MULTIFUNCTIONAL OXIDE HETEROSTRUCTURES

Multifunctional Oxide Heterostructures

Edited by

Evgeny Y. Tsymbal

Department of Physics and Astronomy
Nebraska Center for Materials and Nanoscience
University of Nebraska, Lincoln

Elbio R. A. Dagotto

Department of Physics and Astronomy
University of Tennessee, Knoxville, and
Materials Science and Technology Division
Oak Ridge National Laboratory, Oak Ridge, Tennessee

Chang-Beom Eom

Department of Materials Science and Engineering
University of Wisconsin, Madison

Ramamoorthy Ramesh

Department of Materials Science and Engineering
University of California, Berkeley

OXFORD
UNIVERSITY PRESS

OXFORD
UNIVERSITY PRESS

Great Clarendon Street, Oxford, OX2 6DP,
United Kingdom

Oxford University Press is a department of the University of Oxford.
It furthers the University's objective of excellence in research, scholarship,
and education by publishing worldwide. Oxford is a registered trade mark of
Oxford University Press in the UK and in certain other countries

First Edition published in 2012

Impression: 1

British Library Cataloguing in Publication Data
Data available

Library of Congress Cataloging in Publication Data
Library of Congress Control Number: 2012940356

ISBN 978–0–19–958412–3

Printed and bound by
CPI Group (UK) Ltd, Croydon, CR0 4YY

Preface

The field of research related to complex oxide materials has been developing vigorously in the last few decades. Oxides exhibit an abundance of macroscopic physical properties involving the interplay between magnetism, ferroelectricity, and conductivity. The strong coupling between the electronic, spin, and structural degrees of freedom in complex oxides leads to novel functional behaviors which may open the door to new technologically important applications. Significant experimental and theoretical efforts have been invested to understand fundamental properties of these materials and to elucidate the origin of such complex behavior.

Even more interesting physical phenomena and an even broader spectrum of functional properties occur if two or more complex oxides are combined with atomic-scale precision in a heterostructure to form a new nanoscale material. Recent advances in thin-film deposition and characterization techniques made possible the experimental realization of such oxide heterostructures, promising novel functionalities and device concepts. Especially notable advances have been made over the past few years, driven by the discovery of fascinating new physical phenomena in oxide heterostructures. The fundamental science underlying these phenomena is rich and exciting and provides a new avenue to create novel devices based on the multifunctional physical properties and advanced performance of oxide electronic materials.

This book is devoted to the rapidly developing field of research on oxide thin films and heterostructures. It consists of a set of chapters on topics that represent some of the key innovations in the field over recent years. The book starts from Fundamentals (Part I) which include two chapters introducing strongly correlated electronic materials (Chapter 1) and magnetoelectric coupling and multiferroic materials (Chapter 2). As emphasized in Chapter 1, electron–electron correlations play a decisive role in the properties of many transition-metal oxides, such as manganites, high-temperature superconductors, and multiferroic compounds. In these strongly correlated materials the coupling between charge, spin, and orbital degrees of freedom may produce a rich variety of electronic phases. The interplay between these phases can lead to unusual physical phenomena and interesting functionalities. Chapter 2 is focused on multiferroic materials. These materials have recently attracted significant interest, in particular, due to the magnetoelectric coupling that allows control of the magnetic properties of these materials by electric fields. This chapter gives an introduction to the physics of magnetoelectric coupling and magnetoelectric multiferroics, including the origin of linear and higher-order couplings and the physics of magnetoelectric composites. It reviews different types of magnetoelectric multiferroics and provides a perspective for future directions in this area of research.

Part II of the book is devoted to the growth and characterization of oxide heterostructures. Four chapters on these subjects comprise important experimental developments, providing demonstrations of what can be accomplished using high quality epitaxial growth using advanced deposition techniques and how properties of the resulting thin-film heterostructures can be characterized. Chapter 3 discusses the role of the substrate in synthesizing well-controlled epitaxial thin films of ferroic oxide materials. Examples are given to demonstrate how termination, surface morphology, and strain can be used to tailor functional properties of oxide thin-film heterostructures. Chapter 4 surveys a variety of hard X-ray scattering techniques useful for the study of oxide heterostructures. These techniques provide real space images of the heterostructure with atomic resolution and allow the investigation of composition, valence, and local environment and their effects on the emergent phenomena found at oxide interfaces. Chapter 5 reviews the state-of-the-art of scanning transmission microscopy and electron energy loss spectroscopy (EELS). Starting from a brief introduction to the techniques it gives a number of examples of the applications to both bulk oxide materials and interfaces in oxide heterostructures. Chapter 6 is devoted to piezoresponse force microscopy (PFM) as a tool to characterize ferroelectric properties of oxide thin-films at the nanoscale. A particular emphasis is made on new developments of advanced modes of PFM, such as resonance-enhanced and stroboscopic PFM, to explore static and dynamic properties of thin-film ferroelectrics.

Part III of the book addresses functional properties of oxide heterostructures. The discovery of two-dimensional electron gases (2DEGs) at oxide interfaces has recently attracted considerable interest due to their interesting physical properties and potential for applications. Electronic reconstruction is believed to play an important role in the formation of the 2DEG. Chapter 7 addresses the important question of the role of electrostatics in oxide heterostructures and how to control the properties of their interfaces. The chapter discusses local-bonding effects, crystalline defects, and lattice polarization on an equal footing, providing a useful insight in to the systems which are the subject of extensive current investigations. Chapter 8 continues this discussion by focusing on strongly correlated electronic behavior which may accompany the electronic reconstruction at oxide interfaces. Chapter 9 considers manganite multilayers as a model system to illustrate the dramatic effects of interfaces on magnetism and transport behaviors. Starting from a brief description of the basic properties of manganites and methods for their synthesis and characterization, this chapter describes recent progress in the field of manganite multilayers, including the effects of strain, cation ordering, phase tuning via charge transfer, metal–insulator transitions, and magnetic interactions at interfaces. A different aspect of the physical behavior of oxide heterostructures, namely thermoelectric phenomena, is addressed in Chapter 10. Thermoelectric properties of representative oxide epitaxial films are reviewed that exhibit extraordinary thermoelectric properties.

Part IV of the book is focused on existing and potential applications of oxide heterostructures. Chapter 11 is devoted to high-κ dielectric materials, which are important for the development of advanced CMOS architectures. Various oxide materials and aspects of device design are discussed, aimed at overcoming fundamental limitations associated with conventional materials. Chapter 12 reviews ferroelectric field effect transistors (FeFET) and ferroelectric random access memories (FeRAM). These devices involve ferroelectric polarization as the key functional property responsible for their operation. The chapter describes the important characteristics of ferroelectric materials needed for these devices and the operational principles of FeFET and FeRAM. Finally, Chapter 13 is focused on new concepts of oxide electronics that involve 2DEGs at oxide interfaces, emphasizing the potential for fundamentally new technologies to emerge.

Overall, this book covers the core principles of oxide electronic materials, describes experimental approaches to fabricating and characterizing oxide thin films and heterostructures, demonstrates new functional properties of these materials, and provides an overview of novel applications, as well as the challenges and opportunities in the field. We hope that this book will be an important reference both for graduate students and experienced researchers, specializing in the fields of condensed matter physics, materials science, device engineering, and nanoscience. Finally, we would like to acknowledge Verona Skomski for providing invaluable assistance in preparing the book for submission to the publisher.

Evgeny Y. Tsymbal
Department of Physics and Astronomy, Nebraska Center for Materials and Nanoscience, University of Nebraska, Lincoln, NE 68588, USA

Elbio R. A. Dagotto
Department of Physics and Astronomy, University of Tennessee, Knoxville, TN 37996 and Materials Science and Technology Division, Oak Ridge National Laboratory, Oak Ridge, TN 37831, USA

Chang-Beom Eom
Department of Materials Science and Engineering, University of Wisconsin, Madison, WI 53706, USA

Ramamoorthy Ramesh
Department of Materials Science and Engineering, University of California, Berkeley, CA 94720, USA

Contents

List of contributors

ANAND BHATTACHARYA
Materials Science Division and Center for Nanoscale Materials, Argonne National Laboratory, Argonne, Illinois 60439, USA

DANIELA F. BOGORIN
Chemical Sciences Divison, Oak Ridge National Laboratory, Oak Ridge, Tennessee 37831, USA

GUSTAU CATALAN
Institut Catala de Recerca i Estudis Avançats (ICREA), Catalunya, Spain; and CIN2 (CSIC-ICN), Campus Universitat Autonoma de Barcelona, Barcelona, Spain

CHENG CEN
Semiconductor Research and Development Center, IBM, Hopewell Junction, NY 12533, USA

E. DAGOTTO
Department of Physics and Astronomy, University of Tennessee, Knoxville, Tennessee 37996, USA; and Materials Science and Technology Division, Oak Ridge National Laboratory, Oak Ridge, Tennessee 32831, USA

SUMAN DATTA
Pennsylvania State University, University Park, PA 16802, USA

SHUAI DONG
Department of Physics, Southeast University, Nanjing 211189, China

CHANG-BEOM EOM
Department of Materials Science and Engineering, University of Wisconsin-Madison, Madison, Wisconsin 52706, USA

DILLON D. FONG
Materials Science Division, Argonne National Laboratory, 9700 S. Cass Avenue, Bldg. 212/C222, Argonne, IL 60439, USA

A. GRUVERMAN
Department of Physics and Astronomy, University of Nebraska, Lincoln, NE 68588, USA

T. HIGUCHI
Department of Applied Physics, University of Tokyo, Hongo, Tokyo 113-8656, Japan

HAROLD Y. HWANG
Department of Applied Physics and Stanford Institute for Materials and Energy Science, Stanford University, Stanford, California 94305, USA; and Correlated Electron Research Group (CERG), RIKEN-ASI, Saitama 351-0198, Japan

PATRICK IRVIN
Department of Physics and Astronomy, University of Pittsburgh, Pittsburgh, PA 15260, USA

Hiroshi ISHIWARA
Department of Physics, Division of Quantum Phases and Devices, Konkuk University, Seoul 143-701, Korea

Kunihito KOUMOTO
Department of Physics, Division of Quantum Phases and Devices, Konkuk University, Seoul 143-1701, Korea

C. LEON
Universidad Complutense de Madrid, Madrid 28040, Spain

Jeremy LEVY
Department of Physics and Astronomy, University of Pittsburgh, Pittsburgh, PA 15260, USA

Hiromichi OHTA
Nagoya University, Graduate School of Engineering, Furo-cho, Chikusa, Nagoya 464-8603, Japan; and PRESTO, Japan Science and Technology Agency, Honcho, Kawaguchi 332-0012, Japan

Satoshi OKAMOTO
Materials Science and Technology Division, Oak Ridge National Laboratory, Oak Ridge, TN 37831, USA

S. J. PENNYCOOK
Oak Ridge National Laboratory, Oak Ridge, TN 37830, USA

J. SANTAMARIA
Universidad Complutense de Madrid, Madrid 28040, Spain

Darrell G. SCHLOM
Department of Materials Science and Engineering, Cornell University, Ithaca, NY 14853, USA

James F. SCOTT
Cavendish Laboratory, Department of Physics, University of Cambridge, United Kingdom

Y. TOKURA
Cross-Correlated Materials Research Group (CMRG), ASI, RIKEN, Wako 351-0198, Japan; Multiferroics Project, ERATO, Japan Science and Technology Agency (JST), Tokyo 113-8656, Japan; and Department of Applied Physics, University of Tokyo, Tokyo 113-8656, Japan

Thomas TYBELL
Department of Electronics and Telecommunications, Norwegian University of Science and Technology, 7491 Trondheim, Norway

M. VARELA
Oak Ridge National Laboratory, Oak Ridge, TN 37830, USA; and Universidad Complutense de Madrid, Madrid 28040, Spain

Rong YU
Department of Physics and Astronomy, Rice University, Houston, Texas 77005, USA

Part I

Fundamentals

1
A brief introduction to strongly correlated electronic materials

E. Dagotto and Y. Tokura

1.1 Motivation

A brief introduction is presented to the very active field of research that addresses the properties of strongly correlated electronic materials. Our focus is on the complex transition-metal oxides, with emphasis on the manganese oxides known as manganites, but also describing several other materials, such as high-temperature superconductors and multiferroic compounds. In these correlated electron materials the interactions between the electronic spins, their charges and orbitals, and the lattice produce a rich variety of electronic phases and self-organization. The competition and/or cooperation among these correlated electron phases can lead to the emergence of surprising electronic phenomena and also of interesting functionalities via their nonlinear responses to external fields, potentially forming the basis for a new type of electronics. Our perspective on future interesting directions in this area of research is also included in this chapter.

1.2 Introduction

The study of Strongly Correlated Electronic (SCE) materials is among the most exciting, and challenging areas of research in condensed matter physics [1, 2]. In these materials the behavior of their conduction electrons cannot be described merely by using non-interacting components, as in the often employed one-electron approximation to good metals such as silver or copper. Other sophisticated approaches, including the local-density approximation to density functional theory, are still one-electron theories, and in general they are not applicable to the proper understanding of SCE compounds, particularly at low temperatures. In this family of materials an electron cannot simply be considered to be immersed in the average "mean field" of the others. A typical example is provided by NiO which has a partially filled *3d* band and, for this reason, it should be a metal. However, the Coulomb repulsion between electrons, that induces nontrivial correlations among them, renders NiO an insulator [3].

In the absence of one-particle approximations, theoretical studies of SCE materials are usually carried out using model Hamiltonians, where the dominant terms often correspond to the electrons interacting with one another via Coulomb interactions (that are often assumed to be local in real space, namely on a lattice site or involving nearest neighbor sites, after consideration of screening effects). Coupling with lattice vibrations can also be very important if it is strong. In these model Hamiltonians the electronic movement involving tunneling of electrons between atoms is incorporated via the so-called tight-binding hopping terms.

The complete models including all of these terms are still deceptively simple to write, but solving them, even approximately, requires an enormous effort since many-body SCE systems typically do not have a small parameter to expand on. For this reason, the use of computational techniques is gradually acquiring increasing importance for the theoretical study of this type of material, as discussed in more detail later in this chapter. The above mentioned difficulty in reaching a theoretical understanding of SCE materials, and even of idealized SCE model Hamiltonians, also arises from the several different physical states that are in close competition in these compounds, often involving metals and insulators that can be so close in energy that small external electric or magnetic fields may induce nonlinear responses caused by transitions from one to the other. So a proper theoretical description not only needs good handling of the ground state under investigation, but also of excited states that may have quite different properties [1, 2].

It is clear that SCE materials represent a challenging and exciting frontier for solid state physics: while the understanding of their properties is difficult, their potential applications could be very interesting. These materials can be found in complex states such as high temperature superconductors and heavy fermions, and they can be very susceptible to small perturbations such as in the case of colossal magnetoresistance found in manganese oxides, where relatively small magnetic fields of a few Teslas, which are small in typical electronic energy scales, induce huge changes of many orders of magnitude in resistivity.

The current chapter provides a brief overview of this vast area of research, known colloquially as "strongly correlated electrons". It is important that the reader should be aware that in a single chapter it would be impossible to address all of the many exciting topics of research contained within this field. As a consequence, the authors have focused on the subset with which they have developed their own scientific area of expertise, and for this reason they will mainly address transition-metal oxides. This focus, with an emphasis on bulk materials, will nevertheless be useful to better understand some of the other chapters in this book that also specialize in transition-metal oxides, although in the form of artificially engineered superlattices. It is also important to remark that topics that are still controversial, such as the actual origin of high temperature superconductivity in cuprates, will be only briefly reviewed here, with sufficient references to allow the reader to arrive at their own conclusions on this subfield. Thus, emphasis will be given to particular areas of research, such as colossal magnetoresistance, where in recent years considerable progress has been made and consensus reached due to a remarkable cross fertilization between theoretical and experimental efforts. The chapter ends with our opinion about the present status of the SCE field of research and its future.

The current chapter is based on a recent review article by the authors [4], but it incorporates considerable additional information and several references not included in that previous publication. As expressed before, ours is certainly not a comprehensive review and, moreover, for simplicity the cited literature is naturally biased towards the publications by the authors. As a consequence, readers that are interested in developing a more profound understanding of SCE materials, beyond the information contained in this chapter, are urged to consult the additional literature that is cited in the list of publications at the end of this chapter.

1.3 Why correlated electrons?

As mentioned in the introduction, one of the most intensively studied areas of research in condensed matter physics is the field of SCE materials [1, 2, 4]. Why are these compounds considered to be such an interesting challenge? Typically, SCE materials are made of simple building blocks, such as transition-metal ions located inside an octahedral oxygen cage forming, by periodic repetition, an overall lattice perovskite structure. However, the large ensemble of these constituents behaves in a nontrivial complex manner, leading to collective responses that are quite difficult to predict a priori based on the properties of those small octahedral building blocks. Typical examples are the high temperature superconductivity that exists in layered copper oxides and the colossal magnetoresistance of manganites, to be described in more detail later in this chapter. As mentioned before, in these SCE systems the collective states cannot be understood based on the one-electron (or even one-quasiparticle) approximation, even if the calculations are fully quantum mechanical. The mere addition of the individual electronic properties is not sufficient to rationalize the subtle phenomena that emerge from the full ensemble, such as the zero resistivity that occurs at low temperatures in dense electronic systems where collisions should dominate, or the resistances that change by several orders of magnitude when some SCE materials are immersed in relatively small magnetic fields. The strong correlations between particles that are present when these phenomena occur render the one-particle approximation useless, and more sophisticated approaches are needed to rationalize the properties of these materials. Understanding, controlling, and predicting the emergent complexity of correlated-electron systems is one of the most pressing challenges in condensed matter physics at present. The rich phase diagrams of these materials, to be described below, make them natural candidates for devices where nonlinear responses can be exploited.

Surprises in the field of SCE research abound. Consider, for instance, the recent explosion of interest in the study of the iron-based layered superconductors known as the pnictides [5,6]. The main efforts to increase the record critical temperatures of layered copper oxides have been traditionally focused on the right-hand side of the $3d$ row of the periodic table where Cu is located. Iron (Fe) is in the middle of that row, and it is usually mainly associated with ferromagnetism, not with superconductivity. However, for reasons that are still still unclear, electrons in layers of Fe and As produce an unexpected superconducting state with a high critical temperature of approximately 50 K [6]. While fair it is to say that it is not yet totally clear whether these Fe-based superconductors are indeed part of the correlated-electron family, since the parent undoped compounds are simultaneously magnetic and metallic, as opposed to magnetic and insulating as in the cuprates, evidence is accumulating that the superconductivity is unconventional and perhaps mediated by spin excitations. Some investigations suggest that pnictides are located at *intermediate* on-site Coulombic repulsion couplings, in between the weak and strong limits [7].

On the theory side, it is difficult to carry out reliable calculations for SCE materials, as explained in the introduction. Standard *ab-initio* methods do not properly incorporate correlations and often lead to metallic states for cases where a Mott insulator is the true ground state. In the absence of reliable *ab-initio* methods, a widely used theoretical approach to SCE materials is based on model Hamiltonians, since they seem to better capture the essence of correlated systems. However, as mentioned before, these models are still very difficult to solve accurately particularly using analytic techniques. For these reasons, it is not surprising that computational studies of complex systems in general, and correlated electrons in particular, are becoming a widely followed path to gather information about models, and this avenue has already led to considerable progress in the area of strongly correlated electrons.

The importance of understanding how complex phenomena emerge from simple ingredients was identified as one of the challenges for the next decade in a recent study by the U.S. National Academy of Sciences that addressed the current status of condensed matter and materials physics [8]. Also, the Basic Energy Sciences Advisory Committee of the U.S. Department of Energy has identified five "grand challenges" for science in order to be able to control matter all the way to atomic and electronic levels [9]. Among these generic challenges is an understanding of how the remarkable complex properties of matter emerge from correlations of their atomic or electronic constituents. Without a doubt, controlling correlated electrons is at present one of the most crucial areas of research in condensed matter and materials science.

1.4 Control of correlated electrons in complex oxides

Among the fundamental parameters for controlling the behavior of correlated electrons are (1) the tunneling rate for electron-hopping between atoms (typically regulated by a so-called hopping amplitude t, or by the one-electron bandwidth W which is linearly related to t) and (2) the density of charge carriers (or to be more precise, the band filling) [10]. The hopping amplitude t competes with the on-site electron–electron Coulomb repulsion energy U (also known as the Hubbard repulsion coupling U) and the outcome of this competition is the Mott transition, namely, an insulator–metal transition that often occurs in correlated-electron systems. As a function of the ratio U/t, the system undergoes several changes in the spin and charge arrangement and dynamics at low temperatures. In the limit of large U/t, every electron localizes on an atomic lattice site when the number of electrons precisely equals the number of those atomic sites (assuming just one active orbital per site in this example). In a Mott insulator, the ground state is often also spin antiferromagnetically ordered. But when U/t decreases and reaches some critical value, a transition to a metallic state occurs, as shown schematically in Figure 1.1. Another route to the Mott transition is by changing the filling (charge doping) of the correlated Mott insulator. Ideally, a tiny deviation from half-filling (or from an integer number of conduction electrons per atomic site in the d-electron system) should be sufficient to induce a transition to a paramagnetic metallic state, with a divergently large effective mass of conduction electrons. In most actual cases, however, a small but finite amount of filling change (i.e. doped charge) is necessary to stabilize a metallic state, free of the self-localization of conduction carriers caused by interactions with the lattice, spins, and the influence of disordering effects.

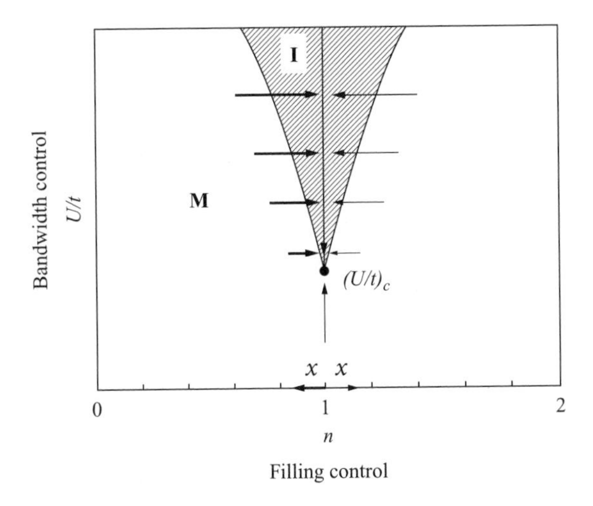

Fig. 1.1 Sketch of the generic low-temperature insulator (I) to metal (M) transition found in correlated-electron systems that are controlled by the bandwidth W (or equivalently by the hopping amplitude t) or the band filling n (doping x). U is the on-site Coulomb repulsion interaction, and $(U/t)_c$ represents the critical value where the transition occurs.

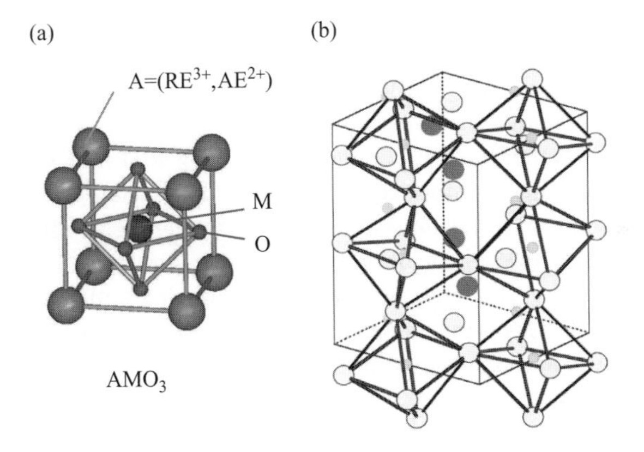

Fig. 1.2 (a) Perovskite structure of compounds with chemical formula (RE, AE)MO_3, with trivalent (3+) rare-earth (RE) ions and divalent (2+) alkaline-earth (AE) ions located at the perovskite A-site, and the transition-metal element at the perovskite B-site. (b) Orthorhombically distorted ($GdFeO_3$-type) structure of the perovskites.

These fundamental, and other related parameters in correlated-electron oxides, can be well controlled typically by the crystal engineering of a perovskite with the formula (RE,AE)MO_3 (Figure 1.2 (a)), where RE, AE, and M represent the trivalent rare-earth ions, the divalent alkaline-earth ions (which can be in solid solution), and the transition-metal element, respectively. The ideal perovskite (AMO_3) exhibits a simple cubic structure, but the lattice distortions (Figure 1.2(b)), usually referred to as $GdFeO_3$-type distortions, are governed by the size mismatch of the ionic radii of the A and M elements, or by the so-called tolerance factor f, defined as

$$f = (r_A + r_O)/\sqrt{2}(r_M + r_O), \tag{1.1}$$

where r_i ($i =$ A, M, or O) represents the ionic size of each element i. When f is close to 1, the cubic perovskite structure is realized. As r_A, or equivalently f, decreases, the lattice structure transforms, first to the rhombohedral and then to the orthorhombic ($GdFeO_3$-type) structure, in which the M-O-M bond is bent and its angle deviates from 180°. This bond-angle distortion decreases the one-electron bandwidth W, because the effective d-electron transfer amplitude t between the neighboring M sites is governed by d-electron hybridization with the intervening O $2p$ state. The impact of the variation in W varying r_A is observed, for example, in the metal insulator transition in the family $RENiO_3$ [11]. In fact, $LaNiO_3$, with a maximum tolerance factor ($f \approx 0.96$) close to unity, is a paramagnetic metal with one conduction d electron per Ni atom, whereas other $RENiO_3$ materials with smaller f values (or decreasing RE ionic size) exhibit antiferromagnetic insulating ground states and undergo a thermally induced insulator metal transition (IMT) with increasing temperature.

Another important advantage of perovskites or related structures (e.g. layered structures, termed the Ruddlesden–Popper series) is the ease of chemical control of the band filling. Using the solid solution $RE_{1-x}AE_x$ at the perovskite A site (AMO_3, see Figure 2(a)), the effective valence of the transition metal (M) becomes $3 + x$. In analogy to doped semiconductors, the increase (decrease) of x is customarily called "hole doping" ("electron doping"). These changes in x reflect a decrease (or increase) of the band filling or of the chemical potential.

A well-known example of a band-filling- (doping-) controlled Mott transition is the case of copper oxide high-temperature superconductors, such as $La_{2-x}Sr_xCuO_4$, in which the effective Cu^{2+} valence state, or more correctly the $[Cu\text{-}O]^0$ state, is doped with holes and becomes a $Cu^{(2+x)+}$ ($[Cu\text{-}O]^x$) state. It is well known that after the insulator to metal transition, that occurs at around $x = 0.06$, the ground state becomes superconducting [12, 13]. Another prototypical example is $La_{1-x}Sr_xTiO_3$, where hole doping of $x \approx 0.05$ drives the ground state from an antiferromagnetic Mott insulator to a paramagnetic metal with a high carrier density (band filling $n = 1 - x$), but with a large enhancement of the carrier mass, as compared to the calculated band mass [14].

Figure 1.3 presents a schematic of the overall electronic phase diagram of $RE_{1-x}AE_xTiO_3$ under the control of both the one-electron bandwidth W and the band filling (i.e. $1 - x$) in terms of varying the RE ionic radius and doping level x [15]. The "undoped" Mott insulator, for example $La_{1-y}Y_yTiO_3$, shows an increasing charge gap (Mott–Hubbard gap) with decreasing W (or increasing U/W), whereas the antiferromagnetic order turns to ferromagnetic through the influence of orbital ordering (to be discussed in more detail below). The critical doping level of the insulator-metal transition is increased from $x_c \approx 0.05$ for $La_{1-x}Sr_xTiO_3$, with a maximum W value (W_{La}), up to $x_c \approx 0.4$ for $Y_{1-x}Ca_xTiO_3$, with a relatively small W value (about 80% of W_{La}). The rather high concentration of holes needed to realize the metallicity of the small-W systems, such as $Y_{1-x}Ca_xTiO_3$, is common in the $3d$ transition-metal oxides with perovskite-like structures. There should be short-range/long-range polaronic ordering in such a highly doped insulating ground state, because of the combined effect of strong electron correlation and electron–lattice coupling. The charge–orbital ordered insulating state

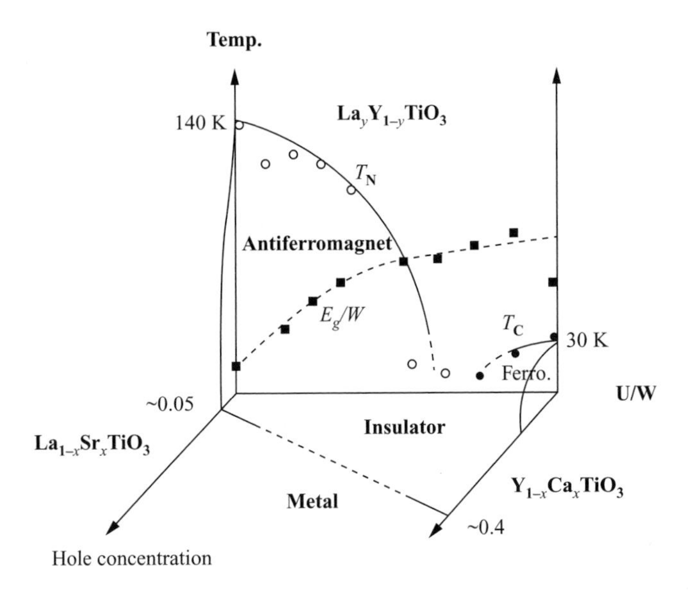

Fig. 1.3 Metal–insulator phase diagram of $RE_{1-x}AE_xTiO_3$ [14, 15]. In the end plane (temperature versus electron correlation U/W) with $x = 0$, the antiferromagnetic (T_N) and ferromagnetic (T_C) transition temperatures are plotted for the RE=La_yY_{1-y} system, where the variation of the normalized Mott–Hubbard energy gap (E_g/W) is also shown.

in the perovskite manganites is one such example where the state changes substantially when subjected to an external stimulus such as a magnetic field, electric field, or light irradiation, as described in the following sections.

Figure 1.3 also illustrates a typical characteristic of materials belonging to the correlated electron family. By varying the bandwidth, electronic concentration, and temperature, a rich variety of interesting phases can be stabilized. These phases present quite different properties, such as behavior like a metal or insulator in transport, or being antiferromagnetic versus ferromagnetic with regard to the spin order at low temperatures. The appearance of quite different tendencies in the same material by varying the above-mentioned parameters is typical of SCE systems. Figure 1.4 illustrates a similar qualitative behavior but for the cases of (a) Ru-oxides such as $Ca_{2-x}Sr_xRuO_4$ (known as the "ruthenates") [2, 16], (b) Mn-oxides such as $Nd_{1-x}Sr_xMnO_3$ (known as the "manganites", to be extensively discussed below in this chapter) [17], (c) the famous high temperature superconductors based on copper (the cuprates) including both hole and electron doping [13], and (d) the Co-oxides such as Na_xCoO_2 [18]. Future work will clarify if the new Fe-based superconductors (the pnictides) are located at strong or intermediate couplings, but they are already known to display a competition between magnetic and superconducting phases [5, 6, 7]. The phenomenon of presenting a very rich phase diagram has considerable importance particularly when close to the boundaries between phases, since a state with characteristics quite different from those of the ground state (say, a metal) may be located close in energy to the actual ground state (say, an insulator). As a consequence, small "perturbations" such as those introduced by relatively small external fields may unbalance

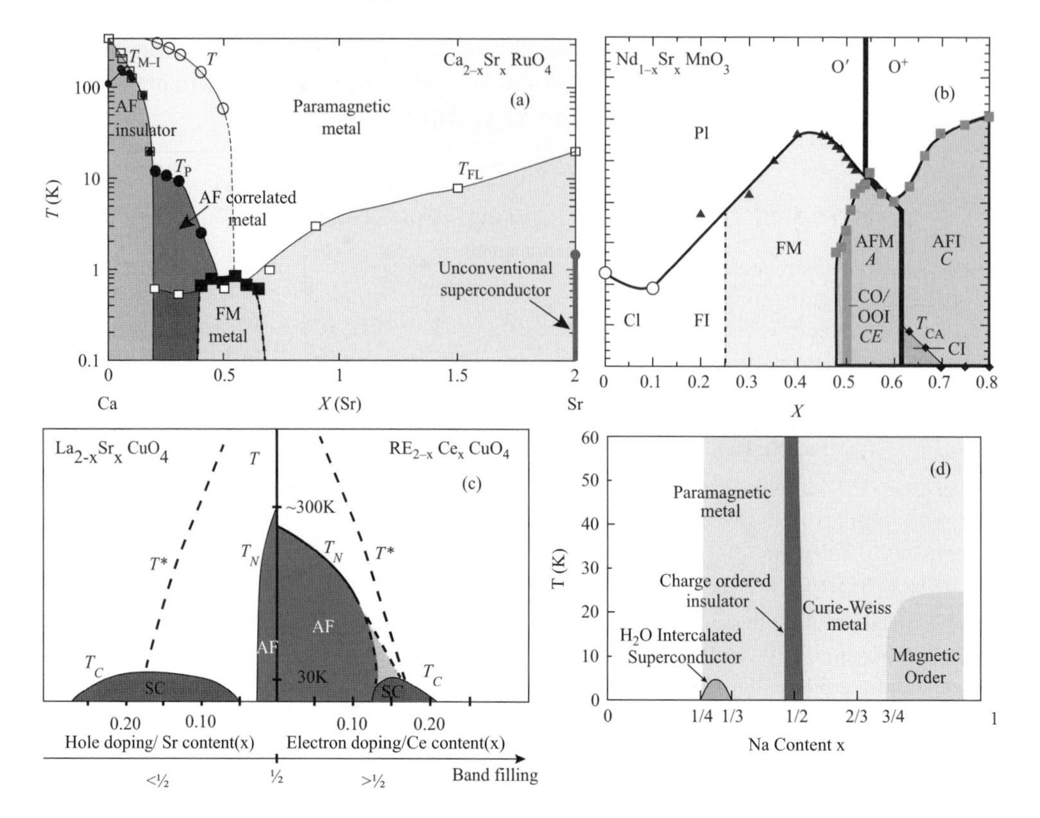

Fig. 1.4 (a) Phase diagram of $Ca_{2-x}Sr_xRuO_4$, a member of the family of compounds known as the ruthenates, reproduced from [2, 16]. Note the remarkably quite different properties obtained in this system by varying the temperature T as well as bandwidth W via the chemical substitution of Sr with Ca. The material can display properties corresponding to a good metal (paramagnetic metal) or a magnetically ordered metal (AF correlated metal and FM metal), or in a narrow range of temperatures, even a superconductor for 100% Sr. (b) Phase diagram of $Nd_{1-x}Sr_xMnO_3$, a member of the family of compounds known as the manganites, which display colossal magnetoresistance. Varying the concentration of holes x, via chemical substitution of Sr for Nd, the material presents, at low temperatures, a variety of metallic (ferromagnetic) and insulating (antiferromagnetic and ferromagnetic) phases. For details and a better description of the many phases see [17]. (c) Phase diagram of the hole-doped $La_{2-x}Sr_xCuO_4$and electron-doped $RE_{2-x}Ce_xCuO_4$ high critical temperature cuprates, showing the antiferromagnetic insulating and superconducting phases, as well as other characteristic temperatures such as T^* which defines the pseudogap phase [13]. (d) Phase diagram of Na_xCoO_2, reproduced from [18]. Note the presence of several very different phases, such as a superconducting phase, a magnetic regime, and a prominent charge-ordered state at $x = 1/2$. (c) Reprinted with permission from [13]. Copyright 2010 by the American Physical Society. (d) Reprinted with permission from [18]. Copyright 2004 by the American Physical Society.

the ground state and transform it into a quite different new ground state. This nonlinear effect of "perturbations" that are naively considered to be small a priori has profound implications for possible applications of these materials in actual devices.

1.5 Ordering of charge, spin, and orbital degrees of freedom

Consider a transition-metal ion (M) in a crystal with a perovskite structure. M is surrounded by six doubly negative oxygen ions (O^{2-}) which produce a crystal field potential that partly lifts the degeneracy of the d-electron levels (Figure 1.5). To understand this effect, simply consider the electronic wave functions. Those pointing towards the negatively charged O^{2-} ions ($d_{x^2-y^2}$ and $d_{3z^2-r^2}$, called e_g orbitals) must have higher energy than those pointing between the ions (d_{xy}, d_{yz}, and d_{zx}, called t_{2g} orbitals). In the Mott insulating state, almost all of the d electrons are localized on their respective atoms, making the spin and orbital degrees of freedom simultaneously active. The combination of these degrees of freedom produces a variety of complex spin–orbital ordering patterns. Prototypical cases are shown in Figure 1.6 for perovskites of $LaVO_3$ and YVO_3, which are t_{2g} electron systems, and $LaMnO_3$ and $BiMnO_3$, which are e_g electron systems.

In $LaVO_3$, for example, the antiferromagnetic spin ordering with ferromagnetic chains along the z direction is known to induce an orbital-ordered state with alternate occupancy of d_{yz} and d_{zx} in the x, y, and z directions, in addition to the commonly occupied d_{xy} orbital [19] (in Figure 1.6 (a), the d_{xy} orbital on each V site is omitted for clarity). In spite of its nearly cubic character, the electronic structure is highly anisotropic because of the spin–orbital ordering. On the other hand, in YVO_3[19] (Figure 1.6(b)), with a larger orthorhombic ($GdFeO_3$-type) lattice distortion, staggered spin order and a d_{yz} or d_{zx} orbital-ordered state along the z axis are known to exist in the ground state. Thus, the relationships between the spin and orbital orders for these two compounds are exactly opposite, indicating the strong coupling between the spin and orbital degrees of freedom.

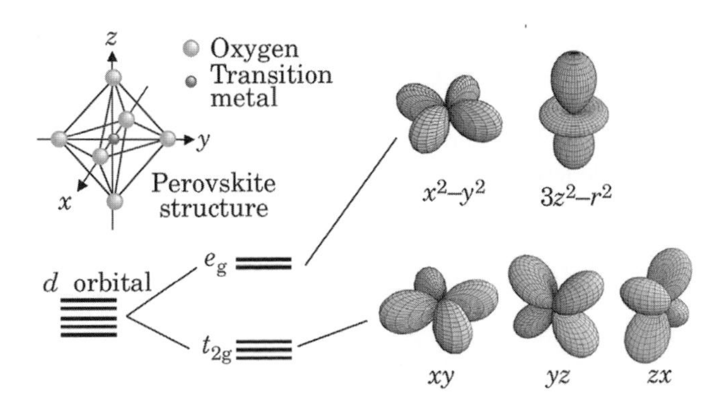

Fig. 1.5 Sketch illustrating how orbitals that are degenerate at the atomic level are split in a crystal field environment. Shown are the d-electron quantum mechanical orbitals and their level splitting in the octahedral coordination of O^{2-}.

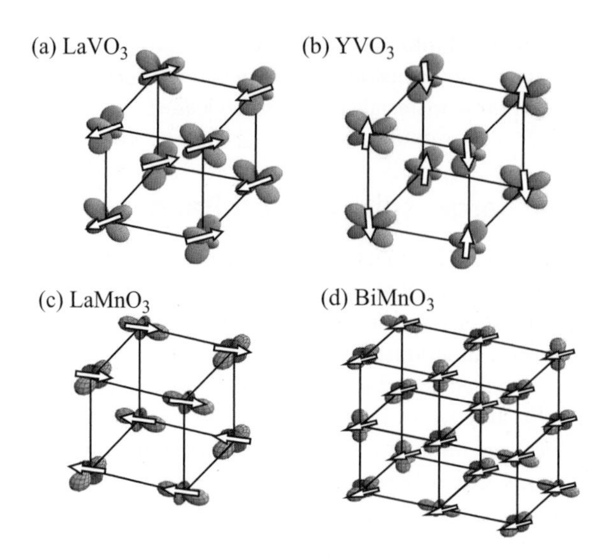

Fig. 1.6 The spin and charge ordering of the d electrons in (a) $LaVO_3$ [19]. (b) YVO_3[19], reprinted with permission from [19]. (c) $LaMnO_3$ [20], and (d) $BiMnO_3$ [21], all with (distorted) perovskite structures. (a) and (b) reprinted with permission from [19]. Copyright 1996 by the American Physical Society. (c) Reprinted from [20], with permission from Elsevier. (d) Reprinted with permission from [21]. Copyright 2002 by the American Physical Society.

In the perovskite manganites, the orbital degeneracy is easily lifted by coupling with the local deformation of the MnO_6 octahedron (called the Jahn–Teller effect), such as the elongation or compression of the octahedron along the c axis, compensated by compression or elongation along the a and b axes to keep the volume constant, favoring the occupation of the $d_{3z^2-r^2}$ and $d_{x^2-y^2}$ orbitals, respectively. Therefore, orbital ordering coupled with the collective Jahn–Teller distortion first emerges with decreasing temperature, and then regulates the spin ordering pattern at lower temperatures. In $LaMnO_3$, a local linear combination of the $d_{3z^2-r^2}$ and $d_{x^2-y^2}$ orbitals produces orbital states elongated along the x and y axes, which alternate on the Mn sites in the ab (xy) plane, as shown in Figure 1.6(c). In this case, the Jahn–Teller distortion produces a macroscopic lattice strain, compressing the c axis and expanding the ab plane. The spins couple ferromagnetically on the ab plane, whereas they are stacked antiferromagnetically along the c axis, producing the so-called A-type layered antiferromagnetic state [20]. In $BiMnO_3$, with a lower crystal symmetry because of the lone-pair on Bi^{3+} (Figure 1.6(d)), the complex orbital order can give rise to a Mott insulating ferromagnetic ground state [21].

In some relatively common cases, the compound remains electrically insulating or marginally metallic over a broad range of band fillings, in which a periodic array of doped holes or electrons appears. This phenomenon is called charge ordering. Figure 1.7 exemplifies several cases of such charge ordering in some quasi-two-dimensional transition-metal (M) oxides with chemical formula $(RE,AE)_2MO_4$, with the K_2NiF_4-type structure (Figure 1.7(a)). In the isolated MO sheet of this crystal form, the spin, charge, and/or orbital tend to form

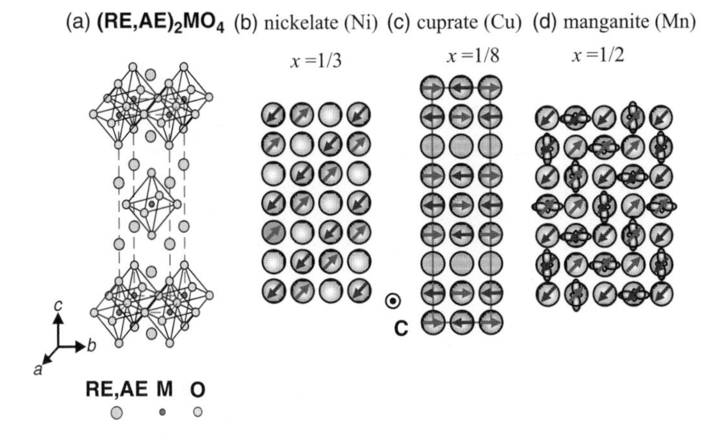

Fig. 1.7 Charge order in the form of "stripes" in hole-doped two-dimensional metal oxide sheets of layered perovskite materials with (a) K_2NiF_4-type structure; (b) $La_{2-x}Sr_xNiO_4$ ($x \approx 1/3$) [22], (c) $La_{2-x}Ba_xCuO_4$ ($x \approx 1/8$) [23, 24], and (d) $La_{1-x}Sr_{1+x}MnO_4$ ($x \approx 1/2$) [25]. This figure is reproduced in color in the color plate section.

"stripes", namely the extra charge (or in their absence, the extra holes) self-organize into quasi-one dimensional arrays. In $La_{2-x}Sr_xNiO_4$ (Figure 1.7(a), with RE = La and AE = Sr), for example, the hole doping, x, in the parent Mott insulator La_2NiO_4 cannot induce the insulator–metal transition up to $x = 0.9$ and instead forms a charge (hole) and spin stripe pattern running parallel to the diagonal direction of the NiO_4 squares ("diagonal stripe"), that is, along (110) in the tetragonal lattice [22]. At $x = 1/3$, in particular, the charge and orbital stripe state is the most stable, as shown in Figure 1.7 (b). This clearly illustrates that doping a Mott insulator does not always lead to a metal.

A similar hole stripe is also known to exist in some of the high-temperature superconducting copper oxides, as exemplified in Figure 1.7(c) for the $x = 1/8$ hole-doped La_2CuO_4 [23, 24] In this case, the one-quarter-filled (i.e. 50% hole-doped) stripes run along the (100) or (010) directions ("horizontal or vertical stripes") in the half-filled CuO_2 background. The incommensurate spin order observed in the underdoped region (say, $x < 0.12$) of La_2CuO_4 might originate from this exotic charge order. Reflecting the one-quarter-filled nature of the vertical/horizontal stripes, the electrical conduction along these stripes appears to survive and manifests itself as a one-dimensional charge dynamics and Fermi-surface structure.

Even more complex features arise in the doped manganites as a consequence of the close interplay among the spin, charge, orbital, and lattice degrees of freedom, as exemplified in Figure 1.7(d) for hole-doping levels of $x = ½$[25]. In a classical picture, Mn^{4+} (with three t_{2g} electrons as the local $S = 3/2$ spin) and Mn^{3+} (with the $S = 3/2$ local spin plus one e_g electron) should coexist (more details are provided in the next section). The charge ordering shows a checkerboard pattern, whereas the orbital on the same Mn^{3+} sites shows the larger unit-cell ordering with approximately alternating $d_{3x^2-r^2} - d_{3y^2-r^2}$ (or $d_{y^2-z^2} - d_{z^2-x^2}$) orbital occupancy. The observed spin ordering pattern arises from the subtle compromise between the antiferromagnetic superexchange interaction among the t_{2g} local spins and the ferromagnetic double-exchange interaction (also to be addressed in more detail below) mediated by the e_g electron hopping between the Mn^{3+} and Mn^{4+} sites. The orbital ordering regulates the anisotropic e_g electron hopping, and hence, the Mn^{4+} sites adjacent to the lobe of the e_g orbital on the nearest Mn^{3+} site are linked through a ferromagnetic interaction. As a result, interesting ferromagnetic "zigzag" chains appear along the diagonal direction in the ground state. These complex forms of charge–orbital–spin ordering are ubiquitous in the highly (moderately) doped nonmetallic manganese oxides with perovskite-related structure.

1.6 Model Hamiltonians

In several transition-metal oxides, particularly those in the middle of the *3d* row of the periodic table, there are two effectively separated electronic degrees of freedom originating from the same ion. Figure 1.8(a) illustrates, similarly to the explanation for Figure 1.5, the case of the Mn oxides where the lattice environment splits the originally five-folded degenerate *3d* orbitals into two e_g levels and three t_{2g} levels. The valence of Mn in, e.g. $LaMnO_3$ is 3+ and, consulting the periodic table, this immediately shows that four electrons must be accommodated in the active five *3d* levels. In practice, the crystal field splitting between these levels is not sufficiently large to prevent Hund's rule of atomic physics from dominating and, as a consequence, the e_g levels carry one electron while the three other electrons are located into the t_{2g} levels, and they all orientate their spins in the same direction. This effectively creates a separation between the electrons at the e_g levels, that become mobile after hole-doping is introduced via the replacement of La by Sr or Ca, and the well-localized spins residing at the t_{2g} levels that are

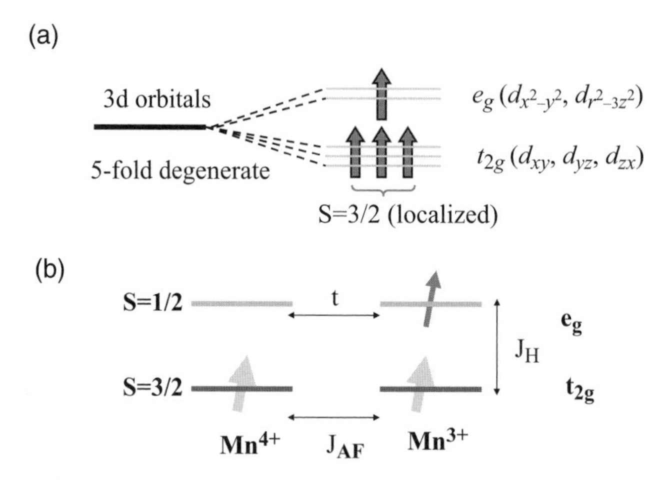

Fig. 1.8 (a) Sketch illustrating the splitting of levels in a manganese ion due to the crystal field, and the effective formation of two types of degrees of freedom: mobile e_g electrons and localized t_{2g} spins (with a sufficiently large total spin that they are often approximated by a classical spin). (b) Sketch illustrating some of the interactions that appear in model Hamiltonians for manganites, as well as the double-exchange mechanism for ferromagnetism in these compounds.

not affected by chemical doping in the ranges usually studied. In addition, it has been shown that a t_{2g} spin value of 3/2 can be considered to be large enough to approximate those t_{2g} spins by classical spins. Thus, effectively there is an intriguing coexistence of quantum itinerant and quasi-classical localized degrees of freedom in these materials. This separation is important for many reasons. In particular, it provides a procedure to carry out computational studies of model Hamiltonians for Mn oxides [26].

After establishing the effective separation between mobile and localized degrees of freedom at each Mn ion, bringing two of those Mn ions together illustrates the typical interactions that need to be incorporated into model Hamiltonians to properly describe their physics. Figure 1.8(b) shows schematically the previously discussed spin 1/2 and 3/2 degrees of freedom of a manganese ion in valence state 3+, together with another manganese that has lost the e_g electron (due to the influence of chemical doping) and is now in a valence state 4+. With regard to the mobility of the e_g electrons, a "hopping" term for those electrons must be incorporated to account for the tunneling of an e_g electron from one Mn to the other (in this approximation the intermediate oxygen merely acts as a bridge). The amplitude for this process to occur is t. In addition, the localized spins 3/2 can interact like any pair of localized spins via a Heisenberg interaction which is regulated here by a coupling called J_{AF}. At each manganese ion, the Hund rule is enforced by including a ferromagnetic term that links the spins of the localized and mobile electrons, with strength J_H. Finally, not shown in the figure but still very important, is the Jahn–Teller coupling of the electrons to the oxygen vibrations that can split the doubly degenerate e_g levels. The orientation of the spins in Figure 1.8 (b), with the two t_{2g} spins pointing in the same direction, illustrates the reason for the presence of a so-called "double exchange" ferromagnetic phase in these compounds: in order for the e_g electrons to be able to move easily and gain kinetic energy without expending an energy as large as the Hund coupling ($\sim$eV), the t_{2g} spins must be oriented ferromagnetically. This simple mechanism, that fails at large enough J_{AF} or e-phonon couplings, is, for historical reasons usually referred to as a double exchange.

To avoid technicalities in this introductory chapter, the actual mathematical form of the model Hamiltonian for manganites will not be explicitly presented here. Interested readers should consult [26] for details (or [12] for models for the cuprates). In [26], a description of the phase diagrams of these models for manganites can also be found, as well as several references and a description of the many-body techniques used to obtain those phase diagrams. Typically they are complex, presenting many phases with competing tendencies involving metals and insulators, and a variety of spin, charge, and orbital orders, similar in several respects to the phase diagrams of the real materials shown in Fig. 1.4.

1.7 Intrinsically inhomogeneous states

A variety of experimental probes and theoretical calculations aimed at understanding materials of the strongly correlated electron family have consistently unveiled a tendency to form states that are not homogeneous, in the sense that rather different spin, charge, and orbital arrangements can coexist in the same sample [2, 26, 27]. It appears that these tendencies originate in the competition between the several phases that are close in energy in some SCE materials. These issues have attracted considerable attention, particularly in the context of the manganites and the cuprates. In manganites, early Monte Carlo and dynamical mean field calculations revealed

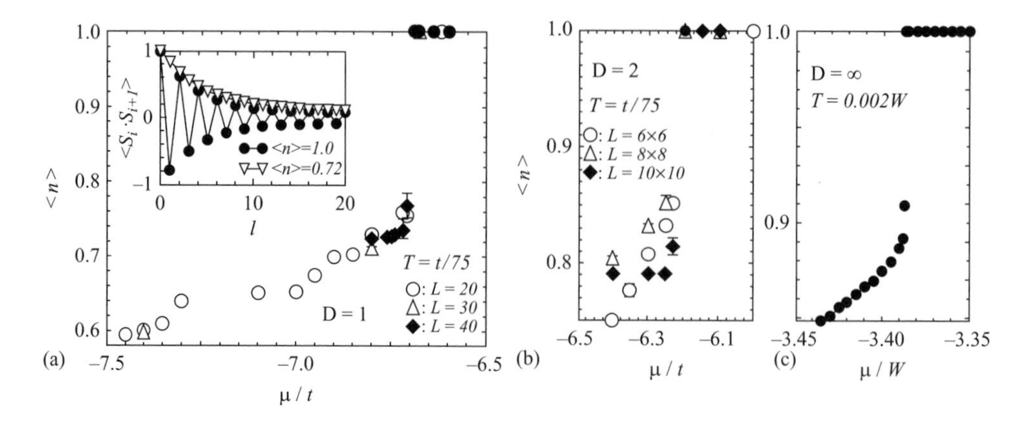

Fig. 1.9 Electronic density <n> of the e_g electrons versus chemical potential, reproduced from [27]. These results were obtained via a Monte Carlo simulation of models for manganites, involving an Exact Diagonalization of the fermionic sector and Monte Carlo sampling of the classical spins [27]. The left panel was obtained in one dimension and it shows a clear discontinuity in the curve, signaling the presence of unstable density regions where phase separation occurs [27]. The middle panel are similar results but for two dimensions. The right panel illustrates similar physical results, but using the dynamical mean-field approximation that is exact at infinite dimensions. For more details see [26, 27, 28].

a tendency towards a phenomenon called phase separation [26, 27, 28]. This manifests itself as a discontinuity in the curve electronic density <n> versus chemical potential [27], as shown in Figure 1.9. In this regime, a cluster that is nominally prepared at an electronic density $<n>_{\text{nom}}$ which is not allowed, e.g. a density in the discontinuity range between 1.0 and ~0.78, if Figure 1.9 (a) is considered, will spontaneously separate into two regions with densities at the extremes of the jump, each occupying a volume in the proper proportion such that the average density is $<n>_{\text{nom}}$. A crude sketch of this phenomenon is shown in Figure 1.10(a), where the white and gray tones indicate the two macroscopic regions involved in the process, with different electronic densities. However, note that such a phase-separated state cannot be stable because there is a macroscopic separation of charge that is not balanced (the positive ions that keep the ensemble neutral are not mobile enough to migrate and compensate for the electronic charge separation). When the Coulombic $1/r$ long-range interactions are taken into account in the calculations, then complex inhomogeneous states are typically formed involving, for example, bubbles of one state embedded into the other, or striped arrangements as previously discussed (see Figures 1.10 (b) and (c)), with typical characteristic lengths in the nanometer scale region. This whole phenomenon is sometimes referred to as "electronic phase separation" and it was first widely discussed in the context of the cuprates, to rationalize the appearance of the stripes [23, 24]. In manganites, inhomogeneous states have also been widely reported [26, 28]. Thus, in both manganites and cuprates, a variety of calculations and experiments suggest that this tendency to separate into two states at the nanometer scale must be seriously considered. Readers interested in this popular subject should consult [23, 24, 26, 28] and the cited literature therein for details.

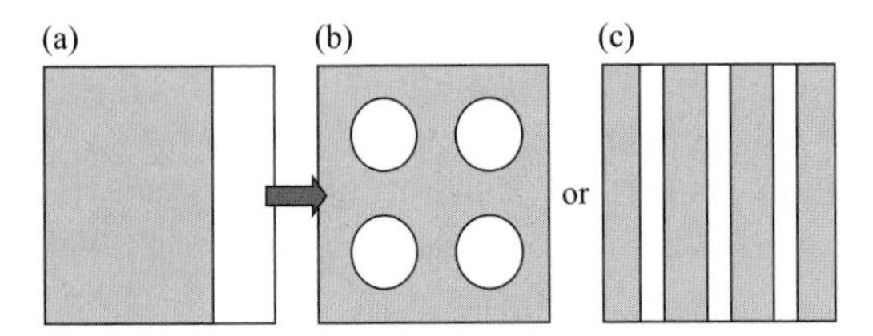

Fig. 1.10 Qualitative sketch of a macroscopically phase-separated state (a), where the gray and white regions represent two phases of the same compound but with different electronic densities. (b) and (c) represent possible nanometer scale arrangements of these phases obtained after proper consideration of the long-range $1/r$ Coulomb repulsion between the electrons. This figure is reproduced from [28] where more details can be found. Reprinted from [28] with kind permission from Springer Science+Business Media.

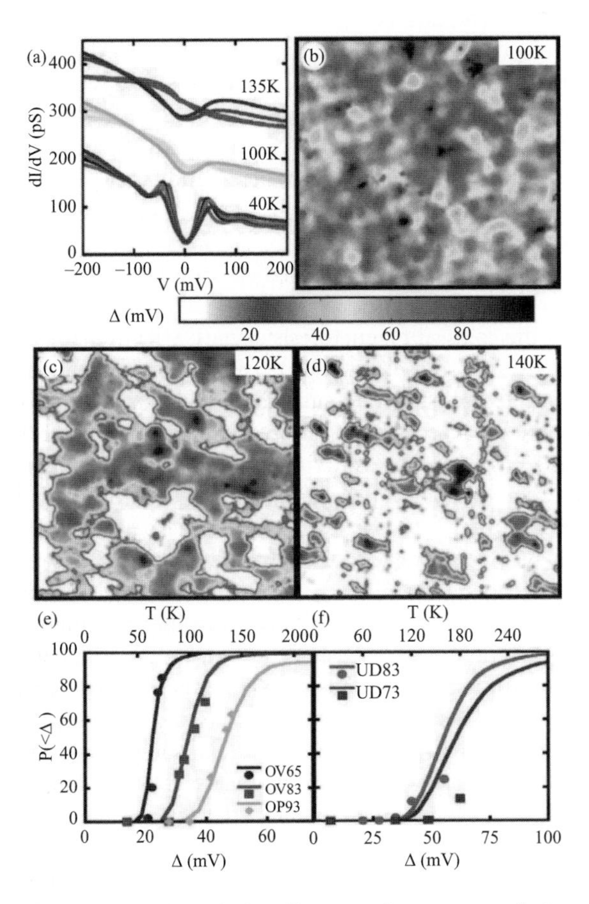

Fig. 1.11 Example of an experimental result that illustrates the presence of inhomogeneous states in a transition-metal oxide. Shown are results reported in [29], containing scanning tunneling microscopy (STM) data gathered for a high-temperature superconductor material with a critical (optimal) temperature of 93 K. The colored regions have a d-wave STM spectrum, and the remarkable result is that these superconducting areas are observed even at temperatures as high as 120 K, well above the critical temperature, contrary to the behavior of typical BCS low-temperature superconductors. Reprinted by permission from Macmillan Publishers Ltd [29], copyright 2007. This figure is reproduced in color in the color plate section.

Recently, Scanning Tunneling Microscopy (STM) experiments have revealed a variety of interesting inhomogeneous states that relate to the previous discussion. For instance, STM experiments for a high critical temperature superconductor [29] with a critical temperature $T_c \sim 93$ K revealed regions where the local density of states closely resembles that of a d-wave superconductor, but at temperatures considerable higher than T_c. Figure 1.11 contains some of these STM results. Note that at 120 K, namely almost 30 K above T_c, as much as half the sample still has local d-wave characteristics, which is a remarkable result. The d-wave regions have random shapes instead of stripes or other more uniform patterns. Some of the theoretical proposals for understanding these effects rely on the influence of quenched disorder over the competition between superconducting and antiferromagnetic states [30, 31].

1.8 Giant responses in correlated electron systems

The several competing tendencies in correlated-electron materials lead to rich phase diagrams, containing a variety of phases with spin, charge, and orbital order; superconducting states; metals and insulators; multiferroics, and other states (as discussed before, see Figures 1.3 and 1.4). These phases are interesting by themselves but, in addition, their presence in the same material suggests that the energies of each phase must be similar, since small changes in composition, temperature, pressure, external fields, and other parameters tend to induce a transition from one to another. These transitions are often of first order, which is natural for states with such different properties.

In several materials of the strongly correlated electron family, giant responses to external fields have been observed in the region of phase competition. A typical and widely studied example is the case of the colossal magnetoresistance (CMR) in manganites [32, 33]. At particular chemical compositions, these materials present a ferromagnetic metallic ground state, but they display unexpected resistivity versus temperature curves. The resistivity has insulating characteristics, namely it increases with decreasing temperature, above the ferromagnetic ordering temperature T_C, followed by a sharp peak at T_C upon cooling (sometimes involving a first-order transition, as discussed below), and finally it drops to a metallic state as the temperature is lowered further (see Figure 1.12 for the case of $La_{1-x}Ca_xMnO_3$). The shape of a typical CMR resistivity versus temperature curve is considered exotic because it is unusual to have an insulator as the high-temperature "normal" state of a low-temperature metallic ferromagnet. However, the most curious property is obtained by immersing the material into an external magnetic field of just a few Teslas. In this case, the resistivity peak is found to be rapidly suppressed, generating as a consequence a large and negative magnetoresistance effect, which was dubbed "colossal" magnetoresistance in the early days of its discovery in 1994 [32, 33]. Although technological applications of this effect in computer-read sensors will need T_C to be raised by a substantial factor to comfortably reach room temperature, nevertheless the CMR phenomenon already defines a fundamental science puzzle that many groups have been trying to decipher for several years.

Present understanding of the CMR effect is based on phase competition. Of the several manganites obtained by varying the trivalent and divalent ions in the chemical formula, only those with relatively small bandwidths show the most prominent CMR effect (the CMR effect in large bandwidth manganites is still present but with a reduced magnitude). Concomitant with a small bandwidth, insulating states that compete with the ferromagnetic metal are often found in

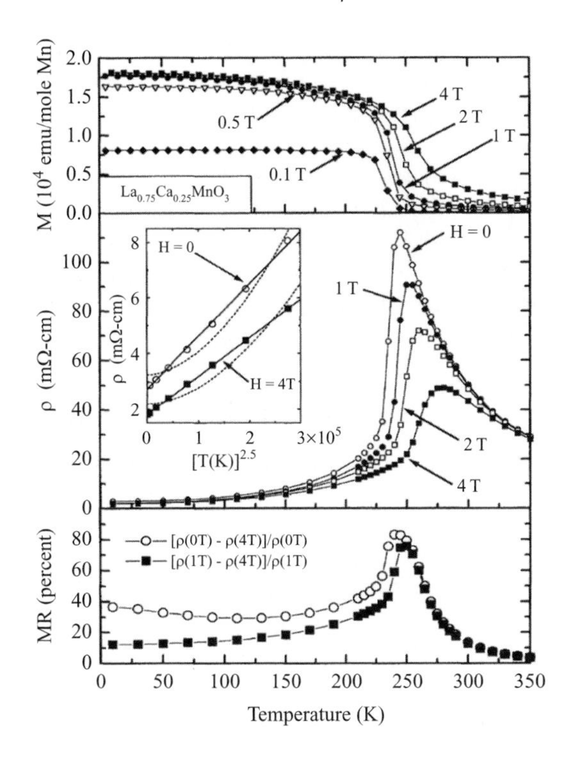

Fig. 1.12 Magnetization, resistivity, and magnetoresistance of the manganite $La_{0.75}Ca_{0.25}MnO_3$, as a function of temperature and for various magnetic fields, illustrating the presence of the CMR effect in this compound. Reprinted with permission from [33]. Copyright 1995 by the American Physical Society.

phase diagrams. The reason is that having a small bandwidth enhances the relative importance of electron–phonon couplings, and complex charge, orbital, and spin-order arrangements with overall insulating properties are found at low temperatures, as discussed earlier in this chapter.

Theoretically, it was argued early on that the region of phase competition between the ferromagnetic metal and the insulator is where CMR is found [26]. This hypothesis was confirmed by recent state-of-the-art computer simulations based on the double-exchange model for manganites, supplemented by a robust coupling to Jahn–Teller lattice displacements [34]. Some representative theoretical curves are shown in Figure 1.13. It is clear that these results, which were obtained using Monte Carlo simulations employing large computer clusters or even supercomputers, display resistivity curves that much resemble those obtained in experiments, including first-order characteristics in the CMR transition under some circumstances, such as when being in close proximity to the competing insulator in the clean limit. It can be said that the CMR effect has now been "trapped" in a box, in the sense that it appears in controlled numerical studies of small clusters of about 100 Mn atoms, and it is now up to theorists to "ask" the proper questions of the computers in order to understand qualitatively its origin at a deeper level.

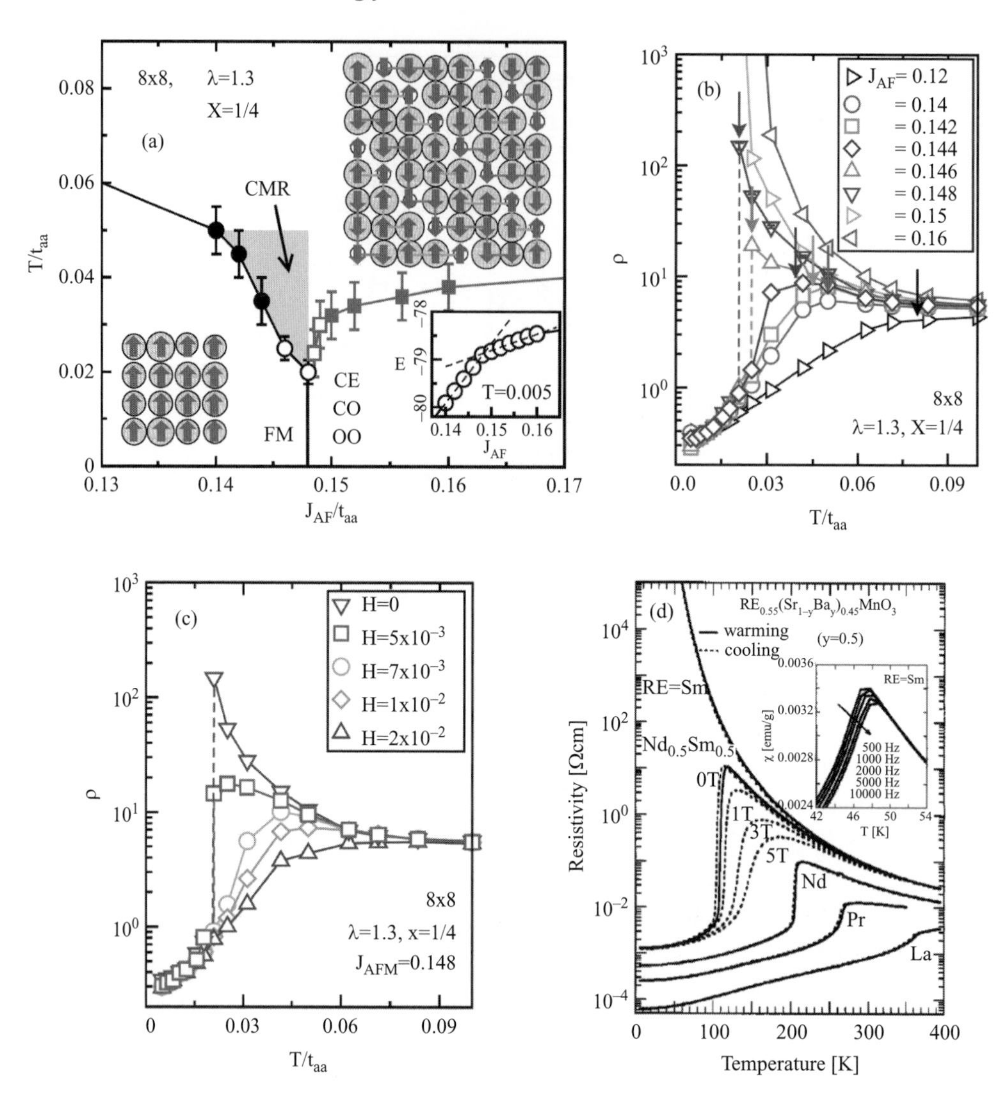

Fig. 1.13 Results of a computer simulation corresponding to a double-exchange two-orbital model for the CMR manganites, adapted from [34]. Shown here are computationally calculated resistivities versus temperature curves, parametric with the couplings in the model, as described in more detail in the original publication. (a) contains the phase diagram at hole doping $x = 1/4$, with the presence of two competing states (FM metal and CE/CO/OO insulator). The characteristic resistivity peak of materials exhibiting colossal magnetoresistance is observed in this study, see panel (b). In panel (c) the rapid reduction of the resistivity with increasing magnetic fields H is shown. Panel (d) contains experimental results [38] to illustrate how similar theory and experiment are.

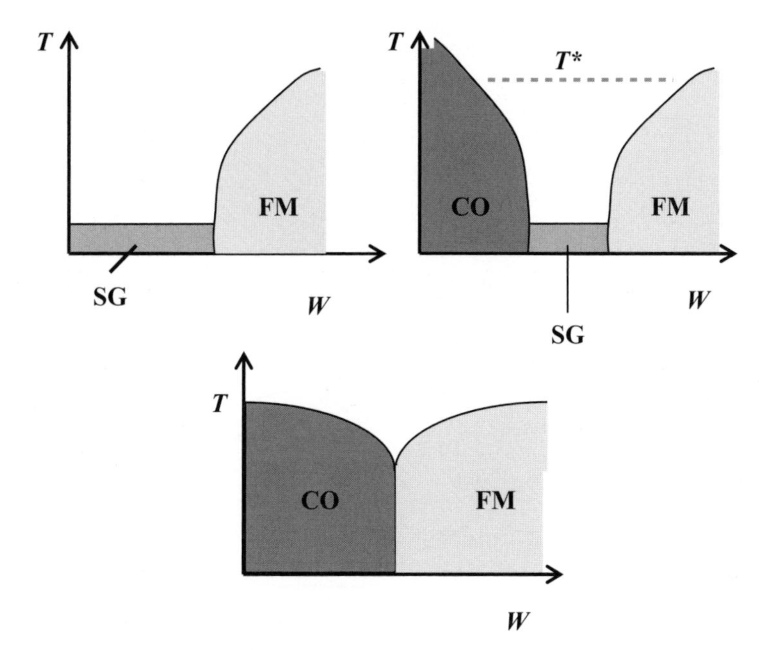

Fig. 1.14 Generic temperature (T)–bandwidth (W) phase diagram of manganites in the region of competition between the FM metallic state and a charge-ordered (CO) (typically also orbitally ordered (OO) and spin antiferromagnetic (AF)) insulator. The lower panel is the case of the "clean limit" without quenched disorder. Here the transition FM–CO is first order. The CMR effect can be observed, as shown in [34, 35], but only in a narrow region close to the FM–CO transition. The upper right panel illustrates the case of weak/moderate quenched disorder, where a spin-glass (SG) region is created in between the two competing phases. T^* is a remnant of the clean limit transitions, and it is the characteristic temperature where clusters of the competing phases appear upon cooling. The upper left panel corresponds to large quenched disorder, where the results of Figure 1.15 indicate that the CO phase disappears. For more details see [2, 26, 28, 36] and references therein.

The same theoretical/computational investigations that lead to Figure 1.13 have also suggested [34, 35] that in the insulating region above T_C the short-distance charge and spin order closely resemble those of the insulating state that competes with ferromagnetism. Perhaps induced by large entropy effects, at temperatures higher than T_C, the system behaves as though the insulator were the dominant state. However, T_C is where the energy prevails over the entropy, and the true metallic ground state is stabilized upon cooling. Moreover, it has been observed [26, 28, 34, 35] both theoretically and experimentally, that quenched disorder caused by chemical doping enhances the window in parameter space where these effects are observed. A schematic phase diagram, including both the bandwidth and disorder strengths, was predicted by theory and confirmed by experiments. In this phase diagram, shown in Figures 1.14 and 1.15, an original first-order transition is smeared by disorder, and it is in the intermediate region, which has glassy properties and mixed phase characteristics, where the colossal effects can be observed [36].

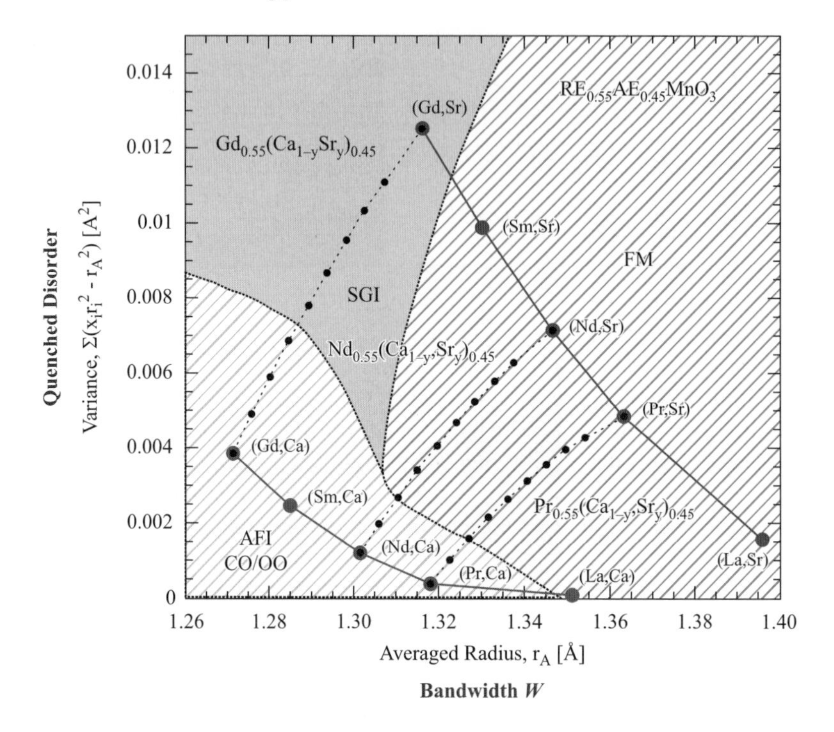

Fig. 1.15 Experimental phase diagram of manganites obtained by varying the bandwidth W and quenched disorder strength (variance). Note the presence of three regimens: the FM metallic phase induced by double exchange, the CO insulating phase (also OO and AF), and a spin-glass insulator (SGI) induced by quenched disorder. Figure adapted from [38].

Let us discuss in more detail Figure 1.13 which illustrates the exciting results obtained in recent calculations, where the properties of the double-exchange model for manganites were studied using computational techniques (more specifically, the Monte Carlo technique supplemented by the Exact Diagonalization method for the e_g electronic sector [27, 34, 35, 37]), employing a small cluster with only 64 manganese sites. These small clusters may seem indeed very tiny compared with the huge number of atoms in a real material. However, it has been extensively shown, via a variety of examples, that the qualitative aspects of the physics of these model Hamiltonians can still be captured, at least qualitatively, using small systems with about 100 active sites (unless very subtle properties such as critical exponents near critical temperatures are needed). Shown in Figure 1.13 (a) is the phase diagram of the analyzed model, varying the coupling J_{AF} among the localized spins and the temperature T, at a fixed value of the electron–phonon coupling. For relatively small values of J_{AF}, the ground state at hole concentration ¼ (equivalent to $x = 0.25$ in the $La_{1-x}Ca_xMnO_3$ language) is ferromagnetic, as expected due to the double exchange mechanism discussed before in this chapter. However, increasing J_{AF} (or at a fixed J_{AF} increasing the electron–phonon coupling), eventually an insulating state is always reached, with a complicated pattern of charge, spin, and orbital order (see sketch in Figure 1.13(a)). But finding an insulating state that is competing with the ferromagnetic metallic state is not the main result. Much more important is the behavior of the resistivity versus temperature in these small clusters, when tuning the J_{AF} coupling to be close to the transition from one phase to the other. Figure 1.13(b) shows typical examples. These results show that inside the FM phase and far from the region of competition, the curve is metallic as it should be. Reciprocally, inside the insulating phase, and also far from the competing phase, the curve is clearly insulating, as it should be. The surprises are in the middle: rather than simply switching from metal- to insulator-like resistivity curves by varying J_{AF}, there is a narrow range of couplings where, above T_C the resistivity is insulating (namely increasing with decreasing temperatures), but when T_C is reached upon cooling then a first-order (i.e. discontinuous) transition transforms the insulating-like behavior into a metal. Thus, in a narrow regime of couplings, the low-temperature ground state is FM and metallic, but the state above T_C much resembles that of the competing insulator. These results are in excellent agreement with experiments, such as those shown for instance in Figure 1.12.

Figure 1.13(c) contains the results when an external magnetic field is added. In this case, a rapid suppression of the resistivity can be observed, even for relatively small magnetic fields, producing a very large negative magnetoresistance similar to that observed experimentally (see Figure 1.12). Figure 1.13(d) contains experimental data showing how similar those results are when compared with theory [38]. In conclusion, via intensive numerical calculations (which use the resources of clusters of computers with hundreds of nodes, or supercomputers with thousands of nodes, see [34, 35]) the complex behavior of models for CMR manganites is starting to be unveiled, and thus far agreement with experiments is remarkable. So it is not difficult to imagine that similar techniques could also be used for the study of superlattices involving manganites, as described elsewhere in this book, bringing theory and experiments closer together in this promising area of research.

The giant nonlinearity with magnetic fields observed in CMR manganites, widely believed to originate in phase competition, is merely one example of the potentially enormous responses that several correlated-electron systems could display. In fact, the sketch in Figure 1.14 actually does not depend in any detail on the properties of the competing states, but only depends on the existence of the competition itself. Note, however, that if one of the phases is not ferromagnetic, then the external "perturbation" that triggers the giant response becomes less obvious than those generated by a uniform magnetic field as in manganites. In fact, recent investigations have suggested that giant responses could occur in underdoped cuprates as well. In this case, the external perturbation is conjectured to be due to the proximity to a well-developed superconductor, which may induce superconductivity in an otherwise non-superconducting underdoped cuprate [30], which is believed to contain superconducting clusters where the superconducting amplitude is developed but the phases between clusters are random. More generally, it might occur that the exotic properties of the famous pseudogap regime of the high-temperature superconductors could originate in phase competition between the parent compound, which is an antiferromagnetic insulator, and the superconducting state itself [30]. In addition, many other correlated-electron materials could potentially present similar exotic nonlinear responses to special perturbations. Surprises of this kind might be found in several compounds of the correlated-electron family.

Note that the nonlinearities described here tend to occur concomitantly with nanometer-length-scale inhomogeneities [2, 26, 28, 36]. For example, a variety of experiments in hole-doped manganites have shown convincingly that, in the CMR regime there are nanometer length-scale clusters with characteristics similar to those of the charge-ordered state that is stable at half-doping $x = 0.5$ in the LCMO phase diagram, even though the ground state at low temperatures at the values of x investigated is a ferromagnetic metal [26, 28]. Even before the experimental investigations in this context established these results, theoretical calculations had already predicted strong tendencies towards inhomogeneous states, a phenomenon widely known as phase separation, as previously described in this chapter. This kind of self-organization of electronic degrees of freedom also occurs in some high-temperature superconductors, and also in nickel oxides where states containing striped arrangements have been identified, as previously mentioned. As also discussed previously, experiments [29] have shown that other kinds of inhomogeneous states, with irregular patterns, could occur in Cu-oxide superconductors above the critical temperature (Figure 1.11).

1.9 Importance of quenched disorder and strain

A variable often overlooked in the consideration of the physics of materials, including correlated electron systems, is the strength of *quenched disorder* in the sample under study. This disorder may merely be caused by impurities or imperfections in the growing process. But it may be intrinsic as well, such as in the cases where chemical doping is needed to add or remove carriers from a particular material, such as when a trivalent ion is randomly replaced by a divalent ion. Not only do these two types of ions have different valences but they also have different sizes which tends to induce lattice distortions, as already discussed when addressing how the bandwidths of material can be altered by chemical substitution.

In the area of manganites, experiments for the case of $Ln_{1/2}Ba_{1/2}MnO_3$, which admits both ordered and disordered crystal forms with regard to the location of the Ln and Ba ions, have unveiled the profound influence of disorder on the properties of these materials (and by extension, in many other SCE compounds as well). Figure 1.16 shows the phase diagrams for both the ordered and disordered versions of this compound (for details see [39]). In the ordered case, a first-order transition separates the FM and CO/OO phases, similarly to Figures 1.13(a), 1.14, and 1.15 for the clean limit. However, in the disordered case, the Curie temperatures for ferromagnetism are substantially reduced, and a spin glass (SG) new phase is created between the FM and CO/OO phases (the latter not shown in the figure). Even more dramatic and crucial for the issues discussed here, are the transport properties of these two versions of the same material. Figure 1.16 shows that, while the ordered version has a standard insulating behavior, the disordered one actually presents the CMR effect. Thus in this case, experiments show that disorder is *needed* for giant nonlinear behavior to appear in these manganites. Presumably, the existence of nonlinearities originating in phase competition may be present in other similar compounds as well.

Also, Cu-oxide high-temperature superconductors may be influenced substantially by quenched disorder. Figure 1.17 contains a revised phase diagram recently proposed for the cuprates [40], including not only the canonical temperature and hole-doping variables, but also the strength of quenched disorder as a "new" axis. In this phase diagram, the material called YBCO corresponds to a fairly clean case, and in this situation the transition from the AF state to the superconducting phase appears to occur via a weak first-order transition or a quantum critical point (this issue is still under discussion). In the other extreme, some one-layer cuprates, such as the 214 compounds, have the strongest disorder and an intermediate spin glass phase

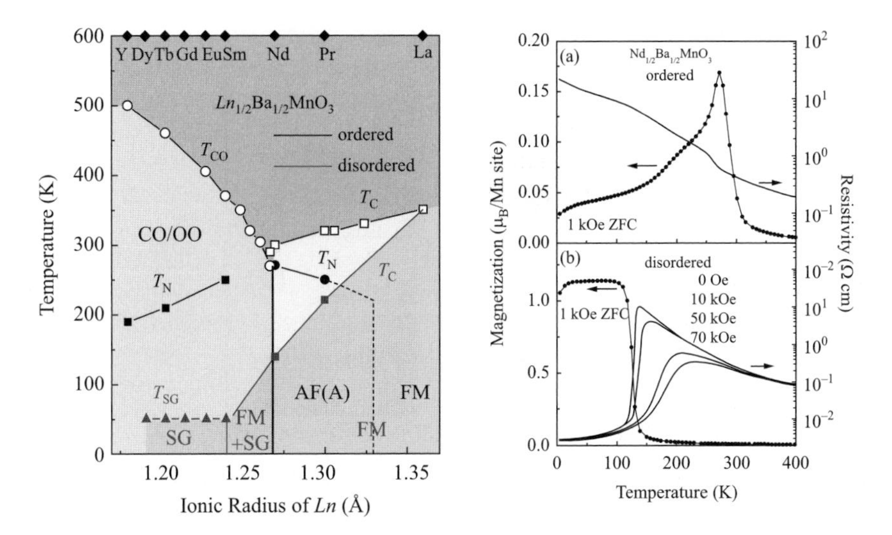

Fig. 1.16 *(Left panel)* Phase diagram for the A-site ordered (black symbols) and disordered (gray triangles and squares symbols) perovskites with half doping $Ln_{1/2}Ba_{1/2}MnO_3$, as a function of the ionic radius of Ln. CO/OO, FM, and SG are the charge/orbital ordered, ferromagnetic, and spin-glass states, respectively. Note that for the case of the A-site ordered compound, the CO/OO and FM phases are in contact and the critical temperatures are between 300 K and 500 K. However, in the disordered version, the critical temperatures are much reduced, the SG phase appears, and the CO phase has disappeared in the range of chemical doping investigated. *(Right panel)* Magnetization and resistivity of ordered and disordered $Nd_{1/2}Ba_{1/2}MnO_3$. Note that the A-site disordered material shows the CMR effect near the Curie critical temperature, while the ordered version does not, illustrating the importance of quenched disorder in this context. All these results are reproduced from [39] where more information can be found.

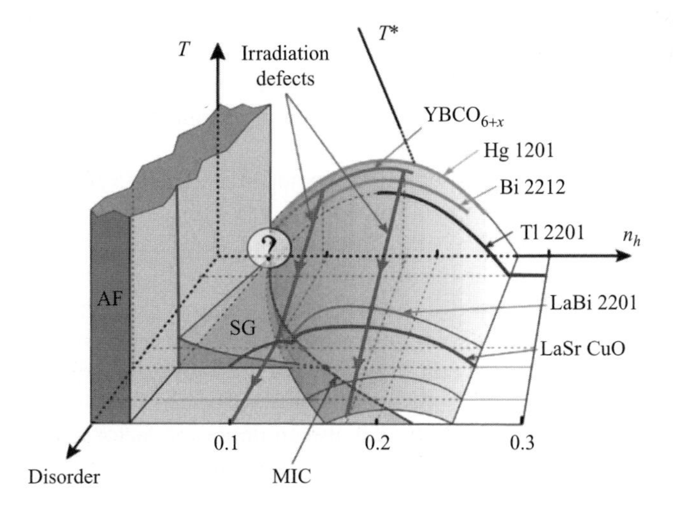

Fig. 1.17 Schematic phase diagram of the high-temperature superconducting cuprates as a function of temperature, hole doping n_h, and disorder strength, reproduced from [40]. Shown are: the AF phase in the undoped case, the superconducting phase upon doping, and the spin-glass (SG) state obtained by increasing the amount of disorder. These experimental results were obtained via resistivity measurements and with the disorder induced by electron irradiation [40].

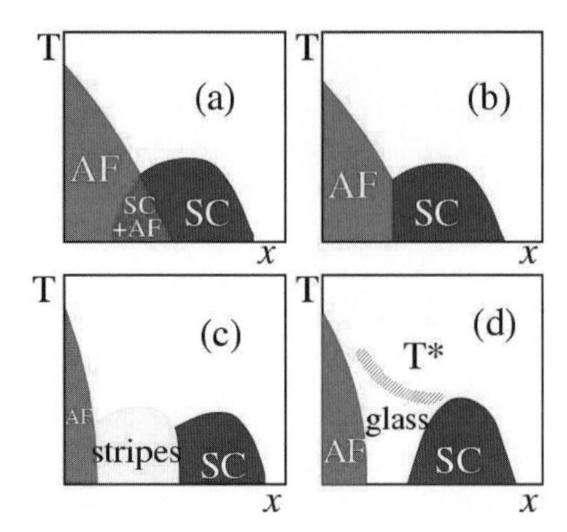

Fig. 1.18 Schematic phase diagram illustrating the possible ways in which an antiferromagnetic (AF) state can compete with a superconducting (SC) state (for details see [30, 31]). In panel (a), the transition from AF to SC states occurs via a region of homogeneous coexistence of both order parameters. In (b), a first-order transition separates both phases. In (c), a region with stripes made of the competing phases is located in between the competing states. These three cases (a,b,c) can be found in the clean limit [30, 31]. In (d), the modified phase diagram of (a) is shown after quenched disorder is introduced. In this case, a spin-glass phase is generated in between the AF and SC states, as in the widely discussed experimental phase diagram of the single-layer "214" cuprate compound. In this context, T^* is a remnant of the original clean limit phase transitions. For additional details, the reader should consult [30, 31].

is formed, similarly to the case of manganites with disorder. Theoretical calculations [30, 31] using phenomenological models have suggested that the phase diagram of cuprates in the clean limit could have a region of uniform coexistence of AF and SC states, or a first-order transition separating those states, or even stripes made out of AF and SC phases. But it is only after quenched disorder is incorporated that a spin glass is formed and a new temperature scale T^* is obtained, a remnant of the original clean-limit phase transitions, reproducing the famous 214 phase diagram (Figure 1.18).

As a consequence, the possibility of producing doped materials in artificial superlattices via the transfer of charge from one component to another (for instance from $LaMnO_3$to $SrMnO_3$ as in the cases described by Bhattacharya, Dong, and Yu, elsewhere in this book) is interesting because this procedure avoids the intrinsic randomness introduced by chemical substitution. Thus, it could potentially be expected that new phase diagrams may emerge employing oxide superlattices for the electronic doping procedure, as opposed to chemical doping via ionic substitution, which could be different from those of the alloy counterparts. This is certainly an exciting area of research worth pursuing.

In addition to quenched disorder *strain* is also an important factor to be considered in the study of transition metal oxides. For instance, by growing thin films of $La_{1-x}Sr_xMnO_3$ on different substrates, the lattice constants of LSMO can be modified, since they tend to adjust to the values corresponding to the substrate during the growing process. By this procedure the ratio c/a, where $c(a)$ is the lattice constant along the z axis (the x or y axes), can be

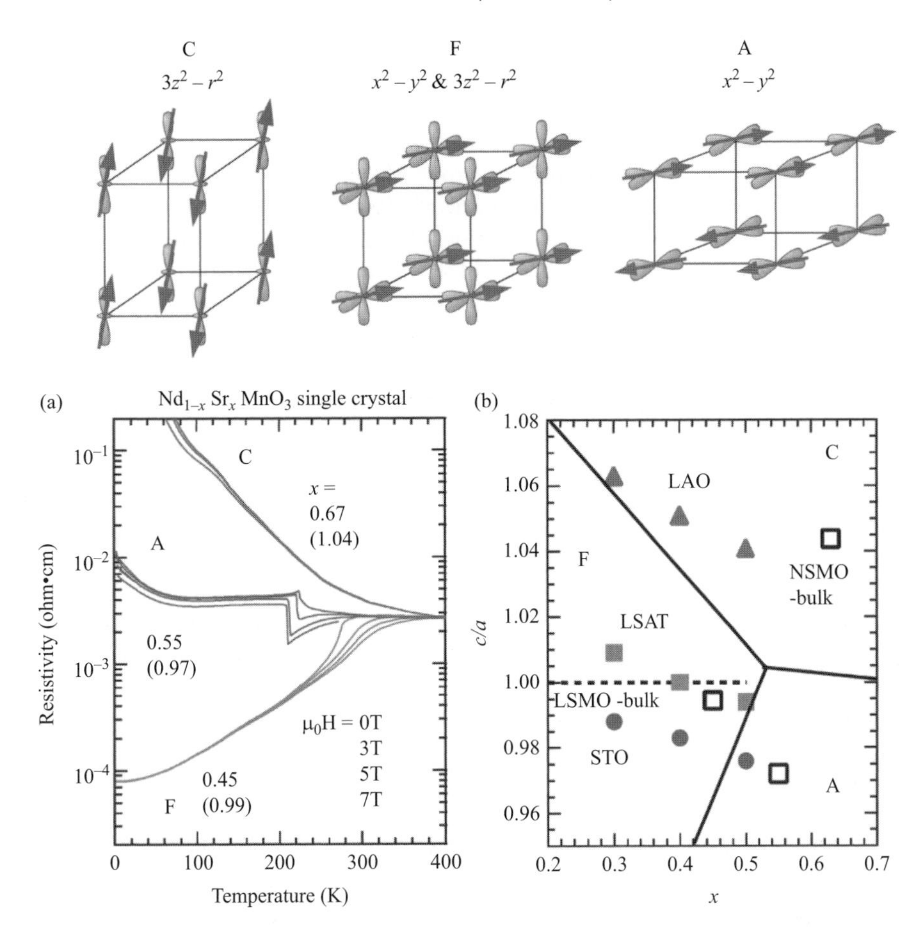

Fig. 1.19 A phase diagram of perovskite manganese oxides, reproduced from [1]. The top panel shows schematically the orbital and spin order realized in the hole-doped manganese oxides corresponding to C, F, and A-type phases. Lower panel (a) contains the temperature dependence of the resistivity in various magnetic fields for the case of $Nd_{1-x}Sr_xMnO_3$, and for the three phases at the top of the figure. The numbers in parentheses represent the uniaxial lattice strain, more specifically the c/a ratio, indicating the coupling of the magnetism to the orbital order. Lower panel (b) is the schematic phase diagram varying the lattice strain c/a and the doping level x. The data labelled LAO, LSAT, and STO represent results for the coherently strained epitaxial thin films of $La_{1-x}Sr_xMnO_3$ grown on the perovskite single-crystal substrates of $LaAlO_3$, (La,Sr)(Al,Ta)O_3, and $SrTiO_3$, respectively. LSMO-bulk and NSMO-bulk stand for the results for the bulk single crystals of $La_{1-x}Sr_xMnO_3$ and $Nd_{1-x}Sr_xMnO_3$, respectively. For details and references see [1].

altered and the relative stability of the competing FM, A-type AF, C-type AF, and G-type AF phases can be modified [41]. *Ab-initio* theoretical calculations [42] are in agreement with the experimental results. The importance of strain has also been remarked on in several other more recent theoretical [43, 44] and experimental [45, 46, 47] studies, and it provides an additional degree of freedom to control the properties of SCE materials. In fact, the phase diagrams of these compounds, which are routinely provided only in terms of the temperature and electronic concentration, should also include other axes, such as the strength of quenched disorder previously discussed, and the degree of strain, to fully understand the physics of these materials, as in Figure 1.19.

1.10 Outlook for correlated-electron technology: spintronics, double perovskites, multiferroics, orbitronics, resistance switching

The use of both the spin and charge degrees of freedom is an essential feature for the emerging field of spin-electronics, or spintronics [48]. Its straightforward application is control of the electrical current by an external magnetic field. In fact, invention of both the giant-magnetoresistive magnetic multilayer (composed of transition metals) and tunneling magnetoresistance (TMR) devices were crucial milestones for the field of spintronics and its industrial applications.

For general spintronic use, a half-metal, meaning a metallic ferromagnet with perfect spin polarization in the ground state, is of considerable importance. Several of the ferromagnetic metallic transition-metal oxides with strong electron correlation are expected to show such a half-metallic ground state, because of the potentially strong spin-charge coupling. The hole-doped manganites with a perovskite structure, such as $La_{1-x}Sr_xMnO_3$ $(0.2 < x < 0.5)$, provide a typical example of a half-metal mediated by the previously discussed strong Hund coupling between the e_g itinerant electron spin and the t_{2g} localized spins. However, the TMR characteristics for the junction that uses $La_{1-x}Sr_xMnO_3$ are known to show unexpected degradation with increasing temperature, say, up to 200 K, despite the fact that their Curie temperatures are much higher (330–370 K). This seemingly rapid fading out of the spin polarization is believed to be due to modification of the interface magnetism of $La_{1-x}Sr_xMnO_3$ $(0.2 < x < 0.5)$ when facing the insulating barrier. To attack this problem, alternative engineered magnetic interfaces are being explored extensively [49], as discussed in the remainder of the book. Moreover, the emerging novel properties and possible functionalities at the interfaces between correlated oxides [49, 50, 51] represent a new and important area for investigations, thus it is of considerable importance to study this issue. In the remainder of this book, several chapters will address this type of research at interfaces.

More robust and higher-T_C half-metals have also been sought for the family of perovskites to improve their performance in spintronic devices. A consequence of such investigations was discovery of the ordered double perovskite family with half-metallic characteristics, represented by $Sr_2B_1B_2O_6$ (B_1= Fe or Cr, B_2= Mo or Re) [52, 53]. In this class of compounds, the perovskite B (transition metal) sites are alternately occupied by two elements B_1 and B_2 in a rock-salt form. The valence of Fe (or Cr) is 3+, corresponding to the maximum spin state $S = 5/2(3/2)$. The Mo^{5+} $(4d^1)$ or Re^{5+} $(5d^2)$ ion provides the conduction electrons, partially hybridizing with

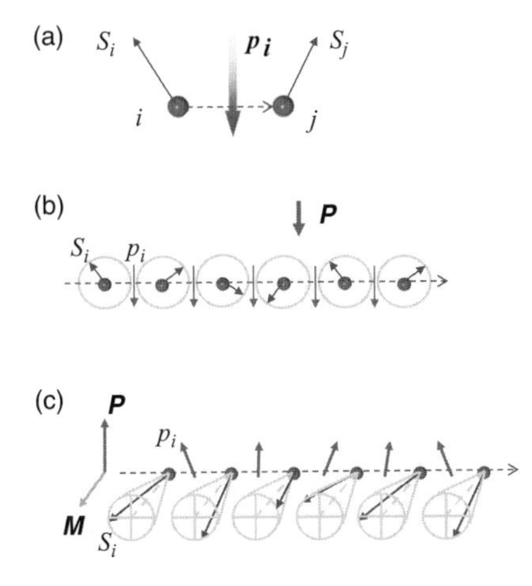

Fig. 1.20 Possible spin superstructures in multiferroics that can present a strong coupling between magnetism and electricity. (a) Canted spins on neighboring atomic sites. (b) Spin spiral structure producing a uniform polarization (P). (c) Conical spin structure allowing both uniform magnetization (M) and P.

the spin-down state of Fe or Cr, and strongly coupling antiferromagnetically with the local spins (spin-up states) on Fe or Cr. Thus, the states near the Fermi level are composed only of the spin-down electrons, forming the half-metallic ground state. The ferromagnetic transition temperature can be very high, for example, 420 K for Sr_2FeMoO_6 [52], and as high as 615 K for Sr_2CrReO_6 [53]. Possible antiferromagnetic metals with a half-metallic ground state, which is a highly spin-polarized state without any magnetization, have also been sought in the family of ordered perovskites [54, 55], but so far attempts have not been successful because of the correlation-induced charge-gap opening.

Another important group of compounds related to spintronic oxides are the multiferroics, materials with coexisting ferroelectric and magnetic orders, which can potentially host a gigantic magnetoelectric (ME) effect. Concerning the coupling between magnetism and ferroelectricity, a new simple scheme has recently been theoretically [56, 57, 58] and experimentally [59] demonstrated: mutually canted spins (such as in Figure 1.20(a)), that may be generated by spin frustration effects [60], can produce an electronic polarization through the spin–orbit interaction. When the spins form a transverse-spiral (cycloidal) modulation along a specific crystallographic direction (Figure 1.20(b)), then every nearest-neighbor pair produces a unidirectional polarization, p_i, which is proportional to the cross product between those nearest-neighbor spins, and, hence, a macroscopic polarization P should be generated. The direction of polarization may be fully determined by the clockwise or counterclockwise rotation of the spins that are proceeding along the spiral propagation axis, called the spin helicity.

In these cycloidal spin compounds, typically, perovskite $TbMnO_3$ and $DyMnO_3$, spontaneous polarization can easily be controlled using an external magnetic field of a specific direction (since the magnetic field can control the orientation of the spin helicity), inducing generation and/or flipping of the spontaneous polarization [59], which can be viewed as a gigantic ME effect. Multiferroics based on this mechanism have also been realized in the conical

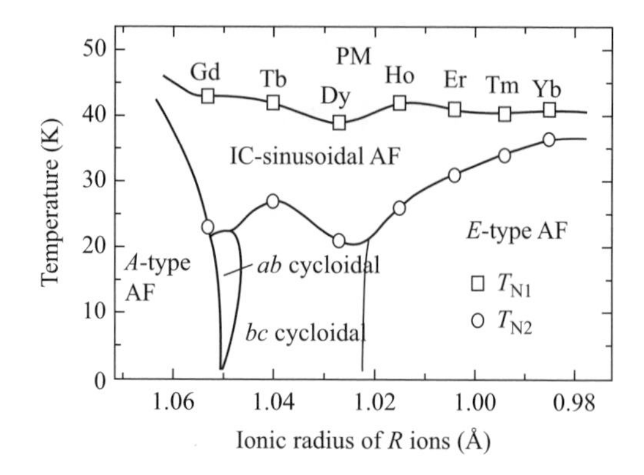

Fig. 1.21 Experimental phase diagram of the multiferroic manganites showing the A-type antiferromagnetic phase (that appears e.g. in $LaMnO_3$) followed, by reducing the rare earth radius *R*, by spin spiral (sinusoidal and cycloidal) phases, where ferroelectricity was found at low temperatures for the cycloidal arrangement, and finally ending with the E-type phase (which is a complex form of an antiferromagnetic state, involving zigzag chains) as the *R* radius is reduced further. More details can be found in [59] (see also [60] for a theoretical discussion and references).

spin state (Figure 1.20 (c)) where the transverse spiral component coexists with the uniform (spontaneous or field-induced) magnetization component along the cone axis, and hence, flexible control of the polarization vector is possible using external magnetic fields. These cycloidal and transverse-conical spin states are often observed in complex transition-metal compounds such as spinels and perovskites, where competing exchange interactions of neighboring spins can cause such a periodically modulated spin structure. As a natural extension of these studies, electrical control of the magnetization vector is currently being explored by several groups as a new and potentially important spintronic function.

To complete the discussion about multiferroics and in order to properly visualize these materials in the context of phase competition, Figure 1.21 shows the phase diagram of the undoped manganite multiferroics with generic chemical formula $REMnO_3$ [59]. Starting with decreasing the ionic radius of the rare earth element from La down to Gd, it is known that the ground state presents A-type antiferromagnetic characteristics. However, as chemical substitution of La by other RE elements proceeds further, the critical temperature toward the A-type AF phase clearly decreases and eventually a state with non-collinear spin order, and concomitant ferroelectric properties in view of the previous discussion, is stabilized. Finally, for the cases of Ho to Yb, another exotic state with so-called E-type antiferromagnetic order is stabilized. Thus, the spiral state that is needed for multiferroicity emerges from the competition between the A- and E-type antiferromagnetic states [60].

Note that hole- or electron-doped manganites may also induce the ferroelectric state as argued for their charge-orbital ordered states theoretically [61] or experimentally [62, 63]. Furthermore, recent computational studies have suggested that for sufficiently low bandwidth, the hole doping of the manganites may lead to a complex exotic state made from stripes that could be multiferroic [64]. While this prediction still remains to be tested experimentally, here

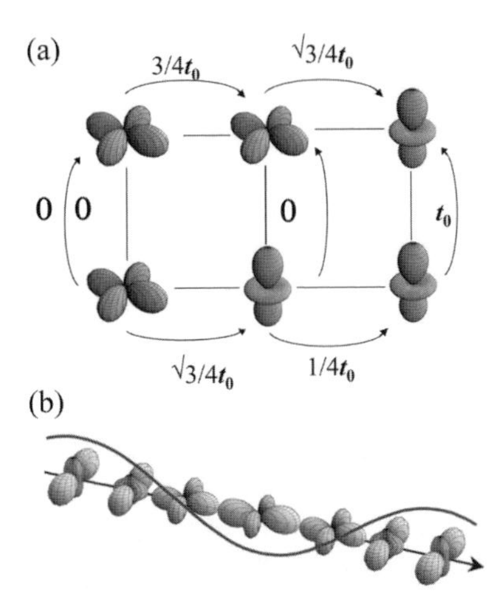

Fig. 1.22 (a) Electron hopping amplitudes between the nearest-neighbor sites involving the various orbital states shown. t_0 stands for the transfer integral unit. (b) Schematic representation of the quantal orbital wave, or orbiton, of the orbital-ordered state.

what it is important to remark is that the field of doped manganite multiferroics deserves considerable attention due to the potential for interesting surprises that still remains.

In analogy with spintronics, one can also consider the possibility of controlling electric currents through the *d*-electron *orbital* state [1]. This potential correlated-electron technology is called orbital-electronics or orbitronics [65]. In a broad context, the CMR phenomenon itself amounts to a field-induced modification of orbital correlation, and hence, it could provide one such example. By analogy to the magnetoresistance effect in the spin-ordered state, one can utilize the orbital degree of freedom to regulate the electrical conduction. For example, when the orbital ordering composed of $d_{x^2-y^2}$ orbitals is realized in the nearly cubic perovskite lattice, the charge dynamic is highly anisotropic, more specifically it is confined within the xy plane and insulating along the z direction, as can be seen from the transfer hopping values shown in Figure 1.22(a). In reality, such uniform orbital ordering is present ubiquitously in overdoped manganites, and a highly two-dimensional charge motion is observed despite the nearly cubic lattice structure [66].

The key idea of orbitronics is ultrafast switching of the orbital state, and hence of the related spin-charge state, by means of electric fields and/or light irradiation. Since the orbital shape, rod-like or planar, represents the electron's probability-density distribution, the orbital degree of freedom can inherently couple with the electric field through its anisotropic polarizability. In this context, the orbital manipulation might bear analogy to liquid-crystal technology, in which rod-shaped or planar molecules can respond to the electric field through their anisotropic polarizability.

To describe the dynamical response of the orbital to external fields, it is necessary to define the orbital wave or orbiton in the orbital-ordered state (Figure 1.22(b)). An advantage of the use of the orbital degree of freedom for the ultrafast control of electronic and magnetic states is that the orbiton frequency is high [67], 10–100 THz, as compared to the typical spin precession frequency (i.e. the $\mathbf{k} = \mathbf{0}$ magnon energy in a ferromagnet) of 1–100 GHz. In fact, in the initial process of photoinduced insulator–metal transition in the perovskite manganite ($Pr_{0.7}Ca_{0.3}MnO_3$), the orbiton mode of 30 THz is observed in ultrafast optical spectroscopy [68].

Correlated-electron science is thus exploiting a broad range of materials and electronic properties, in addition to the famous high-T_c and CMR-related phenomena that have received so much attention. Noteworthy is the recent advance of epitaxial-growth technology for transition-metal oxide thin films and superlattices, as well as discovery of many intriguing properties at the hetero-interfaces, described in more detail in several chapters of this book. At the interface, the spin and orbital, as well as charge, states are greatly modified, and they may be subject to external magnetic and/or electric fields that can induce large nonlinear effects. One of the potential applications of such phenomena is the *resistance switching* memory effect of the correlated-oxide interface with metal electrodes [69, 70]; the mechanisms are still to be clarified, but the possible application of this effect to high-density, fast, nonvolatile memory devices (resistive random-access memory, or ReRAM) is now being investigated extensively. Another interesting example is the creation of a polar ferromagnet with broken inversion symmetry at the interface [71, 72], in which many new intriguing magnetoelectronic functions, such as nonlinear/nonreciprocal magneto-optical and dynamical magnetoelectric effects, can be observed because of the simultaneous breaking of space-inversion and time-reversal symmetries.

1.11 Conclusions

The study of correlated electronic materials continues at a fast pace. Close interaction between theory and experiment has been crucial to progress in this field. The complexity of these compounds, manifested in their rich phase diagrams, self-organization, and nonlinear responses, suggests their potential use in devices. Even leaving aside technological applications, which will take considerable time to realize, the interesting properties of these materials clearly define an exciting field of fundamental scientific research that is full of surprises, and will surely continue to provide exciting and challenging phenomena in the near future.

Acknowledgments

E. Dagotto is supported by the U.S. Department of Energy, Office of Basic Energy Sciences, Materials Sciences and Engineering Division and by the National Science Foundation under Grant No. DMR-11-04386. The work of Y. Tokura was partly supported by the Funding Program for World-Leading Innovative R&D on Science and Technology (FIRST) on "Quantum Science on Strong-correlation".

References

[1] Tokura Y. and Nagaosa N. (2000). Orbital Physics in Transition-Metal Oxides. *Science* **288**, 462–468.

[2] Dagotto E. (2005). Complexity in Strongly Correlated Electronic Systems. *Science* **309**, 257–262.

[3] Mott N.F. (1990). *Metal-Insulator Transitions*. second edition. Taylor and Francis, London.

[4] Dagotto E. and Tokura Y. (2008). Strongly Correlated Electronic Materials: Present and Future. *MRS Bulletin* **33**, 1037–1045.
[5] Kamihara Y., Watanabe T., Hirano M., and Hosono H. (2008). Iron-Based Layered Superconductor $La[O_{1-x}F_x]FeAs$ (x = 0.05–0.12) with T_c = 26 K. *J. Am. Chem. Soc.* **130**, 3296–3297.
[6] Lynn J.W. and Dai P. (2009). Neutron studies of the iron-based family of high Tc magnetic superconductors. *Physica C* **469**, 469–476; and references therein; Lumsden M.D. and Christianson A.D. (2010). Magnetism in Fe-based superconductors. *J. Phys.: Condens. Matter* **22**, 203203; and references therein; Johnston D.C. (2010). The puzzle of high temperature superconductivity in layered iron pnictides and chalcogenides. *Adv. Phys.* **59**, 803–1061 and references therein.
[7] Yu R., Trinh K.T., Moreo A., *et al.* (2009). Magnetic and metallic state at intermediate Hubbard *U* coupling in multiorbital models for undoped iron pnictides. *Phys. Rev. B* **79**, 104510 (16 pages).
[8] Committee on CMMP 2010. (2007). *Condensed-Matter and Materials Physics: The Science of the World Around Us*. National Academies Press, Washington, DC.
[9] Basic Energy Sciences Advisory Committee (BESAC) Grand Challenges Subcommittee. (2007). *Directing Matter and Energy: Five Challenges for Science and the Imagination*. U. S. Department of Energy, Washington, DC.
[10] Imada M., Fujimori A., and Tokura Y. (1998). Metal-insulator transitions. *Rev. Mod. Phys.* **70**, 1039–1263.
[11] Torrance J.B., Lacorre P., Nazzal A.I., Ansaldo E.J., and Niedermayer Ch. (1992). Systematic study of insulator-metal transitions in perovskites $RNiO_3$ (R=Pr,Nd,Sm,Eu) due to closing of charge-transfer gap. *Phys. Rev. B* **45**, 8209–8212.
[12] Dagotto E. (1994). Correlated electrons in high-temperature superconductors. *Rev. Mod. Phys.* **66**, 763–840; and references therein.
[13] Armitage N.P., Fournier P., and Greene R.L. (2010). Progress and perspectives on electron-doped cuprates. *Rev. Mod. Phys.* **82**, 2421–2487; and references therein.
[14] Tokura Y., Taguchi Y., Fujishima Y., *et al.* (1993). Filling dependence of electronic properties on the verge of metal–Mott-insulator transition in $Sr_{1-x}La_xTiO_3$. *Phys. Rev. Lett.* **70**, 2126–2129.
[15] Okimoto Y., Katsufuji T., Okada Y., Arima T., and Tokura Y. (1995). Optical spectra in $(La,Y)TiO_3$: Variation of Mott–Hubbard gap features with change of electron correlation and band filling. *Phys. Rev. B* **51**, 9581–9588; and references therein. See also [14].
[16] Nakatsuji S., private communication. See also Nakatsuji S. and Maeno Y. (2000). Quasi-Two-Dimensional Mott Transition System $Ca_{2-x}Sr_xRuO_4$. *Phys. Rev. Lett.* **84**, 2666–2669; Nakatsuji S., Dobrosavljević V., Tanasković D., *et al.* (2004). Mechanism of Hopping Transport in Disordered Mott Insulators. *Phys. Rev. Lett.* **93**, 146401 (4 pages).
[17] Tokura Y. (2006). Critical features of colossal magnetoresistive manganites. *Rep. Prog. Phys.* **69**, 797–852.
[18] Foo M.L., Wang Y., Watauchi S., *et al.* (2004). Charge Ordering, Commensurability, and Metallicity in the Phase Diagram of the Layered Na_xCoO_2. *Phys. Rev. Lett.* **92**, 247001 (4 pages).
[19] Sawada H., Hamada N., Terakura K., and Asada K.T. (1996). Orbital and spin orderings in YVO_3 and $LaVO_3$ in the generalized gradient approximation. *Phys. Rev. B* **53**, 12742–12749.

[20] Kanamori J. (1959). Superexchange interaction and symmetry properties of electron orbitals. *J. Phys. Chem. Solids* **10**, 87–98.

[21] Moreira dos Santos A., Cheetham A.K., Atou T., *et al.* (2002). Orbital ordering as the determinant for ferromagnetism in biferroic $BiMnO_3$. *Phys. Rev. B* **66**, 064425 (4 pages).

[22] Chen C.H., Cheong S.-W., and Cooper A.S. (1993). Charge modulations in $La_{2-x}Sr_xNiO_{4+y}$: Ordering of polarons. *Phys. Rev. Lett.* **71**, 2461–2464.

[23] Tranquada J.M., Sternlieb B.J., Axe J.D., Nakamura Y., and Uchida S. (1995). Evidence for stripe correlations of spins and holes in copper oxide superconductors. *Nature* **375**, 561–563.

[24] Kivelson S.A., Fradkin E., and Emery V.J. (1998). Electronic liquid-crystal phases of a doped Mott insulator. *Nature* **393**, 550–553.

[25] Murakami Y., Kawada H., Kawata H., *et al.* (1998). Direct Observation of Charge and Orbital Ordering in $La_{0.5}Sr_{1.5}MnO_4$. *Phys. Rev. Lett.* **80**, 1932–1935.

[26] Dagotto E., Hotta T., and Moreo A. (2001). Colossal Magnetoresistant Materials: The Key Role of Phase Separation. *Phys. Rep.* **344**, 1–153.

[27] Yunoki S., Hu J., Malvezzi A., *et al.* (1998). Phase Separation in Electronic Models for Manganites. *Phys. Rev. Lett.* **80**, 845–848. For experimental studies also addressing phase separation see Uehara M., Mori S., Chen C.H., and Cheong S.-W. (1999). Percolative phase separation underlies colossal magnetoresistance in mixed-valent manganites. *Nature* **399**, 560–563.

[28] Dagotto E. (2002). *Nanoscale Phase Separation and Colossal Magnetoresistance*. Springer-Verlag, Berlin.

[29] Gomes K., Pasupathy A.N., Pushp A., *et al.* (2007). Visualizing pair formation on the atomic scale in the high-Tc superconductor $Bi_2Sr_2CaCu_2O_{8+\delta}$. *Nature* **447**, 569–572. For other STM results related with inhomogeneous states see for instance Lang K., Madhavan V., Hoffman J.E., *et al.* (2002). Imaging the granular structure of high-Tc superconductivity in underdoped $Bi_2Sr_2CaCu_2O_{8+\delta}$. *Nature* **415**, 412–416.

[30] Alvarez G., Mayr M., Moreo A., and Dagotto E. (2005). Areas of superconductivity and giant proximity effects in underdoped cuprates. *Phys. Rev. B* **71**, 014514 (7 pages); Mayr M., Alvarez G., Moreo A., and Dagotto E. (2006). One-particle spectral function and local density of states in a phenomenological mixed-phase model for high-temperature superconductors. *Phys. Rev. B* **73**, 014509 (15 pages).

[31] Alvarez G. and Dagotto E. (2008). Fermi Arcs in the Superconducting Clustered State for Underdoped Cuprate Superconductors. *Phys. Rev. Lett.* **101**, 177001 (4 pages).

[32] Jin S., Tiefel T.H., McCormack M., *et al.* (1994). Thousandfold Change in Resistivity in Magnetoresistive La-Ca-Mn-O Films. *Science* **264**, 413–415.

[33] Schiffer P., Ramirez A.P., Bao W., and Cheong S.-W. (1995). Low Temperature Magnetoresistance and the Magnetic Phase Diagram of $La_{1-x}Ca_xMnO_3$. *Phys. Rev. Lett.* **75**, 3336–3339.

[34] Sen C., Alvarez G., and Dagotto E. (2010). First Order Colossal Magnetoresistance Transitions in the Two-Orbital Model for Manganites. *Phys. Rev. Lett.* **105**, 097203 (4 pages).

[35] Sen C., Alvarez G., and Dagotto E. (2007). Competing Ferromagnetic and Charge-Ordered States in Models for Manganites: The Origin of the Colossal Magnetoresistance Effect. *Phys. Rev. Lett.* **98**, 127202 (4 pages); and references therein.

[36] Burgy J., Mayr M., Martin-Mayor V., Moreo A., and Dagotto E. (2001). Colossal Effects in Transition Metal Oxides Caused by Intrinsic Inhomogeneities. *Phys. Rev. Lett.* **87**, 277202 (4 pages).

[37] For other techniques to handle this problem without diagonalizing exactly the fermionic sector at each step, see Kumar S. and Majumdar P. (2006). Insulator–Metal Phase Diagram of the Optimally Doped Manganites from the Disordered Holstein-Double Exchange Model. *Phys. Rev. Lett.* **96**, 016602 (4 pages); and references therein,

[38] Tomioka Y. and Tokura Y. (2004). Global phase diagram of perovskite manganites in the plane of quenched disorder versus one-electron bandwidth. *Phys. Rev. B* **70**, 014432 (5 pages); and references therein.

[39] Akahoshi D., Uchida M., Tomioka Y., *et al.* (2003). Random Potential Effect near the Bicritical Region in Perovskite Manganites as Revealed by Comparison with the Ordered Perovskite Analogs. *Phys. Rev. Lett.* **90**, 177203 (4 pages).

[40] Rullier-Albenque F., Alloul H., Balakirev F., and Proust C. (2008). Disorder, metal–insulator crossover and phase diagram in high-Tc cuprates. *EPL* **81**, 37008 (6 pages); and references therein.

[41] Konishi Y., Fang Z., Izumi M., *et al.* (1999). Orbital-State-Mediated Phase-Control of Manganites. *J. Phys. Soc. Jpn.* **68**, 3790–3793.

[42] Fang, Z., Solovyev I.V., and Terakura K. (2000). Phase Diagram of Tetragonal Manganites. *Phys. Rev. Lett.* **84**, 3169–3172.

[43] Ahn K.H., Lookman T., and Bishop A.R. (2004). Strain-induced metal–insulator phase coexistence in perovskite manganites. *Nature* **428**, 401–404.

[44] Dong S., Yunoki S., Zhang X., *et al.* (2010). Highly anisotropic resistivities in the double-exchange model for strained manganites. *Phys. Rev. B* **82**, 035118 (6 pages).

[45] Wu W., Israel C., Hur N., *et al.* (2006). Magnetic imaging of a supercooling glass transition in a weakly disordered ferromagnet. *Nature Mater.* **5**, 881–886; Yamada H., Kawasaki M., Lottermoser T., Arima T., and Tokura Y. (2006). $LaMnO_3/SrMnO_3$ interfaces with coupled charge-spin-orbital modulation. *Appl. Phys. Lett.* **89**, 052506 (3 pages); Dhakal T., Tosado J., and Biswas A. (2007). Effect of strain and electric field on the electronic soft matter in manganite thin films. *Phys. Rev. B* **75**, 092404 (4 pages); Dekker M.C., Rata A.D., Boldyreva K., *et al.* (2009). Colossal elastoresistance and strain-dependent magnetization of phase-separated $(Pr_{1-y}La_y)_{0.7}Ca_{0.3}MnO_3$ thin films. *Phys. Rev. B* **80**, 144402 (7 pages); Ding Y., Haskel D., Tseng Y.-C., *et al.* (2009). Pressure-Induced Magnetic Transition in Manganite ($La_{0.75}Ca_{0.25}MnO_3$). *Phys. Rev. Lett.* **102**, 237201 (4 pages).

[46] Ward T.Z., Budai J.D., Gai Z., *et al.* (2009). Elastically driven anisotropic percolation in electronic phase-separated manganites. *Nature Phys.* **5**, 885–888; and references therein.

[47] Yamada H., Ogawa Y., Ishii Y., *et al.* (2004). Engineered Interface of Magnetic Oxides. *Science* **305**, 646–648.

[48] Zutic I., Fabian J., and Das Sarma S. (2004). Spintronics: Fundamentals and applications. *Rev. Mod. Phys.* **76**, 323–410; and references therein.

[49] Ohtomo A. and Hwang H.Y. (2004). A high-mobility electron gas at the $LaAlO_3/SrTiO_3$ heterointerface. *Nature* **427**, 423–426.

[50] Reyren N., Thiel S., Caviglia A.D., *et al.* (2007). Superconducting Interfaces Between Insulating Oxides. *Science* **317**, 1196-1199. For additional references see also Dagotto E. (2007). When Oxides Meet Face to Face. *Science* **318**, 1076–1077.

[51] Okamoto S., and Millis A. (2004). Electronic reconstruction at an interface between a Mott insulator and a band insulator. *Nature* **428**, 630–633.
[52] Kobayashi K.I., Kimura T., Sawada H., Terakura K., and Tokura Y. (1998). Room-temperature magnetoresistance in an oxide material with an ordered double-perovskite structure. *Nature* **395**, 677–680.
[53] Kato H., Okuda T., Okimoto Y., Tomioka Y., Takenoya Y., Ohkubo A., Kawasaki M., and Tokura Y. (2002). Metallic ordered double-perovskite Sr_2CrReO_6 with maximal Curie temperature of 635 K. *Appl. Phys. Lett.* **81**, 328 (3 pages).
[54] Pickett W.E. (1996). Single Spin Superconductivity. *Phys. Rev. Lett.* **77**, 3185–3188.
[55] Pardo V. and Pickett W.E. (2009). Compensated magnetism by design in double perovskite oxides. *Phys. Rev. B* **80**, 054415 (6 pages).
[56] Katsura H., Nagaosa N., and Balatsky A.V. (2005). Spin Current and Magnetoelectric Effect in Noncollinear Magnets. *Phys. Rev. Lett.* **95**, 057205 (4 pages).
[57] Sergienko I.A., and Dagotto E. (2006). Role of the Dzyaloshinskii-Moriya interaction in multiferroic perovskites. *Phys. Rev. B* **73**, 094434 (5 pages).
[58] Sergienko I.A., Sen C., and Dagotto E. (2006). Ferroelectricity in the Magnetic E-Phase of Orthorhombic Perovskites. *Phys. Rev. Lett.* **97**, 227204 (4 pages); Picozzi S., Yamauchi K., Sanyal B., Sergienko I. A., and Dagotto E. (2007). Dual Nature of Improper Ferroelectricity in a Magnetoelectric Multiferroic. *Phys. Rev. Lett.* **99**, 227201 (4 pages).
[59] Ishiwata S., Kaneko Y., Tokunaga Y., Taguchi Y., *et al.* (2010). Perovskite manganites hosting versatile multiferroic phases with symmetric and antisymmetric exchange strictions. *Phys. Rev. B* **81**, 100411(R) (4 pages); and references therein. See also Kimura T., Goto T., Sintani H., *et al.* (2003). Magnetic control of ferroelectric polarization. *Nature* **426**, 55–58.
[60] Dong S., Yu R., Yunoki S., Liu J.-M., and Dagotto E. (2008). Origin of multiferroic spiral spin order in the $RMnO_3$ perovskites. *Phys. Rev. B* **78**, 155121 (6 pages); and references therein.
[61] Efremov D.V., van den Brink J., and Khomiskii D. (2005). Bond centered versus site-centered charge ordering: ferroelectricity in oxides. *Physica B* **359**, 1433–1435.
[62] Tokunaga Y., Lotteroser Th., Lee Y.S., *et al.* (2006). Rotation of orbital stripes and the consequent charge-polarized state in bilayer manganites. *Nature Mat.* **5**, 937–941.
[63] Lopes A.M.L., Araujo J.P., Amaral V.S., *et al.* (2008). New Phase Transition in the $Pr_{1-x}Ca_xMnO_3$ System: Evidence for Electrical Polarization in Charge Ordered Manganites. *Phys. Rev. Lett.* **100**, 155702 (4 pages).
[64] Dong S., Yu R., Liu J.-M., and Dagotto E. (2009). Striped Multiferroic Phase in Double-Exchange Model for Quarter-Doped Manganites. *Phys. Rev. Lett.* **103**, 107204 (4 pages); and references therein.
[65] Tokura Y. (2003). Correlated-Electron Physics in Transition-Metal Oxides. *Physics Today* **56**, 50–55.
[66] Kuwahara H., Okuda T., Tomioka Y., Asamitsu A., and Tokura Y. (1999). Two-Dimensional Charge-Transport and Spin-Valve Effect in the Layered Antiferromagnet $Nd_{0.45}Sr_{0.55}MnO_3$. *Phys. Rev. Lett.* **82**, 4316–4319.
[67] Saitoh E., Okamoto S., Takahashi K.T., *et al.* (2001). Observation of orbital waves as elementary excitations in a solid. *Nature* **410**, 180–183.
[68] Polli D., Rini M., Wall S., *et al.* (2007). Coherent orbital waves in the photo-induced insulator–metal dynamics of a magnetoresistive manganite. *Nature Mat.* **6**, 643–647.

[69] Lu S.Q., Wu N.J., and Ignatiev A. (2000). Electric-pulse-induced reversible resistance change effect in magnetoresistive films. *Appl. Phys. Lett.* **76**, 2749 (3 pages).

[70] Sawa A., Fujii T., Kawasaki M., and Tokura Y. (2004). Hysteretic current–voltage characteristics and resistance switching at a rectifying $Ti/Pr_{0.7}Ca_{0.3}MnO_3$ interface. *Appl. Phys. Lett.* **85**, 4073 (3 pages).

[71] Yamada H., Ogawa Y., Kawasaki M., and Tokura Y. (2002). Perovskite oxide tricolor superlattices with artificially broken inversion symmetry by interface effects. *Appl. Phys. Lett.* **81**, 4793 (3 pages).

[72] Kida N., Yamada H., Sato H., *et al.* (2007). Optical Magnetoelectric Effect of Patterned Oxide Superlattices with Ferromagnetic Interfaces. *Phys. Rev. Lett.* **99**, 197404 (4 pages).

2
Magnetoelectric coupling and multiferroic materials

GUSTAU CATALAN AND JAMES F. SCOTT

2.1 Introduction: magnetoelectric coupling and multiferroic materials

The subject of magnetoelectric coupling (coupling between dielectric polarization and magnetization) is not new. In the nineteenth century [1], Pierre Curie had already discussed the symmetry requirements required for electric and magnetic fields to couple statically (dynamic coupling whereby an oscillating magnetic field causes a voltage and vice versa was, of course, already enshrined in Maxwell's laws). Nevertheless, it was not until the second half of the twentieth century that Dzyaloshinskii [2] and Moriya [3] described the physical mechanism whereby breaking space inversion symmetry (which dielectric polarization does) can lead to weak (canting) magnetization in antiferromagnets. This type of bilinear ME coupling, henceforth known as the Dzyaloshinskii–Moriya interaction, was invoked by Igor Dzyaloshinskii to predict that the bilinear ME effect (electrically induced magnetization or magnetically induced polarization) should exist in Cr_2O_3 [4]. Experimental verification of this prediction was provided by Astrov shortly afterwards [5, 6].

Though the first experimental verification of the ME effect was for an oxide (Cr_2O_3), other families of materials were subsequently discovered to show linear ME coupling. Hans Schmid and co-workers, in Geneva, reported strong coupling between polarization and magnetization in nickel-iodine boracide $Ni_3B_7O_{13}I$ in 1966 [7] (see Figure 2.1), while Astrov and co-workers reported magnetoelectric effects in fluorides [8, 9, 10, 11, 12], later shown by Fox and Scott to be caused by ferroelectrically induced spin canting [13, 14, 15]. In spite of the early interest, however, it was soon recognized that the physics of the Dzyaloshinskii–Moriya effect is such that linear ME effects are intrinsically small (more on this later), and this probably led to the ME effect remaining a backwater of solid state research for several decades. In the last decade, however, this situation has changed dramatically, and magnetoelectric multiferroics (materials that are simultaneously ferroelectric and magnetic, and where both order parameters are coupled) have taken off exponentially [16]. This "revival of the magnetoelectric effect", to use the expression coined in the already classic review by Fiebig [16], probably has three interlinked causes:

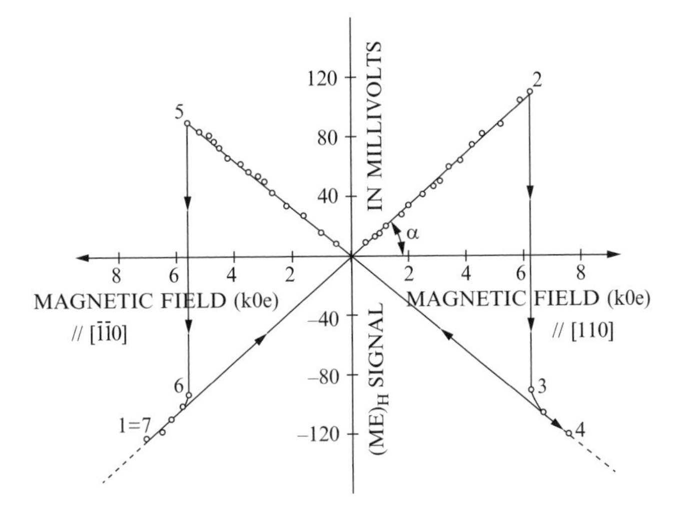

Fig. 2.1 Magnetically induced voltage caused by magnetoelectric switching of the ferroelectric polarization in nickel-iodine boracite. Adapted from [7]. Copyright 1966, American Institute of Physics.

1. One is the publication of two different reports, in 2003, on multiferroic properties in perovskite oxides $TbMnO_3$ [17] and $BiFeO_3$ [18] (while the latter was known as a multiferroic, its intrinsic properties, and in particular its high polarization, were only uncovered when high quality epitaxial thin films were made by the Ramesh group in Berkeley). Perovskites are a favourite of solid-state research due to their different compositions being able to accommodate a miriad of different functional properties such as ferroelectricity, superconductivivity, or colossal magnetoresistance. Hence, many scientists already familiar with these oxides on account of their other properties find it easy to make the transition to magnetoelectricity.
2. Perovskite oxides also have the added bonus that their structure is rather simple and lends itself easily to epitaxial thin film growth techniques, so that novel heterostructures can be made that may be useful in nanoelectronics. In connection with this, ways to exploit ME coupling for thin film-based spintronic applications have been proposed [19]. In fact, the area of thin film magnetoelectrics is so important in its own right that is discussed in a separate chapter in this book.
3. Finally, the importance of magnetoelectric couplings other than linear (such as strain-mediated biquadratic coupling, for example) has also been recognized, opening the field for new practical applications. These, in fact, may end up reaching the market long before applications based on linear coupling.

The present chapter will be divided as follows: In Section 2.2 we begin with an introduction to the physics of magnetoelectric coupling and magnetoelectric multiferroics, including the origin and relative magnitude of linear and higher-order couplings, and the physics of magnetoelectric composites. This will be followed by a catalogue of the different families of magnetoelectric multiferroics (Section 2.3). Special attention will be paid to fluorides as there are no reviews about these in the modern literature. There are also two appendices. Appendix A deals with technical aspects of measuring magnetoelectric multiferroics, paying attention to the kind of effects that can easily be mistaken for true magnetoelectricity; this should be particularly useful for those venturing into magnetoelectric research for the first time. Appendix B, on the other hand, will discuss in some detail the critical exponents involved in isostructural phase transitions, an issue of relevance in connection with recent reports of such transitions in $BiFeO_3$, the most popular magnetoelectric multiferroic.

2.2 Magnetoelectric coupling

Magnetoelectricity is the cross-coupling between electric polarization and magnetism in a material. The polarization and magnetism of the crystal may be remanent (as in magnetoelectric multiferroics) or field-induced. Thermodynamically, this is described by adding a cross-coupling term to the free energy electrically-induced magnetization [16]:

$$F = F_0 + \varepsilon E^2 + \mu H^2\text{-}EP\text{-}MH\text{-}\alpha EH\text{-}\beta EH^2 - \delta E^2H\text{-}\gamma E^2H^2 + \ldots O(E^4, H^4) \tag{2.1}$$

The last four terms in this expansion describe different types of magnetoelectric coupling. Of these, the two most studied are bilinear coupling (αEH) and biquadratic coupling (γE^2H^2). The former was the first to be theoretically described and also the first to be found experimentally (in Cr_2O_3, see next section). The latter, on the other hand, has the advantage of being allowed for all crystal symmetries, and moreover, it can lead to bigger magnetoelectric voltages than linear coupling, thanks to the clever use of strain mediating mechanisms in composites. At present, there appears to be little research on linear-quadratic coupling terms.

Minimization of free energy leads to expressions for the magnetically induced polarization and electrically-induced magnetization:

$$P_{ME} = \alpha H + \beta H^2 + 2\delta EH + 2\gamma EH^2 + \ldots \tag{2.2}$$

$$M_{ME} = \alpha E + 2\beta EH + \delta E^2 + 2\gamma E^2H + \ldots \tag{2.3}$$

In practical terms, it is often better to describe magnetoelectric coupling, not in terms of magnetically induced polarization, but magnetically induced voltage (since the latter can be measured directly while polarization can only be indirectly inferred [21]). In order to do this, we divide the polarization by the permittivity to obtain the magnetically-induced electric field. Restricting ourselves to the linear coupling term, then, $E_{ME}(H) = \frac{\alpha}{\varepsilon}H = \alpha' H$, where $\alpha' \equiv \frac{\alpha}{\varepsilon_0 \varepsilon_r}$, where ε_0 is the permittivity of a vacuum and ε_r is the relative dielectric constant of the material. In terms of units, α has dimensions of time divided by space, and is typically given in ps m^{-1}. The magnetovoltaic coefficient α' has dimensions of voltage divided by magnetic field, and is typically given in units of mV Oe^{-1} (or mV Oe^{-1} cm^{-1} for the magnetoelectric coefficient).

2.2.1 Linear coupling: Dzyaloshinskii–Moriya effect, electrically induced spin canting, and Shtrikman limit

Dielectric polarization breaks space inversion symmetry. Consider a system polarized along z: if we invert the polarization, the crystal will look as though we had turned it upside down, $P(-z) = -P(z)$. Magnetization, on the other hand, breaks *time* inversion symmetry. This concept may sound at first a little more confusing, but it is not: imagine a very small coil through which a current circulates. This current will generate a magnetic field. If we invert the sense of the current (which is the same as imagining the original current running backwards in time), then the sense of the magnetic field will be inverted. Although of course magnetic spins are not currents, the symmetry principle still stands (and indeed that is why the spin is called spin: it was originally thought to originate from the rotation of electrons!). Thus, $M(-t) = -M(t)$. Therefore, a multiferroic where the polarization P and the magnetization M are bilinearly coupled needs to break both time and space inversion symmetry simultaneously. One physical mechanism that allows this to happen is the Dzyaloshinskii–Moriya effect [2–4]. This effect is caused by an energy contribution to the Hamiltonian of a crystal whereby the spins in an otherwise antiferromagnetic system can be canted; in a magnetoelectric, the amount of canting is directly proportional to the size of an intermediate electric dipole. Mathematically, the Dzyaloshinskii–Moriya energy term is expressed as $E_{DM}=\mathbf{D}\bullet[\mathbf{S}_1 \times \mathbf{S}_2]$, and it is depicted schematically in Figure 2.2. Here $\mathbf{D}_{12}$ is a term proportional to the size of the electric dipole and to the distance between the spins, that is, $\mathbf{D}_{12} \propto \mathbf{P} \times \mathbf{r}_{12}$.

The magnetoelectric Dzyaloshinskii–Moriya effect allows for a voltage-induced electric dipole to cause spin canting, or for magnetically induced spin canting to cause an electric dipole, in otherwise paraelectric and antiferromagnetic Cr_2O_3 [4, 5, 6]. There are also systems where a permanent (ferroelectric) dipole causes permanent spin canting, like $BaMnF_4$ [14, 15], and, *ab-initio* calculations predict that this mechanism may allow certain perovskite oxides to show bilinear multiferroic coupling [20]. A variation on the theme of DM-based magnetoelectric coupling is that dielectric polarization can be induced by a spin cycloid, and vice versa. The physics behind "spiral magnetoelectrics" have been discussed in papers by Sergienko *et al.* [22], and Mostovoy [23], as well as a review co-authored by the latter and Sang-Wook Cheong [24].

The most important use of bilinear coupling is the possibility of reversing the sign of the magnetization using a voltage, as this would allow new types of memory devices. Although the net magnetization in most of these systems is weak (it arises from canting), it can be magnified by the use of exchange bias, which allows control of the magnetization of thin ferromagnetic layers grown on top of a multiferroic, an idea that was first demonstrated by Borisov *et al.*

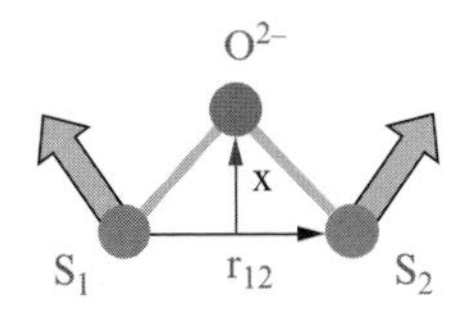

Fig. 2.2 Schematic illustration of the Dzyaloshinskii–Moriya interaction and how the electric dipole causes spin canting in an antiferromagnet. Figure courtesy of M. Mostovoy, University of Groningen.

in Duisburg using magnetoelectric Cr_2O_3 [25]. This material, however, is not ferroelectric and therefore the induced changes in magnetism were not permanent. Implementation of this idea using a switchable (multiferroic) material, $YMnO_3$, was first achieved by Laukhin and co-workers in Barcelona [26].

The Dzyaloshinskii–Moriya interaction can only exist under very restrictive symmetry constraints, however, and thus the number of crystals that are able to show it are rather limited. Moreover, the magnetization arises from spin canting, so it can never be too large. Using thermodynamics, it was shown by Brown, Hornreich, and Shtrikman that the maximum bilinear magnetoelectric coupling is limited by the magnetic and dielectric susceptibilities of the material [27]; if the linear magnetoelectric energy term in the thermodynamic potential is $\alpha_{ij}E_iH_j$, then the magnetoelectric coupling coefficient must necessarily fulfill $\alpha_{ij}^2 < \varepsilon_{ii}\mu_{jj}$, where ε and μ are the dielectric constant and magnetic susceptibilities, respectively. In practice, the largest linear magnetoelectric coefficients experimentally measured fall many orders of magnitude below the Brown–Hornreich–Shrtikman (BHS) limit, with the current record being 30×10^{-12} sm^{-1} [16]. In magnetovoltaic units, this is of the order of 0.3 V m^{-1} Oe^{-1} (3mV cm^{-1} Oe^{-1}). As we shall see, this is not only well short of the linear BHS limit, but also several orders of magnitude smaller than what can be achieved through biquadratic magnetoelectric coupling. It has also been recently emphasized by Dzhialoshinskii ["Magnetoelectric to multiferroic transitions", EPL 96 (2011) 17001] that in multiferroic phase transitions the linear magnetoelectric coefficient can diverge in spite of the Shtrikman limit.

2.2.2 Biquadratic (strain mediated) coupling

Biquadratic coupling between polarization and magnetization places no constraints on inversion symmetry (its sign is inversion invariant) and is thus allowed in all materials. Moreover, there are no intrinsic limitations to the size of this type of coupling. This is not only a very general type of coupling, but is in fact the one that is the most commonly exploited, and so far also the most commercially viable. The reason is that biquadratic magnetoelectricity can be easily achieved and maximized using elastic deformation (strain) as the intermediate step.

The idea behind strain-mediated biquadratic coupling is that electrostriction (coupling between deformation and electric polarization) coupled to magnetostriction (coupling between strain and magnetization) provides a natural bridge between magnetism and electricity. Crucially, for this coupling to take place, the electrostriction and magnetostriction need not reside in the same material. Thus, composites or thin-film heterostructures combining a good piezoelectric material, such as perovskite $Pb(Zr,Ti)O_3$, and a good magnetostrictive one, such as Terfenol-D, provide very large effective magnetoelectric couplings, which can be further enhanced by operating the device at mechanical resonance frequencies. Reported magnetoelectric coefficients for such devices are as large as 90 V cm^{-1} Oe^{-1}, [28] that is, 30 000 times bigger than those measured in the best single-compound linear magnetoelectric materials.

This strain-mediated coupling is not only much bigger but, importantly, the materials are usually more chemically stable and easier to make than single-compound magnetoelectrics, offering excellent performances at room temperature [29]. Variants on the strain-coupling idea seek to maximize the effect by changing the architecture of the heterostructure, e.g. columns embedded in a matrix (1-2 composites) [30] or heteroepitaxial thin films (2-2 composites) [31].

Fig. 2.3 Commercial multilayer capacitor consisting of interspersed Ni electrodes and $(Ba,Sr)TiO_3$. The picture captures both the size of such capacitors and their current price, which is only a fraction of a cent [32]. Figure courtesy of Nature Publishing Group.

It is also worth noticing that the capacitor industry has already—unwittingly—been making very adequate and extremely cheap 2-2 composite magnetoelectric transducers for some years. When the erstwhile standard palladium electrodes were replaced by nickel in multilayer capacitors, a nonmagnetic metal was being swapped for a magnetostrictive one. The dielectric material in most of these capacitors is a doped $(Ba,Sr)TiO_3$, itself a ferroelectric composition with electromechanical properties. Thus biquadratic strain-mediated coupling was readily achieved between the magnetostrictive Ni electrodes and the electrostrictive $(Ba,Sr)TiO_3$, as pointed out by Israel *et al.* [32], see Figure 2.3.

A drawback—for some applications—of biquadratic coupling is that it cannot be exploited for switching devices. Because the coupling term is $\propto P^2M^2$, it is mathematically—and, of course, physically—impossible to gain energy by inverting the sign of either P or M. This means that magnetoelectric composites based on magnetostrictive/electrostrictive coupling cannot be used for making switchable memory devices. Of course that is not the only possible application of a magnetoelectric material and indeed they remain extremely useful for magnetoelectric transducing devices (sensors, etc.) [29].

Finally, although piezoelectric/magnetostrictive composites are a very convenient way of obtaining biquadratic magnetoelectricity, they are by no means the only way. Very large biquadratic coupling has also been measured in a family of materials that are not composites in the conventional sense, but that have intrinsic heterogeneity at the nanoscale, showing both dipolar nanoclusters and spin clusters. These are the so-called relaxors, and their properties will be discussed in greater detail in a later section.

2.2.3 Perovskite oxides: why are they seldom multiferroic?

Nicola Hill (now Spaldin) has asked the important question of why do there seem to be so few magnetic ferroelectrics among ABO_3 perovskites [33]? Some reasons are phenomenological, such as the fact that ferromagnetism in ABO_3 perovskites sometimes comes about from double exchange, an itinerant electron mechanism that leads to metallic behaviour and thus precludes ferroelectric switching. But she also points to a more subtle and arguably more important effect, which is that ferroelectricity in ABO_3 perovskites is usually caused by covalent bonding between the B site and one or more of the neighboring oxygens. In this case d^0 ions such as Ti^{+4} or Zr^{+4} are favored, as the empty d^0 orbital easily bonds covalently with the *p* orbitals of the oxygen. The archetypal example of this is Ti–O bonding in $BaTiO_3$. On the other hand, magnetism in perovskites comes from electrons in the d orbital of the B site ions, i.e. they must be d^n ($n \neq 0$). This idea is powerful, but it is worth remembering that not all oxide ferroelectrics are covalent perovskites: another archetypal ferroelectric, $PbTiO_3$ for example, makes an important contribution to the polarization from the hybridization between the lone-pair s^2 orbital in the A site (Pb^{+2}) and the oxygen [34]. Therefore the argument that there will be very few multiferroic oxides should be stated as "there will be few covalently bonded perovskites that are multiferroic." In the formulation of this statement lies also the solution, already put forward by Spaldin herself [33]: in searching for perovskite multiferroics, the magnetism and the ferroelectricity should ideally reside in different sites: lone-pair polarization in the A site and d^n magnetization in the B site is presently the most popular route, and examples will be given below. Other strategies revolve around alternative routes for ferroelectricity (as in spiral multiferrroics). A few examples of these will also be given. Finally, it is also worth keeping in mind that there are many magnetic fluoride ABF_3 perovskites, such as $KMnF_3$ and $RbCoF_3$, and $KNiF_3$. The field of multiferroic fluorides is likely to lead to important advances, and will also be discussed at length.

2.3 Magnetoelectric multiferroics

2.3.1 Perovskites with ferroelectricity caused by lone-pair polarization: $BiFeO_3$

A solution to the incompatibility problem between the d^0 requirement for B-site polarization and the d^n requirement for B-site magnetization in perovskites is to have polarization coming from the A-site instead. This can be achieved by having Pb^{+2} or Bi^{+3} in the A-site: the last electronic shell of these ions is s^2, and the lone-pair of electrons in the *s* shell easily hybridizes with the oxygens, thereby producing polarization. The most famous example of this mechanism is $BiFeO_3$, which shall be discussed in some detail here, but there are other examples such as $BiMnO_3$ [33, 35, 36] or $Bi(Fe_{1/2}Cr_{1/2})MnO_3$[37, 38]. A related compound, $BiCrO_3$, turns out to be antiferroelectric rather than ferroelectric [39].

Bismuth ferrite $BiFeO_3$ is without a doubt the most important multiferroic at the moment, on account of (i) being virtually the only one that is multiferroic at room temperature, having a ferroelectric Tc of 825°C and an antiferromagnetic $T_N = 370$ °C, and (ii) having one of the largest ferroelectric polarizations measured for any material, nearly 100 μC/cm^2 along the polar direction, $<111>_{pseudocubic}$. [40] The Berkeley report from 2003 which triggered interest in this material was the first to identify the very large polarization [18], later also measured in high quality single crystals [41] and ceramics [42]. Intriguingly, a rather large saturation magnetization (150 emu/ccc) was also reported in the original 2003 report [18], which was surprising because $BiFeO_3$ is thought of as an antiferromagnet.

In bulk, $BiFeO_3$ has a basic G-type antiferromagnetic order (each ion spin is surrounded by six antiparallel spins in the nearest neighbors), slightly modified on two counts: (i) there is a weak canting between neighbouring spins, and (ii) the canting moment (and indeed the whole magnetic sublattice) rotates in a cycloid with a huge inconmensurate period (63 nm) [43, 44, 45]. Both magnetic distortions are due to magnetoelectric coupling to the polarization,

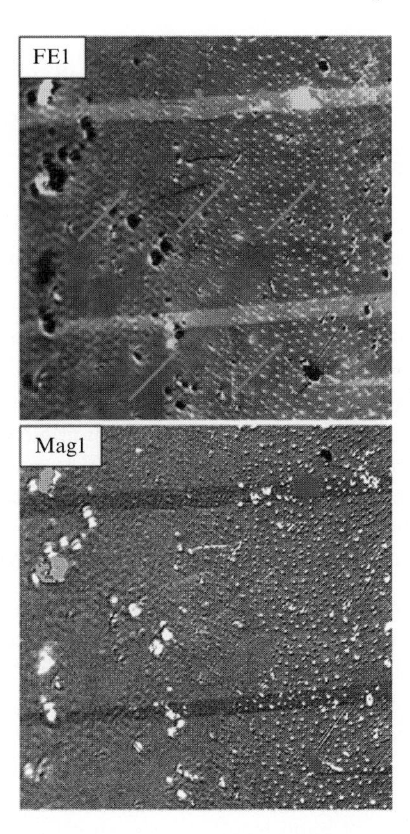

Fig. 2.4 Correlation between ferroelectric domains in a BFO single crystal (above) and exchange-biased ferromagnetic domains in a permalloy film grown on top of it (below). Reprinted with permission from [51]. Copyright 2009 American Physical Society.

and the existence of the cycloid ensures that even the weak canting moment (8 emu/c.c.) is averaged out in any macroscopic measurement, i.e. bulk $BiFeO_3$ has no net magnetization at all. The origin of the magnetization reported in the 2003 paper [18] has been the subject of much controversy ever since, with hypotheses ranging from intrinsic, such as strain-induced unwinding of the spiral and enhancing of the canting moment [46], to extrinsic, such as magnetic impurities [47, 48]. Recently, a third alternative has been proposed, whereby at least some of the magnetization may arise from the domain walls [49].

In spite of its essentially antiferromagnetic nature, $BiFeO_3$ remains a workhorse of magnetoelectric spintronics. This is because the antiferromagnetic sublattice can be coupled to thin ferromagnetic layers grown on top of $BiFeO_3$, via the mechanism of exchange bias. Moreover, in $BiFeO_3$ the antiferromagnetic sublattice can be flopped between different orientations by switching the polarization [44, 45]. Thus, indirectly, a voltage used to switch ferroelectric polarization also causes the magnetization of an exchange-biased ferromagnetic layer to switch [50, 51]. The correlation between ferroelectric domains in $BiFeO_3$ and the magnetization of an exchange-biased layer grown on top of it is illustrated in Figure 2.4 [51]. For further details,

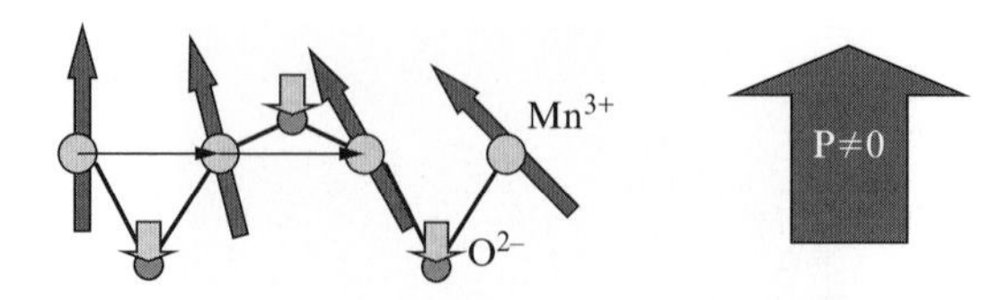

Fig. 2.5 Schematic illustration of how the spin cycloid of a manganese oxide induces ferroelectricity. Figure courtesy of Dennis Meier and Manfred Fiebig, University of Bonn.

the reader is referred to monographic reviews about the general properties of $BiFeO_3$ [40] and its spintronic applications [19].

Because $BiFeO_3$ is the most popular multiferroic, and because it exhibits not one but two phase transitions that do not change symmetry (orthorhombic Pbnm-Pbnm at high T and monoclinic Cc-Cc under epitaxial stress), it is also useful to review the critical phenomena expected at these transitions. This is discussed as an appendix at the end of this chapter.

2.3.2 Oxides with ferroelectricity caused by spins spirals: $TbMnO_3$, $TbMn_2O_5$

Spin cycloids induce polarization thanks to the Dzyaloshinskii–Moriya interaction: this interaction depends on a cross product between spins [see section 2.2.1], so a chiral spin arrangement renders a net dipole moment, as depicted in figure 2.5.

Some important oxides showing ferroelectricity caused by a spin cycloid are $TbMnO_3$[17] and CuO [52]. Another multiferroic compound with a spin cycloid is $TbMn_2O_5$ [53, 54], but in this latter case, as in others with the formula RMn_2O_5 (R = rare earth), it appears that the ferroelectricity may not be dominated by the spin cycloid, but by exchange-striction (a magnetostrictive distortion of bond lengths) between canted spins in the ab-plane [55]. Although it has had a massive revival since its rediscovery in $TbMnO_3$, it is worth emphasizing that weak ferroelectricity induced by a spin spiral is *not* new and had already been discovered and explained by the team of Newnham and Cross in Penn State, who reported weak pyroelectricity in Cr_2BeO_4, and linked it to the onset of the spiral antiferromagnetic state of this compound [56].

Although bilinear coupling effects are rather spectacular (see e.g. Figure 2.6 for an illustration of the inversion of polarization that can be achieved with magnetic fields), it must nevertheless be pointed out that the absolute magnitude of ferroelectric polarization in these types of "spiral" multiferroics tends to be very small, (nC/cm^2), about 1000 times smaller than in proper ferroelectrics such as $BaTiO_3$, $PbTiO_3$, or indeed multiferroic $BiFeO_3$. Moreover, the ferroelectric phase transition typically takes place below room temperature, and therefore these materials must presently be regarded as conceptually important, but not yet suitable candidates for most practical applications.

2.3.3 Hexagonal multiferroics: $YMnO_3$

Hexagonal rare-earth manganites have an intrinsically unstable magnetic ordering due to the geometrical frustration imposed by a triangular lattice: if two spins are antiparallel, the third one must be parallel to one of the other two. The resulting magnetic order results from the uneasy compromise between the antiferromagnetic exchange and the geometric frustration,

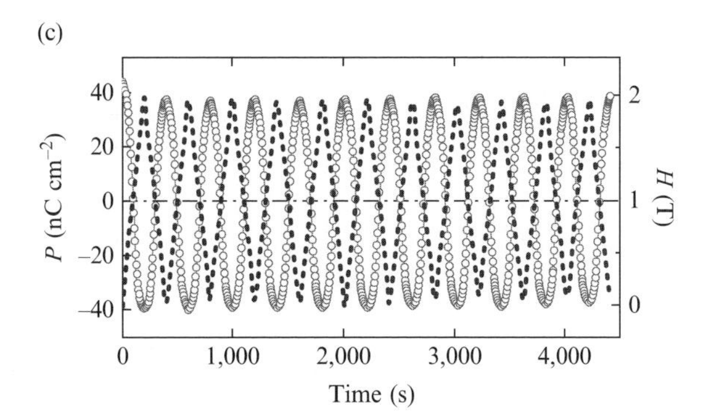

Fig. 2.6 In multiferroic $TbMn_2O_5$, the orientation of ferroelectric polarization can be switched by switching the magnetic field. Note, however, the small absolute value of the polarization, around 1000 times smaller than in conventional or "proper" ferroelectrics. Reprinted by permission from Macmillan Publishers Ltd: Nature Materials [54], copyright 2004.

and it is quite sensitive to lattice distortions, such as those caused by ferroelectricity. In the specific case of $YMnO_3$, ferroelectric ordering takes place at very high temperature, $T_C \approx 1000K$, and antiferromagnetic order takes place at $T_{Neel} = 71$ K, with strong concomitant magnetoelastic effects [57]. The exact nature and origin of the ferroelectricity in this compound has been the subject of attention from several groups. First principles calculations by Van Aken *et al.* [58] suggest that the polarization results from tilting of the Mn-centered oxygen octahedra and buckling of the Y–O planes; this mechanism is sometimes termed "geometric ferroelectricity", as opposed to the more conventional off-centering mechanism. Later studies by Fennie and Rabe suggest, on the other hand, that this is an improper ferroelectric with a transition caused by a zone-tripling phonon instability [59]. It is interesting that, in spite of being an improper ferroelectric (meaning that the primary order parameter of the transition is not the polarization itself), the ferroelectric polarization is still rather high, 6 $\mu c/cm^2$.

An important feature of hexagonal multiferroics is the strong coupling between antiferromagnetic and ferroelectric order parameters at the domain walls. This has been amply studied by Fiebig and co-workers, who point out the existence of a net magnetization at the center of the antiferromagnetic domain walls (Figure 2.7) [60] and show also that domain walls play a strong role in the magnetoelectric properties of $HoMnO_3$ [61]. The theory of domain wall multiferroic coupling was first examined by Privatska and Janovec, who listed the space groups that might allow net magnetization to develop in the domain walls of materials that are not ferromagnetic [62], while Daraktchiev and co-workers have more recently developed the quantitative phenomenological description of domain wall magnetoelectric coupling [49].

2.3.4 Unconfirmed oxide multiferroics: $RNiO_3$ (R = rare earth or Bi)

Another family of perovskite oxides that have been proposed to be multiferroics are the nickelates. There are two different arguments for proposing multiferroicity in these compounds, depending on the composition:

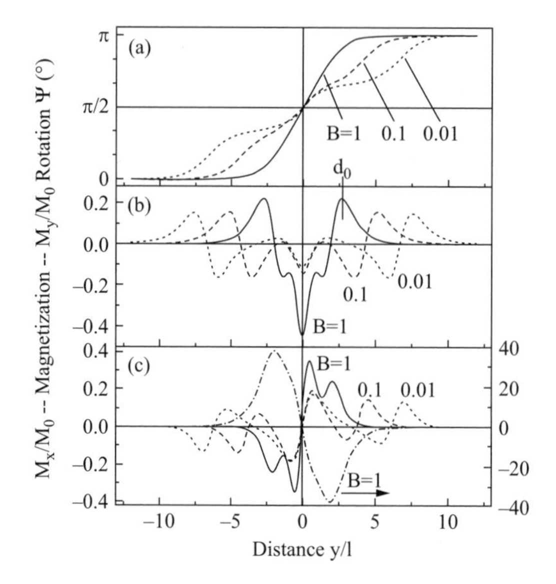

Fig. 2.7 Sublattice rotation (a) and net magnetization (b and c) across the antiferromagnetic domain walls of $YMnO_3$. Although the total integrated magnetization is zero, the local magnetization at the centre of the wall is nonzero. Reprinted with permission from [60]. Copyright 2003 by the American Physical Society.

(1) Rare earth nickelates have simultaneous charge order and spin order, but while the charge order (a disproportionation of the valence of the nickel) takes place every other ion, the magnetic order takes place every second Ni ion. The resulting exchange-striction between unequal charge and unequal spin breaks the centre of symmetry and has been predicted to lead to important ferroelectric polarizations [63, 64].
(2) $BiNiO_3$ has a lone-pair in the A-site, and it is therefore possible that this material could have ferroelectric polarization by the same mechanism as $BiFeO_3$ [65].

In spite of these predictions, however, the ferroelectric polarization of perovskite nickelates has never been measured. Of course, this may be due to the predictions being wrong, but it is probably more likely that the inability to measure hysteresis loops may also be due to the fact that these materials are rather conductive (in fact, most of them show a metal–insulator transition as a function of temperature). As is further discussed in the appendix, low conductivity (high resistivity) is an essential requisite for measuring P(E) hysteresis loops. It is therefore quite possible that these materials may be multiferroic but unswitchable due to their high conductivity.

2.3.5 Magnetoelectric relaxors

General view

In these materials, nanoscopic spin clusters and dipolar clusters coexist, but there is no long-range ordering. The cluster polarization fluctuates thermally, slowing down as the temperature cools, and eventually freezing in a glassy process whose dynamics are well described by the Vogel–Fulcher law:

$$\tau = \tau_0 e^{\frac{U}{k(T-T_f)}} \tag{2.4}$$

This equation reflects the slowing down and eventual divergence of the relaxation time (the average time between fluctuations in cluster polarization) as the freezing temperature, T_f, is approached. Below the freezing temperature, static polarization can be achieved and the material may be treated as a proper multiferroic.

The idea of materials that are simultaneously ferroelectric relaxors and magnetic relaxors was initiated by Levstik, Filipic, and Bobnar, who developed the basic idea [66] and also showed a specific example [67] involving $PbFe_{1/2}Nb_{1/2}O_3$ and $PbMg_{1/2}W_{1/2}O_3$. A large amount of research has also been carried out on three lead-ferrite-based relaxors: $PbFe_{2/3}W_{1/3}O_3$ [68–83]; $PbFe_{1/2}Ta_{1/2}O_3$ [84]; and $PbFe_{1/2}Nb_{1/2}O_3$ [85–90]—and on their solid solutions with PZT (lead zirconate-titanate). The Ta- and Nb- compounds have high Neel temperatures (*c.* 150 K) and ferroelectric T_c well above ambient (310 K and 380 K, respectively) and when combined with PZT are both ferroelectric and somewhat magnetic at room temperature, with evidence of coupling between the polarization and magnetization. The mechanisms are somewhat unclear as yet, and spin clustering may be involved. $PbFe_{2/3}W_{1/3}O_3$ has a much higher Neel temperature (T_N = 360–380 K), and it is important that it lies well above room temperature; hence, mixtures of PFW with PZT can gain improved long-range ferroelectricity while remaining magnetic at ambient temperatures. The PZT-mixed compounds are good single-phase materials, generally with a tetragonal $BaTiO_3$ structure.

A final word of caution is in order, however, in that relaxational dynamics are often seen but are not always caused by intrinsic relaxor behavior. As shall be discussed in the appendix, systems with poor resistivity can show relaxational behaviour due to the changes in their RC time constant. It is imperative that these be ruled out before claiming that a system is a relaxor.

$PbFe_{2/3}W_{1/3}O_3$ ("PFW") and its solid solutions with PZT

The aim in magnetoelectric memories is to have a magnetoelectric with large coupling at room temperature. In this goal two constraints, already mentioned, must be circumvented. The first is Spaldin's requirement that ABO_3 ferroelectric perovskites have d^0 B-site electron configurations, whereas the magnetic properties of the same material require d^n ($n \neq 0$). Some solutions have been discussed in previous sections; but another strategy is, for example, to have roughly half of the B-site ions as Fe and half as Ti, as in $PbFe_{2/3}W_{1/3}O_3/PbZr_{0.4}Ti_{0.6}O_3$ [PFW/PZT], discussed below.

Lead iron tungstate has been known since the days of Smolensky to be a multiferroic with a "diffuse" phase transition (now termed a "relaxor"). Very recently, a theory has been developed that gives very specific predictions for magnetic field dependence of electrical properties in such materials [91]. Based upon the earlier work of Pirc and Blinc, in their new paper Pirc *et al.* calculate that, for systems like PFW that are both magnetic relaxors (nano-regions of oriented spins) and ferroelectric relaxors (polar nano-regions), the coupling of magnetization and polarization through strain s, i.e. via magnetostrictio sM^2 and electrostriction sP^2 can be very large—much larger in fact than the direct biquadratic coupling P^2M^2. This kind of indirect strain-coupled magnetoelectric effect is not subject to the mathematical limit on magnetoelectric susceptibilities imposed by the Brown–Hornreich–Shtrikman constraint, that $\chi_{ME} < [\chi_{elec}\chi_{mag}]^{1/2}$.

The conclusion of this theory [91] is that materials near the instability limit between short-range electric relaxor ordering and long-range ferroelectricity can be driven to the relaxor state by application of a magnetic field H. This occurs as a continuous dynamic process in which the polarization relaxation time τ is increased with applied field H. The explicit dependence of τ (H) is given by

$$\tau(\mathrm{H}) = \text{constant} \times \exp[-\mathrm{H}_c^2/(\mathrm{H}_c^2\text{-}\mathrm{H}_2)], \tag{2.5}$$

where the critical field H_c is estimated to be $0.4T < H_c < 4T$, using averages of magnetostriction and electrostriction tensor components from related compounds (those for PFW/PZT are as yet unmeasured). Readers will recognize the unusual algebraic dependence in (2.6) as that of a Vogel–Fulcher equation (known to fit ordinary ferroelectric relaxors) in which the freezing temperature T_f has been replaced with a critical magnetic field H_c. This formula fits the observed data (Figure 2.8) in PFW/PZT very well [92], and the critical field H_c is evaluated as $H_c = 0.92$ T, rather close to the theoretical estimate. The drive of a long-range ferroelectric to a relaxor requires that the product of magnetostriction tensor and electrostriction tensor be negative; for a positive product, the applied magnetic field will, conversely, drive the relaxor to long-range ferroelectric ordering.

When the applied magnetic field H becomes close to H_c, the hysteresis loop becomes unmeasurable, as shown in Fig. 2.8, and the material exhibits only the small extrinsic (space-charge) loop of a very lossy linear dielectric [92]. If the time constant of the measuring apparatus were unlimited, this would occur at H_c; but in reality it will occur at a slightly lower field, as the polarization relaxation time moves out of the frequency range of the detector. In the data

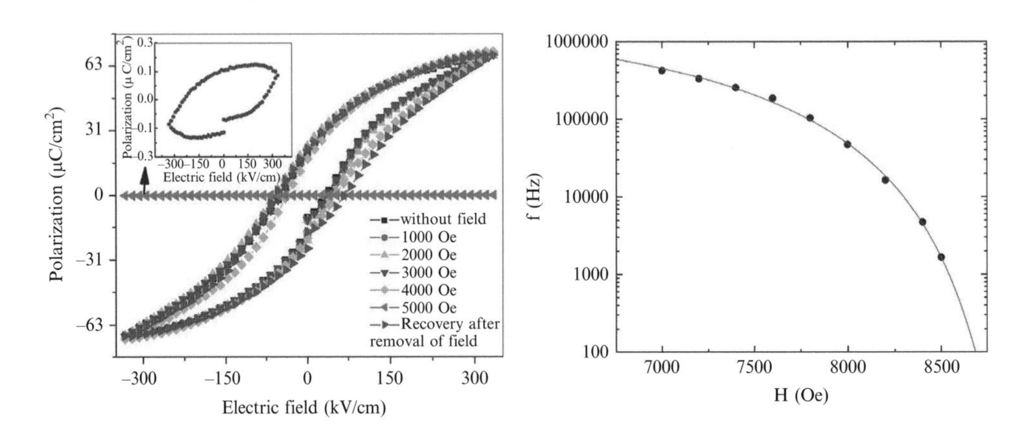

Fig. 2.8 Ferroelectric hysteresis in PFW/PZT as a function of magnetic field. Beyond a critical magnetic field, the ferroelectricity disappears. (right) Vogel–Fulcher-type fitting of the relaxation frequency as a function of magnetic field, showing that the dynamics are slowed down as the magnetic field is increased. Figure from [92], courtesy of the Institute of Physics.

shown in Fig. 2.8 $H_c = 0.92T$ and the polarization relaxation time increases from *c.* 200 ns at $H = 0$ to 100 μs at $H = 0.5T$. Longer relaxation times exceed the measuring window of the apparatus employed; but independent of the kit used, the hysteresis loop should vanish before $H = 0.92T$.

These PFW/PZT films also exhibit positive temperature coefficients of resistivity (PTCR) [93]. The data fit the model of Dawber and Scott [93b], which involves a relationship between negative differential resistivity and PTCR. PTCR is of commercial device importance because it eliminates the problem of thermal-runaway shorts: as the samples heat up from increased current flow, PTCR is a self-limiting process that raises resistivity and allows temperature to stabilize. In related compounds it has been found that the linear magnetoelectric coupling term can also be important [Sanchez Dilsom A.; Ortega N.; Kumar Ashok; Katiyar RS, and Scott JF; "Symmetries and multiferroic properties of novel room-temperature magnetoelectrics: Lead iron tantalate-lead zirconate titanate (PFT/PZT)" AIP ADVANCES 1, 2011 (042169)], leading to magnetic control of ferroelectric domain walls [Evans D M, Gregg J M, Kumar R S, Sanchez D, Ortega N, and Scott J F, Domain Wall Motion in Ferroelectrics with Applied Magnetic Fields (submitted, 2012)].

2.3.6 Ferromagnetic ferroelectric fluorides

Early work on magnetoelectric fluorides

Although oxides have clearly taken a more prominent role in recent magnetoelectric research, early work devoted just as much attention to fluorides. Shortly after their experimental verification [5, 6] of bilinear magnetoelectric coupling in Cr_2O_3, Astrov *et al.* made similar measurements [8–12] in $BaMnF_4$ and $BaCoF_4$. These results were however different from those in chromia; note that the effect is a frequency-dependent ac response, not a dc electrostatic response. In fact these data have never been explained, although the role of domain walls is suggested. A detailed frequency-dependent response was calculated for this family of multiferroics by Tilley and Scott [13], but that only included bulk terms and not domain walls, and it does not correspond closely to the data of Astrov *et al.* [8–12]. The structure of these compounds

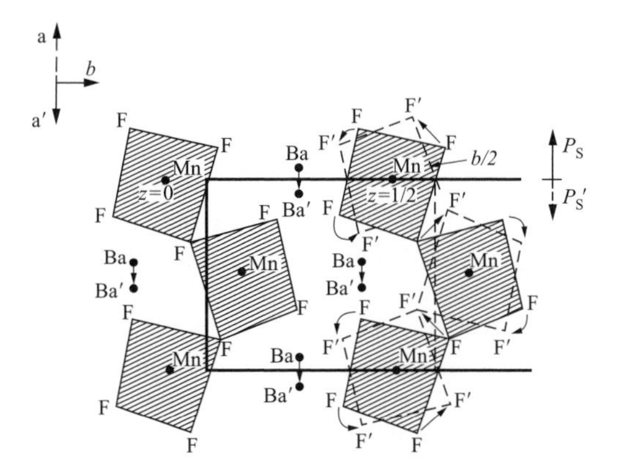

Fig. 2.9 Structure of $BaMnF_4$. Adapted from [94].

Table 2.1 Requirements for ferroelectrically-induced-ferromagnetism (after Birss [101])

Magnetic point group symmetry	Relationship required between **P** and M	Spin-canting via ferroelectricity
1	none	allowed
m′	P parallel m′; M parallel m′	allowed
m	P parallel m; M perpendicular m	allowed
2′	P perpendicular M; P parallel 2′	allowed
M′m2′	P parallel 2′; M perpendicular m	allowed
2 or m′m′2	P parallel M parallel 2	forbidden
3 or 3m′ or 4; 4m′m′ or 6 or 6m′m′	P parallel M parallel principal axis	forbidden

is illustrated in Figure 2.9 [94, 95]. Note in this diagram that the magnetic Mn (Ni or Co or Fe) ions do not move in the ferroelectric soft mode eigenvector; rather, the motion consists of Ba-ion displacements coupled to nearly rigid rotation of MF_6 octahedra. This makes it less than obvious why there should be strong coupling between polarization P and magnetization M. (The same kind of argument can be made for $BiFeO_3$, where the ferroelectricity arises from the Bi-ion and the magnetism from the Fe-ion.) That understanding required a more detailed recent *ab-initio* model [96, 97]. The basic mechanism is that of Dzyaloshinskii–Moriya anisotropic exchange, but the models and experiments are both subtle. In 1975, Venturini and Morgenthaler at MIT showed [98a] that $BaMnF_4$ has weak ferromagnetism (3 mrad canting angle), and that the axis of sublattice magnetization is also tilted (9 degrees off the b-axis), but both of these results remained controversial, with both Kizhaev and Prozorova in Leningrad and Moscow respectively [99, 100] finding no ferromagnetism, presumably due to the spatial averaging of domains in their specimens. Only very recently Zvezdin *et al.* confirmed the 9-degree tilt of sublattice magnetization [98b].

It is especially interesting (and surprising) that $BaMnF_4$ and $BaCoF_4$ exhibit similar magnetoelectric data, because as shown in Table 2.1 [101, 102] and the more detailed model of Fox *et al.* [14], the Mn compound manifests ferroelectrically induced (weak) ferromagnetism [15], whereas the Co isomorph cannot, since its spins and polarization are collinear, yielding magnetic point group 2′ rather than 2, as in $BaMnF_4$.

Rather simpler effects in $BaMnF_4$ arising from coupling of polarization P and magnetization M are the dielectric anomalies. This change in dielectric constant $\varepsilon(T)$ near $T_N = 29\,K$ is small (<1%), negative, and proportional [103, 104] to the sublattice magnetization squared $M^2(T)$, in accord with the general theory of Gehring [105]. All of these characteristics are expected for intrinsic effects, whereas extrinsic Maxwell–Wagner space charge effects are usually much larger, often positive, and not proportional to any power of M(T)

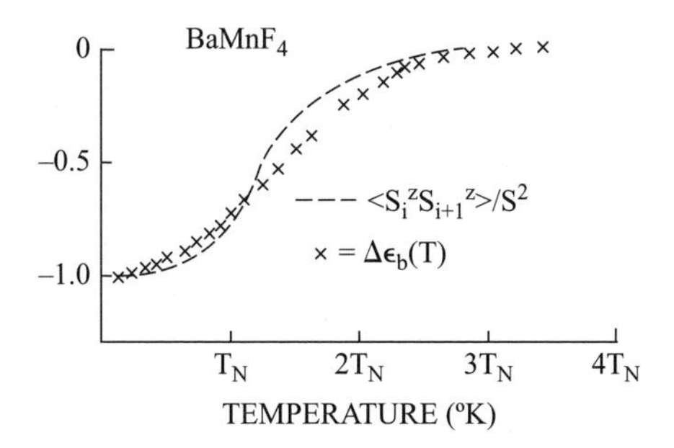

Fig. 2.10 Dielectric anomaly $\Delta\varepsilon_b$ (T) for $BaMnF_4$, compared with the nearest-neighbor magnetic energy [103]. Figure courtesy of the American Physical Society.

[106–108], see also Appendix 2.1. Although these interpretations have remained contentious in the context of other materials [109–114], recent studies discriminate carefully between intrinsic and extrinsic effects [107, 115, 116]. Note that an earlier theory by Albuquerque and Tilley [117] incorrectly predicted that the change in ε(T) near T_N must be positive; this was due to neglect of higher-order terms in the free energy they used; these terms are necessarily positive definite and can be larger than the lowest-order negative term. A very similar model was independently published by Negran and Glass *et al.* [118] and by Negran [119] to fit data on $BaNiF_4$. However, a more comprehensive model was detailed by Fox *et al.* [14], which explains the magnitude, temperature dependence, and rather importantly, the signs of the dielectric constant changes at both T_N in $BaMF_4$ and also at T_{2D} (a higher temperature at which the spins order in-plane).

The dielectric data in Figure 2.10 are for the magnetic b-axis in $BaMnF_4$. Here the anomaly is larger, still negative, and goes well above $T_N = 29$ K. It is maximal near 90 K, which is the temperature at which magnetic in-plane ordering sets in (Zorin *et al.* [10, 12]). It arises from the linear-quadratic term in the free energy $<P_i\ M_j\ M_k>$, which is nonzero well above T_N in any antiferromagnet, since $<M^2\gg 0$, but is very large (nearly equal to $<M>^2$ in-plane) in any magnet that orders in two dimensions.

This is a particularly fine example of the linear-quadratic magnetoelectric coupling first discovered by Hou and Bloembergen [120] because it permits very quantitative analysis of $<M^2>$ in-plane. This quadratic magnetoelectric effect is also strong in $BaMnF_4$ [121, 122].

In 1977 Fox and Scott showed [15] that it is plausible that the weak ferromagnetism in $BaMnF_4$ measured by Venurini and Morgenthaler [98] arises from ferroelectricity. The key term in the free energy developed explicitly by that group [14] is of form

$$G(P,M,L,T) = b_1\mathbf{P}(T)\cdot(\mathbf{M}\times\mathbf{L}) \qquad (2.6)$$

where P is the polarization; L, the sublattice magnetization; and M, the weak ferromagnetic moment. b_1 is a Dzyaloshinskii–Moriya anisotropic exchange term; it signifies physically that the ferroelectric displacement of the magnetic ion will modify the quantum mechanical exchange in particular directions. More recently Perez-Mato [123] and Fennie [20] have shown independently that an interaction of this algebraic form is generic and is required for ferroelectrically induced ferromagnetism in all crystals, and not just the $BaMF_4$ family. We can see from (2.1) why such a ferroelectric canting of spins is favored in $BaMnF_4$, since P is along the a-axis; L, along the b-axis; and M, along c. However, in $BaCoF_4$, P and L are both along the a-axis, and this spin flop makes the interaction term vanish. The space group requirements for ferroelectrically induced (weak, canted) ferromagnetism are given in Table 2.1, from Scott [102], based upon the original theory of Birss [101].

Using the definition of the magnetoelectric tensor a'_{ij}, Fox and Scott write

$$M_c = 4\pi a'_{ac} P_a \tag{2.7}$$

for the magnetization along the c-axis. But by the usual model of spins with spin S = 5/2 for Mn^{+2} and g approximately 2.0, with $N = 1.1 \times 10^{22}$ cm^{-3} [94] and $P = 11.5 \mu C/cm^2$ [94], we can also express M_c as

$$M_c = \mu_B g\, N\, S\, \sin\phi \tag{2.8}$$

where ϕ is the canting angle of the spins. Solving (2.7) and (2.8) together for ϕ, using the known values of off-diagonal a′ magnetoelectric tensor components for other Mn^{+2} systems, Fox and Scott estimated ϕ as 2.1 ± 1.0 mrad, in good agreement with the measured value of 3.0 mrad [98].

In addition to explaining the induced ferromagnetism, a magnetoelectric theory of $BaMnF_4$ should be able to explain the dielectric anomalies. The assumed free energy of Fox *et al.* was

$$\begin{aligned} G(P,M,L) = {} & \tfrac{1}{2} A L^2 - \tfrac{1}{2} aL_z^2 + (1/4)B'L^4 + \tfrac{1}{2} BM^2 + (b_0 + b_1 p + b_2 p^2)M_x L_z - \gamma M_z L_x \\ & + \tfrac{1}{2} D(L \cdot M)^2 + \varepsilon p^2 - P \cdot E - M \cdot H \end{aligned} \tag{2.9}$$

where P(total) = $P_r + p$; P_r, the remnant polarization; and p, the part of the polarization induced by applied magnetic field H.

Note that this is a mean-field (Landau) theory; this is quite allowed for weak ferromagnets, because the order parameter (the canting) is small at all temperatures. Single-ion anisotropy (originally favored by Rado in some materials) has been ignored.

More importantly, no electrostriction or magnetostriction have been included; these are of paramount importance in the discussion of $PbFe_{2/3}W_{1/3}O_3$ at the end of this section.

Minimizing this free energy showed that the dielectric anomaly at the Neel temperature is of form

$$\Delta\varepsilon(T) = (b_1^2 + b_0 b_2)L^2(T)/(B\varepsilon_0\varepsilon^2). \tag{2.10}$$

This shows several important things: firstly, the dielectric anomaly varies as magnetization squared; secondly, it can be positive or negative, depending upon the sign of $b_0 b_2$; thirdly, it is not very large (numerically of order 1% of the background dielectric constant). Therefore, when we see publications about other materials in which the dielectric anomaly is 20–30% of the background value and is not proportional to the magnetization or its square, we can be sceptical [107, 109–114]. See, however, the next paragraph.

The theory of Fox *et al.* explains very well the small negative anomaly in the a-axis dielectric constant in $BaMnF_4$ at and below $T_N = 29$ K, but it cannot explain the larger dielectric anomaly along the b-axis near T_{2D} = c. 90 K. This anomaly arises because of the

Hou–Bloembergen linear-quadratic term <P_i M_j M_k>, and is due to the in-plane ordering up to *c.* T = $3T_N$. Two-dimensional systems cannot be modelled via mean-field theories, simply because integrating over all three dimensions gives zero. In this particular case, the correct details are given elsewhere [103, 118, 119], but the main conclusion is that the dielectric anomaly in this case is proportional not to $L^2(T)$, but to the magnetic energy. In [107, 109–114] the observed dielectric anomaly is not proportional to either; hence it may arise from extrinsic Maxwell–Wagner space charge, as discussed in the Appendix 2.1.

A general *ab-initio* theory of ferroelectricity in this $BaMF_4$ family has been published by Ederer and Spaldin [124].

There is an added complication for careful analysis of $BaMnF_4$, which is that it is structurally incommensurate below its antiferroelectric phase transition [125], near 254 K. Eibschutz *et al.* reported [126] that the initial value of the wave vector for the soft mode is at q = (½, ½, 0.39)a*, and that the incommensurate 0.39 value remains independent of temperature down to 4 K. This is perhaps true only in their specimen, however, and probably due to incommensurate antiphase boundary pinning by defects (perhaps fluorine vacancies). In other samples, several transitions are found via dielectric studies [127], by piezoelectric resonance [128], a double peak in specific heat [129], and most importantly, neutron scattering (Barthes-Regis *et al.* [130, 131]), and it would appear that a Devil's staircase of wave vectors of probable form $q_n = (5+2n)/(13+5n)$ exists, starting at $5a^*/13 = 0.385a^*$ and asymptotically reaching a lock-in at q = (½, ½, 2/5) a*, i.e. with a unit cell of five MnF_6 octahedra, compared with two in the paraelectric phase. Some modeling of this has been done [132].

2.3.7 Ferrimagnetic ferroelectrics

Ferroelectricity exists below the Vervey metal–insulator transition of magnetite Fe_3O_4, [133, 134]. This is a famous ferrimagnetic material: historically, this was the very first magnetic material identified by mankind. The ferroelectric polarization and magnetization of magnetite thin films is shown in Figure 2.11.

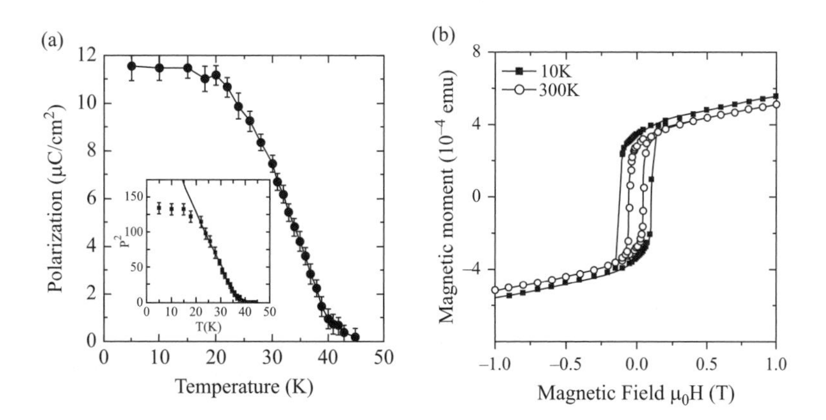

Fig. 2.11 Polarization as a function of temperature and magnetization hysteresis loops of ferrimagnetic multiferroic magnetite, Fe_3O_4. Figure adapted from [134], courtesy of Wiley Interscience.

Ferrimagnetic ferroelectrics also feature among fluorides and oxifluorides. The study of ferroelectric magnetic fluorides has recently been centered at two locations: Groningen [135] and Ljubljana [68–72]. The Ljubljana effort has emphasized the family $K_3Fe_2^{+3}Fe_3^{+2}F_{15}$. This is in fact an unusual ferromagnetic ferroelectric, with two Fe^{+3} ions and three Fe^{+2} ions per formula group (4 Fe^{+3} and 6 Fe^{+2} ions per primitive unit cell). It is possible to substitute separately for the Fe^{+2} ions, giving for example $K_3Fe_2^{+3}Cu_3F_{15}$, and for the Fe^{+3} ions, giving $K_3Fe_3^{+2}Cr_2F_{15}$. All of these are magnetoelectric, and the latter exhibits two phase transitions at low temperatures, possibly signifying the spin ordering of Fe (higher T) and Cr (lower T) ions. A more general discussion of the possibility of magnetoelectricity in the wider family of $A_3Fe_5F_{15}$ has been given by Abrahams [73, 74]. Here the A-ion can be Na, Li, K, etc. No lattice dynamical modeling of these structures has been published.

Other fluoride and oxyfluoride magnetoelectrics

In addition to the $PbFeWO_3$, $PbFeTaO_3$, $PbFeNbO_3$ family of oxides, there are a number of other fluoride magnetoelectrics [137a] and some oxyfluoride crystals that are magnetoelectric. The latter are reviewed by Ravez [137b]; $Na_5W_3O_9F_5$ is a good example. Most of these are magnetic and ferroelectric above 100 K. Table 2.2 lists a few of the fluoride multiferroics that have been studied. $Sr_3(FeF_6)_2$ is of academic interest because its ferroelectric phase $I4_1$ is not a subgroup of its paraelectric phase I4/m symmetry. Since the phase transition is first-order, this is allowed, but it is an unusual case, since the transition is not "reconstructive" in the sense of requiring large ionic displacements.

Finally, other fluorides and oxyfluoride families of interest are listed below:

(1) $Pb_5Cr_3F_{19}$ and $Pb_5Fe_3F_{19}$ family [139ab, 144]
(2) $Sr_3(FeF_6)_2$ family of antiferromagnetic ferroelectrics [145]

Table 2.2 Structures of some multiferroic fluorides and oxyfluorides

Formula	Space group	T_c	Formula groups/ primitive cell Z
$Pb_5Cr_3F_{19}^a$	I4cm	555 K	4
$Pb_5W_3O_9F_{10}^b$	I4	785 K	4
$Sr_3(FeF_6)_2$	$I4_1$	1160 K	4
$Pb_3(MF_6)_2^c$			
$A_2KMO_3F_3^d$			
$A_3MO_3F_3^e$			
$K_3Fe_5F_{15}^f$	Pba2	T_N =123 K; T_c=490K	2
$K_3Fe_2Cu_3F_{15}^g$	P4/mbm	T_N =85 K	2
$K_3Fe_3Cr_2F_{15}^h$	T_N = 17 K and 37 K		2
$(NH_4)_2FeF_5^i$	Pbcm-Pnma	T_N =13 K;T_c =168 K	4/8
$NaPbFe_2F_9^j$	C2/c		4

[a] There are ten other well-characterized members of this family, [138, 139]. [b] M = Ti, Mo, or W
[c] M = Ti, V, Cr, Fe, or Ga: [140] [d] A = K, Rb, or Cs; M = Mo or W [e] A = K, Rb, or Cs; M = Mo or W
[f] [141] [g] [142] [h] [70] [i] [143] [j] [144–147]

(3) $(NH_4)_2FeF_6$ family [146, 147]
(4) Oxyfluorides [137abc]

An added wrinkle of some academic interest is that, as emphasized by Fischer and Hertz in their text [148], published spin-glass theories generally do not apply to crystals such as ferroelectrics that lack inversion centers. In particular, they stress that such systems cannot be Ising-like. Since some multiferroics exhibit properties resembling spin-glasses [149], this means that existing theories are apt to be inapplicable.

Appendix 2.1 Magnetoelectric measurements

The measurement of magnetoelectric coupling implies measuring a conjugated property, either magnetization induced by an electric field, or polarization induced by a magnetic field. Often, however, it is easier—though less direct—to measure either an electrically induced change in susceptibility or, more often, a magnetically induced change in dielectric constant; the latter is called magnetodielectric effect or magnetocapacitance.

The typical configuration for a magnetoelectric measurement will involve placing the magnetoelectric sample between two electrodes, which may either be used to measure magnetically induced voltage or permittivity, in the direct magnetoelectric effect, or to apply a voltage in order to induce changes in magnetization or susceptibility, in the converse magnetoelectric effect. The magnetic field is supplied or sensed by placing the capacitor structure inside a coil.

Unfortunately, all of these measurements are susceptible to artifacts: magnetoelectric effects that do not reflect the intrinsic magnetoelectric properties of a material. Below we depict (Figure A2.1.1) and summarize some of these artifacts.

(1) Magnetoelectric induction/Faraday effect
These derive directly from Maxwell's equations. Any oscillating magnetic field is equivalent to an electric field, and likewise any oscillating electric field is equivalent to a magnetic field. The solution is rather trivial, however: one simply has to make sure that the external electric field is continuous (DC) rather than oscillating. If the external field is an oscillating one (as is sometimes required in order to improve sensitivity using lock-in amplifiers), then one has to make sure that the maximum magnetoelectric effect takes place coinciding with the peaks in the applied field and NOT coinciding with the maximum time-derivative in applied field. If maximal magnetoelectricity coincides with maximal time-derivative then the measured magnetoelectric effect is almost certainly an artifact: in the case of a sinusoidal field, the maximum time-derivative in fact takes place where the external field is zero, which directly rules out true magnetoelectricity.

(2) Magnetostrictive magnetoelectricity
This is a subtle and small effect, yet true magnetoelectric coupling is often also a small effect, so measurements should be corrected for strain effects. This artifact affects only magnetodielectric (magnetocapacitance) measurements. The capacitance (of a parallel-plate capacitor) is $C = \varepsilon \frac{S}{d}$, where ε is the dielectric constant, S is the surface area of the electrodes, and d is the thickness of the sample. When applying a magnetic field to the sample, it may affect the dielectric constant (if there is true magnetodielectric coupling) but, if the material is magnetostrictive (as is the case for all magnetic materials), then it will also affect the dimensions (S and d) of the capacitor, hence giving a change in capacitance that is only due to geometric factors and not to true ME coupling. This

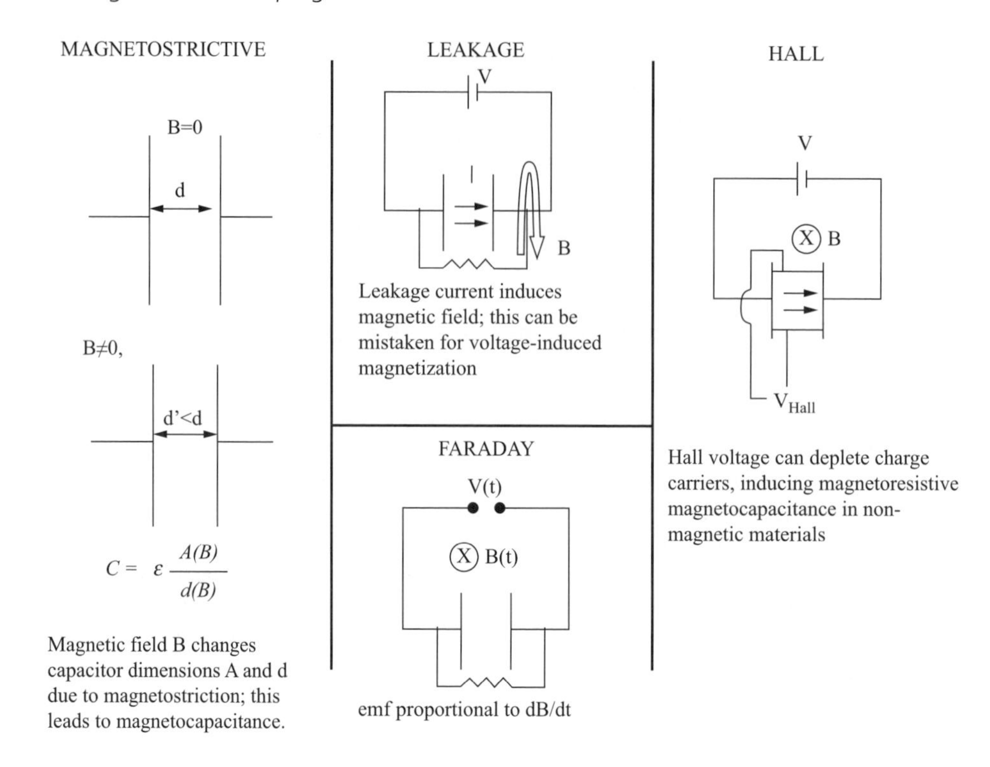

Fig. A2.1.1 Schematic depiction of different effects that may be mistaken for genuine magnetoelectric coupling.

effect is easy to correct if the magnetostriction of the material is known. At any rate, the magnetically induced strains are rarely bigger than 0.1%, so these effects are generally quite small—unless measured in the vicinity of a phase transition.

(3) Lenz's law
Again, this artifact is a direct consequence of the intrinsic coupling between electric and magnetic fields in Maxwell's equations. Specifically, any moving charge—any current—induces a magnetic field around it. If a voltage is applied to a material that is not a very good insulator, the leakage current will induce a magnetic field that may be mistaken for a change in the magnetization of the material, when it is nothing of the sort. Ruling out such an artifact is trickier, and it can only really be done by ensuring that the sample is sufficiently insulating. Good electrical insulation, in fact, is a requirement of virtually any characterization of dielectric and ferroelectric materials, and not just magnetoelectricity. Unfortunately, for reasons that are discussed elsewhere, most magnetoelectrics tend to be rather poor insulators.

(4) Hall effect
The Hall effect is another classic example of magnetoelectric coupling: a magnetic field applied to a current will divert the carriers hence inducing a voltage. This effect is well known and it is in fact the basis for most commercial magnetic field sensors. It has recently been pointed out as well that the Hall effect can lead to magnetoresistive effects,

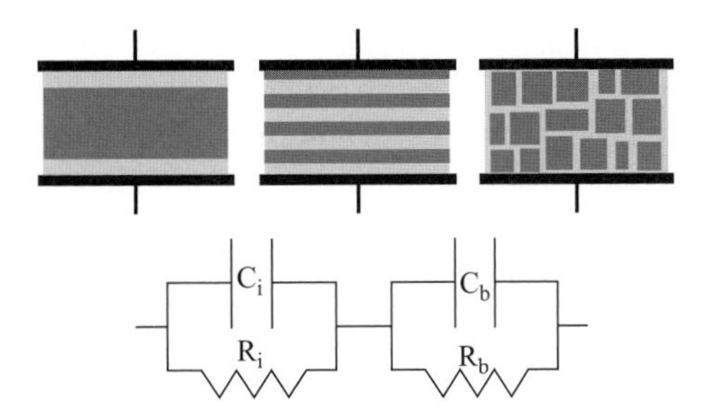

Fig. A2.1.2 Different types of heterogeneous dielectric where finite resistivity can lead to an artificially inflated value of the dielectric constant, as given by the Maxwell–Wagner equations. Figure reprinted with permission from [106]. Copyright 2006, American Institute of Physics.

even in non-magnetic materials [150]. The relevance of this is that magnetoresistance often masquerades as magnetocapacitance, as explained in the next section.

(5) Magnetoelectric magnetoresistance
This is one of the most pervasive artifacts in the literature, and is therefore the one to look out for most carefully in measurements. Again, its basis is in poor electrical insulation of the sample, combined with the existence of heterogeneity in the impedance characteristics, see Figure A2.1.2. The finite resistivity of the dielectric means that its relaxation time constant, $\tau = RC$, is also finite. Measurements of capacitance performed at a frequency slower than $1/\tau$ will therefore have an important contribution from leakage charge (as opposed to pure displacement charge in a dielectric). This leakage artificially inflates the capacitance value, a phenomenon known as Maxwell–Wagner relaxation; the equivalent circuit equations for the real and imaginary parts of the dielectric constant are, respectively:

$$\varepsilon'(\omega) = \frac{1}{C_0 (Ri + R_b)} \frac{\tau_i + \tau_b - \tau + \omega^2 \tau_i \tau_b \tau}{1 + \omega^2 \tau^2}$$

$$\varepsilon''(\omega) = \frac{1}{\omega C_0 (R_i + R_b)} \frac{1 - \omega^2 \tau_i \tau_b + \omega^2 \tau (\tau_i + \tau_b)}{1 + \omega^2 \tau^2}$$

Importantly, if the finite-conductivity dielectric happens to be magnetoresistive (or if the geometry is such that Hall-based magnetoresistance can take place [150]), then application of a magnetic field will result in a change of this artificial capacitance; in other words, pure magnetoresistance can easily be mistaken for magnetocapacitance [106, 107, 108].

Appendix 2.2 Critical exponents in isostructural phase transitions

Recent studies on $BiFeO_3$ have revealed two first-order phase transitions that do not change symmetry (Pbnm-Pbnm at *c.* 1200 K and Cc-Cc under uniaxial stress). Twenty years ago Ishibashi and Hidaka showed [151] that such systems generally exhibit a phase diagram with tricritical points and critical end points giving the unusual mean-field critical exponents $\alpha = 2/3$, $\beta = 1/3$, and $\gamma = 2/3$ at the end points. In the present note we extend that to show $\delta = 3$, $\nu = 1/3$, and $\eta = 0$, and suggest some experimental tests.

There are relatively few crystals that exhibit phase transitions that retain the full space-group symmetry. Half a dozen were listed in an earlier paper [152] to which we can add $(NH_4)_3H(SO_4)_3$ [153] and both di-calcium strontium propionate and its lead isomorph [154, 155] The analytical Landau–Devonshire theory for these situations has been well described by Ishibashi and Hidaka, using a free energy eighth order in polarization P [151]. Although these iso-symmetric phase transitions are necessarily first-order (discontinuous change in both unit cell volume and in c/a lattice constant ratio), they also gave a full phase diagram, showing that in general such systems exhibit a line of phase boundaries that includes both a tricritical point (where the transition goes from first-order to second-order) and a critical end point. A similar critical end point (as a function of applied electric field E and temperature T) has been reported recently in the relaxor ferroelectric lead magnesium niobate (PMN) [156] This remains somewhat controversial in that the theoretical models of PMN predict tricritical points but do not yield a critical end point [157]; however, the experimental data are convincing. The iso-symmetric transitions in $BiFeO_3$ are monoclinic-monoclinic Cc-Cc (under electric field or epitaxial stress in thin films) [158, 159] and orthorhombic-orthorhombic Pbnm-Pbnm at *c.* 1200 K in ceramics [160].

Calculations: If we write the Landau–Devonshire free energy as

$$G(P,T) = a(T\text{-}Tc)P^2 + bP^4 + CP^6 + dP^8$$

then the coefficients b, c, d can vary as functions of field E or pressure P. Depending upon their signs and magnitudes, one can have a second-order transition, a tricritical point, a first-order transition, or a critical endpoint (as in fluids). Ishibashi and Hidaka show directly by differentiating their free energy that the critical exponent describing the divergence of specific heat is,

$$C_V(t) = A\, t^{-\alpha}, \text{ where } \alpha = 2/3. \tag{2.2.1}$$

Here t is reduced temperature $[|T\text{-}T_c|/T_c]$, and T_c is the Curie temperature.

Similarly, the polarization P is,

$$P(t) = P_0 t^{\beta} \text{ where } \beta = 1/3. \tag{2.2.2}$$

And the isothermal electric susceptibility is,

$$C(t) = C\, t^{-\gamma} \text{ where } \gamma = 2/3. \tag{2.2.3}$$

To these we can add the dependence of displacement vector D (or polarization P) upon E as $\delta = 3$ from the Widom inequality expressed under the assumption of scaling:

$$P(E) = p\, E^{1/\delta} \tag{2.2.4a}$$

$$\text{whence } \gamma = \beta(\delta - 1) \text{ gives } \delta = 3. \tag{2.2.4b}$$

Finally, from hyperscaling [161, 162] we have for v, the exponent in the correlation function, and h, the exponent in the structure factor S(q) (d is dimensionality, taken as d = 4 for mean field):

Table 2.2.1 Mean-field exponents in iso-symmetric phase transitions near second-order transitions, tricritical points, and critical end points

Exponent	Second-order	Tricritical	Critical endpoint
α	0	½	2/3**
β	½	¼	1/3**
γ	1	1	2/3**
δ	3	5	3
ν	½	½*	1/3
η	0	0	0

*From [163]
**From [151]

$$\gamma - 2\beta = \nu\,(4 - d - 2\eta) \tag{2.2.5a}$$

$$\gamma = (2 - \eta)\nu \tag{2.2.5b}$$

$$\text{and} \quad d - 2 + \eta = 2\beta/\nu, \tag{2.2.5c}$$

$$\text{whence} \quad \nu = 1/3 \tag{2.2.5d}$$

$$\text{and} \quad \eta = 0. \tag{2.2.5e}$$

These "critical" exponents are all mean field. They are compared in Table 2.2.1 with mean field exponents at second-order phase transitions and at tri-critical points. We note in passing that these values do NOT satisfy the relationship

$$\alpha = 2 - 3\nu \tag{2.2.6}$$

which Holwerda *et al.* claim [162] is required of all hyperscaling exponents. The correct expression is $\alpha = 2 - d\nu$, where $d = 4$ for mean field, not 3.

Experiments: It can be seen from Table 2.2.1 that the specific heat exponent α is unusually large (2/3) and the susceptibility exponent γ unusually small (2/3). Values of γ are normally 1 or larger in models. Therefore specific heat and dielectric measurements might be revealing.

References

[1] Curie, P. (1894). Sur la symétrie dans les phénomènes physiques, symétrie d'un champ électrique et d'un champ magnétique. *Journal de physique* **3**, 393.
[2] Dzialoshinski. I. (1958). A thermodynamic theory of "weak" ferromagnetism of antiferromagnetics. *J. Phys. Chem. Solids* **4**, 241.
[3] Moriya, T. (1960). Anisotropic superexchange interaction and weak ferromagnetism. *Physical Review* **120**, 91.
[4] Dzyaloshinskii, I.E. (1960). On the magneto-electrical effects in antiferromagnets. *Sov. Phys. JETP* **10**, 628.

[5] Astrov, D.N. (1960). The magnetoelectric effect in antiferromagnets. *Sov. Phys. JETP* **11**, 708–709.
[6] Astrov, D.N. (1961). Magnetoelectric effect in chromium oxide. *Sov. Phys. JETP* **13**, 729–733.
[7] Ascher, E., Rieder, H., Schmid, H., and Stossel, H. (1966). Some Properties of Ferromagnetoelectric Nickel-Iodine Boracite, $Ni_3B_7O_{13}I$. *J. Appl. Phys.* **37**, 1404.
[8] Alshin, B.I., Astrov, D.N., and Tischen, A.V. (1970). Magnetoelectric effect in $BaCoF_4$. *JETP Lett.* **12**, 142.
[9] Astrov, D.N., Alshin, B.I., Zorin, R.V., and Drobyshe, L.A. (1969). Spontaneous magnetoelectric effect. *Sov. Phys. JETP* **28**, 1123.
[10] Zorin, R.V., Tischen, A.V., and Astrov, D.N. (1972). 2-dimensional magnetic ordering in $BaMnF_4$. *Fiz. Tverd. Tela* **14**, 3103.
[11] Alshin, B.I., Astrov, D.N., and Zorin, R.V. (1972). Low-frequency magnetoelectric resonances in $BaMnF_4$. *Zh. Eksp. Teor. Fiz.* **63**, 2198–2204.
[12] Zorin, R.V., Astrov, D.N., and Alshin, B.I. (1972). Low-frequency magnetoelectric resonances in $BaCoF_4$. *Zh. Eksp. Teor. Fiz.* **62**, 1201.
[13] Tilley, D.R. and Scott, J.F. (1982). Frequency Dependence of Magnetoelectric Phenomena in $BaMnF_4$. *Phys. Rev. B* **25**, 3251–3260.
[14] Fox, D.L., Tilley, D.R., Scott, J.F., and Guggenheim, H.J. (1980). Magnetoelectric Phenomena in $BaMnF_4$ and $BaMn_{0.99}Co_{0.01}F_4$. *Phys. Rev. B* **21**, 2926–2936.
[15] Fox, D.L. and Scott, J.F. (1977). Ferroelectrically Induced Ferromagnetism. *J. Phys. C: Sol. St. Phys.* **10**, L329–L331.
[16] Fiebig, M. (2005). Revival of the Magnetoelectric Effect. *J. Phys. D: Appl. Phys.* **38**, R123–R152.
[17] Kimura, T., Goto, T., Shintani, H., *et al.* (2003). Magnetic control of ferroelectric polarization. *Nature* **426**, 55–58.
[18] Wang, J., Neaton, J.B., Zheng, H., *et al.* (2003). Epitaxial $BiFeO_3$ Multiferroic Thin Film Heterostructures. *Science* **299**, 1719–1722.
[19] Bea, H., Gajek, M., Bibes, M., and Barthelemy, A. (2008). Spintronics with Multiferroics. *J. Phys.: Condens. Matter* **20**, 434221.
[20] Ederer, C. and Fennie, C. J. (2008). Electric–field switchable magnetization via the Dzyaloshinskii–Moriya interaction: $FeTiO_3$ versus $BiFeO_3$. *J. Phys.: Condens. Matter* **20**, 434219.
[21] Scott, J.F. (2008). Ferroelectrics go bananas. *J. Phys.: Condens. Matter* **20**, 021001.
[22] Sergienko, I.A. and Dagotto, E. (2006). Role of the Dzyaloshinskii-Moriya interaction in multiferroic perovskites. *Phys. Rev. B* **73**, 094434.
[23] Mostovoy, M. (2006). Ferroelectricity in spiral magnets. *Phys. Rev. Lett.* **96**, 067601.
[24] Cheong, S.-W. and Mostovoy, M. (2007). Multiferroics: a magnetic twist for ferroelectricity. *Nature Mat.* **6**, 13–20.
[25] Borisov, P., Hochstrat, A., Chen, X., Kleemann, W., and Binek, Ch. (2005). Magnetoelectric Switching of Exchange Bias. *Phys. Rev. Lett.* **94**, 117203.
[26] Laukhin, V., Skumryev, V., Martí, X. *et al.* (2006). Electric-Field Control of Exchange Bias in Multiferroic Epitaxial Heterostructures. *Phys. Rev. Lett.* **97**, 227201.
[27] Brown, W.F., Hornreich, R.M., and Shtrikman, S. (1968). Upper bound on magnetoelectric susceptibility. *Phys. Rev.* **168**, 574.

[28] Laletsin, U., Padubnaya, N., Srinivasan, G., and Devreugd, C.P. (2004). Magnetoelectric Effects in Ferromagnetic Metal-Piezoelectric Oxide Layered Structures. *Appl. Phys. A* **78**, 33.

[29] Nan, C.W., Bichurin, M.I., Dong, S.X., Viehland, D., and Srinivasan, G. (2008). Multiferroic magnetoelectric composites: Historical perspective, status, and future directions. *J. Appl. Phys.* **103**, 031101.

[30] Zheng, H., Wang, J., Lofland, S.E. *et al.* (2004). Multiferroic $BaTiO_3$-$CoFe_2O_4$ Nanostructures. *Science* **303**, 661–663.

[31] Thiele, C., Dörr, K., Bilani, O., Rödel, J., and Schultz, L. (2007). Influence of strain on the magnetization and magnetoelectric effect in $La_{0.7}A_{0.3}MnO_3$/PMN-PT(001) (A=Sr,Ca). *Phys. Rev. B* **75**, 054408.

[32] Israel, C., Mathur, N.D., and Scott, J.F. (2008). A one-cent room-temperature magnetoelectric sensor. *Nature Mat.* **7**, 93–94.

[33] Hill, N.A. (2000). Why are there so few magnetic ferroelectrics? *J. Phys. Chem. B* **104**, 6694–6709.

[34] Cohen, R.E. (1992). Origin of ferroelectricity in perovskite oxides. *Nature* **358**, 136–138.

[35] Gajek, M., Bibes, M., Fusil, S., *et al.* (2007). Tunnel junctions with multiferroic barriers. *Nature Mat.* **6**, 296–302.

[36] Grizalez, M., Martinez, E., Caicedo, J., Heiras, J., and Prieto, P. (2008). Occurrence of ferroelectricity in epitaxial $BiMnO_3$ thin films. *Microelectronics Journal* **39**, 1308–1310.

[37] Baettig, P. and Spaldin, N.A. (2005). *Ab initio* prediction of a multiferroic with large polarization and magnetization. *Appl. Phys. Lett.* **86**, 012505.

[38] Nechache, R., Harnagea, C., Carignan, L.-P. *et al.* (2009). Epitaxial thin films of the multiferroic double perovskite Bi_2FeCrO_6 grown on (100)-oriented $SrTiO_3$ substrates: Growth, characterization, and optimization. *J. Appl. Phys.* **105**, 061621.

[39] Kim, D.H., Lee, H.N., Varela, M., and Christen, H.M. (2006). Antiferroelectricity in multiferroic $BiCrO_3$ epitaxial films. *Appl. Phys. Lett.* **89**, 162904.

[40] Catalan, G. and Scott, J.F. (2009). Physics and Applications of Bismuth Ferrite. *Advanced Materials* **21**, 2463–2485.

[41] Lebeugle, D., Colson, D., Forget, A., and Viret, M. (2007). Very large spontaneous electric polarization in $BiFeO_3$ single crystals at room temperature and its evolution under cycling fields. *Appl. Phys. Lett.* **91**, 022907.

[42] Shvartsman, V.V., Kleemann, W., Haumont, R., and Kreisel, J. (2007). Large bulk polarization and regular domain structure in ceramic $BiFeO_3$. *Appl. Phys. Lett.* **90**, 172115.

[43] Sosnowska, I. Peterlinneumaier, T., and Steichele, E. (1982). Spiral magnetic-ordering in bismuth ferrite. *Journal of Physics C-Solid State Physics* **15**, 4835–4846.

[44] Lebeugle, D., Colson, D., Forget, A., *et al.* (2008). Electric-Field-Induced Spin Flop in $BiFeO_3$ Single Crystals at Room Temperature. *Phys. Rev. Lett.* **100**, 227602.

[45] Lee, S., Ratcliff, W., Cheong, S.W., and Kiryukhin, V. (2008). Electric field control of the magnetic state in $BiFeO_3$ single crystals. *Appl. Phys. Lett.* **92**, 192906.

[46] Bai, F., Wang, J., Wuttig, M., *et al.* (2005). Destruction of spin cycloid in (111)c-oriented $BiFeO_3$ thin films by epitiaxial constraint: Enhanced polarization and release of latent magnetization. *Appl. Phys. Lett.* **86**, 032511.

[47] Eerenstein, W., Morrison, F.D., Dho, J., *et al.* (2005). Comment on "Epitaxial $BiFeO_3$ Multiferroic Thin Film Heterostructures". *Science* **307**, 1203a.

[48] Bea, H., Bibes, M., Barthelemy, A., *et al.* Influence of parasitic phases on the properties of $BiFeO_3$ epitaxial thin films. *Appl. Phys. Lett.* **87**, 072508 (2005).

[49] Daraktchiev, M., Catalan, G., and Scott, J.F. (2010). Landau Theory of Domain Wall Magnetoelectricity. *Phys. Rev. B* **81**, 224118.

[50] Chu, Y.-H., Martin, L.W., Holcomb, M.B. *et al.* (2008). Electric-field control of local ferromagnetism using a magnetoelectric multiferroic. *Nature Mat.* **7**, 478–482.

[51] Lebeugle, D., Mougin, A., Viret, M., Colson, D., and Ranno, L., (2009). Electric Field Switching of the Magnetic Anisotropy of a Ferromagnetic Layer Exchange Coupled to the Multiferroic Compound $BiFeO_3$. *Phys. Rev. Lett.* **103**, 257601.

[52] Kimura, T., Sekio, Y., Nakamura, H., Siegrist, T., and Ramirez, A.P. (2008). Cupric oxide as an induced-multiferroic with high-T_C. *Nature Mat.* **7**, 291.

[53] Saito, K. and Kohn, K. (1995). Magnetoelectric effect and low-temperature phase transitions of $TbMn_2O_5$. *J. Phys.: Condens. Matter* **7**, 2855.

[54] Hur, N., Park, S., Sharma, P.A., *et al.* (2004). Electric polarization reversal and memory in a multiferroic material induced by magnetic fields. *Nature* **429**, 392.

[55] Radaelli, P.G., Chapon, L.C., Daoud-Aladine, A. *et al.* (2008). Electric Field Switching of Antiferromagnetic Domains in YMn_2O_5: A Probe of the Multiferroic Mechanism. *Phys. Rev. Lett.* **101**, 067205.

[56] Newnham, R.E., Kramer, J.J., Schulze, W.A., and Cross, L.E. (1978). Magnetoferroelectricity in Cr_2BeO_4. *J. Appl. Phys.* **49**, 6088.

[57] Lee, S., Pirogov, A., Kang, M. *et al.* (2008). Giant magneto-elastic coupling in multiferroic hexagonal manganites. *Nature* **451**, 805–808.

[58] Van Aken, B.B., Palstra, T.T.M., Filippetti, A., and Spaldin, N.A. (2004). The origin of ferroelectricity in magnetoelectric $YMnO_3$. *Nature Mat.* **3**, 164.

[59] Fennie, C.J. and Rabe, K.M. (2005). Ferroelectric transition in $YMnO_3$ from first principles. *Phys. Rev. B* **72**, 100103.

[60] Goltsev, A.V., Pisarev, R.V., Lottermoser, Th., and Fiebig, M. (2003). Structure and Interaction of Antiferromagnetic Domain Walls in Hexagonal $YMnO_3$. *Phys. Rev. Lett.* **90**, 177204.

[61] Lottermoser, T. and Fiebig, M. (2004). Magnetoelectric behavior of domain walls in multiferroic $HoMnO_3$. *Phys. Rev. B* **70**, 220407(R).

[62] Privratska, J. and Janovec, V. (1997). Pyromagnetic domain walls connecting antiferromagnetic non-ferroelastic magnetoelectric domains. *Ferroelectrics* **204**, 321–331.

[63] Sergienko, I.A., Sen, C., and Dagotto, E. (2006). Ferroelectricity in the Magnetic E-Phase of Orthorhombic Perovskites. *Phys. Rev. Lett.* **97**, 227204.

[64] Giovannetti, G., Kumar, S., Khomskii, D., Picozzi, S., and van den Brink, J. (2009). Multiferroicity in Rare-Earth Nickelates $RNiO_3$. *Phys. Rev. Lett.* **103**, 156401.

[65] Catalan, G. (2008). Progress in perovskite nickelate research. *Phase Transitions* **81**, 729.

[66] Levstik, A., Bobnar, V., Filipic, C. *et al.* (2007). Magnetoelectric relaxor. *Appl. Phys. Lett.* **91**, 012905.

[67] Levstik, A., Filipic, C., Bobnar, V. *et al.* (2008). $0.3Pb(Fe_{1/2}Nb_{1/2})O_3$-$0.7Pb(Mg_{1/2}W_{1/2})O_3$, A magnetic and electric relaxor. *J. Appl. Phys.* **104**, 054113.
[68] Blinc, R., Tavčar, G., Žemva, B., *et al.* (2009). Electron paramagnetic resonance and Mossbauer study of antiferromagnetic $K_3Cu_3Fe_2F_{15}$. *J. Appl. Phys.* **106**, 023924.
[69] Blinc, R., Tavčar, G., Žemva, B. *et al.* (2008). Weak Ferromagnetism and Ferroelectricity in $K_3Fe_5F_{15}$. *J. Appl. Phys.* **103**, 074114.
[70] Blinc, R., Zalar, B., Cevc, P. *et al.* (2009). K-39 NMR and EPR study of multiferroic $K_3Fe_5F_{15}$. *J. Phys.: Condens. Matter* **21**, 045902.
[71] Levstik, A., Filipic, C., Bobnar, V., *et al.* (2009). Polarons in magnetoelectric $(K_3Fe_3Cr_2F_{15})$. *J. Appl. Phys.* **106**, 073720.
[72] Blinc, R., Zalar, B., Cevc, P. *et al.* (2009). K-39 NMR and EPR study of multiferroic $K_3Fe_5F_{15}$. *J. Phys.: Condens. Matter* **21**, 045902.
[73] Abrahams, S.C. (1999). Systematic prediction of new inorganic ferroelectrics in point group 4. *Acta Cryst.* **B55**, 494–506.
[74] Abrahams, S.C. (1989). Structurally based predictions of ferroelectricity in 7 inorganic materials with space group Pba2 and two experimental confirmations. *Acta Cryst.* **B45**, 228.
[75] Kumar, A., Katiyar, R.S., Premnat, R.N., *et al.* (2009). Strain-induced artificial multiferroicity in $Pb(Zr_{0.53}Ti_{0.47})O_3/Pb(Fe_{0.66}W_{0.33})O_3$ layered nanostructure at ambient temperature. *J. Mater. Sci.* **44**, 5113–5119.
[76] Kumar, A., Murari, N.M., and Katiyar, R.S. (2009). Investigation of dielectric and electrical behavior in $Pb(Fe_{0.66}W_{0.33})_{0.50}Ti_{0.50}O_3$ thin films by impedance spectroscopy. *J. Alloys & Comp.* **469**, 433–440.
[77] Peng, W., Lemee, N., Dellis, J.L. *et al.* (2009). Epitaxial growth and magnetoelectric relaxor behavior in multiferroic $0.8Pb(Fe_{1/2}Nb_{1/2})O_3$-$0.2Pb(Mg_{1/2}W_{1/2})O_3$ thin films. *Appl. Phys. Lett.* **95**, 132507.
[78] Scott, J.F., Palai, R., Kumar, A. *et al.* (2008). New phase transitions in perovskite oxides: $BiFeO_3$, $SrSnO_3$, and $Pb[Fe_{2/3}W_{1/3}]_{1/2}Ti_{1/2}O_3$. *J. Am. Ceram. Soc.* **91**, 1762–1768.
[79] Kumar, A., Rivera, I., Katiyar, R.S. *et al.* (2008). Multiferroic $Pb(Fe_{0.66}W_{0.33})_{0.80}Ti_{0.20}O_3$ thin films: A room-temperature relaxor ferroelectric and weak ferromagnetic. *Appl. Phys. Lett.* **92**, 132913.
[80] Choudhary, R.N.P., Pradhan, D.K., Tirado, C.M. *et al.* (2007). Effect of La substitution on structural and electrical properties of $Ba(Fe_{2/3}W_{1/3})O_3$ nanoceramics. *J. Mater. Sci.* **42**, 7423–7432.
[81] Kumar, A., Murari, N.M., Katiyar, R.S. *et al.* (2007). Probing the ferroelectric phase transition through Raman spectroscopy in $Pb[Fe_{2/3}W_{1/3}]_{1/2}Ti_{1/2}O_3$thin films. *Appl. Phys. Lett.* **90**, 262907.
[82] Choudhary, R.N.P., Pradhan, D.K., Tirado, C.M. *et al.* (2007). Impedance characteristics of $Pb(Fe_{2/3}W_{1/3})O_3$-$BiFeO_3$ composites. *Phys. Stat. Sol.* **244**, 2254–2266.
[83] Kumar, A., Murari, N.M., and Katiyar, R.S. (2007). Diffused phase transition and relaxor behavior in $Pb(Fe_{2/3}W_{1/3})O_3$ thin films. *Appl. Phys. Lett.* **90**, 162903.
[84] Choudhury, R.N.P., Rodriguez, C., Bhattacharya, P., Katiyar, R.S., and Rinaldi, C. (2007). Low-frequency dielectric dispersion and magnetic properties of La, Gd modified $Pb(Fe_{1/2}Ta_{1/2})O_3$ multiferroics. *J. Magn. Magn. Mater.* **313**, 253–260.

[85] Peng, W., Lemee, N., Holc, J. *et al.* (2009). Epitaxial growth and structural characterization of $Pb(Fe_{1/2}Nb_{1/2})O_3$ thin films. *J. Magn. Magn. Mater.* **321**, 1754–1757.
[86] Peng, W., Lemee, N., Karkut, M. *et al.* (2009). Spin-lattice coupling in multiferroic thin films. *Appl. Phys. Lett.* **94**, 012509.
[87] Kumar, A., Katiyar, R.S., Rinaldi, C., Lushnikov, S.G., Shaplygina, T.A. (2008). Glasslike state in $PbFe_{1/2}Nb_{1/2}O_3$ single crystal. *Appl. Phys. Lett.* **93**, 232902.
[88] Correa, M., Kumar, A., Katiyar, R. S. *et al.* (2008). Observation of magnetoelectric coupling in glassy epitaxial $Pb(Fe_{1/2}Nb_{1/2})O_3$ thin films. *Appl. Phys. Lett.* **93**, 192907.
[89] Varshney, D., Choudhary, R.N.P., Rinaldi, C. *et al.* (2007). Dielectric dispersion and magnetic properties of Ba-modified $Pb(Fe_{1/2}Nb_{1/2})O_3$. *Appl. Phys. A* **89**, 793–798.
[90] Varshney, D., Choudhary, R.N.P., Katiyar, R.S. *et al.* (2006). Low frequency dielectric response of mechano-synthesized $(Pb_{0.9}Ba_{0.1})(Fe_{0.50}Nb_{0.50})O_3$ nanoceramics. *Appl. Phys. Lett.* **89**, 172901.
[91] Pirc, R., Blinc, R., and Scott, J.F. (2009). Mesoscopic model of a system possessing both relaxor ferroelectric and relaxor ferromagnetic properties. *Phys. Rev. B* **79**, 214114.
[92] Kumar, A., Sharma, G.L., Katiyar, R.S., and Scott, J.F. (2009). Magnetic control of large room-temperature polarization. *J. Phys.: Condens. Matter* **21**, 382204.
[93] (a) Kumar, A., Katiyar, R.S., and Scott, J.F. (2009). Positive temperature coefficient of resistivity and negative differential resistivity in lead iron tungstate-lead zirconate titanate. *Appl. Phys. Lett.* **94**, 212903; (b) Dawber, M. and Scott, J.F. (2004). Negative differential resistivity and positive temperature coefficient of resistivity effect in the diffusion-limited current of ferroelectric thin-film capacitors. *J. Phys.: Condens. Matter* **16**, L515-L521.
[94] Keve, E.T., Abrahams, S.C., and Bernstein, J.L. (1969). Crystal Structure of pyroelectric paramagnetic barium manganese fluoride, $BaMnF_4$. *J. Chem. Phys.* **51**, 4928; (1971). *Ferroelectrics* **2**, 129.
[95] Keve, E.T., Abrahams, S.C., and Bernstein, J.L. (1970). Crystal Structure of pyroelectric paramagnetic barium cobalt fluoride, $BaCoF_4$. *J. Chem. Phys.* **53**, 3279.
[96] Ederer, C. and Spaldin, N.A. (2005). Recent progress in first-principles studies of magnetoelectric multiferroics. *Current Opinion Sol. St. & Mater. Sci.* **9**, 128–139.
[97] Ederer, C. and Spaldin, N.A. (2006). Electric-field-switchable magnets: The case of $BaNiF_4$. *Phys. Rev. B* **74**, 020401.
[98] (a) Venturini, E.L. and Morgenthaler, F.R. (1975). AFMR versus orientation in weakly ferromagnetic $BaMnF_4$. *AIP Conf. Proc.* **24**, 168; (b) Zvezdin, A.K., Vorob'ev, G.P., Kadomsteva, A.M., *et al.* (2009). Quadratic Magnetoelectric Effect and the Role of the Magnetocaloric Effect in the Magnetoelectric Properties of Multiferroic $BaMnF_4$. *J. Exp. Theor. Phys.* **109**, 221–226.
[99] Kizhaev, S.A. (1981). Ioffe Institute (private communication); Kizhaev, S.A. and Pisarev, R.V. (1984). Low-temperature dielectric and magnetic properties of BaMnF. *Fiz. Tverd. Tela* **26**, 1669–1674.
[100] Prozorova, L.A. (1981). Kapitza Institute (private communication).
[101] Birss, R.R. (1964). Magnetic symmetry and forbidden effects. *Am. J. Phys.* **32**, 142.
[102] Scott, J.F. (1979). Phase Transitions in $BaMnF_4$. *Rep. Prog. Phys.* **42**, 1055–1084.
[103] Scott, J.F. (1977). Mechanisms of Dielectric Anomalies in $BaMnF_4$. *Phys. Rev. B* **16**, 2329–2331.

[104] Samara, G. and Scott, J.F. (1977). Dielectric Anomalies in $BaMnF_4$ at Low Temperatures. *Sol. St. Commun.* **21**, 167–170.
[105] de Alcantara Bonfim, O.F. and Gehring, G. (1980). Magnetoelectric effect in antiferromagnetic crystals. *Adv. Phys.* **29**, 731–789.
[106] Catalan, G. (2006). Magnetocapacitance without magnetoelectric coupling. *Appl. Phys. Lett.* **88**, 102902.
[107] Scott, J.F. (2007). Electrical Characterization of Magnetoelectrical Materials. *J. Mater. Res.* **22**, 2053–2062.
[108] Catalan, G. and Scott, J.F. (2007). Is $CdCr_2S_4$ a multiferroic relaxor? *Nature* **448**, E4–E5.
[109] Staresinic, D., Lunkenheimer, P., Hemberger, J., and Loidl, A. (2006). Giant dielectric response in the one-dimensional charge-ordered semiconductor $(NbSe_4)_3I$. *Phys. Rev. Lett.* **96**, 046402.
[110] Lunkenheimer, P., Fichtl, R., Hemberger, J., and Loidl, A. (2005). Relaxation dynamics and colossal magnetocapacitive effect in $CdCr_2S_4$. *Phys. Rev. B* **72**, 060103.
[111] Hemberger, J., Lunkenheimer, P., Fichtl, R., *et al.* (2006). Relaxor ferroelectricity and colossal magnetocapacitive coupling in ferromagnetic $CdCr_2S_4$. *Nature* **434**, 364–367.
[112] Krohns, S., Schrettle, F., Lunkenheimer, P., and Loidl, A. (2008). Colossal magnetocapacitive effect in differently synthesized and doped $CdCr_2S_4$. *Physica* **B403**, 4224–4227.
[113] Hemberger, J., Lunkenheimer, P., Fichtl, R., and Loidl, A. (2007). Magnetoelectrics - Is $CdCr_2S_4$ a multiferroic relaxor? Reply, *Nature* **448**, E5–E6.
[114] Weber, S., Lunkenheimer, P., Fichtl, R., and Loidl, A. (2006). Colossal magnetocapacitance and colossal magnetoresistance in $HgCr_2S_4$. *Phys. Rev. Lett.* **96**, 157202.
[115] Kliem, H. and Martin, B. (2008). Pseudo-ferroelectric properties by space charge polarization. *J. Phys.: Condens. Matter* **20**, 321001.
[116] Ortega, N., Kumar, A., Katiyar, R.S., and Scott, J.F. (2007). Maxwell-Wagner space charge effects on the $PbZrTiO_3$ - $CoFe_2O_4$ multilayers. *Appl. Phys. Lett.* **91**, 102902.
[117] Albuquerque, E.L. and Tilley, D.R. (1978). Mode mixing and dielectric function. *Sol. St. Commun.* **26**, 817–821.
[118] Glass, A.M., Lines, M.E., Eibschutz, M., Hsu, F.S.L., and Guggenheim, H.J. (1977). Observation of anomalous pyroelectric behavior in $BaNiF_4$ due to cooperative magnetic singularity. *Commun. Phys.* **2**, 103–107.
[119] Negran, T.J. (1981). Measurement of the thermal diffusivity in $BaMnF_4$ by means of its intrinsic pyroelectric response. *Ferroelectrics* **34**, 285–289.
[120] Hou, S.L. and Bloembergen, N. (1965). Paramagnetoelectric effects in $NiSO_4{\cdot}6H_2O$. *Phys. Rev.* **138**, A1218–A1226.
[121] Sciau, P., Clin, M., Rivera, J.P., and Schmid, H. (1990). Magnetoelectric measurements on $BaMnF_4$. *Ferroelectrics* **105**, 201–206.
[122] Zvezdin, A.K., Vorob'ev, G.P., and Kadomsteva, A.M. (2009). Quadratic Magnetoelectric Effect and the Role of the Magnetocaloric Effect in the Magnetoelectric Properties of Multiferroic $BaMnF_4$. *J. Exp. Theor. Phys.* **109**, 221–226.
[123] Perez-Mato, J.M. (2008). Summer School on Multiferroics, Girona, Spain.
[124] Ederer, C. and Spaldin, N.A. (2006). Origin of ferroelectricity in the multiferroic barium fluorides $BaMF_4$: A first principles study. *Phys. Rev. B* **74**, 024102.
[125] Scott, J.F. (1979). Phase Transitions in $BaMnF_4$. *Rep. Prog. Phys.* **42**, 1055.

[126] Cox, D.E., Shapiro, S.M., Nelmes, R.J., and Cowley, R.A. (1983). X-ray diffraction and neutron diffraction measurements on $BaMnF_4$. *Phys. Rev. B* **28**, 1640–1643.
[127] Levstik, A., Blinc, R., Kadaba, P., *et al.* (1976). Dielectric properties of the 250 K phase transition in $BaMnF_4$ and of CsH_2PO_4 and CsD_2PO_4. *Ferroelectrics* **4**, 703.
[128] Hidaka, M., Nakayama, T., and Scott, J.F. (1985). Piezoelectric Resonance Study of Structural Anomalies in $BaMnF_4$. *Physica B/C* **133**, 1–9.
[129] Scott, J.F., Habbal, F., and Hidaka, M. (1982). Phase transition in $BaMnF_4$ – Specific Heat. *Phys. Rev. B* **25**, 1805–1812.
[130] Barthes, M., Almairac, R., St Grégoire, P. *et al.* (1983). Temperature dependence of the wave vector of the incommensurate modulation in two $BaMnF_4$ crystals grown by different techniques; neutron and X-ray measurements. *J. Physique Lettres* **44**, 829.
[131] Barthes-Regis, M., Almairac, R., St. Gregoire, P. (1983). Temperature-dependence of the wave vector of the incommensurate modulation in two $BaMnF_4$ crystals grown by different techniques – neutron and X-ray measurements. *J. Physique Lettres* **44**, L829–L835.
[132] Lavrencic, B.B. and Scott, J.F. (1981). Dynamical model for the polar-incommensurate transition in $BaMnF_4$. *Phys. Rev. B* **24**, 2711–2717.
[133] Kato, K. and Iida, S. (1982). Observation of Ferroelectric Hysteresis Loop of Fe_3O_4 at 4.2 K. *J. Phys. Soc. Jpn.* **51**, 1335.
[134] Alexe, M., Ziese, M., Hesse, D., *et al.* (2009). Ferroelectric Switching in Multiferroic Magnetite (Fe_3O_4) Thin Films. *Advanced Materials* **21**, 4452–4455.
[135] Nenert, G. and Palstra, T.T.M. (2007). Prediction for new magnetoelectric fluorides. *J. Phys.: Condens. Matter* **19**, 406213.
[136] Levstik, A., Filipic, C., and Holc, J. (2008). The magnetoelectric coefficients of $PbFe_{1/2}Nb_{1/2}O_3$ and 0.8 $PbFe_{1/2}Nb_{1/2}O_3$-$0.2PbMg_{1/2}W_{1/2}O_3$. *J. Appl. Phys.* **103**, 066106.
[137] (a) Abrahams, S.C. (1992). Dielectric and related properties of fluorine-octahedra ferroelectrics. *Ferroelectrics* **135**, 21; (b) Ravez, J. (1997). The Inorganie Fluoride and Oxyfluoride Ferroelectrics. *J. Physique III* **7**, 1129–1144; (c) Abrahams, S.C., Marsh, P., and Ravez, J. (1989). Coupling of Ferroelasticity to ferroelectricity in $Na_5W_3O_9F_5$ and the structure at 295 K. *Acta Cryst.* **B45**, 364–370.
[138] Ravez, J., Arquis, S., Grannec, J., Simon, A., and Abrahams, S.C. (1987). Phase transitions and ferroelectric attributes in $A_5M_3F_{19}$ compounds with A=Sr, Ba and M=Ti, V, Cr, Fe, Ga. *J. Appl. Phys.* **62**, 4299–4301.
[139] Abrahams, S.C., Albertsson, J., Svensson, C., and Ravez, J. (1990). Structure of $Pb_5Cr_3F_{19}$, polarization reversal, and the 555K phase transition. *Acta Cryst.* **B46**, 497.
[140] Abrahams, S.C., Bernstein, J.L., and Ravez, J. (1981). Paraelectric-paraelastic $Rb_2KMoO_3F_3$ structure at 343 and 473 K. *Acta Cryst.* **B37**, 1332–1336.
[141] Blinc, R., Tavčar, G., Žemva, B., *et al.* (2008). Weak Ferromagnetism and Ferroelectricity in $K_3Fe_5F_{15}$. *J. Appl. Phys.* **103**, 074114.
[142] Blinc, R., Tavčar, G., Žemva, B., *et al.* (2009). Electron paramagnetic resonance and Mossbauer study of antiferromagnetic $K_3Cu_3Fe_2F_{15}$. *J. Appl. Phys.* **106**, 023924.
[143] Forquet, J.L., LeBail, A., Duroy, H., and Moron, M.C. (1989). $(NH_4)_2FeF_5$: crystal structures of its α and β forms. *Eur. J. Sol. St. & Iorg. Chem.* **26**, 435–443.

[144] Sarraute, S., Ravez, J., Von der Mühll, R., *et al.* (1996). Structure of ferroelectric $Pb_5Al_3F_{19}$ at 160 K, polarization reversal and relationship to ferroelectric $Pb_5Cr_3F_{19}$ at 295 K. *Acta Cryst.* **B52**, 72–77.

[145] Kroumova, E., Aroyo, M.I., Pérez-Mato, J.M., and Hundt, M.R. (2001). Ferroelectric-paraelectric phase transitions with no group-supergroup relation between the space groups of both phases? *Acta Cryst.* **B57**, 599–601.

[146] Lorient, M., Von der Mühll, R., Ravez, J., and Tressaud, A. (1980). Etude de la transition de phase de $(NH_4)\ _3FeF_6$ par mesures dielectriques et de thermocourant. *Sol. St. Commun.* **36**, 383–385.

[147] Lorient, M., Von der Mühll, R., Tressaud, A., and Ravez, J. (1981). Polarisation remanente dans les varietes de basse temperature de $(NH_4)\ _3AlF_6$ ET $(NH_4)\ _3FeF_6$. *Sol. St. Commun.* **40**, 847–852.

[148] Fischer, K.H. and Hertz, J.A. (1991). *Spin Glasses*, p.123. Cambridge Univ. Press, Cambridge.

[149] Singh, M.K., Prelier, W., Singh, M.P., Katiyar, R.S., and Scott, J.F. (2008). Spin glass transition in single-crystal $BiFeO_3$. *Phys. Rev. B* **77**, 144403.

[150] Parish, M.M. and Littlewood, P.B. (2008). Magnetocapacitance in Nonmagnetic Composite Media. *Phys. Rev. Lett.* **101**, 166602.

[151] Ishibashi, Y. and Hidaka, Y. (1991). On an Isomorphous Transition. *J. Phys. Soc. Jpn.* **60**, 1634.

[152] Scott, J.F. (2010). Iso-Structural Phase Transitions in $BiFeO_3$. *Adv. Mat.* **22**, 2106–2107.

[153] Hosokawa, T., Kobayashi, J., Uesu, Y., and Miyazaki, H. (1978). Crystal symmetry of $Ca_2Sr(C_2H_5CO_2)_6$ in the low temperature phase. *Ferroelectrics* **20**, 201–202.

[154] Gesi, K. and Ozawa, K. (1974). Critical points of the II-III transitions in $Ca_2Pb(C_2H_5COO)_6$ and $Ca_2Sr(C_2H_5COO)_6$. *Phys. Lett. A* **49**, 283.

[155] Gesi, K. (1977). Pressure-Induced Ferroelectricity in $(NH_4)_3H(SO_4)_2$. *J. Phys. Soc. Jpn.* **43**, 1941.

[156] Kutnjak, Z., Petzelt, J., and Blinc, R. (2006). The giant electromechanical response in ferroelectric relaxors as a critical phenomenon. *Nature* **441**, 956.

[157] Wu, Z. and Cohen, R.E. (2005). Pressure-induced anomalous phase transitions and colossal enhancement of piezoelectricity in $PbTiO_3$. *Phys. Rev. Lett.* **95**, 037601.

[158] Lisenkov, L., Rahmedov, D., and Bellaiche, L. (2009). Electric-Field-Induced Paths in Multiferroic $BiFeO_3$ from Atomistic Simulations. *Phys. Rev. Lett.* **103**, 047204.

[159] Zeches, R.J., Rossell, M.D., Zhang, J.X., *et al.* (2009). A Strain-Driven Morphotropic Phase Boundary in $BiFeO_3$. *Science* **326**, 977.

[160] Arnold, D., Morrison, F.D. *et al.* (2010). The β-to-γ Transition in $BiFeO_3$: A Powder Neutron Diffraction Study. *Adv. Func. Mat.* **20**, 2116.

[161] Scott, J.F., Pirc, R., Levstik, A., Filipic, C., and Blinc, R. (2006). Resolving the quantum criticality paradox in O-18 isotopic $SrTiO_3$. *J. Phys.: Condens. Matter* **18**, L205.

[162] Holwerda, M.J., van Neerven, W.L., and van Royen, R.P. (1979). Hyperscaling and the critical exponent ν. Nuov. *Cim.* **52**, 77.

[163] Ocko, B.M., Birgeneau, R.J., Litster, J.D., and Neubert, M.E. (1983). Critical and Tricritical Behavior at the Nematic to Smectic-A Transition. *Phys. Rev. Lett.* **52**, 208.

Part II

Oxide films and interfaces: growth and characterization

3
Synthesis of epitaxial multiferroic oxide thin films

Thomas Tybell and Chang-Beom Eom

3.1 Introduction

Over the last two decades there has been tremendous advancement in epitaxial thin-film growth of perovskite based material systems. In the first part of the 1990s, high temperature superconductors were a driving force towards the synthesis of high quality epitaxial thin films. Since the basic structure of $YBa_2Cu_3O_{7-\delta}$ resembles the simple perovskite unit cell, it was rapidly realized that the same techniques could be used to synthesize other perovskite material systems. Today's control of functional properties is largely due to two areas of continuous advancement: thin-film deposition techniques, and development of high quality single crystalline substrate materials with single surface termination and well-controlled miscut. Physical deposition techniques of choice today are often either 90° off-axis sputtering [1], pulsed laser deposition (PLD), or molecular beam epitaxy (MBE). *In-situ* monitoring by high pressure reflection high-energy electron diffraction (RHEED) [2] has enabled sub-unit cell control during deposition, making it possible to control the termination layer and deposition rate with high precision.

Perovskite materials exhibit strong structure–property coupling, opening for epitaxy as a tool to control the functional properties in thin-film systems. Taking advantage of the recent advancements in physical deposition and substrate preparation, it is now not only possible to control, but also to tune and enhance physical properties during synthesis. The basis for obtaining thin films with desired properties is hence a combination of precise control of deposition parameters during synthesis and choice of substrate. An epitaxial approach allows synthesis of single crystalline thin films of materials that are only accessible in polycrystalline form in bulk, for example $Pb(Zr,Ti)O_3$ [3]. This opens new avenues to study both fundamental aspects of the materials, as well as a gateway to high performance thin films for device applications. Furthermore, advanced thin film synthesis offers not only single crystalline thin films, but also provides a means of controlling the microscopic, at unit cell level, and macroscopic, e.g. crystalline domain, structure of thin films.

The correct choice of substrate is key in this regard. For example, single crystalline thin films with different orientations can be obtained by controlling the orientation of the substrate [4]. Since the properties of perovskite materials are often anisotropic, the control of crystalline direction enables control of macroscopic physical responses, such as ferroelectric polarization and fatigue [4]. Furthermore, macroscopic structure can be controlled by miscut substrates. The symmetry of the substrate can be employed to control the unit cell symmetry of the thin film, such as for $SrRuO_3$ grown on $SrTiO_3$ [5]; and hence, the properties. Another important role of the substrate is to introduce a coherent strain state in the films. Since many functional properties are sensitive to the distortions and volume of the unit cell, epitaxial strain can be used to tailor and enhance their properties, as demonstrated, for example, in ferroelectric thin films [6]. Figure 3.1 displays a schematic of the degrees of freedom that substrates offer for controlling the properties of perovskite thin films. The choice of substrate, its crystalline symmetry, orientation, and surface morphology, are hence central in order to ensure a high level of control of the outcome during thin-film deposition.

In this chapter we will focus on how the use of substrates allows thin-film growers to both control and enhance functional properties of perovskite based systems. Two important aspects for thin-film growth that we will discuss are how strain engineering and structural domain control by miscut substrates can be used to tailor thin-film systems. We will use examples from our own research throughout the chapter to elucidate on the current possibilities. In Section 3.2 some common substrates will be introduced, and the last two sections cover how state-of-the-art deposition can be used to tune and control physical properties of thin films. Section 3.3

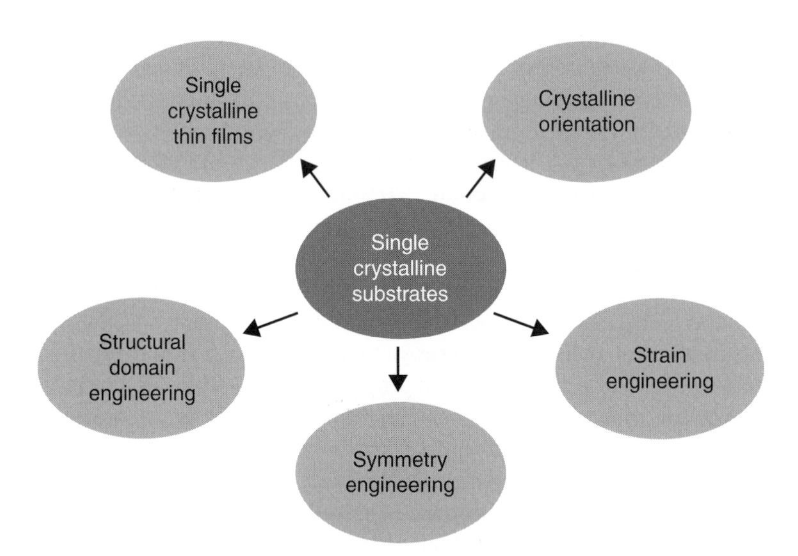

Fig. 3.1 A schematic of the degrees of freedom that substrates offer for controlling the properties of perovskite thin films. Substrates enable control of both microscopic and macroscopic properties of thin films by varying lattice parameters, symmetry, miscut, crystallographic orientation, and surface termination of substrates.

discusses how strain imposed by the substrate on the thin film can be utilized to control physical properties, and the effect of substrate surface will be discussed in Section 3.4.

3.2 Substrates

A key parameter in thin-film synthesis is the substrate, an important tool used to enhance and control functional properties of perovskite thin films. This section will briefly introduce common substrates, and some important concepts of strain and domain engineering which will be discussed in Sections 3.3 and 3.4.

The available selection of single crystalline substrates is large, but the most widely used substrate material for perovskite thin-film synthesis is $SrTiO_3$. This has a cubic unit cell with a lattice parameter of 3.905 Å at room temperature [7]. $SrTiO_3$ is well lattice-matched to a wide selection of perovskite oxides, enabling epitaxial growth of various materials systems. For example, epitaxial growth of high temperature superconductors [3, 8], metallic perovskites [9], antiferromagnetics [10], ferroelectrics [3], ferromagnetics [11], and multiferroics [12] have been demonstrated on $SrTiO_3$. It is important to note that the structure of $SrTiO_3$ does not change between room temperature and typical deposition temperatures of 700–900°C. However, $SrTiO_3$ undergoes a structural phase transition, and goes from cubic to tetragonal when the temperature is lowered below 105 K [7]. From a deposition point of view the thermal expansion coefficient is important since a large thermal expansion mismatch can lead to thermal strain. $SrTiO_3$ has a thermal expansion coefficient of 9.4×10^{-6} C^{-1} at room temperature [7].

There are a variety of substrates to choose from, with a large span of lattice parameters and crystalline symmetry. Table 3.1 shows a list of some of the most common substrate materials, together with their lattice parameters and crystal symmetry at room temperature. The list is not extensive and can for instance be complemented with compositional effects, such as chemical doping. For example, a metallic/conducting substrate can often be beneficial for electrical characterization of multiferroic materials, and it is possible to obtain conducting oxide substrates by doping. Doping $SrTiO_3$ with either Nb or La makes it conducting, thus enabling direct growth on top of conducting substrates. Typical doping levels are 0.1–0.5 wt%.

3.2.1 Strain, orientation, and symmetry control by choice of substrate

It is the large variety in lattice parameters among the various substrates that allows for strain engineering of functional materials. That is, thin films with a controlled biaxial in-plane strain state. By growing the thin film coherently on the substrate, below the critical thickness where strain relaxation occurs, the thin film will have the in-plane parameters dictated by the substrate values. Hence, due to the wide spectrum of substrate lattice parameters available, both tensile and compressive strain states can often be obtained for a given material system. The use of strain engineering in order to control and enhance physical properties will be discussed in more detail in Section 3.3.

The properties of multiferroic oxides are often anisotropic, and coupled to certain crystalline directions, as for example, ferroelectric polarization [17]. The use of substrates having specific crystalline orientations, for example (100) or (110) terminated crystalline surfaces, is thus beneficial. Figure 3.2 shows a schematic of $SrTiO_3$ having a (001), (110), and (111) top surfaces respectively. Based on such surface it is then possible to synthesize thin films which have various

Table 3.1 Common perovskite substrates and their structural properties at room temperature. Pseudo cubic (pc) lattice parameters are defined by $a_{pc} = 0.5^*$ square root $(a^2 + b^2)$.

Substrate	Crystal structure	Lattice parameters	References
$SrTiO_3$	cubic	$a = b = c = 3.905$ Å	7
$LaAlO_3$	rhombohedral	$a = b = c = 5.356$ Å $a_{pc} = 3.787$ Å	13
$NdGaO_3$	orthorhombic	$a = 5.428$ Å $b = 5.498$ Å $c = 7.708$ Å $a_{pc} = 3.863$ Å	14
$LaGaO_3$	orthorhombic	$a = 5.520$ Å $b = 5.490$ Å $c = 7.770$ Å $a_{pc} = 3.893$ Å	15
$DyScO_3$	orthorhombic	$a = 5.440$ Å $b = 5.713$ Å $c = 7.887$ Å $a_{pc} = 3.944$ Å	16
$GdScO_3$	orthorhombic	$a = 5.488$ Å $b = 5.746$ Å $c = 7.934$ Å $a_{pc} = 3.973$ Å	16
$TbScO_3$	orthorhombic	$a = 5.466$ Å $b = 5.727$ Å $c = 7.915$ Å $a_{pc} = 3.958$ Å	16

crystalline directions. This can be utilized in controlling the physical and structural properties of the oxide thin films [4].

Many functional oxides belong to point-groups having low-symmetry unit cells, lower than cubic, which is the symmetry of $SrTiO_3$. The possibility of growing samples with a well-defined crystalline domain structure is therefore important. Often heterostructures may be required, for example if epitaxial electrodes are needed, or if materials with different symmetry properties are to be combined, thus making the choice of substrate more delicate. As can be seen in Table 3.1, there is not only a wide range of available lattice parameters of the various substrates, but they also have different symmetries. One possibility for control of the structure is to match the symmetry of the substrate with that of the film material. The orthoscandate family of substrates is interesting in this regard; they possess an orthorombic crystal structure, and are very stable with respect to defects due to the stable Sc^{3+} valence [18]. Another possibility is to control the symmetry of the thin-film unit cell through the choice of substrate, i.e. symmetry

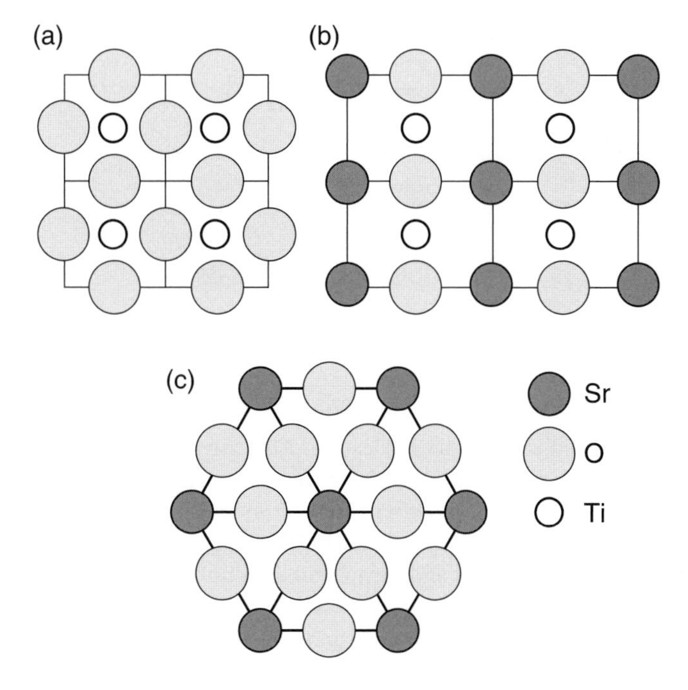

Fig. 3.2 Schematic of (a) (001), (b) (110), and (c) (111) $SrTiO_3$ surfaces.

engineering. In the case of different unit-cell symmetry between the thin film and the substrate, the symmetry of the substrate can couple into the thin film, close to the interface, as predicted theoretically by He *et al.* [19]. The oxygen octahedra in a thin film can be susceptible to deformation due to the substrate symmetry creating a novel phase at the interface. That is, the substrate can not only influence lattice parameters due to strain, but it can actually affect the detailed symmetry of the thin film.

3.2.2 Substrate termination and surface quality

It is advantageous to be able to work with substrates which have an atomically perfect termination, with respect to both chemistry and morphology. That is, atomically engineered substrates should be beneficial, and there has been a large focus on this topic over the last 15 years. One important tool in this respect, in order to control the physical properties of perovskite thin films, is the degree of miscut of the substrate, that is the angle between the average surface normal and the crystalline surface normal. Figure 3.3 is a schematic of a miscut substrate showing terraces separated by one unit-cell-high steps. The degree of miscut, the larger the miscut angle the smaller the terrace width, can be used to control the detailed growth mode for a given deposition process [20]. For a given set of deposition parameters the ad-atoms will have a given surface diffusivity, and the relation between a typical diffusion length and terrace width is important. Ideally one would like to control where nucleation takes place, for example at the step edges. However, it is important to impose a sufficiently large diffusivity to enable the

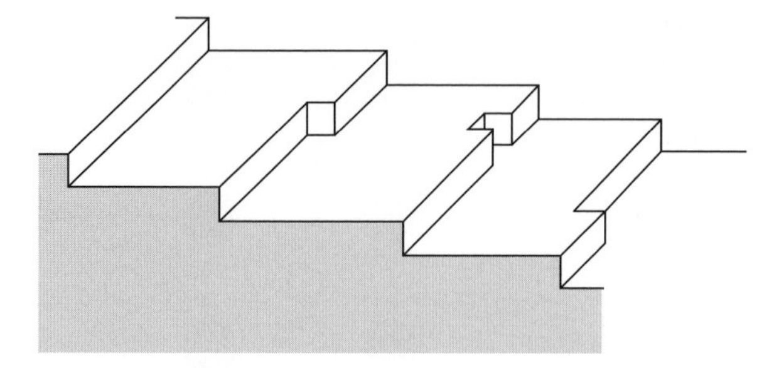

Fig. 3.3 An example of the step and terrace structure.

ad-atoms to reach the step edge before nucleating. Also the miscut direction is an important parameter, [21] since this can be used to tailor the structural domain structure of the thin films, and this is especially true for growth of low-symmetry materials. This topic with regard to multiferroics will be discussed in more detail in Section 3.4.

Not only are the degree of miscut and the miscut direction important, but also the detailed atomic surface structure of the substrate plays a role in determining thin-film quality. The termination of the substrate, if the perovskite unit-cell closest to the surface terminates with

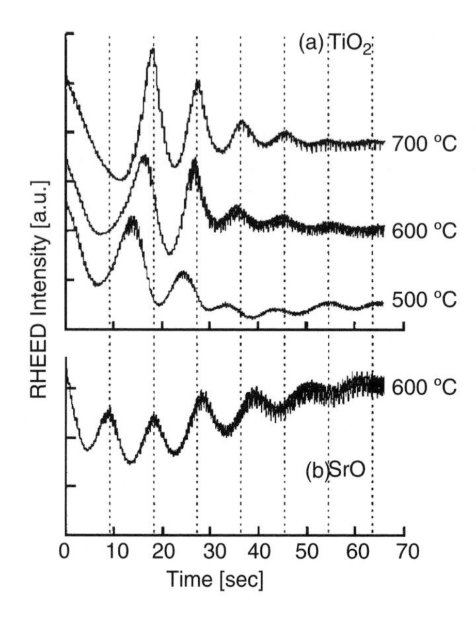

Fig. 3.4 RHEED intensity for both TiO_2-terminated (a) and SrO-terminated (b) $SrTiO_3$ as a function of time after the start of deposition. As can be seen, TiO_2 termination leads to a switch in terminating layer. Reprinted with permission from [23]. Copyright 2004, American Institute of Physics.

Table 3.2 Common substrates with reported single terminations.

Substrate	*Reported single termination*	*Reference*
$SrTiO_3$	SrO	25
	TiO_2	26, 27
$GdScO_3$	ScO_2	28
$DyScO_3$	ScO_2	28
$NdScO_3$	ScO_2	28
$NdGaO_3$	NdO	29

an AO or BO_2 layer, plays a role during thin-film growth. It has, for example, been reported that TiO_2 termination is required for layer-by-layer growth of $SrRuO_3$ [22]. The deposition sequence can also be altered by the substrate termination. For example, by depositing $SrRuO_3$ on SrO-terminated $SrTiO_3$, the main perovskite sequence is not altered, and RuO_2 layers are alternated with SrO layers. However, when a TiO_2-terminated substrate is used, a switch in termination is obtained during the growth of the first unit-cell, effectively producing a SrO-terminated $SrRuO_3$ unit-cell [23, 24]. This is illustrated in Figure 3.4 which shows RHEED data during growth, clearly demonstrating a switch in termination of $SrRuO_3$ when grown on TiO_2-terminated $SrTiO_3$. These two examples exemplify that having access to the correct substrate termination is important for a high degree of control during epitaxial growth to be achieved. Hence, single terminated substrates are desirable, and there has been significant progress in developing single terminated substrates. Most of the focus has been on $SrTiO_3$, with several reports on methods for obtaining either A-site termination [25] or B-site termination [26, 27]. There has also been a focus on obtaining single terminated substrates from other perovskite substrates. Recently, it was demonstrated that a combination of initial annealing at 1000 °C, followed by chemical etching in 12M NaOH-DI water solution enables ScO_2 termination of scandate substrates, as verified by time-of-flight mass spectroscopy [28]. Table 3.2 lists reported terminations for various substrates of interest for epitaxial growth of perovskite oxides.

3.3 Strain engineering as a tool for controlling functional oxide thin films

Many functional properties of perovskites, such as strength of ferroelectric polarization, depend on the distortion of the unit-cell [17]. If the thin-film deposition process can be used to tune the lattice parameters of thin films, tailoring of the physical properties through strain engineering is possible. We have discussed previously that there exists a large range of substrates with varying in-plane lattice parameters. In this section, we will focus on the effect of lattice mismatch between the thin-film material and the substrate. We will especially discuss the possibility of using controlled biaxial in-plane strain as a tool to control and enhance the functional properties of ferroic and multiferroic perovskite thin films.

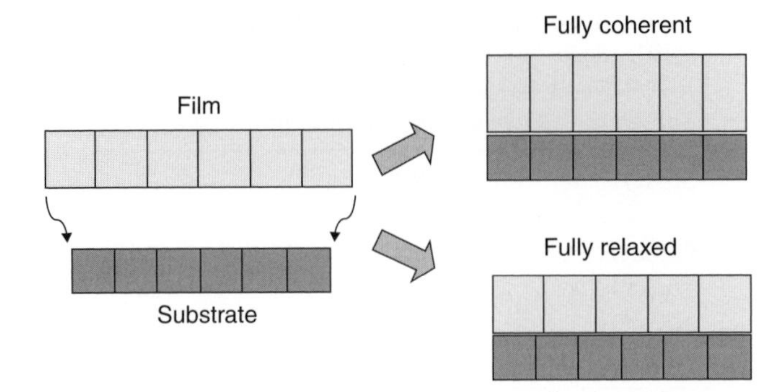

Fig. 3.5 Schematic showing the difference between fully strained film for which the in-plane lattice parameters of the thin film is dictated by the substrate, and a fully relaxed thin film having the materials bulk lattice parameters.

When a thin film, with an in-plane lattice parameter different from the substrate's in-plane lattice parameter is grown on a substrate, the film can either be fully epitaxial strained or it can start to relax. If the film is fully strained, the in-plane lattice parameter of the thin film will be equal to the substrate value. At the opposite end, a fully relaxed thin film would have its bulk lattice parameter, and, ideally, misfit dislocations at the interface. This is illustrated schematically in Figure 3.5. For fully commensurate material systems, thin films with a thickness below the critical thickness for which relaxation starts, one can define the in-plane strain as $\varepsilon = (a - a_0)/a_0$, where a_0 and a are the in-plane lattice constant of the thin film material and substrate, respectively. Based on the range of available substrates, having pseudo-cubic in-plane lattice parameters ranging from 3.79Å ($LaAlO_3$) to 3.97Å ($GdScO_3$), a $SrTiO_3$ thin film would have a biaxial strain state ranging from 2.94% compressive strain to 1.66% tensile strain. This indicates that there is quite a large range of strain that can be applied to a thin film system via the substrate, making it an effective contributor to the epitaxial toolbox. For a small difference in lattice parameter between the thin film and the substrate, it is possible to grow thick commensurate thin films. In the case of $SrRuO_3$ on $SrTiO_3$ it is possible to grow more than 1000 Å-thick films that are fully strained [30]. However, for Ce-doped $BaTiO_3$, which has an in-plane lattice parameter of approximately 4.15 Å, the thin films begin to relax after two to three unit-cells in thickness when grown on $SrTiO_3$. Hence, there is a maximum amount of strain that can be employed. As the degree of strain increases, relaxation processes become more effective and misfit dislocations will be introduced, limiting the maximum film thickness for which one can obtain fully commensurate thin films. However, with this in mind, strain is nevertheless a tremendous tool, as we shall see in the coming examples.

3.3.1 $SrRuO_3$—a case study of strain engineering

Here we will first exemplify the effect of strain in a fully commensurate film by focusing on $SrRuO_3$. $SrRuO_3$ is a ferromagnetic metallic perovskite with a Curie temperature of 160 K. It is a material that is widely used as an electrode layer in multiferroic epitaxial heterostructures [9, 31], it is therefore important to understand how different substrates, and hence strain,

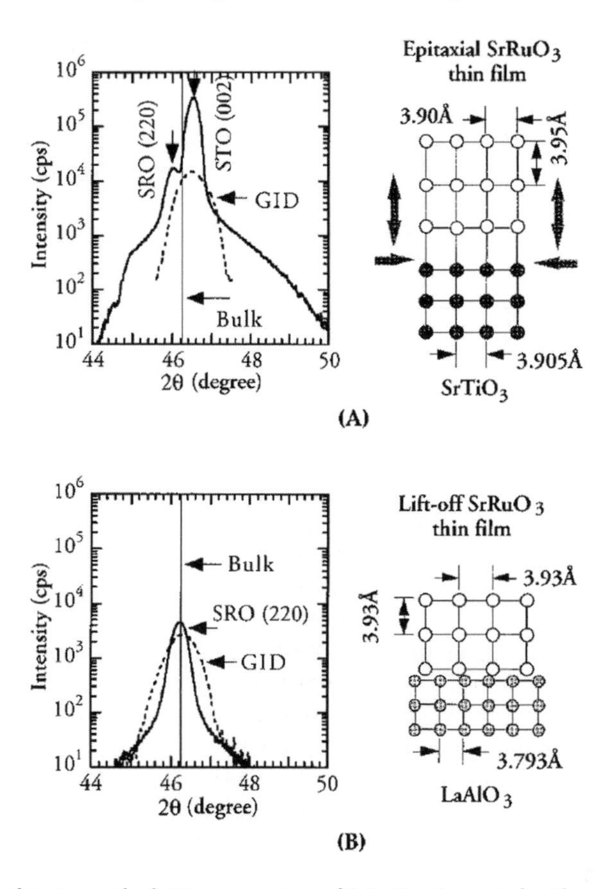

Fig. 3.6 The effect of strain on the lattice parameters of $SrRuO_3$. Reprinted with permission from [30]. Copyright 1998, American Institute of Physics.

state affects its properties. $SrRuO_3$ is orthorhombic with a pseudo-cubic lattice parameter of 3.93Å, and when grown on $SrTiO_3$ the lattice mismatch is small, of the order of -0.64%, but sufficiently large to alter the properties of a thin film. $SrRuO_3$ can be grown coherently on $SrTiO_3$ using both 90° off-axis sputtering [9], pulsed laser deposition [24], and molecular beam epitaxy [32]. In the case of 90° off-axis RF magnetron sputtering, typical deposition parameters are: a 6 to 4 Ar to oxygen ambient at a total pressure of 200 mTorr, and a deposition temperature of ~680 °C. For pulsed laser deposition, one typically uses a temperature of ~700 °C.

As will be discussed in the next section, by carefully choosing the substrates, and having a focus on miscut angle and direction of miscut, it is possible to grow mono-domain thin films of low-symmetry materials. Thin films of $SrRuO_3$ can easily have domain formation due to the orthorhombic symmetry of the unit-cell. By relying on single domain thin films, Gan *et al.* demonstrated the effect of the compressive strain on both the Curie temperature and magnetization of the $SrRuO_3$ epilayer [30]. The $SrRuO_3$ thin films were grown by 90°

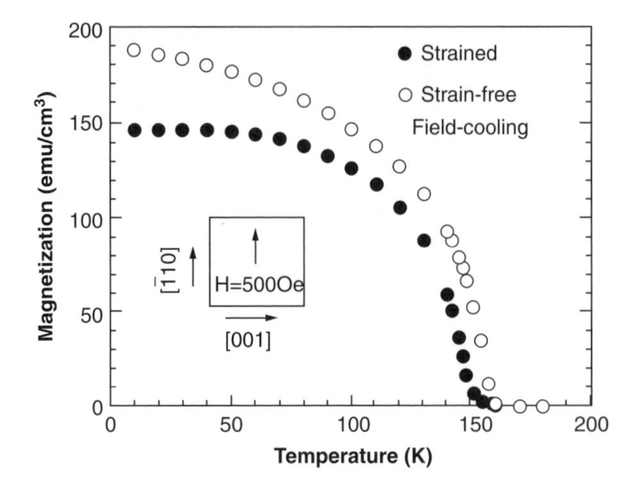

Fig. 3.7 The effect of strain on magnetization of $SrRuO_3$. Reprinted with permission from [30]. Copyright 1998, American Institute of Physics.

off-axis magnetron sputtering on miscut $SrTiO_3$ substrates. The as-grown 1000 Å-thick samples were fully coherent with the substrate. In order to investigate the effect of the biaxial strain imposed by the substrate Gan *et al.* relied on the fact that stoichiometric $SrRuO_3$ is chemically stable, enabling removal of the substrate by selectively etching $SrTiO_3$ using a $HF+HNO_3+H_2O$ solution. In order to rule out other extrinsic effects the samples were cut into dice in order to compare the structure and properties of the same as-grown and substrate etched sample. Figure 3.6 shows grazing incident X-ray diffractograms of the as-grown sample (A) and the free-standing $SrRuO_3$ layer on top of a support (B). The data clearly show that the free-standing film has the same pseudo-cubic lattice parameters as bulk $SrRuO_3$, 3.93 Å. The coherent sample, on the other hand, is clamped to the $SrTiO_3$ substrate having an in-plane lattice parameter of 3.905 Å, and exhibits an out-of-plane lattice parameter of 3.95 Å. This gives a good demonstration of the effect of the biaxial strain on the out-of-plane lattice parameter. It is also interesting to note that, since the thin $SrRuO_3$ layer recovered the bulk lattice parameter, although being coherently compressively strained to the substrate prior to the removal, epitaxial strain imposed by a substrate can be regarded as elastic and reversible. If the strain imposed by the substrate were plastic, or permanent, a reversible process to the bulk value after removal would not take place. The elastic nature of the epitaxial strain is helpful towards this, since it allows for more simple theoretical implementation of the effects of strain on various physical properties. That is, one should be able to treat a coherently strained sample theoretically, based only on its measured lattice parameters, including the real change of the lattice in the course of simulations, neglecting any difficulties with plasticity.

The epitaxial strain does not only affect the lattice parameters. Physical investigations of the strained $SrRuO_3$ layers have shown that the compressive strain increases the saturation magnetization by 20%, as compared to the relaxed sample with bulk lattice parameters. Also, the Curie temperature was affected; the compressive strain resulted in a negative 10 K shift of the Curie temperature, as shown in Figure 3.7.

Bulk $SrRuO_3$ undergoes an orthorhombic to tetragonal, and a tetragonal to cubic phase transition at 820 K and 950 K, respectively [33]. Figure 3.8 depicts how the out-of-plane lattice parameter (a) and in-plane parameters (b) vary as a function of temperature for a 500 Å-thick, coherently grown $SrRuO_3$ film on $SrTiO_3$ [5]. The phase transition temperatures for bulk $SrRuO_3$ are also shown for reference. As can be seen, for the coherently grown thin films, phase transitions are altered as compared to bulk. The X-ray analysis reveals that below 553 K the thin film has an orthorhombic unit-cell, which is in agreement with the bulk structure. Since the thin film is coherently grown on top of the substrate, the in-plane lattice structure is commensurate with the substrate. This is actually true for the entire temperature interval investigated. In Figure 3.8 (b) the orthorhombic a and b lattice parameters are shown, as deduced from 400_o and 040_o X-ray diffraction reflections. The effect of the compressive in-plane strain from the substrate is to lower the bulk phase transition temperature. Moreover, above the phase transition temperature the thin film is still under the influence of the substrate. The biaxial strain exerted by the substrate makes the thin film strained-cubic at higher temperatures, and hence the tetragonal phase in bulk is not observed.

By carefully controlling the deposition process, strain engineering not only changes the magnitude of physical properties, and critical temperatures for the phase transitions, it is also possible to alter the phases above and below the critical temperatures, as compared to bulk. The ideal cubic perovskite structure can be distorted by cation distortions, oxygen octahedral tilting and/or rotations, resulting in materials having other crystal structures such as tetragonal, common among ferroelectrics, orthorhombic, rhombohedral, and monoclinic structures. Hence, one can expect that coherently grown films, exhibiting full biaxial in-plane strain from the substrate, can affect the symmetry of epitaxial thin films. If this is the case, thin-film deposition can be used to control not only the magnitude of a functional property, but also the class of functionality of a thin film, since the symmetry of the unit-cell is often

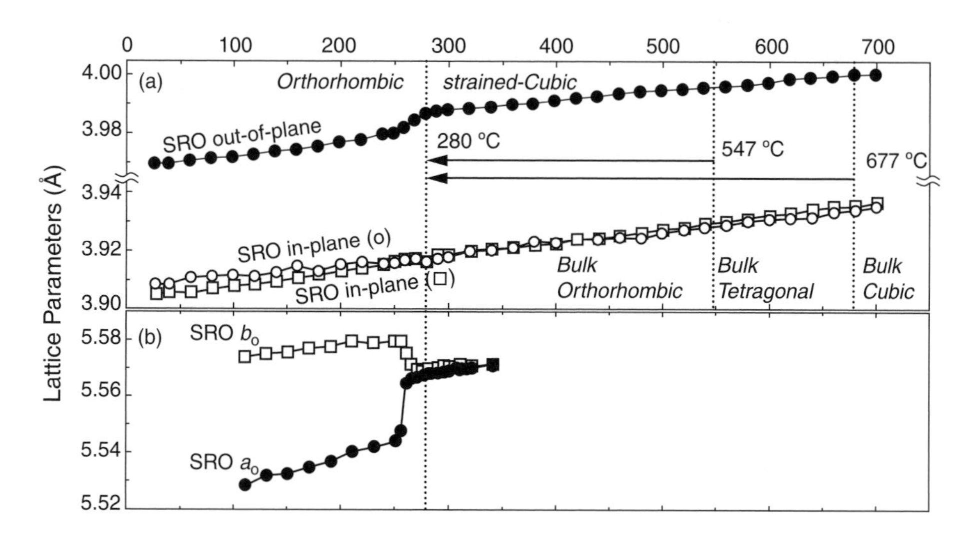

Fig. 3.8 The lattice parameters as a function of temperature of an epitaxial $SrRuO_3$ thin film, demonstrating the shift in structural phase transitions as compared to bulk. [5].

coupled to the functionality. Since the thin films are clamped to the substrates and their in-plane lattice constants are now controlled by the substrate, only the out-of-plane parameter can change at a phase transition. However, internal degrees of freedom can also be allowed, such as oxygen octahedral rotation, which can be used to tune the properties. Going back to the $SrRuO_3$ case, by replacing $SrTiO_3$ with $GaScO_3$, the resulting strain changes from a -0.64% compressive strain state to a 0.97% tensile strain state. This change in strain effectively changes the resulting symmetry of epitaxial $SrRuO_3$ thin films, possibly imposing a cubic symmetry at room temperature [5]. This demonstrates that strain engineering and symmetry engineering can take place simultaneously, and that the interplay between the two governs the final symmetry of the thin film.

As discussed, perovskite oxides undergo structural phase transition as a function of temperature. If such a transition of the substrate material is coupled to a change in in-plane lattice parameter, the resulting strain state of an epitaxial thin film will be altered. Hence, one has to choose a substrate, based not only on the required strain, but also on the temperature interval of interest. Here we will use ferromagnetic/ferroelectric $SrRuO_3/BaTiO_3$ heterostructures as a test case to discuss this effect. Lee *et al.* [34] used $BaTiO_3$ substrates in order to study the effect of crystalline phase transitions in a substrate on a coherently grown thin film. $BaTiO_3$ undergoes three phase transitions between 393 K and 183 K when cooled: first a cubic to tetragonal transition at 393 K, followed by a tetragonal to monoclinic transition at 278 K, and lastly a monoclinic to rhombohedral transition at 183 K, as shown in Figure 3.9. For a coherently grown $SrRuO_3$ thin film on $BaTiO_3$ the strain will thus change abruptly at the phase transition

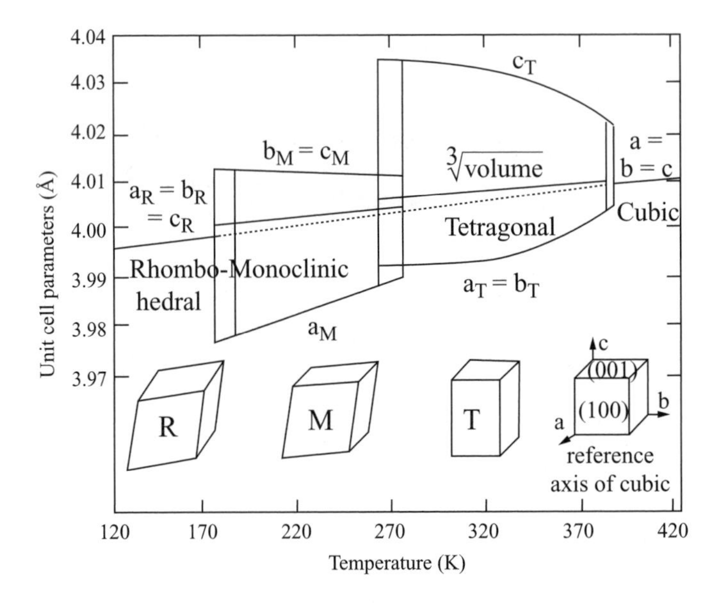

Fig. 3.9 The evolution of lattice parameters of $BaTiO_3$ as a function of temperature, clearly displaying three phase transitions. Reprinted with permission from [34]. Copyright 2000, American Institute of Physics.

temperatures, and based on the discussion above, one would expect a change of properties at these temperatures. Lee *et al.* reported abrupt changes in the resistivity of the $SrRuO_3$ epilayer at the phase transition temperatures.

One advantage enabled by strain engineering is to compare how different degrees of strain affect the physical properties of multiferroic materials, the basis for strain engineering of enhanced properties. This can be achieved by either choosing different oxide substrate materials as the starting point or a-making it strained-tuned buffer layers [35]. However, one prerequisite is that the thin films can be grown coherently on the various substrates; so strain relaxation starts at different thicknesses depending on the degree of strain. As we have seen, the substrate onto which the thin films are deposited can alter the bulk symmetry of the unit-cell and strongly affect the physical properties of a thin film. In this section we will briefly focus on how strain engineering can be used to optimize ferroelectric materials. The application of strain engineering to ferroelectric materials is a large focus of research, and there is a vast amount of literature on the subject [6, 36]. By synthesizing $BaTiO_3$ on $GdScO_3$ and $DyScO_3$, both substrates impose a compressive in-plane strain state of -1 and -1.7%, respectively, Choi *et al.* [36] demonstrated that the Curie temperature of the thin ferroelectric film was drastically raised as compared to bulk. In Figure 3.10, evolution of the lattice parameters as a function of temperature for the two cases is shown. As can be seen, the temperature at which the thin film becomes paraelectric is drastically larger for a thin film as compared to bulk.

Any perovskite oxide material having properties susceptible to strain can, in principle, be tuned by strain engineering. The recently discovered multiferroic materials are interesting in this respect, and they have attracted wide attention [37, 38]. One example of the effect of

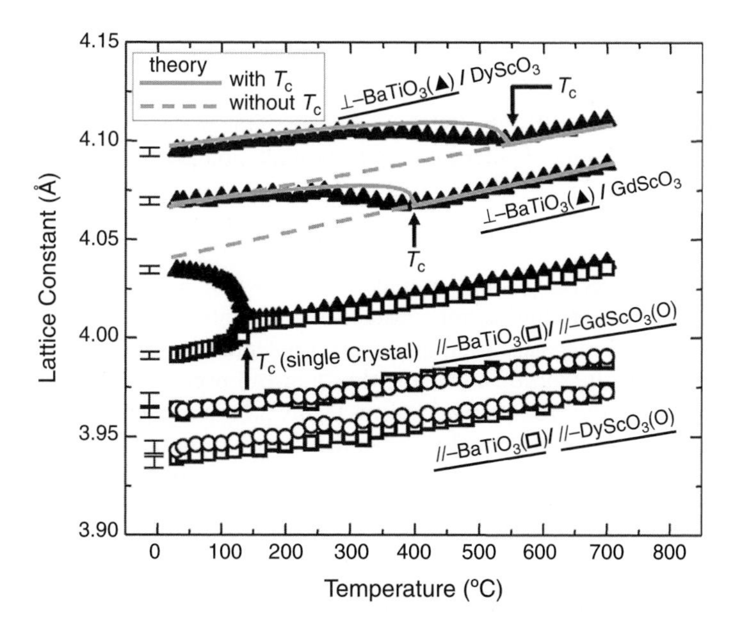

Fig. 3.10 A comparison of how strain affects the ferroelectric–paraelectric phase transition of $BaTiO_3$, as measured by X-ray diffraction. Reprinted with permission from [36].

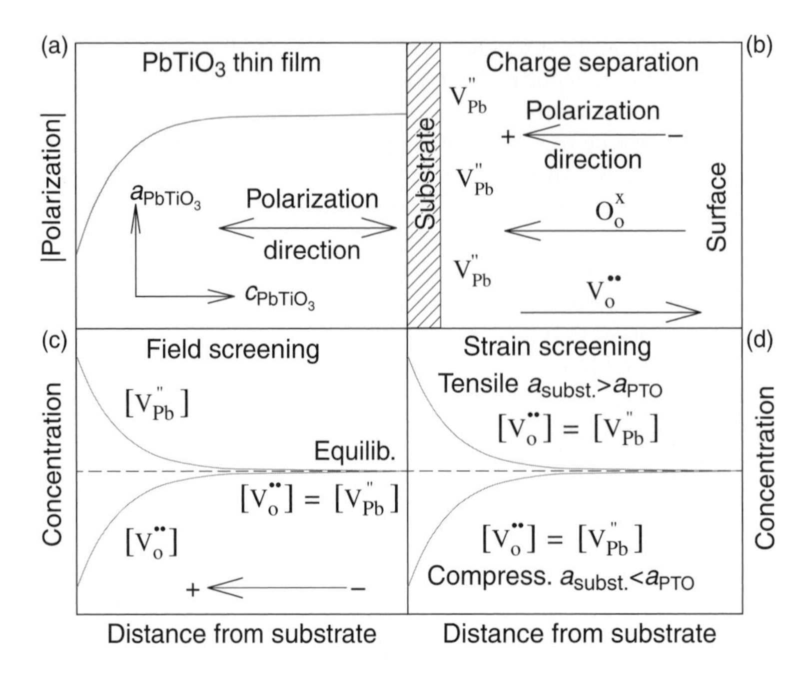

Fig. 3.11 A schematic explaining how strain and cation defect concentration can interplay at a ferroelectric / dielectric interface enabling defect engineering in order to control strain in epitaxial thin films and heterostructures. Reprinted with permission from [43]. Copyright 2011, American Institute of Physics.

epitaxial strain on $BiFeO_3$ is that the (001) component of the remnant polarization is strongly tunable due to a strain-induced polarization direction rotation [39].

3.3.2 Effect of defects

We note finally that defects can locally modify the lattice [40], and affect physical properties [41]. If the defect concentration is not constant in a thin film, for example due to internal fields from ferroelectric depolarization or a contact potential [42], a defect gradient close to the interface between the substrate and the thin film can occur. Such defect gradients can effectively act as a strain relaxation mechanism. Figure 3.11 is a schematic representation for a ferroelectric/electrode interface and possible defect–strain states [43]. This allows for the combination of strain and defect engineering in the synthesis of well-controlled functional multiferroic thin films for device applications.

3.4 Vicinal control of functional properties

In this section we discuss how vicinal substrate surfaces can be used to control the structure and phases of epitaxial thin films. The focus is on the synthesis of materials having low-symmetry unit-cells, deviating from ideally cubic materials, with special focus on how miscut substrates can be a key parameter to control the domain structure of the thin films during deposition.

3.4.1 $SrRuO_3$—a case study of vicinal control of orthorhombic domain structure

We start our discussion focusing on $SrRuO_3$, an orthorhombic material. Six different crystallographic domain variants are possible for $SrRuO_3$ on (001) $SrTiO_3$ surfaces, as displayed in Figure 3.12. The (001), (110), or (1-10) planes of $SrRuO_3$ can epitaxially align with the (001) $SrTiO_3$ surface [44]. If synthesized on exact substrates, that is substrates with large terraces due to minimal or no miscut, the $SrRuO_3$ epitaxial thin film consist of equal amounts of the [001] $SrRuO_3$ // [010] $SrTiO_3$ family, and the [001] $SrRuO_3$ // [100] $SrTiO_3$ family of domains [45].

By imposing a miscut on the substrate surface it is possible to lower the number of possibilities, and hence possibly obtain control of structural domain formation during synthesis. Two parameters of importance are the miscut angle (α) and the miscut direction along the surface (β). The miscut direction, β, is often defined by the projection of the surface normal on the (001) $SrTiO_3$ plane, and the in-plane [010] direction, as shown in Figure 3.13.

As seen in Figure 3.13, the miscut locally exposes (010) and (100) planes at the steps, effectively breaking the fourfold symmetry of the cubic substrate. By relying on step-flow growth, characterized by an ad-atom diffusion length larger than the terrace width, two possible in-plane organizations can take place, named A and B in the figure. For A [-110]° $SrRuO_3$ will be parallel to [010] $SrTiO_3$ and [001]° $SrRuO_3$ parallel to [100] $SrTiO_3$. In case B, [001]° $SrRuO_3$ will be parallel to [100] $SrTiO_3$ and [-110]° $SrRuO_3$ parallel to [100] $SrTiO_3$. This would correspond to two structural variances in the films. If the flow of ad-atoms could be arranged preferentially in one direction, a mono domain sample would be the result. Single domain samples are obtained if large miscut angles, $\alpha > 1.9°$ and small β values are used, effectively suppressing one of the two types of growth. For smaller α values, corresponding to larger terraces, the growth becomes more two-dimensional layer by layer growth, supporting both variances, and by having a larger β the step surface structure allows step-flow growth in both directions, A and B, hence resulting in two domain samples [46]. The possibility of

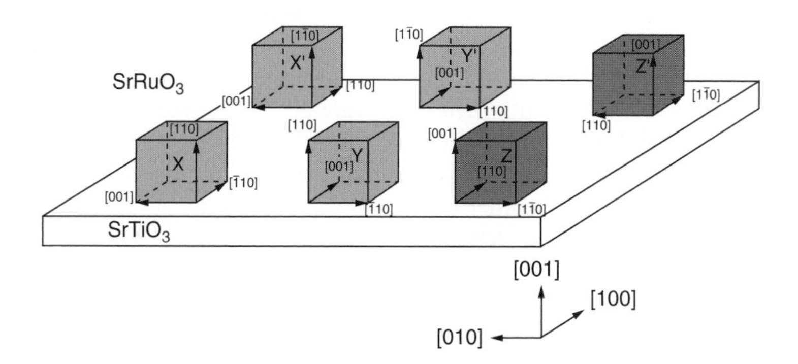

Fig. 3.12 A schematic of six different crystallographic domain variants of orthorhombic $SrRuO_3$ on cubic $SrTiO_3$, displayed in a pseudo-cubic fashion. Reprinted with permission from [44]. Copyright 1998, American Institute of Physics.

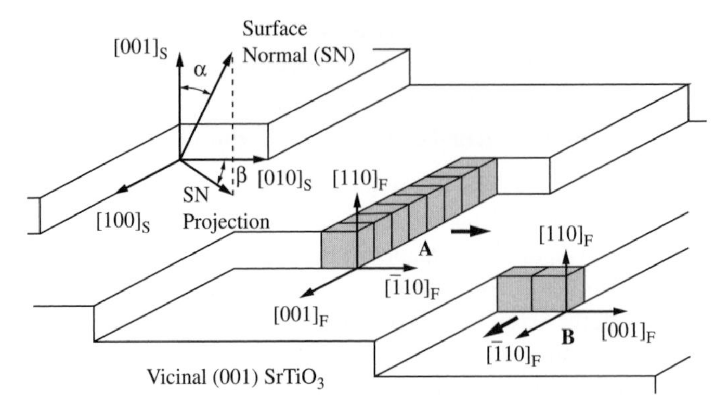

Fig. 3.13 Schematic depicting a miscut (001) $SrTiO_3$ surface defined by the angles α and β. A and B represent two types of step-flow growth on the surface. Reprinted with permission from [46]. Copyright 1997, American Institute of Physics.

controlling the domain structure in an electrode material such as $SrRuO_3$ is a good starting point when different ferroic systems are to be developed. This will enable a well-defined domain geometry as the starting point.

3.4.2 $BiFeO_3$—domain control of a prototype rhombohedral material by substrate miscut

$BiFeO_3$ is a multiferroic material, being both ferroelectric, T_c=820 °C, and antiferromagnetic, T_N = 370 °C. It has a rhombohedral perovskite crystal structure, and can be described by the pseudo-cubic lattice parameters a_r = 3.96 Å and α_r = 0.6° [47]. The rhombohedral distortions, being directed along one of the four pseudo-cubic (111) directions (Figure 3.14), are named r_1, r_2, r_3, and r_4. This results in a complex domain structure when deposited on cubic $SrTiO_3$. Four different domain variances are possible, having either 100 or 101 twin walls [48]. This is illustrated in Figure 3.15, showing transmission electron microscopy data for a 600 nm $BiFeO_3$/100 nm $SrRuO_3$ heterostructure on non-miscut (001) $SrTiO_3$ [49]. As can be seen, there are both 100 (corresponding to 109° domain walls) and 101 (corresponding to 71° domain walls) walls present. The selected area electron diffraction patterns display a splitting/elongation of the higher-order diffraction spots along the [100] direction for 109° domains (b), and along the [101] direction for 71° domains (c).

The complicated domain structure of $BiFeO_3$ on non-miscut substrates can be a disadvantage if $BiFeO_3$ is to be used in factual devices, possibly eroding the response of the material to external stimuli. As for $SrRuO_3$, it is possible to rely on miscut substrates for domain engineering of $BiFeO_3$ [50]. But before discussing how domain control can be used to tailor monodomain or two-variant samples, we will briefly mention the possibility of using structurally adopted substrates based on symmetry, as introduced earlier. It has been shown that it is possible for two-variant $BiFeO_3$ thin films to be grown on orthorhombic substrates [51, 52]. Figures 3.16 and 3.17 exemplify this for $BiFeO_3$ grown on (110) $TbScO_3$ substrates. As can

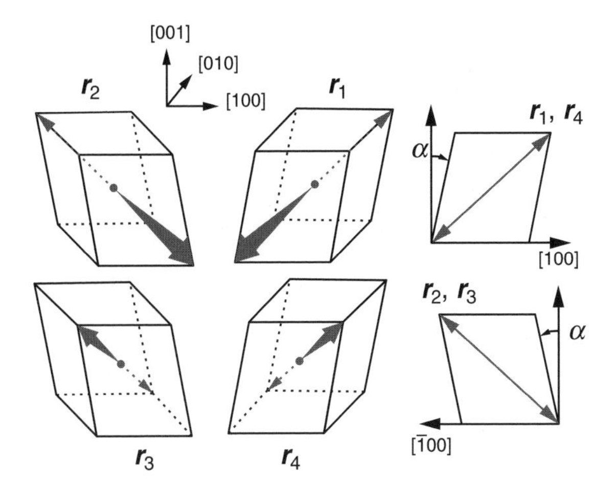

Fig. 3.14 A schematic displaying the four ways a rhombohedral unit-cell can align on cubic $SrTiO_3$. r_1, r_2, r_3, and r_4 are the four variances possible with the [111] elongated axis in four different directions.

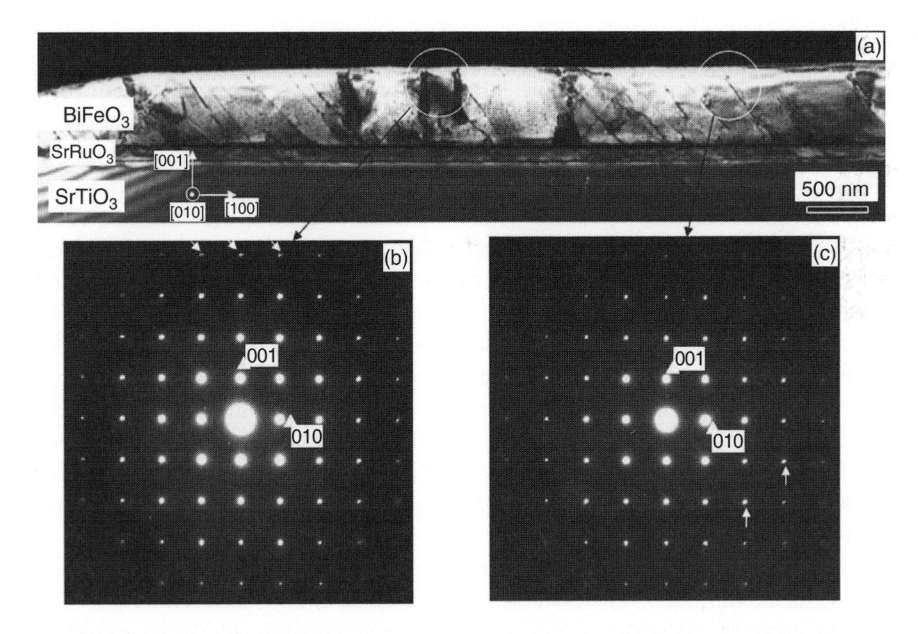

Fig. 3.15 (a) Dark field image of a 600 nm $BiFeO_3$/100 nm $SrRuO_3$, heterostructure on non-miscut (001) $SrTiO_3$, displaying both 001 and 101 types of domain walls. (b) Selected area electron diffraction taken from a region with 109° domains, and (c) for a region with 71° domains. Reprinted with permission from [49]. Copyright 2007, American Institute of Physics.

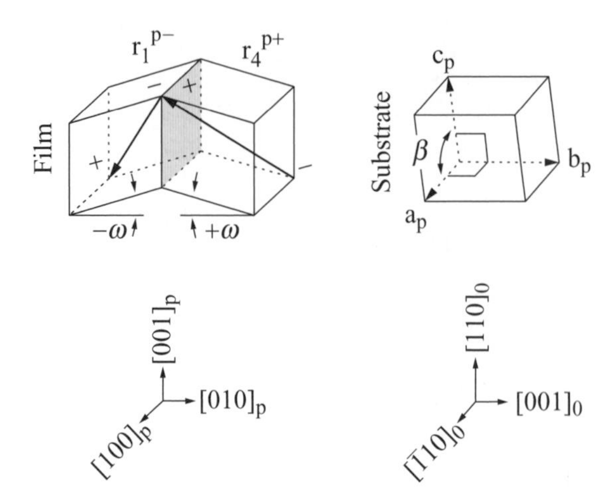

Fig. 3.16 A schematic showing the resulting ferroelastic domain structure of $BiFeO_3$ when deposited on $TbScO_3$. Only two variants were obtained having (010)-type domain walls. The arrows to the left denote the direction of the ferroelectric polarization. Reprinted with permission from [52]. Copyright 2009, American Institute of Physics.

be seen, the [010] $BiFeO_3$ aligns with the [001] $TbScO_3$ direction, resulting in two variances having (010)-type domain walls (Figure 3.16). The regularity of the twin domain pattern is demonstrated by the AFM and dark field TEM data in Figure 3.17. Formation of the two ferroelastic domains is controlled by the rectangular (110) surface of $TbScO_3$ [52]. That is, by relying on orthoscandate substrates, it is possible to control stripe domain patterns.

Now, we will discuss the possibility of using miscut to control the domain structure of $BiFeO_3$ [53]. As mentioned above, when synthesizing $BiFeO_3$ on exact substrates four variances, named r_1, r_2, r_3, and r_4 respectively, are obtained. This is illustrated schematically in Figure 3.18 (a), where nucleation takes place randomly on the terraces. This is typical for large terraces, inhibiting step-flow growth, and leading to 3D islands. On the other hand, when miscut substrates with small terrace width are used, step-flow growth is possible. Because of the mechanical conditions imposed by the step edges, r_2 and r_3 domains are inhibited from forming due to the unfavorable energetics of having the rhombohedral distortions pointing towards the edges. The resulting domain structure tilts away from the miscut direction, favoring a rhombohedral distortion r_1 and r_4 [54], see Figure 3.18 (b). This tilt angle increases as a function of thickness.

The miscut substrate not only changes the domain structure, but also affects the lattice parameters of the thin film. When deposited on exact (001) $SrTiO_3$ substrates, the BFO in-plane lattice parameter increases, and the out-of-plane lattice parameter decreases, as the biaxial compressive strain is relaxed, as shown in Figure 3.19. However, on vicinal substrates there is a faster strain relaxation parallel to the miscut direction, compared to the vertical direction. Furthermore, the out-of-plane lattice parameter decreases faster as a function of thickness, compared to when an exact substrate is used.

By combining miscut control of the ferroelastic variances with different crystalline surfaces of the substrate it is possible to synthesize mono-domain $BiFeO_3$, as shown in Figure 3.20

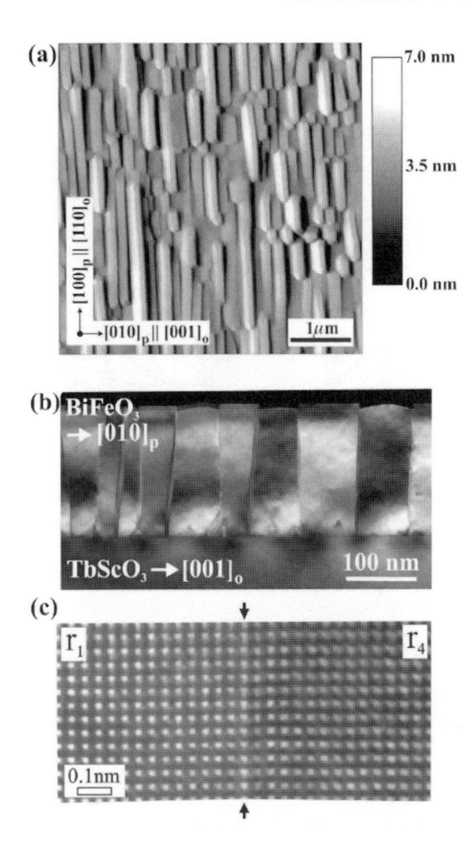

Fig. 3.17 AFM (a), and dark field TEM (b) data of $BiFeO_3/TbScO_3$. (c) High resolution TEM showing the (010) domain wall. Reprinted with permission from [52]. Copyright 2009, American Institute of Physics.

[4, 53, 55, 56]. It has been shown that mono-domain $(001)_{pc}$ $BiFeO_3$ can be synthesized on highly miscut (001) $SrTiO_3$ ($\alpha \approx 4°$) along the [110] direction, mono-domain $(110)_{pc}$ $BiFeO_3$ can be synthesized on 0.3° (110) $SrTiO_3$ along the [1-11] direction, and mono-domain $(111)_{pc}$ $BiFeO_3$ can be synthesized on (111) $SrTiO_3$ [55]. In Figure 3.21, an example of mono-domain $BiFeO_3$ is shown. The AFM data (Figure 3.21(a)) reveal two or three unit-cell high steps due to step bunching from the high miscut substrate. A reciprocal space map taken around the (-103) reflections demonstrates only one variance (Figure 3.21(b)), and TEM analysis along the [-110] zone axis, and electron diffraction, are consistent with a single domain sample (Figure 3.21(c)).

3.4.3 Mono-domain samples—enabling fundamental studies and enhanced properties of $BiFeO_3$

The possibility of synthesizing epitaxial single-domain rhombohedral thin film emerges for well-controlled fundamental studies of intrinsic physical properties, eliminating extrinsic

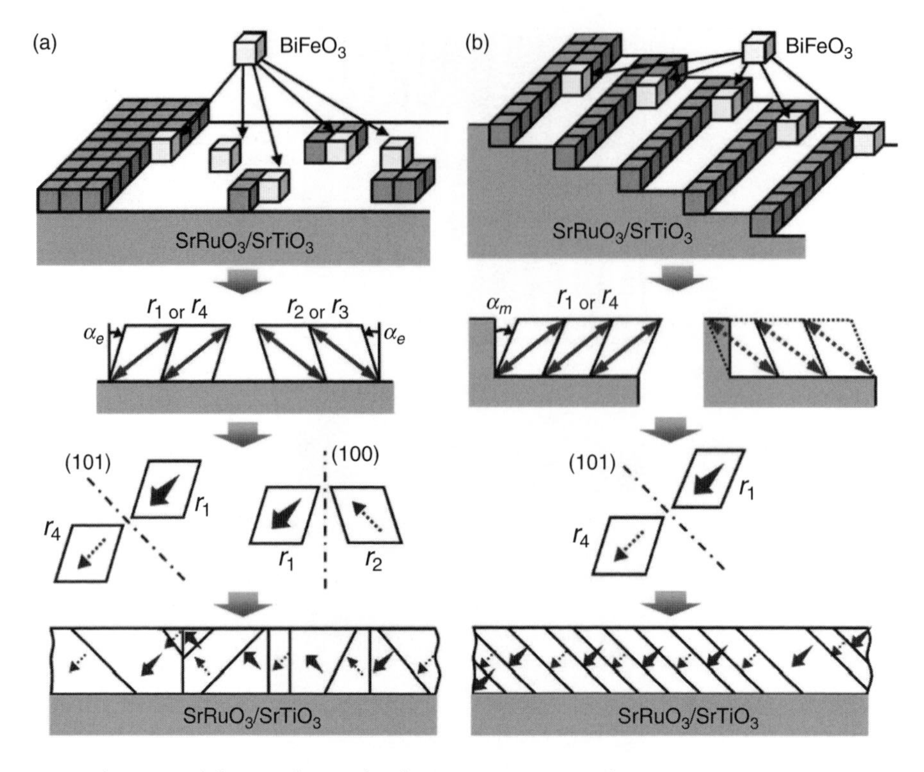

Fig. 3.18 Schematic of the initial growth of $BiFeO_3$ on exact and miscut $SrRuO_3/SrTiO_3$. On exact substrates, all four ferroelastic variances are formed (a). When vicinal substrates are used (b) two of the domains, named r_2 and r_3, have rhombohedral distortions that are unfavorable at the step edges where the nucleation takes place. Reprinted with permission from [54].

influence from multi-domain structure and twin walls. One example of such a study is shown in Figure 3.22. If multiple domains are present in a sample, it is difficult to study the intrinsic mechanism for fatigue, since it is hard to separate the contributions from the various switching events from the different ferroelastic domains. Also, domain walls and twin walls can pin domain wall motion, and hence hamper polarization switching. However, if mono-domain samples are used such problems can be avoided, and the intrinsic response from each variance can be obtained. As can be seen in the figure, both mono-domain $(110)_{pc}$, and $(001)_{pc}$ samples do not fatigue up to 10^6 switching cycles, however the mono-domain $(111)_{pc}$ oriented sample starts to fatigue after $\sim 10^4$ switching cycles. This switching path-dependent fatigue is due to domain wall pinning from the (111) switching event, enabled by free carrier stabilization of head to head domain walls. The possibility of forming such domain walls is much smaller for both the (001) and (110) orientations, hence they are more fatigue resistant. It is interesting to note that it is the thin-film orientation with the largest remnant polarization that actually fatigues the most. [4]

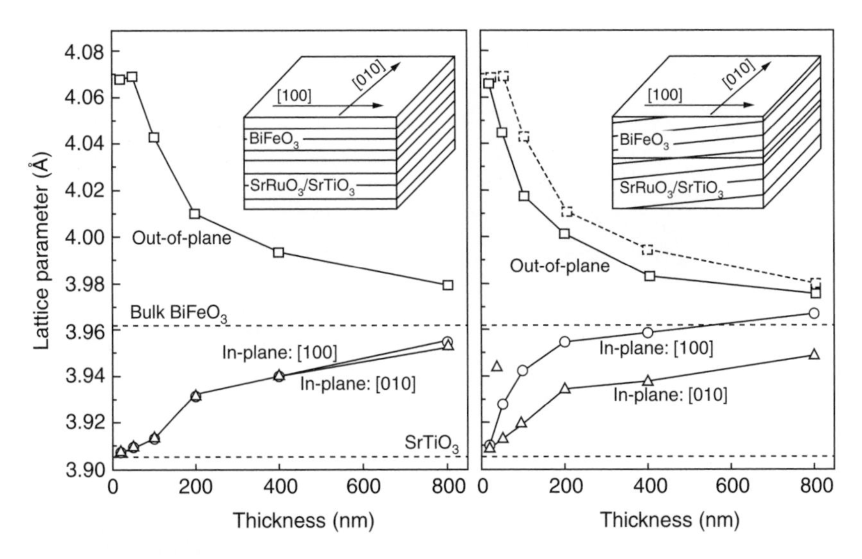

Fig. 3.19 The effect of vicinal substrates and film thickness on the lattice parameters of $BiFeO_3$. When miscut substrates are used, the [100] and [010] lattice parameters are different, and the [001] parameter is lower, compared to $BiFeO_3$ grown on exact substrates. Reprinted with permission from [54].

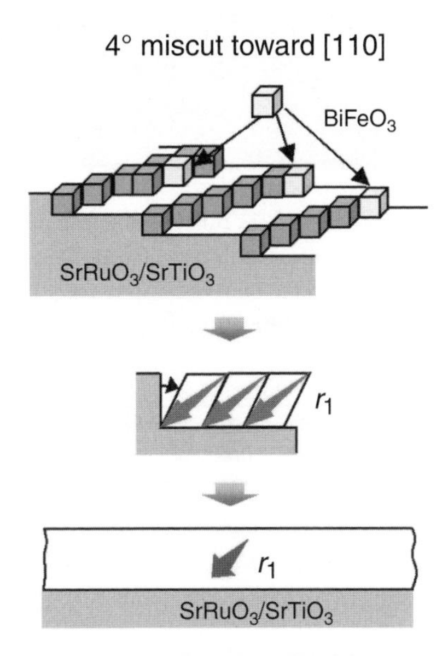

Fig. 3.20 Schematic of the initial growth of mono-domain $BiFeO_3$ on highly miscut $SrRuO_3/SrTiO_3$ along the [110] direction. r_1 is the favorable distortion at the step edges where the nucleation takes place. Reprinted with permission from [56].

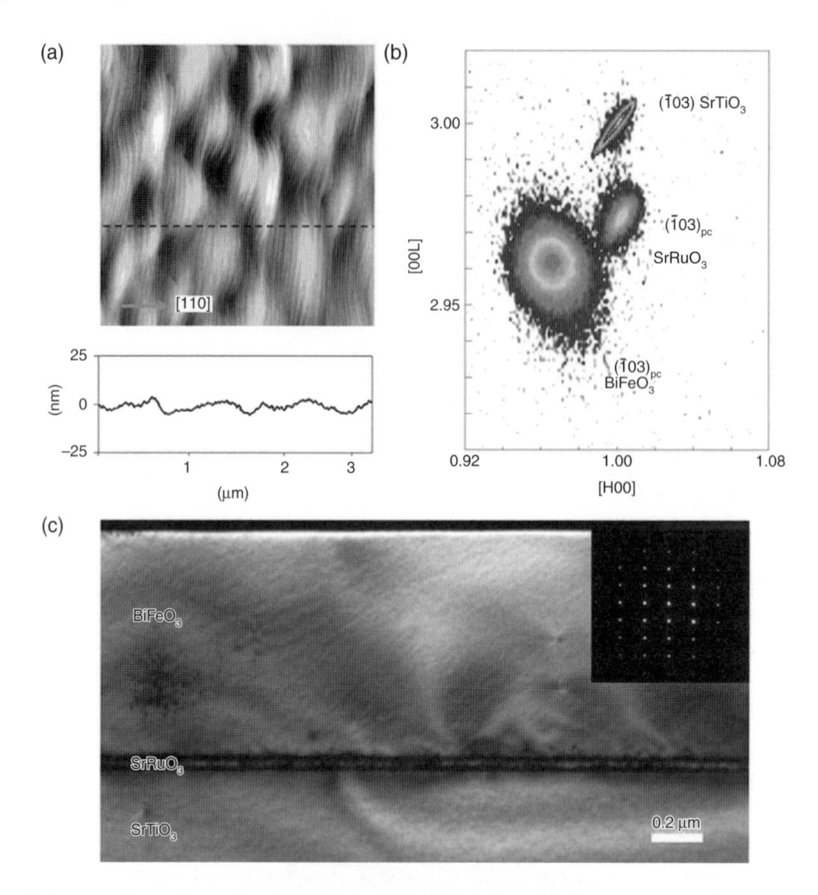

Fig. 3.21 (a) AFM data of a mono-domain (001) $BiFeO_3$ film exhibiting two or three unit-cell high steps due to step bunching from the high miscut substrate. (b) Reciprocal space map around the (-103) reflections demonstrating only one variance. (c) TEM data along the [-110] zone axis demonstrating one domain, the insert is a selected area electron diffraction diagram of $BiFeO_3$. Reprinted with permission from [55].

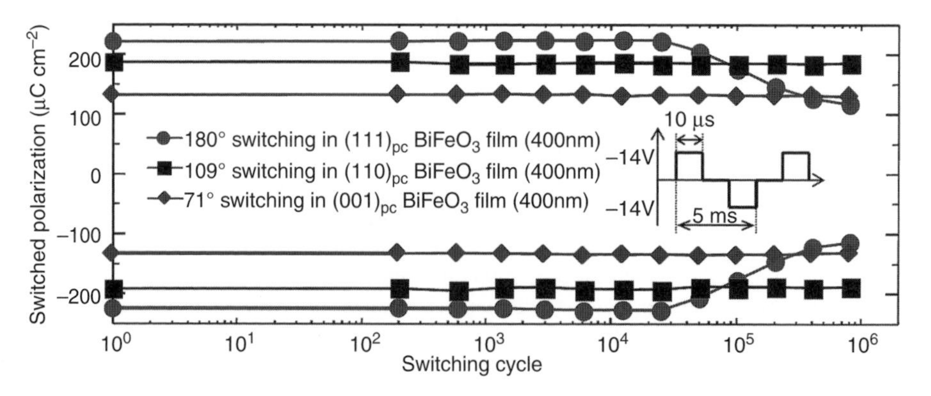

Fig. 3.22 Data revealing how ferroelectric fatigue depends on film orientations in mono-domain $BiFeO_3$ samples. Reprinted with permission from [4].

Domain engineering by miscut substrate also enhances the physical properties of ferroic materials. By relying on mono $(001)_{pc}$ oriented mono-domain $BiFeO_3$ samples, enabled by the use of 4° miscut (001) $SrTiO_3$ substrates along the [110] direction, Baek *et al.* [56] have studied how technologically important parameters such as the remnant polarization, coercive field, and leakage current correlate compared with samples showing a structural multi-domain structure. In the thickness range investigated, 200–800 nm, the two-variant sample showed a smaller remnant polarization, as compared to mono-domain samples, Figure 3.23 (a). The strain state is different between mono-domain and multi-domain samples, since the domain walls of the multi-domain samples can accommodate strain. Also, when the effect of possible strain induced polarization is taken into account, mono-domain films exhibit a larger effective polarization. In Figure 3.23 (b), the effect of domain geometry and film thickness on the coercive field is shown. For all thicknesses the mono-domain samples display a lower coercive field as compared to the two-variant samples, scaling with the effective strain state [56]. That is, it is possible to enhance functional properties by domain control of materials having symmetry lower than cubic.

3.5 Conclusions

In this chapter we have discussed the importance of the substrate on the synthesis and structural and physical properties of epitaxial multiferroic thin films. By relying on a combination of strain engineering and vicinal substrates, it is possible to control and tune the physical properties of thin films. Correct choice of substrate enables tuning of the symmetry of the unit-cell of the deposited materials, and hence control of the properties or production of novel phases. It is also possible to engineer samples with a controlled number of structural variances for materials having symmetry lower than cubic by relying on either orthoscandate substrates, or domain engineering by vicinal substrates. Such domain engineering opens a new avenue for synthesizing materials with given properties for both electronic applications and fundamental studies.

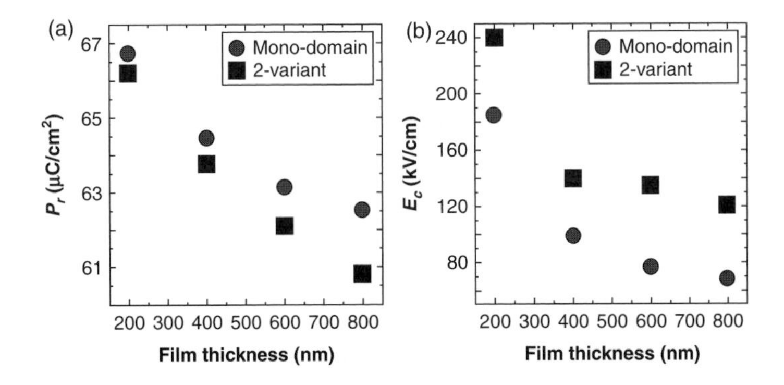

Fig. 3.23 A comparison of mono-domain versus two-variant $BiFeO_3$ with regard to film thickness for: (a) remnant polarization, and (b) coercive field. Adapted with permission from [56].

Acknowledgments

The authors would like to thank S.H. Baek, C.W. Bark, J. E. Boschker, C. Folkman, T. Grande, and S.M. Selbach for fruitful discussions. This work is supported by the Army Research Office under Grant No. W911NF-10-1-0362. Work in Norway was supported by The Research Council of Norway (proj. # 162874 and proj. # 10239707).

References

[1] C.B. Eom *et al.* (1989) *In-situ* Grown $YBa_2Cu_3O_{7-d}$ Thin Films from Single Target Magnetron Sputtering, Appl. Phys. Lett. **55**, 595–597.
[2] G. Rijnders, G. Koster, G, D.H. Blank, and H. Rogalla. (1997) In situ monitoring during pulsed laser deposition of complex oxides using reflection high energy electron diffraction under high oxygen pressure. Appl. Phys. Lett. **70**, 1888–1890.
[3] R. Ramesh *et al.* (1991) Epitaxial cuprate superconductor ferroelectric heterostructures. Science **252**, 944–946.
[4] S.H. Baek, *et al.* (2011) The nature of polarization fatigue in $BiFeO_3$. Adv. Mat. **23**, 1621–1625.
[5] K.J. Choi, S.H. Beak, H.W. Jang, *et al.* (2010) Phase transition temperatures of strained single crystal $SrRuO_3$ thin films. Adv. Mat. **22**, 759–762.
[6] D.G. Schlom, L.-Q. Chen, C.-B. Eom, *et al.* (2007) Strain tuning of ferroelectric thin films. Annu. Rev. Mater. Res. **37**, 589–626.
[7] F.W. Lytle. (1964) X-ray diffraction of low temperature phase transformations in strontium titanate. J. Appl. Phys. **35**, 2212–2215.
[8] D. Dijkkamp *et al.* (1997) Preparation of Y-Ba-Cu oxide superconductor thin films using pulsed laser evaporation from high Tc bulk material. Appl. Phys. Lett. **51**, 619–621.
[9] C.B. Eom *et al.* (1992) Single-Crystal Epitaxial Thin Films of the Isotropic Metallic Oxides $Sr_{1-x}Ca_xRuO_3$ $(0 \leq x \leq 1)$. Science **258**, 1766–1769.
[10] A. Scholl *et al.* (2000) Observation of antiferromagnetic domains in epitaxial thin films. Science **287**, 1014–1016.
[11] S. Jin, T.H. Tiefel, M. McCormack, *et al.* (1994) Thousandfold change in resistivity in magnetoresistive La-Ca-Mn-O films. Science **264**, 413–415.
[12] J. Wang *et al.* (2003) Epitaxial $BiFeO_3$ multiferroic thin film heterostructures. Science **299**, 1719–1722.
[13] S. Geller and P.M. Raccah. (1970) Phase transitions in perovskite-like compounds of rare earths. Phys. Rev. **B 2**, 1167–1172.
[14] L. Vasylechko, L. Akselrud, W. Morgenroth, *et al.* (2000) The crystal structure of $NdGaO_3$ at 100 K and 293 K based on synchrotron data. Journal of Alloys and Compounds **297**, 46–52.
[15] I.K. Bidkin, I.M. Shmyt'ko, A.M. Balbashov, and A.V. Kazansky. (1993) Twinning of $LaGaO_3$ single crystals. J. Appl. Cryst. **26**, 71–76.
[16] J. Schubert *et al.* (2003) Structural and optical properties of epitaxial $BaTiO_3$ thin films grown on $GdScO_3$ (110). Appl. Phys. Lett. **82**, 3460–3462.
[17] M. Lines and A. Glass (1977) *Principles and Applications of Ferroelectrics and Related Materials.* (Clarendon, Oxford).

[18] G.L. Yuan and A. Uedono. (2009) Behavior of oxygen vacancies in $BiFeO_3$/$SrRuO_3$/ $SrTiO_3$(100) and $DyScO_3$(100) heterostructures. Appl. Phys. Lett. **94**, 132905.
[19] J. He, A. Borisevich, S.V. Kalinin, S.J. Pennycook, and S.T. Pantilides. (2010) Control of octahedral tilts and magnetic properties of perovskite oxide heterostructures by substrate symmetry. Phys. Rev. Lett. **105**, 227203.
[20] Q. Gan, R.A. Rao, and C. B. Eom. (1997) Control of the growth and domain structure of epitaxial $SrRuO_3$ thin films by vicinal (001) $SrTiO_3$ substrates. Appl. Phys. Lett. **70**, 1962–1964.
[21] J.C. Jiang, W. Tian, X.Q. Pan, Q. Gan, and C.-B. Eom. (1998) Domain structure of epitaxial $SrRuO_3$ thin films on miscut (001) $SrTiO_3$ substrates. Appl. Phys. Lett. **72**, 2963–2965.
[22] S. Stoyanov. (1999) Layer growth of epitaxial films and superlattices. Surf. Sci. **199**, 226–242.
[23] G. Rijnders, D.H. Blank, J. Choi, and C.B. Eom. (2004) Enhanced surface diffusion through termination conversion during epitaxial $SRuO_3$ growth. Appl. Phys. Lett. **84**, 505–508.
[24] J. Choi, C.B. Eom, G. Rijnders, H. Rogalla, and D.H.A. Blank. (2001) Growth mode transition from layer by layer to step flow during the growth of heteroepitaxial $SrRuO_3$ on (001) $SrTiO_3$. Appl. Phys. Lett. **79**, 1447–1449.
[25] G. Rijnders, G. Koster, V. Leca, D.H.A. Blank and H. Rogalla. (2000) Imposed layer by layer growth with pulsed laser interval deposition. Appl. Surf. Sci. **168**, 223–226.
[26] M. Kawasaki *et al.* (1994) Atomic control of the $SrTiO_3$ Crystal Surface. Science **226**, 1540–1542.
[27] G. Kostner *et al.* (1998) Quasi-ideal strontium titanate crystal surfaces through formation of strontium hydroxide. Appl. Phys. Lett. **73**, 2920–2922.
[28] J.E. Kleibeuker *et al.* (2010) Atomically defined rare-earth scandate crystal surfaces. Adv. Func. Mat. **20**, 3490–3496.
[29] T. Ohnishi *et al.* (1999) A-site terminated perovskite substrate: $NdGaO_3$. Appl. Phys. Lett. **74**, 2531–2533.
[30] Q. Gan, R.A. Rao, C.-B. Eom, J.L. Garret, and M. Lee. (1998) Direct measurement of strain effects on magnetic and electrical properties of epitaxial $SrRuO_3$ thin films. Appl. Phys Lett. **72**, 978–980.
[31] C.B. Eom *et al.* (1993) Fabrication and Properties of Epitaxial Ferroelectric Heterostructures with (SrRuO3) Isotropic Metallic Oxide Electrodes. Appl. Phys. Lett., **63**, 2570–2572.
[32] C.H. Ahn *et al.* (1996) Ferroelectric field effect in ultrathin $SrRuO_3$ films. Appl. Phys. Lett **70**, 206–208.
[33] B. J. Kennedy and B.A. Hunter. (1998) High-temperature phases of $SrRuO_3$. Phys. Rev. **B 58**, 653–658.
[34] M.K. Lee, T.K. Nath, C.-B. Eom, M.C. Smoak, and F. Tsui. (2000) Strain modification of epitaxial perovskite oxide thin films using structural transitions of ferroelectric $BaTiO_3$ substrate. Appl. Phys. Lett. **77**, 3547–3549.
[35] K. Terai *et al.* (2002) In-plane lattice constant tuning of an oxide substrate with $Ba_{1-x}Sr_xTiO_3$ and $BaTiO_3$ buffer layers. Appl. Phys. Lett. **80**, 4437–4439.
[36] K.J. Choi *et al.* (2004) Enhancement of Ferroelectricity in Strained $BaTiO_3$ Thin Films. Science **306**, 1005–1009.

[37] R. J. Zeches *et al.* (2009) A strain driven morphotropic phase boundary in $BiFeO_3$. Science **326**, 977–980.
[38] I.C. Infante *et al.* (2010) Bridging multiferroic phase transitions by epitaxial strain in $BiFeO_3$. Phys. Rev. Lett. **105**, 057601.
[39] H.W. Jang *et al.* (2008) Strain-induced polarization rotation in epitaxial (110) $BiFeO_3$ thin films. Phys. Rev Lett. **101**, 107602.
[40] M.F. Chisholm, W.D. Luo, M.P. Oxley, S.T. Pantelides, and H.N. Lee. (2010) Atomic-scale compensation phenomena at polar interfaces. Phys. Rev. Lett. **105**, 197602.
[41] C.M. Folkman *et al.* (2010) Study of defect-dipoles in an epitaxial ferroelectric thin film. Appl. Phys. Lett. **96**, 052903.
[42] O. Dahl, J.K. Grepstad, and T. Tybell. (2009) Polarization direction and stability in ferroelectric lead titanate thin films. J. Appl. Phys. **106**, 084104.
[43] S.M. Selbach, T. Tybell, M.-A. Einarsrud, and T. Grande. (2011) PbO-deficient $PbTiO_3$: Mass transport, structural effects and possibility for intrinsic screening of the ferroelectric polarization. Appl. Phys. Lett. **98**, 091912.
[44] J.C. Jiang, W. Tian, X.Q. Pan, and C.-B. Eom. (1998) Domain structure of epitaxial $SrRuO_3$ thin films on miscut (001) $SrTiO_3$ substrates. Appl. Phys. Lett. **72**, 2963–2965.
[45] J.C. Jiang, X.Q. Pan, and C.L. Chen. (1997) Microstructure of epitaxial $SrRuO_3$ thin films on (001) $SrTiO_3$. Appl. Phys. Lett. **72**, 909–911.
[46] Q. Gan, R.A. Rao, and C.-B. Eom. (1997) Control of the growth and domain structure of epitaxial $SrRuO_3$ thin films by vicinal (001) $SrTiO_3$ substrates. Appl. Phys. Lett. **70**, 1962–1964.
[47] F. Kubel and H. Schmid. (1990) Structure of a ferroelectric and ferroelastic monodomain crystal of the perovskite $BiFeO_3$. Acta Crystallogr. Sect. **B 46**, 698–702.
[48] S.K. Streiffer *et al.* (1998) Domain patterns in epitaxial rhombohedral ferroelectric films. I. Geometry and experiments. J. Appl Phys., **83**, 2742–2753.
[49] Y.B. Chen *et al.* (2007) Ferroelectric domain structure of epitaxial (001) $BiFeO_3$ thin films. Appl. Phys. Lett. **90**, 072907.
[50] R.R. Das *et al.* (2006) Synthesis and ferroelectric properties of epitaxial $BiFeO_3$ thin films grown by sputtering. Appl. Phys. Lett. **88**, 242904.
[51] Y.H. Chu *et al.* (2006) Nanoscale domain control in multiferroic $BiFeO_3$ thin films. Adv. Mater. **18**, 2307–2311.
[52] C.M. Folkman *et al.* (2009) Stripe domain structure in epitaxial (001) $BiFeO_3$ thin films on orthorhombic $TbScO_3$ substrates. Appl. Phys. Lett. **94**, 251911.
[53] Y.H. Chu. (2007) Domain control in multiferroic $BiFeO_3$ through substrate vicinality. Adv. Mater. **19**, 2662–2666.
[54] H.W. Jang *et al.* (2009) Domain engineering for enhanced ferroelectric properties of epitaxial (001) BiFeO thin films. Adv. Mater. **21**, 817–823.
[55] S.H. Baek *et al.* (2010) Ferroelastic switching for nanoscale non-volatile magnetoelectric devices. Nat. Materials **9**, 309–314.
[56] S.H. Baek, H.W. Jang, C.M. Folkman, and C.-B. Eom. (2011) Enhanced ferroelectric properties of the mono-domain epitaxial (001) $BiFeO_3$ thin films. Submitted.

4
Synchrotron X-ray scattering studies of oxide heterostructures

DILLON D. FONG

This chapter surveys a variety of hard X-ray scattering techniques useful for the study of oxide heterostructures. We begin with non-resonant surface X-ray diffraction and proceed to techniques reliant on modifications to the non-resonant atomic form factor. In addition to providing real space images of the heterostructure with atomic resolution, these techniques permit investigation into the effects of composition, valence, and local environment on the emergent phenomena found at many oxide interfaces. Resonant scattering, in particular, has proven to be essential for disentangling the complex chemical, structural, and electronic behaviors found in these materials.

4.1 Introduction

There are a multitude of techniques useful for the study of oxide heterostructures. Hard X-ray synchrotron techniques offer the ability to probe the heterostructure *in situ* with sub-Ångstrom resolution, whether during deposition, in reactive environments, or under applied fields [1, 2, 3, 4]. This ability stems from the weak interaction of X-rays with matter, which allows X-rays to penetrate through obstacles and "see" the structure of interest. While highly monochromatic X-rays from third generation sources typically cause little radiation damage, high energy photons can create ozone in certain environments and affect the defect-mediated surface reactivity of some oxides (e.g. TiO_2) [5, 6, 7]. Photoinjected electrons are also known to stimulate the formation and incorporation of atomic oxygen at the surface [8], and uniform UV illumination is a known method of enhancing oxygen incorporation during thin film growth [9]. Damage can occur more readily with a larger energy spread or significantly higher flux.

The great interest in oxide heterostructures originates from both the rich array of properties found in such systems and the ability to form coherent interfaces with each other as well as with various commercially available substrates [10]. This has allowed for the creation of "new" materials such as ferroelectric $SrTiO_3$ [11] or superconducting $LaAlO_3/SrTiO_3$ superlattices [12]. As noted by Glazer [13], such perovskites exhibit strong structure–property relationships, and changes in electronic structure are often correlated with structural distortions. In general, perovskites can deviate from their cubic aristotype by (1) tilting of the oxygen octahedra, (2) displacement of the cations, and (3) distortions of the octahedra. These structural changes can easily be observed with high resolution scattering techniques.

In this chapter, we briefly survey different hard X-ray scattering techniques used for the investigation of oxide heterostructures. For more general information on synchrotron X-ray techniques, we refer readers to [14].

4.2 Surface X-ray diffraction

Within the kinematical approximation, the scattered intensity is related to the electron density, $\rho(\mathbf{r})$, by a Fourier transform:

$$I(\mathbf{q}) \propto |F(\mathbf{q})|^2 = \left| \int \rho(\mathbf{r}) e^{i\mathbf{q}\cdot\mathbf{r}} \, d\mathbf{r} \right|^2 . \tag{4.1}$$

Here, $F(\mathbf{q})$ is the total structure factor for the system in question and represents the sum of independent scattering contributions from all electrons in the system:

$$F(\mathbf{q}) = \sum_j f_j(\mathbf{q}) e^{i\mathbf{q}\cdot\mathbf{r}}. \tag{4.2}$$

The atomic form factor, $f_j(\mathbf{q})$, corresponds to the Fourier transform of the electron density for a single atom. Tabulated values of the form factor as a function of the scattering vector, $\mathbf{q}$, can be found in [15].

In surface X-ray diffraction (SXRD) [16, 17, 18, 19], the sample typically consists of a single crystal heterostructure, and the region probed by X-rays can be as large as several mm^2 (Figure 4.1). With surface roughnesses on the order of nanometers or less, the Fourier transform of the region of study is a three-dimensional grid of reciprocal lattice points, each convolved with a crystal truncation rod (CTR) extending through it along the direction of the surface normal. In the event of a surface reconstruction, the change in in-plane symmetry leads to the appearance of superstructure rods (SRs) in reciprocal space (shown in blue). While the incident angle can be fixed near the material's critical angle for total external reflection to maximize the scattered signal from the surface and minimize the penetration depth, varying the incident angle is not necessary for obtaining depth dependent information from model samples. The full, three-dimensional structure is completely encoded within the CTRs and SRs.

For a thin film heterostructure, the total structure factor can be expressed as the sum of contributions from the bulk (B), which represents the (known) structure factor of the substrate, and contributions from the surface (S), which represents the (unknown) structure factor of the deposited layers:

$$F(\mathbf{q}) = |F(\mathbf{q})| e^{i\phi(\mathbf{q})} = B(\mathbf{q}) + S(\mathbf{q}). \tag{4.3}$$

Since the measured intensity is only related to the modulus, $|F(\mathbf{q})|$, one cannot easily recover the phase, $\phi(\mathbf{q})$, and consequently the desired information, $\rho(\mathbf{r})$. Therefore, analysis has most commonly been performed by fitting the total structure factor to a model with numerous

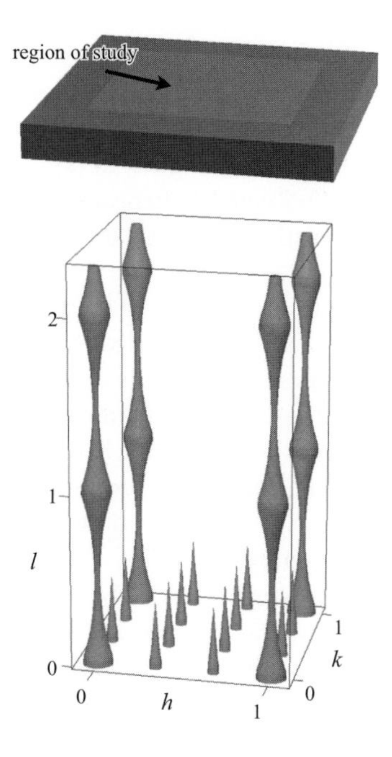

Fig. 4.1 Schematic depiction of surface X-ray diffraction. The real space image of the sample lies above its reciprocal space counterpart. CTRs are in green while SRs are blue. Courtesy of S. O. Hruszkewycz. This figure is reproduced in color in the color plate section.

adjustable parameters [20, 21]. For instance, the bulk and surface structure factors can be represented by

$$B(\mathbf{q}) = \left(\frac{1}{1 - e^{-2\pi i l}}\right) \sum_{j}^{\text{bulk unit cell}} f_j(\mathbf{q}) e^{2\pi i (hx_j + ky_j + lz_j)} \tag{4.4}$$

and

$$S(\mathbf{q}) = \sum_{j}^{\text{surface unit cell}} f_j(\mathbf{q}) e^{2\pi i (hx_j + ky_j + lz_j)}. \tag{4.5}$$

With known atomic form factors and positions for the bulk unit-cell, the measured CTR/SR intensities can be fitted for the atomic positions of the surface unit-cell. This assumes the necessary geometrical corrections have been applied [22, 23].

This technique was used to solve for the equilibrium surface structure of $PbTiO_3$ (001). Using *in-situ* synchrotron X-ray scattering at elevated temperature and PbO partial pressure, Munkholm *et al.* [24] measured the integrated intensities of 14 independent SRs. Equally important to the understanding of the structure is the systematic absence of intensity (here at $h = 0$ or $k = 0$ positions), which indicated that the symmetry of the structure must be at least that of plane group *p*2*gg*. Scans along the out-of-plane (l) direction showed that the intensities decayed by less than 50% up to $q_z = 2.5$ Å^{-1}, indicating that the reconstruction was only one atomic layer thick. Based on size considerations, the atoms considered in the model were restricted to Ti and O, and only three parameters were fit. Comparison between the observed and calculated values of $|S(\mathbf{q})|$ is provided in Figure 4.2(a). The resulting real space structure is shown in Figure 4.2(b). This TiO_2 layer lies beneath the surface PbO layer and consists of an antiferrodistortive structure with oxygen octahedral cages rotated by 10° about the Ti ions.

May *et al.* [25] also used fitting techniques to solve for the structure of epitaxial $LaNiO_3$ (001) films. In bulk form, $LaNiO_3$ is rhombohedral with $R\bar{3}c$ symmetry. The Glazer tilt system is $a^-a^-a^-$ indicating that, with reference to the cubic aristotype, the oxygen octahedra are rotated around the x-, y-, and z-axes by equal amounts. The superscript (-) implies that octahedra along each axis are counter-rotated in each successive unit cell. Since octahedral tilting determines the B-O-B bond angle in ABO_3 perovskites, which is correlated to a variety of electronic and magnetic properties, it is crucial to be able to directly measure octahedral rotations in strained oxide heterostructures.

After growing strained $LaNiO_3$ films on both $SrTiO_3$ and $LaAlO_3$ (001) substrates by ozone-assisted molecular beam epitaxy, May *et al.* [25] used SXRD to measure the integrated intensities of half-order peaks found in reciprocal space. These peaks arise from the tilted oxygen octahedra and are directly related to the $a^-a^-a^-$ tilt system, which causes a $2 \times 2 \times 2$ unit-cell in real space with respect to the cubic aristotype. Since each $2 \times 2 \times 2$ unit-cell contains 24 oxygen atoms, the structure factor for each half-order peak is

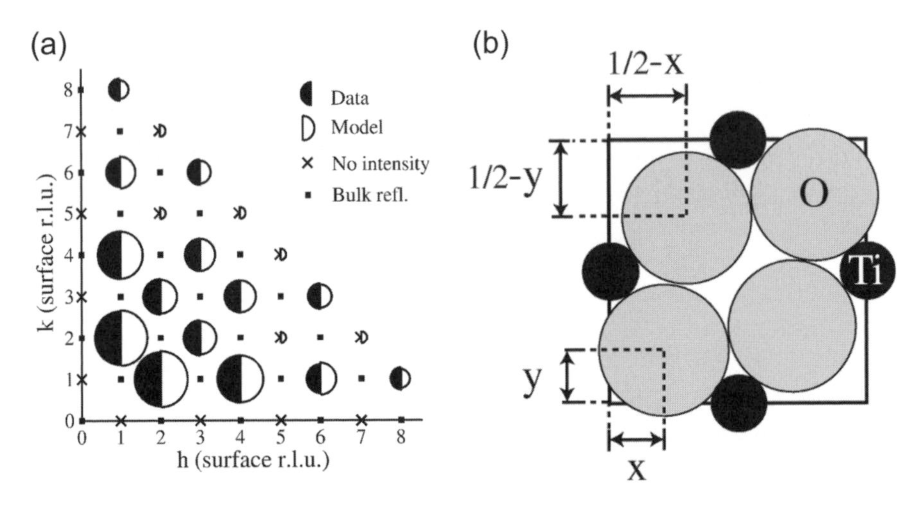

Fig. 4.2 The antiferrodistortive reconstruction of the $PbTiO_3$ (001) surface. (a) Structure factors for the $c(2 \times 2)$ reconstruction, plotted vessus surface Miller indices h and k for the primitive $(\sqrt{2} \times \sqrt{2})R45°$ unit-cell. Areas of filled and open half circles are proportional to observed and calculated structure factors. (b) Schematic of the TiO_2 layer in the antiferrodistortive model for the $c(2 \times 2)$ $PbTiO_3$ (001) reconstruction, showing counter-rotated oxygen cages. Reprinted with permission from [24]. Copyright 2001 by the American Physical Society.

$$S(\mathbf{q}) = f_{O^{2-}}(\mathbf{q}) \sum_{j}^{24} e^{2\pi i(hx_j+ky_j+lz_j)} \tag{4.6}$$

where (x_j, y_j, z_j) is the position of the jth oxygen atom. However, each $2 \times 2 \times 2$ unit-cell can be oriented four different ways, and each domain is expected to be energetically equivalent. If each domain scatters incoherently with each other, the scattered intensity of the half-order peak is given by

$$I(\mathbf{q}) \propto \sum_{j=1}^{4} D_j \left|S(\mathbf{q})\right|^2 \tag{4.7}$$

where D_j is the relative volume fraction of structural domains.

The tilt system of $LaNiO_3$ becomes $a^-a^-c^-$ when strained biaxially. Fitting for the oxygen positions, May *et al.* [25] found that the $LaNiO_3$ film in biaxial tension (on $SrTiO_3$) exhibits larger rotation angles around the x- and y-axes (7.1° versus 5.2° for bulk) and a smaller rotation angle around the z-axis (0.3°). Opposite behavior was found for the film grown in biaxial compression (on $LaAlO_3$), as shown in Figure 4.3. They also discovered that the film grown on

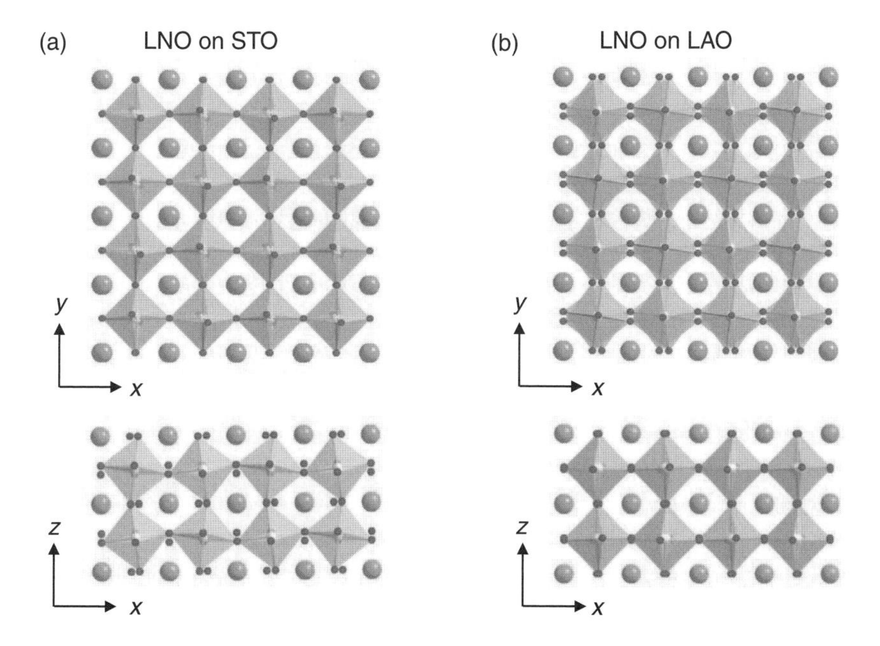

Fig. 4.3 Atomic structure of $LaNiO_3$ (LNO) films grown on either (a) $SrTiO_3$ (STO) or (b) $LaAlO_3$ (LAO). Biaxial tension (LNO/STO) reduces the octahedral rotation around out-of-plane z-axis while enhancing the rotations around the in-plane x and y directions. Biaxial compression (LNO/LAO) produces the opposite effect. Reprinted with permission from [25]. Copyright 2010 by the American Physical Society.

$SrTiO_3$ had equal domain fractions, as expected, but the film grown on $LaAlO_3$ had only two domains, with unequal populations caused by twins in the rhombohedral $LaAlO_3$ substrate.

In these examples, with relatively few integrated intensities and model parameters, fitting methods have proven to be reliable means of determining certain aspects of surface/film structure. However, $\rho(\mathbf{r})$ is encoded within the CTRs/SRs, and recently developed techniques allow us to overcome the phase problem in SXRD and solve directly for the full three-dimensional electron density. By measuring all of the independent CTRs/SRs over a large q-range (which determines the spatial resolution, $\pi/(q_{\text{max}} - q_{\text{min}})$) with a small sampling interval (which determines the window size, $2\pi/\Delta q$), one can analyze this oversampled dataset to recover the missing phase information [26].

The phases are recovered through an iterative process of alternately satisfying constraints in real and reciprocal space [27]. The constraint in reciprocal space demands that the Fourier transform of a trial electron density corresponds to the measured $|F(\mathbf{q})|$. If the correspondence is poor, a "minimum-change" correction is made, and an error function is created to monitor the size of the correction. There are several possible constraints in real space. The *support* constraint demands that the electron density is zero outside a region normal to the surface [28, 29]. This is a very powerful constraint used by most direct methods including "Phase and Amplitude Recovery And Diffraction Image Generation Method" (PARADIGM) [30]. The *positivity* constraint ensures that the electron density is real and positive; the "COherent Bragg Rod Analysis" (COBRA) technique employs both of these constraints. Another possible constraint is that of *atomicity*, which requires that each atom's electron density is well-separated from the others. An analytical technique that employs atomicity is "Difference map using the Constraints of Atomicity and Film shift " or DCAF [31], which also uses an additional constraint to assist that of the support.

As shown in Figure 4.4(a), $F(\mathbf{q}) = B(\mathbf{q}) + S(\mathbf{q})$ can be represented as vector addition on the complex plane. We assume $B(\mathbf{q})$ is known, but only the radius of the circle, $|F(\mathbf{q})|$, is measured. Thus, $S(\mathbf{q})$ can be any vector originating from the perimeter of the circle and ending on the tail of $B(\mathbf{q})$.

In order to improve the rate of convergence to a solution, COBRA employs an additional assumption. If the rate of change of $S(\mathbf{q})$ varies slowly relative to $B(\mathbf{q})$ with respect to q, then we can make the approximation

$$|F_-| = |B_- + S_0| \tag{4.8}$$

$$|F_+| = |B_+ + S_0| \tag{4.9}$$

where the subscripts $(-)$ and $(+)$ refer to adjacent q-points along a CTR. Therefore, for any two successive points along a CTR, F_- and F_+, we have two real equations that can be solved for one complex unknown, S_0, although there will typically be two possible solutions, S_a and S_b. These solutions are displayed in Figure 4.4(b), where it is shown that S must originate from the intersections between the two circles $|F_-|$ and $|F_+|$ and arrive at the origin of the complex plane. The "correct" solution, S_a, is the one that varies more slowly with q and can only be determined after considering the adjacent q-points. Because of this assumption between $B(\mathbf{q})$ and $S(\mathbf{q})$, COBRA can be applied to CTRs but not SRs. Yacoby *et al.* have also shown that for the CTRs, $F(\mathbf{q})$ is the Fourier transform of the electron density laterally "folded" into a substrate-defined

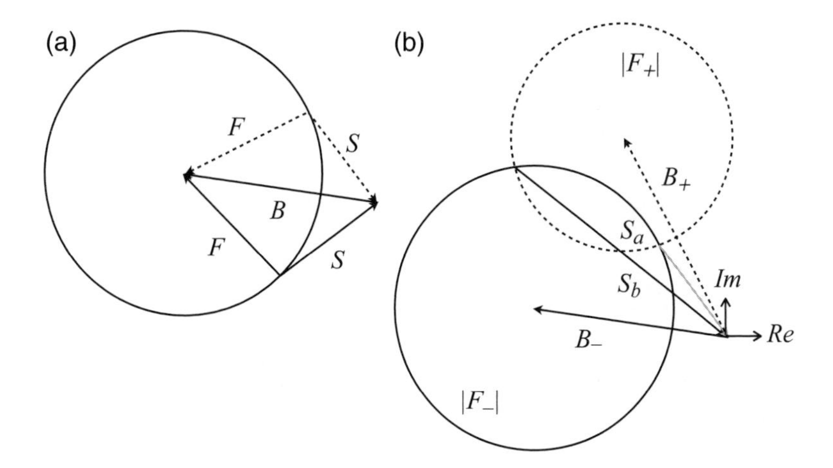

Fig. 4.4 Schematic depiction of $F(\mathbf{q})$, $B(\mathbf{q})$, and $S(\mathbf{q})$ on the complex plane. (a) It is assumed that $B(\mathbf{q})$ is known, but only the magnitude of $F(\mathbf{q})$ is measured. Therefore, $S(\mathbf{q})$ must be a vector drawn from the circle perimeter (with radius $|F(\mathbf{q})|$) to the tail of $B(\mathbf{q})$. There are an infinite number of possible solutions. (b) When it is assumed that the variation of $S(\mathbf{q})$ is slow relative to $B(\mathbf{q})$, there are, in general, only two possible solutions. The two circles represent $F(\mathbf{q})$ changing from F_- to F_+. $B(\mathbf{q})$ varies from B_- to B_+, with the origin of $B(\mathbf{q})$ corresponding to the origin of the complex plane. If $S_- \approx S_+ = S_0$, the two solutions originate from the intersections between the two circles, giving the two vectors S_a and S_b.

unit-cell [32]. This allows the consideration of in-plane atomic displacements and how these vary as a function of distance away from an interface [33].

The COBRA technique was recently used to determine the through-thickness structure of superconducting $La_{2-x}Sr_xCuO_4$ bilayers grown on $LaSrAlO_4$ [34]. Figure 4.5 shows the result for a bilayer consisting of three unit-cells of insulating La_2CuO_4 on two unit-cells of metallic $La_{1.55}Sr_{0.45}CuO_4$. Interestingly, the bilayer is superconducting below 34K. The (100) electron density profile shows that the copper-to-oxygen apical distance increases away from the substrate, as indicated by the elongated CuO_6 octahedra. The (110) profile shows that the La(Sr)-apical oxygen planes become increasingly buckled away from the substrate. By comparing this result with those from metallic and superconducting single layer films, Zhou *et al.* [34] find that the larger copper–apical oxygen bond, which correlates with the superconducting transition temperature, arises from the interface between the insulator and the metal.

4.3 Resonant scattering

We now consider the dependence of X-ray energy on the atomic form factor. Within the dipole approximation, the anomalous dispersion corrections are only dependent on energy so we may write

$$f(\mathbf{q}, E) = f_0(\mathbf{q}) + f'(E) + if''(E). \tag{4.10}$$

Values for the energy-dependent f' and f'' are available online [35]. If this energy-dependent form factor is inserted into the structure factor, equation (4.5) can be rewritten as

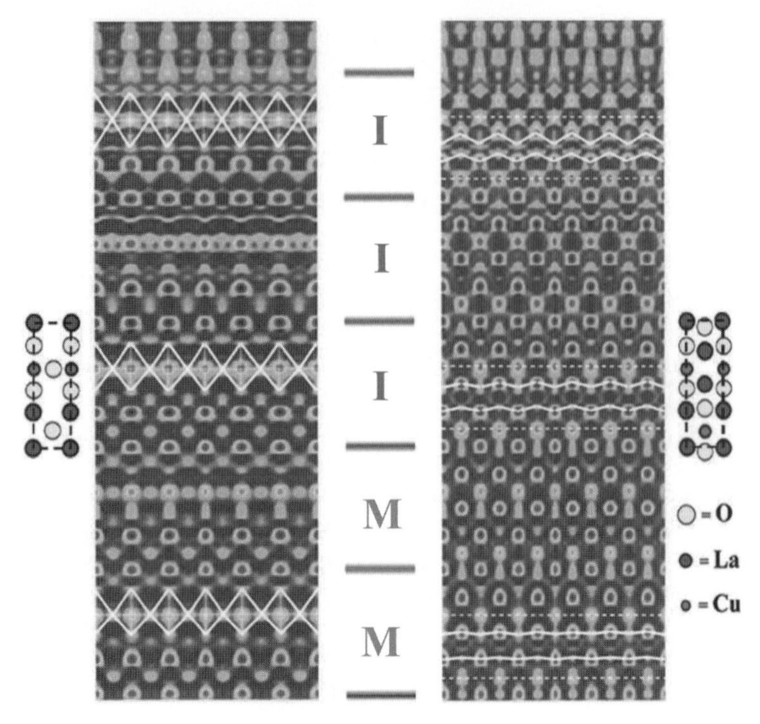

Fig. 4.5 Two-dimensional electron density (ED) profiles along (100) and (110) atomic planes of the M-I bilayer system. The EDs were determined from the experimentally measured diffraction intensities using the COBRA technique. Schematic cross-sections of the complete tetragonal unit cell along each atomic plane are illustrated near the sides of the two respective panels. Left: in (100) plane, the white lines highlight the projected shapes of the CuO_6 octahedra, in particular the elongation near the surface; Right: in (110) plane, the white lines highlight the projected profiles of the La-apical O planes, in particular the corrugation near the surface. Reprinted with permission from [34].

$$S(\mathbf{q}, E) = \sum_j \left(f_{j,0}(\mathbf{q}) + f_j'(E) + if_j''(E)\right) e^{2\pi i(hx_j+ky_j+lz_j)}, \tag{4.11}$$

or, if we separate contributions from both non-resonant and resonant atoms,

$$\begin{aligned} S(\mathbf{q}, E) &= \left[\sum_m S_{N,m}(\mathbf{q}) + \sum_n S_{A,n}(\mathbf{q})\right] \\ &\quad + \sum_a \left(f_a'(E) + if_a''(E)\right) e^{2\pi i(hx_a+ky_a+lz_a)} \\ &= S_T(\mathbf{q}) + S_R(\mathbf{q}, E) \end{aligned} \tag{4.12}$$

where $S_N(\mathbf{q})$ is the non-resonant scattering contribution from non-resonant atoms, $S_A(\mathbf{q})$ is the non-resonant contribution from the single anomalous scatterer, a, $S_T(\mathbf{q}) = \sum_m S_{N,m}(\mathbf{q}) + \sum_n S_{A,n}(\mathbf{q})$, and $S_R(\mathbf{q}, E)$ is the resonant contribution from a. The scattered intensity is then given by [36]

$$\begin{aligned} \left|S(\mathbf{q},E)\right|^2 \propto \left|S_T(\mathbf{q})\right|^2 &+ \left(\frac{f_a'(E)^2 + f_a''(E)^2}{f_{a,0}(\mathbf{q})^2}\right)\left|S_A(\mathbf{q})\right|^2 \\ &+ 2\left(\frac{f_a'(E)}{f_{a,0}(\mathbf{q})}\right)\left|S_T(\mathbf{q})\right|\left|S_A(\mathbf{q})\right|\cos\left(\delta\Phi(\mathbf{q})\right) \\ &+ 2\left(\frac{f_a''(E)}{f_{a,0}(\mathbf{q})}\right)\left|S_T(\mathbf{q})\right|\left|S_A(\mathbf{q})\right|\sin\left(\delta\Phi(\mathbf{q})\right) \end{aligned} \tag{4.13}$$

where $\delta\Phi(\mathbf{q}) = \Phi_T(\mathbf{q}) - \Phi_A(\mathbf{q})$.

The inclusion of energy-dependent absorption allows the distinction between $|S(\mathbf{q}, E)|$ and $|S(-\mathbf{q}, E)|$. For a non-centrosymmetric space group, these reflections are known as a Friedel pair, as long as they are not otherwise symmetry related. A consequence of resonant scattering is that Friedel's law is broken such that $|S(hkl, E)| \neq |S(\bar{h}\bar{k}\bar{l}, E)|$, the latter having an intensity

$$\begin{aligned} \left|S(-\mathbf{q},E)\right|^2 \propto \left|S_T(\mathbf{q})\right|^2 &+ \left(\frac{f_a'(E)^2 + f_a''(E)^2}{f_{a,0}(\mathbf{q})^2}\right)\left|S_A(\mathbf{q})\right|^2 \\ &+ 2\left(\frac{f_a'(E)}{f_{a,0}(\mathbf{q})}\right)\left|S_T(\mathbf{q})\right|\left|S_A(\mathbf{q})\right|\cos\left(\delta\Phi(\mathbf{q})\right) \\ &- 2\left(\frac{f_a''(E)}{f_{a,0}(\mathbf{q})}\right)\left|S_T(\mathbf{q})\right|\left|S_A(\mathbf{q})\right|\sin\left(\delta\Phi(\mathbf{q})\right). \end{aligned} \tag{4.14}$$

As a result, the difference in intensities (called the Bijvoet difference) is simply

$$\left|S(\mathbf{q},E)\right|^2 - \left|S(-\mathbf{q},E)\right|^2 \propto 4\left(\frac{f_a''(E)}{f_{a,0}(\mathbf{q})}\right)\left|S_T(\mathbf{q})\right|\left|S_A(\mathbf{q},E)\right|\sin\left(\delta\Phi(\mathbf{q})\right). \tag{4.15}$$

Although we can obtain two measurements per X-ray energy, we have three unknowns, $\left|S_T(\mathbf{q})\right|$, $\left|S_A(\mathbf{q})\right|$, and $\delta\Phi(\mathbf{q})$. If we wish to solve for them, we need data collected for at least one more X-ray energy. With the Multi-wavelength Anomalous Dispersion (MAD) phasing technique, one constructs Bijvoet Difference Patterson maps to retrieve the positions of the anomalous scatterers ($\Phi_A(\mathbf{q})$), and therefore obtain $\Phi_T(\mathbf{q})$ [37].

For materials with $P4mm$ symmetry, breaking of Friedel's law implies $|S(hkl, E)| \neq |S(hk\bar{l}, E)|$. For the case of ferroelectric $BaTiO_3$ (001), equation (4.13) can be approximated by

$$\begin{aligned} |S(hkl,E)|^2 \propto |S(hkl)|^2 &+ f_{\mathrm{Ba}}''^2 \\ &+ 2f_{\mathrm{Ba}}'' f_{\mathrm{Ti}}\,(-1)^{h+k+l}\sin 2\pi l\Delta z_{\mathrm{Ti}} \end{aligned} \tag{4.16}$$

$$+ 2f''_{\text{Ba}}f_{\text{O}}\,(-1)^{h+k}\sin 2\pi l\Delta z_{\text{O}_1}$$

$$+ 2f''_{\text{Ba}}f_{\text{O}}\left[(-1)^{h+l} + ((-1)^{k+l}\right]\sin 2\pi l\Delta z_{\text{O}_2}$$

when working near the Ba absorption edge and where Δz_j is the ferroelectric displacements for atom j [38]. This energy dependence is shown in Figure 4.6 for both up and down ferroelectric polarization. As seen, the strongest effect is at the Ba L_1 absorption edge for both polarizations. Furthermore, there is a significant difference in intensity between the up and down polarization states, allowing one to distinguish between them.

Do *et al.* [40] exploited this intensity difference for monitoring polarization fatigue in epitaxial Pb(Zr,Ti)O_3 capacitors. Figure 4.7 shows real space images of the Pb(Zr,Ti)O_3 002 intensity, where Do *et al.* used a Fresnel zone plate (to focus 10 keV X-rays down to a 0.8 μm spot) and a sample translator to image the portion of the capacitor near the electrode boundary. As seen, a 30% change in intensity was observed after reversing the polarization direction. The dark region represents a defective portion of the film that does not switch. During high-field cycling (1.2 MV/cm), such defects grow and coalesce, and eventually cover the entire area beneath the electrodes. It should be noted that the oxide need not be ferroelectric for such displacements to be seen. For example, Kolpak *et al.* [41] used resonant scattering near the Sr K-edge to help determine Ti displacements in polar $SrTiO_3$ (001) grown on Si. They found that the chemistry of the $SrTiO_3$/Si interface prevented ferroelectric switching.

Considering only the specular rod, Park and Fenter developed a model-independent technique for recovering the composition profile of an anomalous scatterer [42]. Beginning with equation (4.12), they noted that the resonant contribution could be written as

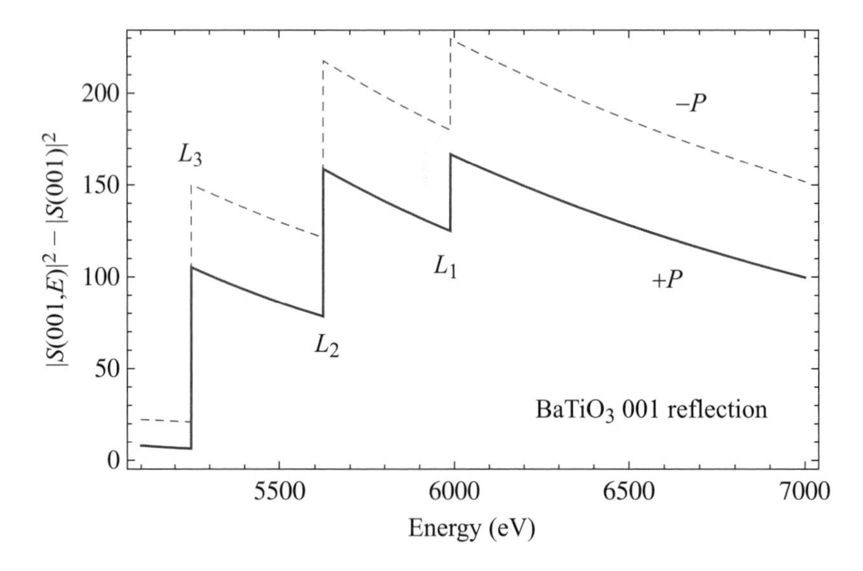

Fig. 4.6 The effect of resonant scattering on the $BaTiO_3$ 001 intensity for both up and down polarization, calculated with Cromer–Liberman values for $f'_{\text{Ba}}(E)$ and $f''_{\text{Ba}}(E)$. Values for atomic displacements were taken from [39].

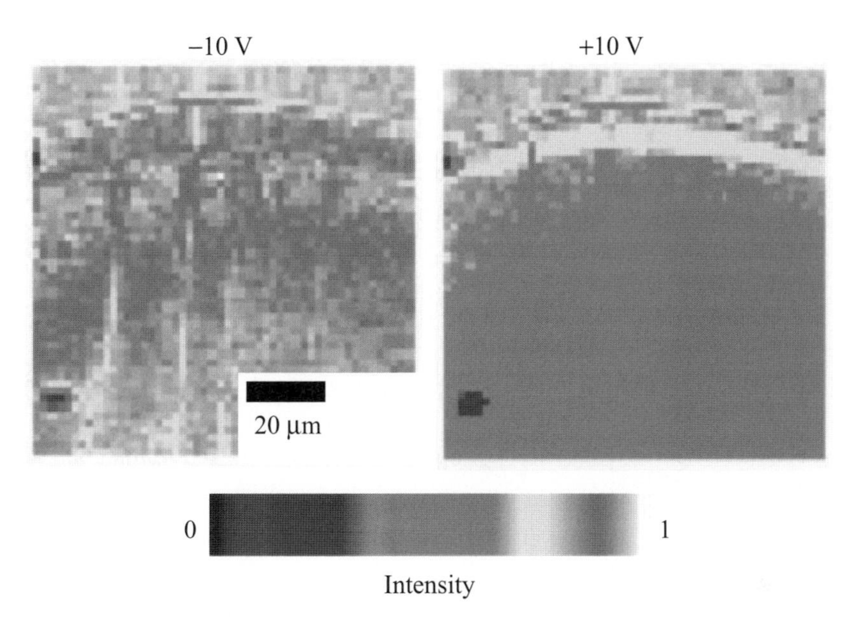

Fig. 4.7 Real space image of 002 intensity from a $Pb(Zr,Ti)O_3$ thin film capacitor after application of -10 V and +10 V pulses to the bottom $SrRuO_3$ electrode. Reprinted by permission from Macmillan Publishers Ltd: Nature Materials [40], copyright 2004.

$$S_R(\mathbf{q}, E) \simeq \left[f_a'(E) + if_a''(E)\right] A_R(\mathbf{q}) e^{2\pi i \phi_R(\mathbf{q})}. \tag{4.17}$$

If $S_T(\mathbf{q})$ could be determined by a separate technique, and $f_a'(E)$ and $f_a''(E)$ were known (e.g. from a separately measured X-ray Absorption Near Edge Structure (XANES) spectrum), then an energy scan across the absorption edge taken at q_0 could be described by $A_R(q_0)$ and $\phi_R(q_0)$. In this formulation, equation (4.13) becomes

$$\begin{aligned} \left|\frac{S(q_0, E)}{S_T(q_0)}\right|^2 &= 1 + \frac{\left|f_a'(E) + if_a''(E)\right|^2 \left|A_R(q_0)\right|^2}{\left|S_T(q_0)\right|^2} \\ &+ 2f_a'(E)\frac{\left|A_R(q_0)\right|}{\left|S_T(q_0)\right|} \cos\left(\Phi_T(q_0) - \Phi_R(q_0)\right) \\ &+ 2f_a''(E)\frac{\left|A_R(q_0)\right|}{\left|S_T(q_0)\right|} \sin\left(\Phi_T(q_0) - \Phi_R(q_0)\right) \end{aligned} \tag{4.18}$$

after normalizing by $\left|S_T(q_0)\right|^2$.

An example electron density profile and the resulting reflectivity are shown in Figures 4.8(a) and (b), respectively. The element-specific density profile (a Gaussian) is assumed small compared to the total electron density, such that the non-resonant reflectivity is insensitive to its location. Potential locations for the resonant atom are labeled in the inset, and the calculated

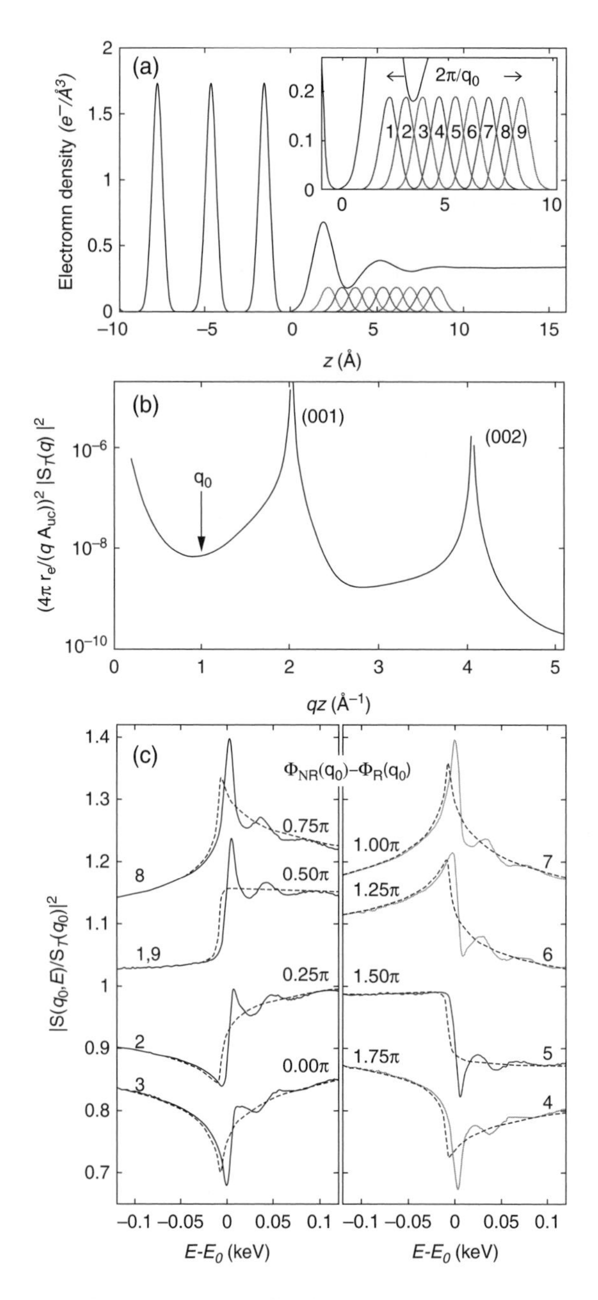

Fig. 4.8 Model scattering and spectroscopic profiles demonstrating the RAXR technique. (a) Simulated 1D solid-liquid electron density profile containing a resonant atom near the interface. Inset shows the possible locations for the resonant atom and are labeled 1–9. (b) Calculated total non-resonant specular X-ray reflectivity for this model interface. (c) RAXR spectra calculated using equation (4.18) at q_0 for each of the numbered resonant atom positions shown in (a). The values for $\Phi_T(q_0) - \Phi_R(q_0)$ are indicated. The dashed lines corresponded to spectra calculated using Cromer–Liberman values for $f'_{Sr^{2+}}(E)$ and $f''_{Sr^{2+}}(E)$, while the solid lines represent those calculated with experimentally derived dispersion corrections for aqueous $Sr(NO_3)_2$. Reprinted with permission from [42].

RAXR profiles (equation (4.18)) for the different locations are shown in Figure 4.8(c). For the best results, $f_a''(E)$ should be taken from a XANES spectrum measured from the same or similar material. The value of $f_a'(E)$ is obtained by a difference Kramers–Krönig transform [43].

Park and Fenter showed that measured RAXR spectra can be fit for $A_R(q_0)$ and $\Phi_R(q_0)$, assuming a linearly varying background [42]. As long as $S_T(\mathbf{q})$ is known (e.g. from COBRA phase retrieval), the electron density profile of the resonant atom can be directly reconstructed by Fourier synthesis:

$$\rho_R(z) = \frac{Z_R}{2\pi A_{\text{uc}}} \sum_m A_R(q_m) \cos\left[2\pi \Phi_R(q_m) - q_m z\right] \Delta q_m \tag{4.19}$$

where Z_R is the atomic number of the resonant atom. They applied this technique to image Rb^+ and Sr^{2+} ions at the muscovite (001)-liquid interface, (for RbCl and $Sr(NO_3)_2$ solutions, respectively). The technique is general, however, and when applied in parallel with a non-resonant phase retrieval technique, should allow direct determination of composition profiles in oxide heterostructures.

The power of this technique is not limited to the specular rod. For example, Specht and Walker performed an energy scan at an off-specular CTR to determine the valence state at a buried interface [44]. The measurement was performed on a mostly relaxed 142-nm-thick Cr_2O_3 film on (0001) Al_2O_3; only the few monolayers adjacent to the substrate remained coherently strained and contributed to the CTR intensity. Figure 4.9 shows an energy scan across the Cr K-edge at an *hkl* of 1 1 $3\frac{1}{4}$. The dotted and solid lines correspond to calculations of the spectrum using Cr and Cr^{3+} form factors, respectively. As expected, the Cr atoms in the monolayers closest to the substrate exhibit 3+ valence.

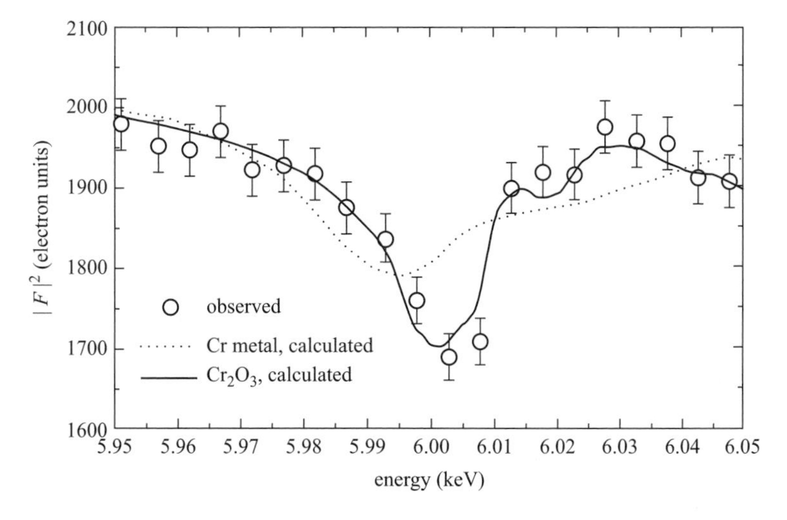

Fig. 4.9 Energy dependence of CTR intensity at (1, 1, 3.25). The solid line is the fit using experimentally determined f' and f'' from either Cr metal (dotted line) or Cr^{3+} (solid line). Reprinted with permission from [44]. Copyright 1993 by the American Physical Society.

We note that composition mapping is also possible using resonant COBRA [45, 46]. Here, CTRs are measured at several different energies, away from and near absorption edges, and COBRA is applied to each energy's dataset. Based on the resulting electron density profiles and the known anomalous dispersion corrections, the composition profile can be modeled in a straightforward manner.

Much can be learned by performing a more careful analysis of the Diffraction Anomalous Fine Structure (DAFS) [47, 48] shown in Figures 4.8(c) and 4.9. Here, we consider another modification to the atomic form factor:

$$f(\mathbf{q}, E) = \left[f_0(\mathbf{q}) + f'(E) + if''(E)\right] + \left[f''(E)\tilde{\chi}(E)\right] \tag{4.20}$$

where the last term accounts for the effects of the local environment on the form factor, and the complex fine structure, $\tilde{\chi}(E)$, can be calculated theoretically [49]. With this term, there is an added *hkl* dependence to the dispersion corrections, $f'_{hkl}(E)$ and $f''_{hkl}(E)$, and their measurement allows the extraction of *hkl*-dependent $\tilde{\chi}$.

DAFS can be used to determine site-specific valence. For example, Cross *et al.* [50] measured DAFS spectra from eight different Bragg peaks across the Cu K-edge for a 200-nm-thick epitaxial $YBa_2Cu_3O_7$ film grown on MgO (001). An iterative Kramer–Krönig algorithm is used to reduce the DAFS intensity data to the desired $f''(E)$ [43, 48]. In order to correct for the effect of fine structure on thin film absorption, fluorescence X-ray Absorption Fine Structure (XAFS) was measured simultaneously. Since each Bragg peak contains different scattering contributions from each of the two types of Cu scatterers, all the intensities were analyzed in parallel for extraction of the two $f''_{\mathrm{Cu}}(E)$ profiles shown in Figure 4.10(a). The energy shift between the two profiles can be used to extract the difference in Cu valence. As seen in Figure 4.10(b), the weighted sum of the two contributions is similar to the fluorescence XAFS spectrum.

More recently, Yang *et al.* [51] measured DAFS from an epitaxial $MnFe_2O_4$ film grown on MgO (111). $MnFe_2O_4$ is a spinel where the cations occupy eight tetragonal sites and 16 octahedral sites in a closed packed oxygen lattice, and its magnetic properties depend on the distribution of cations over these sites. DAFS spectra from the 222 and 422 reflections were measured across both the Mn and Fe K-edges, along with fluorescence XAFS (Figure 4.11(a)). Both $f''_{\mathrm{Mn}}(E)$ and $f''_{\mathrm{Fe}}(E)$ were extracted using the iterative Kramer–Krönig algorithm [43, 48]. Figure 4.11(b) shows the real part of the Fourier transformed Fe and Mn data along with their best fits using the FEFF code [49]. From both the XAFS and DAFS data, Yang *et al.* [51] determined that 9% of the tetragonal sites contained Fe and 18% of the octahedral sites contained Mn. Using the concept of coordination charge to model the shift in energy [52], they determined that the valences were $Mn^{3.38+}$ and $Fe^{2.92+}$ in the octahedral sites, and $Mn^{2.18+}$ in the tetrahedral site. The presence of Mn^{3+} in the octahedral sites increases the magnetic anisotropy of this material. They also calculated a net charge of +8.31 for the Mn and Fe cations, suggesting that the anion sublattice is oxygen deficient.

Resonant scattering has recently been used for studies of charge and orbital ordering in complex oxide systems. As shown in Figure 4.12(a), the layered perovskite manganite $La_{0.5}Sr_{1.5}MnO_4$ can exhibit charge, orbital, and spin ordering. The charge-ordered structure has dimensions of $\sqrt{2}a \times \sqrt{2}a \times c$ relative to the $La_{0.5}Sr_{1.5}MnO_4$ unit-cell, giving rise to weak half-order reflections. Since the absorption edge of Mn^{3+} is slightly different from that of Mn^{4+}, an anomaly in the half-order reflection is expected when run across the Mn K-edge.

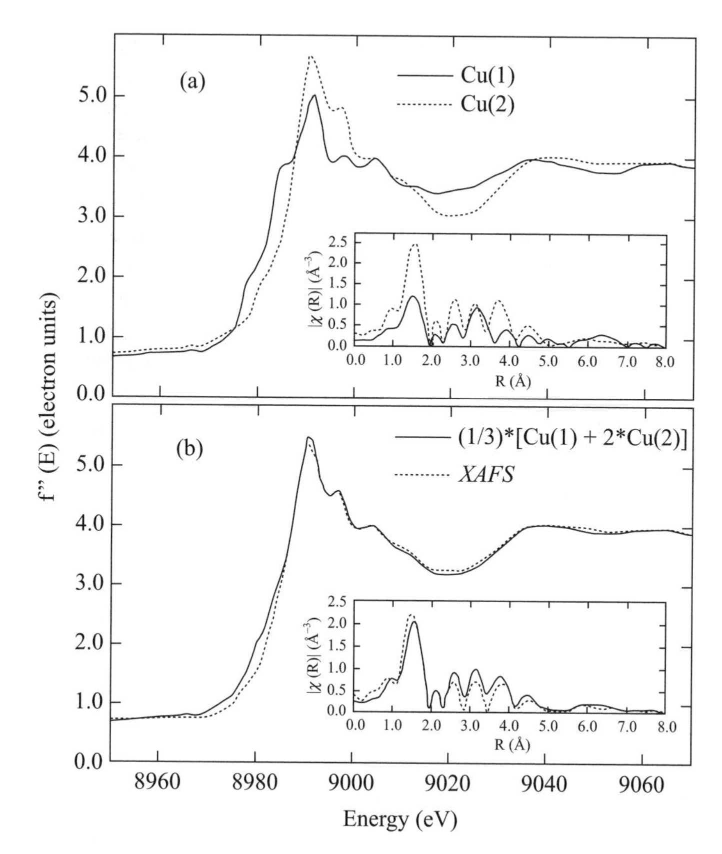

Fig. 4.10 Plots of $f_{Cu''(E)}$ for both the Cu(1) and Cu(2) sites in $YBa_2Cu_3O_7$. (a) The site-separated $f_{Cu''(E)}$ across the Cu K-edge. (b) The weighted sum of the two $f_{Cu''(E)}$ components compared to fluorescence XAFS. Insets show the R-space transforms of the corresponding $\tilde{\chi}$ data. Reprinted with permission from Ref. [50].

This anomaly is shown in Figure 4.12(b), which was observed at $T = 29.6$K. Based on the anomalous dispersion terms of $f_{Mn^{3+}}(E)$ and $f_{Mn^{4+}}(E)$, which were determined by XANES, Murakami *et al.* [53] calcuated the expected intensity (solid line in Figure 4.12(a)) according to the $\frac{h}{2}\frac{h}{2}0$ structure factor,

$$S\left(\frac{h}{2}\frac{h}{2}0, E\right) \propto \left[f'_{Mn^{3+}}(E) - f'_{Mn^{4+}}(E)\right] + i\left[f''_{Mn^{3+}}(E) - f''_{Mn^{4+}}(E)\right], \tag{4.21}$$

which agrees well with the data.

Orbital ordering in $La_{0.5}Sr_{1.5}MnO_4$ is also expected, with the unit-cell $\sqrt{2}a \times 2\sqrt{2} \times c$, as shown by the thick lines in Figure 4.12(a). This gives rise to weak quarter-order reflections. As shown in Figure 4.13(a), the intensity of quarter-order reflection can be significantly enhanced at the Mn absorption edge. Since the surface also truncates orbital ordering in the crystal,

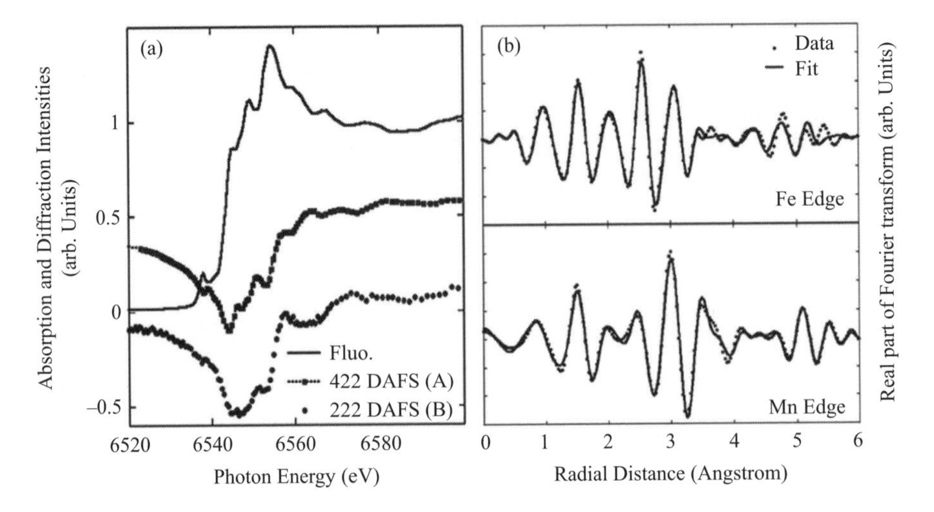

Fig. 4.11 (a) Mn K-edge DAFS for the 422 and 222 Bragg reflections along with EXAFS data. The spectra have been vertically shifted. (b) Fourier transformed Mn and Fe EXAFS data along with the best fit. Reprinted with permission from [51]. Copyright 2008, American Institute of Physics.

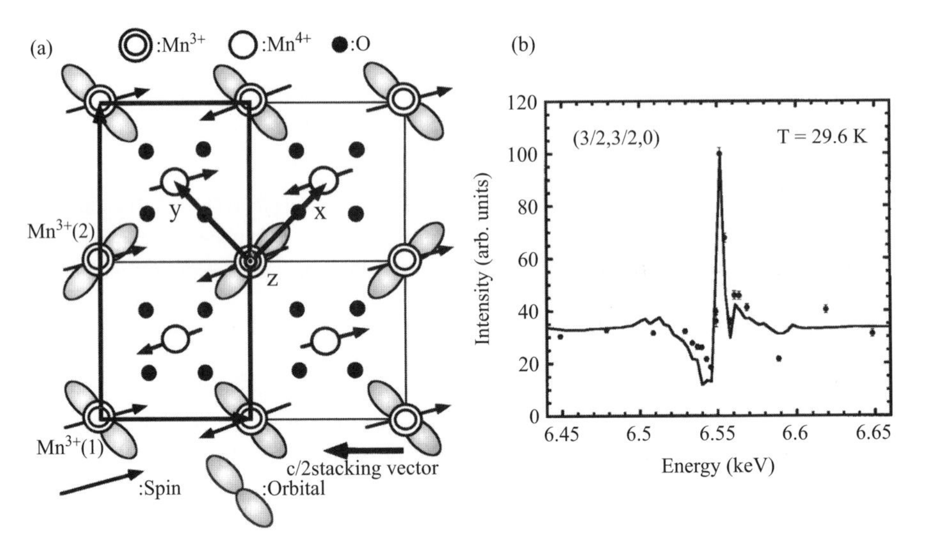

Fig. 4.12 (a) Schematic depiction of charge, spin, and orbital ordering in $La_{0.5}Sr_{1.5}MnO_4$. (b) Intensity of the $\frac{3}{2}\frac{3}{2}0$ peak as a function of energy at 29.6 K. The solid curve was calculated from equation (4.21). Reprinted with permission from [53]. Copyright 1998 by the American Physical Society.

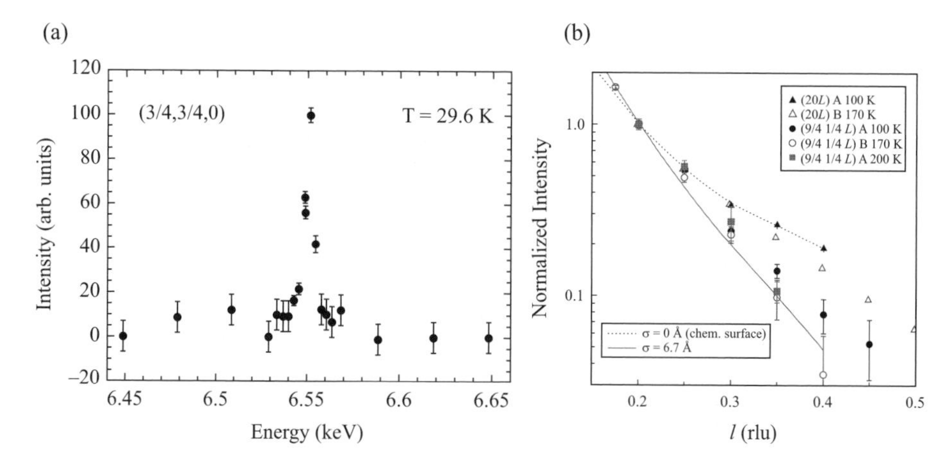

Fig. 4.13 Quarter-order reflections from orbital ordering in $La_{0.5}Sr_{1.5}MnO_4$. (a) Energy dependence of the $\frac{3}{4}\frac{3}{4}0$ peak from orbital ordering at 29.6 K. From [53]. (b) $20l$ and $\frac{9}{4}\frac{1}{4}l$ truncation rod intensities for $La_{0.5}Sr_{1.5}MnO_4$ at 100 K, 170 K and 200 K. Samples A and B were cleaved in He or in air, respectively. The solid line is the calculated intensity for a CTR with 6.7 Å surface roughness. (a) Reprinted with permission from [53]. Copyright 1998 by the American Physical Society. (b) Reprinted by permission from Macmillan Publishers Ltd: Nature Materials [54], copyright 2007.

Wakabayashi *et al.* [54] were able to measure CTRs from such quarter-order reflections. As shown in Figure 4.13(b), the intensity for quarter-order reflection decays faster along l than the intensity from integer reflections, indicating that the "orbital ordered surface" is rougher than the physical/chemical surface.

4.4 Anisotropic effects

Synchrotron x-rays are linearly polarized in the orbital plane, and this can be used to explore the anisotropy of the atomic form factor. The form factor can be expressed as a tensor expansion [55]:

$$f(\mathbf{q},E) = f_0(\mathbf{q}) + e_j e'_k S^{jk} + i e_j e'_k \left(K'_m - K_m\right) T^{jkm} + e_j e'_k K'_m K_n U^{jkmn} + e_j e'_k \left(K'_m K'_n + K_m K_n\right) V^{jkmn} + \ldots \quad (4.22)$$

using Einstein notation, where S^{jk}, T^{jkm}, etc., are tensors, and e_j and e'_k are unit polarization vectors for the incident X-rays and scattered X-rays, respectively. With resonant scattering, the incoming X-ray promotes a core electron to an excited state, which then decays back to the core hole and emits a photon of the same energy. Since the electron is sensitive to the anisotropy of its environment, it causes anisotropy in the form factor. In the $YBa_2Cu_3O_7$ DAFS measurement described above, the rotation axis of the sample was set to be parallel to that of the X-ray polarization to minimize anisotropic contributions to their results [50]. Murakami *et al.* [53]

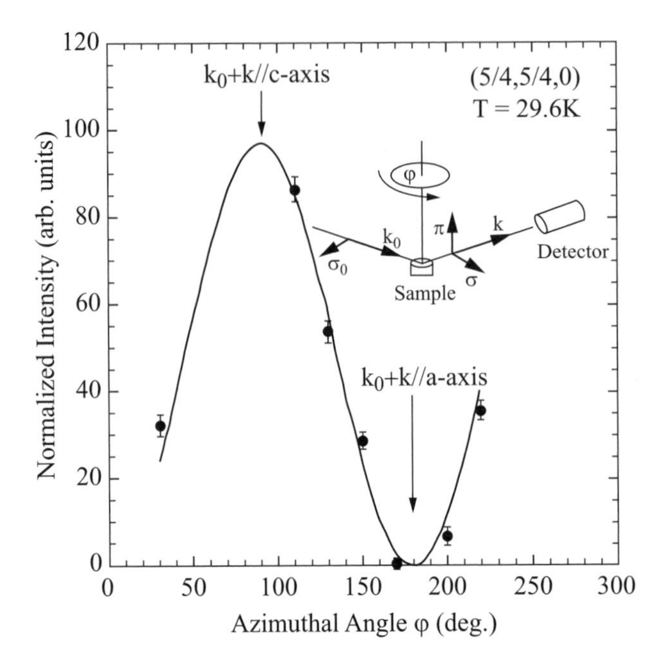

Fig. 4.14 Dependence of the $La_{0.5}Sr_{1.5}MnO_4$ $\frac{5}{4}\frac{5}{4}0$ normalized intensity on azimuthal angle at $E =$ 6.552 keV and 29.6 K. The inset shows the experimental setup and polarization directions. Reprinted with permission from [53]. Copyright 1998 by the American Physical Society.

found that the $\frac{5}{4}\frac{5}{4}0$ orbital ordering peak for $La_{0.5}Sr_{1.5}MnO_4$ exhibits strong anisotropy, as shown in Figure 4.14.

While anisotropic effects can stem from the anisotropy of electronic orbitals, it is now clear that much of the intensity of the fractional order peaks observed by Murakami *et al.* [53] originates from the Jahn–Teller lattice distortions associated with charge and orbital ordering in the manganites [56, 57, 58, 59, 60, 61]. By combining a large dataset (50 reflections over a 120 eV energy range) with first-principles theory, Nazarenko *et al.* [62] could specifically account for the effects of atomic displacements and thereby confirm charge ordering in low temperature Fe_3O_4. Some of their results are shown in Figure 4.15(a), where the solid and dotted lines correspond to models calculated with and without charge ordering, respectively. As seen, the $\bar{1}10$ and $\bar{4}41$ reflections are sensitive to charge ordering, while others like the $\bar{4}42$ are not. They found that the ordering can be characterized by two different charge disproportionations, each associated with its own (uncoupled) ordering: one gives rise to $[001]c$ modulation and the other causes $[00\frac{1}{2}]c$ ordering.

We note finally that the atomic form factor also depends on electron spin, although very weakly. Expressions for the magnetic contributions to the atomic form factor can be found in [64]. The dependence can be enhanced strongly near an absorption edge, particularly the L_2 and L_3 for the rare earth elements. Nandi *et al.* [63] used resonant scattering to examine the low temperature structure of the multiferroic material $HoMnO_3$. Figure 4.15(b) shows resonant scattering from the 009 antiferromagnetic reflection across the Ho L_3-edge below 39 K. With

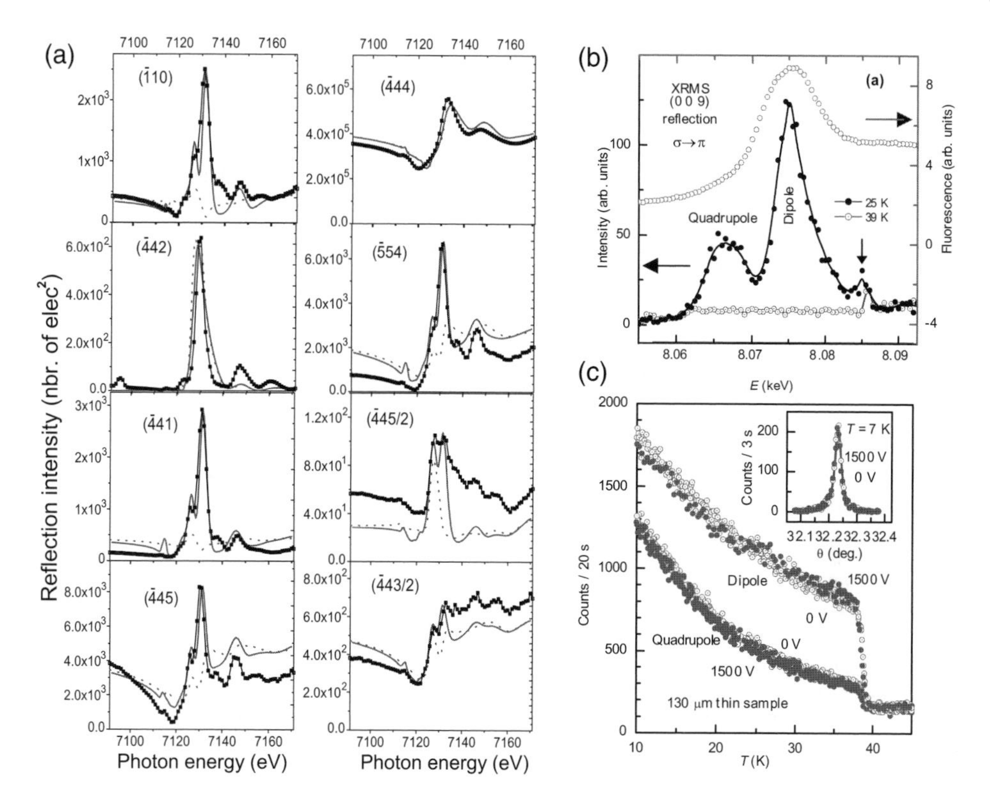

Fig. 4.15 (a) Experimental and calculated resonant scattering in Fe_3O_4 at 50 K. Shown are experimental data (squares), calculated spectra with charge ordering (solid lines), and calculated spectra without charge ordering (dotted lines). The $\bar{1}10$ and $\bar{4}41$ reflections are particularly sensitive to charge ordering. From [62]. (b) Intensity of 009 antiferromagnetic reflection and fluorescence as a function of energy across the Ho L_3-edge. The solid lines are guides to the eye. The small peak on the right is due to multiple charge scattering. From [63]. (c) Intensities for both the dipole and quadrupole 009 peaks as a function of temperature and applied electric field. Inset shows rocking scans at both 0 and 1500 V for the quadrupole peak at 7 K. From [63]. (a) Reprinted with permission from [62]. Copyright 2006 by the American Physical Society. (b) Reprinted with permission from [63]. Copyright 2008 by the American Physical Society. (c) Reprinted with permission from [63]. Copyright 2008 by the American Physical Society.

the incident X-rays polarized perpendicular to the scattering plane, the measurements were carried out in the rotated ($\sigma - \pi$) scattering channel. Two resonance peaks were observed, one from quadrupole resonance and the other from dipole resonance. The temperature dependence of these peaks is shown in Figure 4.15(c). As seen, magnetic ordering of Ho^{3+} begins at 39 K and is unaffected by the applied electric field. Therefore, Ho^{3+} is not responsible for the reported change in $HoMnO_3$'s ordered state from antiferromagnetic to ferromagnetic in an applied electric field.

4.5 Summary

In this chapter, we have provided a brief survey of synchrotron X-ray scattering techniques useful for the study of oxide heterostructures. By necessity, we have omitted a myriad of other important techniques such as soft X-ray scattering and spectroscopy [65, 66, 67, 68, 69, 70], standing wave studies [71, 72], and inelastic scattering [73, 74, 75], and the reader is urged to consult these references. It is clear that many of the intricacies of complex oxides can be examined with the use of synchrotron methods, and progress in both technique development and analysis will go hand-in-hand with improved fundamental understanding of these remarkable materials.

Acknowledgments

This work was supported by the U.S. Department of Energy, Office of Basic Energy Sciences, Division of Materials Sciences and Engineering. Argonne National Laboratory, a U.S. Department of Energy Office of Science laboratory, is operated under Contract No. DE-AC02-06CH11357.

References

[1] Fong, D.D., Thompson, C. (2006). In situ synchrotron X-ray studies of ferroelectric thin films. *Annu. Rev. Mater. Res.* **36**, 431.

[2] Grigoriev, A., Do, D.H., Kim, D.M., *et al.* (2006). Nanosecond domain wall dynamics in ferroelectric $Pb(Zr,Ti)O_3$ thin films Phys. Rev. Lett. **96**, 187601.

[3] Fister, T.T., Fong, D.D., Eastman, J.A., *et al.* (2008). In situ characterization of strontium surface segregation in epitaxial $La_{0.7}Sr_{0.3}MnO_3$ thin films as a function of oxygen partial pressure. Appl. Phys. Lett. **93**, 151904.

[4] Fong, D.D., Lucas, C.A., Richard, M.-I., and Toney, M.F. (2010). X-ray probes for in situ studies of interfaces. MRS Bull. **35**, 504.

[5] Shultz, A., Jang, W., and Hetherington, W. (1995). Comparative second harmonic generation and X-ray photoelectron spectroscopy studies of the UV creation and O_2 healing of Ti^{3+} defects on (110) rutile TiO_2 surfaces. Surf. Sci. **339**, 114.

[6] Horiuchi, T., Ochi, H., Kaisei, K., Ishida, K., and Matsushige, K. (2003). Observations of TiO_2 surfaces using totally reflected X-ray in-plane diffraction under UV irradiation. Mat. Res. Soc. Symp. Proc. **751**, Z3.47.1.

[7] Gutmann, S., Wolak, M.A., Conrad, M., Beerbom, M.M., and Schlaf, R. (2010). Effect of ultraviolet and X-ray radiation on the work function of TiO_2 surfaces. J. Appl. Phys. **107**, 103705.

[8] Kazor, A. (1995). Space-charge oxidant diffusion model for rapid thermal oxidation of silicon. J. Appl. Phys. **77**, 1477.

[9] Tsuchiya, M., Sankaranarayanan, S.K.R.S., and Ramanathan, S. (2009). Photon-assisted oxidation and oxide thin film synthesis: A review. Prog. Mater. Sci. **54**, 981.

[10] Schlom, D.G., Chen, L.Q., Eom, C.B., *et al.* (2007). Strain tuning of ferroelectric thin films. Annu. Rev. Mater. Res. **37**, 589.

[11] Haeni, J.H., Irvin, P., Chang, W., *et al.* (2004). Room-temperature ferroelectricity in strained $SrTiO_3$. Nature **430**, 758.

[12] Reyren, N., Thiel, S., Caviglia, A.D., *et al.* (2007). Superconducting interfaces between insulating oxides. Science **317**, 1196.
[13] Glazer, A.M. (1975). Simple ways of determining perovskite structures. Acta Cryst. A **31**, 756.
[14] Als-Nielsen, J. and McMorrow, D. (2001). *Elements of Modern X-ray Physics*, John Wiley & Sons, Ltd., West Sussex, England.
[15] Waasmaier, D. and Kirfel, A. (1995). New analytical scattering-factor functions for free atoms and ions. Acta Cryst. A **51**, 7673.
[16] Feidenhans'l, R. (1989). Surface structure determination by X-ray diffraction. Surf. Sci. Rep. **10**, 105.
[17] Fuoss, P.H. and Brennan, S. (1990). Surface sensitive X-ray scattering. Annu. Rev. Mater. Sci. **20**, 365.
[18] Robinson, I.K, and Tweet, D.J. (1992). Surface X-ray diffraction. Rep. Prog. Phys. **55**, 599.
[19] Renaud, G. (1998). Oxide surfaces and metal/oxide interfaces studied by grazing incidence X-ray scattering. Surf. Sci. Rep. **32**, 1.
[20] Vlieg, E. (2000). ROD: a program for surface X-ray crystallography. J. Appl. Cryst. **33**, 401 .
[21] Björck, M. and Andersson, G. (2007). GenX: an extensible X-ray reflectivity refinement program utilizing differential evolution. J. Appl. Cryst. **40**, 1174.
[22] Vlieg, E. (1997). Integrated intensities using a six-circle surface X-ray diffractometer. J. Appl. Cryst. **30**, 532.
[23] Schlepütz, C.M., Herger, R., Willmott, P.R., *et al.* 2005. Improved data acquisition in grazing-incidence X-ray scattering experiments using a pixel detector. Acta Cryst. A **61**, 418.
[24] Munkholm, A., Streiffer, S.K., Ramana Murty, M.V., *et al.* (2001). Antiferrodistortive reconstruction of the $PbTiO_3(001)$ surface. Phys. Rev. Lett. **88**, 16101.
[25] May, S.J., Kim, J.-W., Rondinelli, J.M., *et al.* (2010). Quantifying octahedral rotations in strained perovskite oxide films. Phys. Rev. B **82**, 014110.
[26] Sayre, D. (2002). X-ray crystallography: The past and present of the phase problem. Struct. Chem. **13**, 81.
[27] Marchesini, S. (2007). Invited article: A unified evaluation of iterative projection algorithms for phase retrieval. Rev. Sci. Instrum. **78**, 011301.
[28] Gerchberg, R.W., and Saxton, W.O. (1972). A practical algorithm for the determination of the phase from image and diffraction plane pictures. Optik **35**, 237.
[29] Fienup, J.R. (1978). Reconstruction of an object from the modulus of its Fourier transform. Opt. Lett. **3**, 27.
[30] Fung, R., Shneerson, V.L., Lyman, P.F., *et al.* (2007). Phase and amplitude recovery and diffraction image generation method: structure of Sb/Au(110)–$\sqrt{3}$?×?$\sqrt{3}R54.7°$ from surface X-ray diffraction. Acta Cryst. A **63**, 239.
[31] Björck, M., Schlepütz, C.M., Pauli, S.A., *et al.* (2008). Atomic imaging of thin films with surface X-ray diffraction: introducing DCAF. J. Phys.: Condens. Matter **20**, 445006.
[32] Yacoby, Y., Pindak, R., MacHarrie, R., *et al.* (2000). Direct structure determination of systems with two-dimensional periodicity. J. Phys.: Condens. Matter **12**, 3929.
[33] Yacoby, Y., Sowwan, M., Stern, E., *et al.* (2002). Direct determination of epitaxial interface structure in Gd_2O_3 passivation of GaAs. Nat. Mater. **1**, 99.

[34] Zhou, H., Yacoby, Y., Butko, V.Y., *et al.* (2010). Anomalous expansion of the copper-apical-oxygen distance in superconducting cuprate bilayers. Proc. Nat'l. Acad. Sci. **107**, 8103.

[35] http://henke.lbl.gov/optical_constants

[36] Hendrickson, W.A. (1991). Determination of macromolecular structures from anomalous diffraction of synchrotron radiation. Science **254**, 51.

[37] Hodeau, J.-L., Favre-Nicolin, V., Bos, S., *et al.* (2001). Resonant diffraction. Chem. Rev **101**, 1843.

[38] van Reeuwijk, S.J., Karakaya, K., Graafsma, H., and Harkema, S. (2004). Polarization switching in $BaTiO_3$ thin films measured by X-ray diffraction exploiting anomalous dispersion. J. Appl. Cryst. **37**, 193.

[39] Harada, J., Pedersen, T., and Barnea, Z. (1970). X-ray and neutron diffraction study of tetragonal barium titanate. Acta Cryst. A **26**, 336.

[40] Do, D.H., Evans, P.G., Isaacs, E.D., *et al.* (2004). Structural visualization of polarization fatigue in epitaxial ferroelectric oxide devices. Nat. Mater. **3**, 365.

[41] Kolpak, A.M., Walker, F.J., Reiner, J.W., *et al.* (2010). Interface-induced polarization and inhibition of ferroelectricity in epitaxial $SrTiO_3$/Si. Phys. Rev. Lett. **105**, 217601.

[42] Park, C., and Fenter, P.A. (2007). Phasing of resonant anomalous X-ray reflectivity spectra and direct Fourier synthesis of element-specific partial structures at buried interfaces. J. Appl. Cryst. **40**, 290.

[43] Cross, J.O., Newville, M., Rehr, J.J., *et al.* (1998). Inclusion of local structure effects in theoretical X-ray resonant scattering amplitudes using *ab initio* X-ray-absorption spectra calculations. Phys. Rev. B **58**, 11215.

[44] Specht, E.D., and Walker, F.J. (1993). Oxidation state of a buried interface: Near-edge X-ray fine structure of a crystal truncation rod. Phys. Rev. B. **47**, 13743.

[45] Kumah, D.P., Riposan, A., Cionca, C.N., *et al.* (2008). Resonant coherent Bragg rod analysis of strained epitaxial heterostructures. Appl. Phys. Lett. **93**, 081910.

[46] Kumah, D.P., Shusterman, S., Paltiel, Y., Yacoby, Y., and Clarke, R. (2009). Atomic-scale mapping of quantum dots formed by droplet epitaxy. Nat. Nanotech. **4**, 835 .

[47] Stragier, H., Cross, J.O., Rehr, J.J., *et al.* (1992). Diffraction anomalous fine structure: A new X-ray structural technique. Phys. Rev. Lett. **69**, 3064.

[48] Pickering, I.J., Sansone, M., Marsch, J., and George, G.N. (1993). Diffraction anomalous fine structure: a new technique for probing local atomic environment. J. Am. Chem. Soc. **115**, 6302.

[49] Rehr, J.J. (2006). Theory and calculations of X-ray spectra: XAS, XES, XRS, and NRIXS. Rad. Phys. Chem. **75**, 1547.

[50] Cross, J.O., Newville, M., Sorensen, L.B., *et al.* (1997). Separated anomalous scattering amplitudes for the inequivalent Cu sites in $YBa_2Cu_3O_{7-\delta}$ using DAFS. J. Phys. IV France **7**, C2-745.

[51] Yang, A., Chen, Z., Geiler, A.L., *et al.* (2008). Element- and site-specific oxidation state and cation distribution in manganese ferrite films by diffraction anomalous fine structure. Appl. Phys. Lett. **93**, 052504.

[52] Ovsyannikova, I.A., Batsanov, S.S., Nasonova, L.I., Batsanova, L.R., and Nekrasova, E.A. (1967). Bull. Acad. Sci. USSR, Phys. Ser. (Engl. Transl.) **31**, 936.

[53] Murakami, Y., Kawada, H., Kawata, H., *et al.* (1998). Direct observation of charge and orbital ordering in $La_{0.5}Sr_{1.5}MnO_4$. Phys. Rev. Lett. **80**, 1932.

[54] Wakabayashi, Y., Upton, M.H., Grenier, S., *et al.* (2007). Surface effects on the orbital order in the single-layered manganite $La_{0.5}Sr_{1.5}MnO_4$. Nat. Mater. **6**, 972.
[55] Templeton, D.H. (1994). X-ray resonance, then and now. In: Materlik. G., Sparks, C.J., and Fischer, K., eds. *Resonant Anomalous X-Ray Scattering: Theory and Applications*, p. 1-4. Elsevier Science B. V., Amsterdam.
[56] Benfatto, M., Joly, Y., and Natoli, C.R. (1999). Critical reexamination of the experimental evidence of orbital ordering in $LaMnO_3$ and $La_{0.5}Sr_{1.5}MnO_4$. Phys. Rev. Lett. **83**, 636.
[57] Elfimov, I.S., Anisimov, V.I., and Sawatzky, G.A. (1999). Orbital ordering, Jahn-Teller distortion, and anomalous X-ray scattering in manganates. Phys. Rev. Lett. **82**, 4264.
[58] Joly, Y., Di Matteo, S., and Natoli, C.R. (2004). *Ab initio* simulations of resonant X-ray scattering on the insulating phase of V_2O_3 compared with recent experiments. Phys. Rev. B **69**, 224401.
[59] Wakabayashi, Y., Sawa, H., Nakamura, M., Izumi, M., and Miyano, K. (2004). Lack of influence of anisotropic electron clouds on resonant X-ray scattering from manganite thin films. Phys. Rev. B **69**, 144414.
[60] Wilkins, S.B., Di Matteo, S., Beale T.A.W., *et al.* (2009). Critical reexamination of resonant soft X-ray Bragg forbidden reflections in magnetite. Phys. Rev. B **79**, 201102.
[61] Di Matteo, S. (2009). Which orbital and charge ordering in transition metal oxides can resonant x-ray diffraction detect? J. Phys.: Conf. Ser. **190**, 012008.
[62] Nazarenko, E., Lorenzo, J.E., Joly, Y., *et al.* (2006). Resonant x-ray diffraction studies on the charge ordering in magnetite. Phys. Rev. Lett. **97**, 056403.
[63] Nandi, S., Kreyssig, A., Tan, L., *et al.* (2008). Nature of Ho magnetism in multiferroic $HoMnO_3$. Phys. Rev. Lett. **100**, 217201.
[64] Blume, M. (1994). Magnetic effects in anomalous dispersion. In: Materlik, G., Sparks, C.J., and Fischer, K., eds. *Resonant Anomalous X-Ray Scattering: Theory and Applications*, p. 495. Elsevier Science B.V., Amsterdam.
[65] Chakhalian, J., Freeland, J.W., Habermeier, H.-U., *et al.* (2007). Orbital reconstruction and covalent bonding at an oxide interface. Science **318**, 1114.
[66] Freeland, J.W., Kavich, J.J., Gray, K.E., *et al.* (2007). Suppressed magnetization at the surfaces and interfaces of ferromagnetic metallic manganites. J. Phys.: Condens. Matter **19**, 315210.
[67] Wadati, H., Hawthorn, D.G., Geck, J., *et al.* (2009). Resonant soft x-ray scattering studies of interface reconstructions in $SrTiO_3/LaAlO_3$ superlattices. J. Appl. Phys. **106**, 083705.
[68] Gray, A.X., Papp, C., Balke, B., *et al.* (2010). Interface properties of magnetic tunnel junction $La_{0.7}Sr_{0.3}MnO_3/SrTiO_3$ superlattices studied by standing-wave excited photoemission spectroscopy. Phys. Rev. B **82**, 205116.
[69] Valvidares, S.M., Huijben, M., Yu, P., Ramesh, R., and Kortright, J.B. (2010). Native $SrTiO_3$ (001) surface layer from resonant Ti $L_{2,3}$ reflectance spectroscopy. Phys. Rev. B **82**, 235410.
[70] Benckiser, E., Haverkort, M.W., Brück, S., *et al.* (2011). Orbital reflectometry of oxide heterostructures. Nat. Mater. **10**, 189.
[71] Bedzyk, M.B., Kazimirov, A., Marasco, D., *et al.* (2000). Probing the polarity of ferroelectric thin films with x-ray standing waves. Phys. Rev. B **61**, 7873.

[72] Thiess, S., Lee, T.L., Bottin, F., and Zegenhagen, J. (2010). Valence band photoelectron emission of $SrTiO_3$ analyzed with X-ray standing waves. Solid State Communications **150**, 553.

[73] Berner, G., Glawion, S., Walde, J., *et al.* (2010). $LaAlO_3/SrTiO_3$ oxide heterostructures studied by resonant inelastic x-ray scattering. Phys. Rev. B **82**, 241405.

[74] Ellis, D.S., Kim, J., Zhang, H., *et al.* (2011). Electronic structure of doped lanthanum cuprates studied with resonant inelastic x-ray scattering. Phys. Rev. B **83**, 075120.

[75] Fister, T.T., Fong, D.D., Eastman, J.A., *et al.* (2011). Total-reflection inelastic x-ray scattering from a 10-nm thick $La_{0.6}Sr_{0.4}CoO_3$ thin film. Phys. Rev. Lett. **106**, 037401.

5 Scanning transmission electron microscopy of oxides

M. VARELA, C. LEON, J. SANTAMARIA, AND S. J. PENNYCOOK

Complex oxides form a very hot field at the forefront of materials physics. They exhibit some of the most disparate physical behaviors both in bulk and in thin film form, such as colossal magnetoresistance (CMR) or High Tc superconductivity, which are still far from being understood. Many of these properties rely on the presence of small active regions: nanoscale inhomogeneity and phase separation are widely believed to be at the root of CMR [1]. Another example is the metallicity found at the interface between a Mott insulator and a band insulator which is confined within a very thin layer near the interface [2, 3]. In many cases, it has been suggested that these behaviors may be linked to an underlying density of defects, such as O vacancies. In order to achieve a full understanding of the physical processes involved, probes capable of studying these systems in real space and with atomic resolution are very useful tools, which sometimes offer unique information. In this chapter we will review the state-of-the-art of scanning transmission microscopy and electron energy loss spectroscopy (EELS) along with some applications to complex oxides. We will start with a brief introduction to the technique, including explanation of the mechanisms underlying imaging, issues with interpretation, etc. followed by a number of applications to both bulk oxide materials and interfaces in heterostructures.

5.1 Introduction to STEM

The scanning transmission electron microscope (STEM) is unique in its ability to provide two-dimensional maps revealing atomic and electronic structure with sub-Ångstrom spatial resolution and sub-eV energy resolution. The key advantage of STEM is the ability to detect multiple signals simultaneously. While the probe scans a thin specimen, scattered electrons can be used to form a Z-contrast image showing atom locations, while at the same time, spectroscopic signals can be mapped showing elemental compositions, or, from the near edge fine structure, details of electronic band structure. EELS is formally equivalent

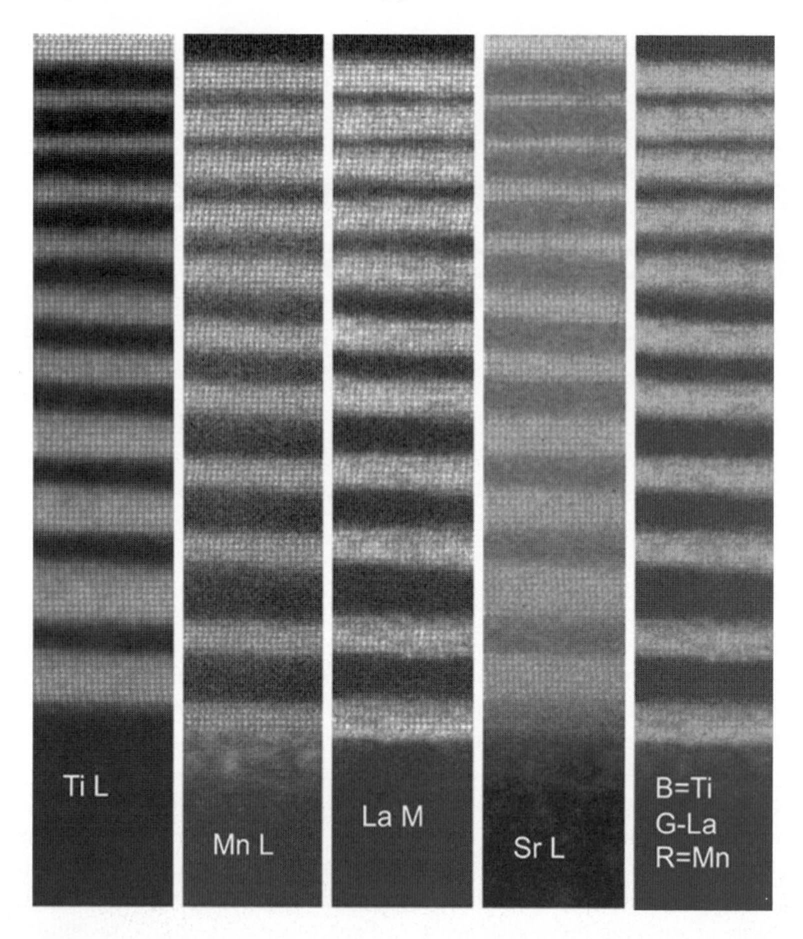

Fig. 5.1 Atomic resolution elemental maps of a $La_{0.7}Sr_{0.3}MnO_3/SrTiO_3$ superlattice with changing $SrTiO_3$ layer thicknesses, obtained in the aberration corrected Nion UltraSTEM 100. From left to right: Ti $L_{2,3}$ map, Mn $L_{2,3}$ map, La $M_{4,5}$ map, and Sr $L_{2,3}$ maps. The RGB is an overlay of the Ti, La, and Mn images. This figure is reproduced in color in the color plate section.

to X-ray absorption spectroscopy, except that STEM can provide such data with atomic resolution. The cost of spatial resolution is of course limited counting statistics. There is a limit to the amount of current that can be focused into one unit-cell before damage will occur through ionization or atomic knock-on events. Nevertheless, the ability to provide two-dimensional maps is extremely complementary to average measurements based on X-rays, revealing any differences in termination or interdiffusion between different interfaces of an oxide thin film or multilayer, as well as any local deviations at defects such as dislocation cores, stacking faults, antiphase boundaries, or grain boundaries. Figure 5.1 shows a beautiful example of elemental mapping in a $La_{0.7}Sr_{0.3}MnO_3/SrTiO_3$ superlattice with varying layer thicknesses.

The capability of STEM has improved dramatically in the last few years with the ability to correct for lens aberrations. Round lenses have inherently high aberrations that for most of the history of electron microscopy have meant an effective maximum usable lens aperture of under 1°. This sets the resolution in the region of about 50 wavelengths, so that for an electron accelerated to 100 kV, despite having a wavelength of only 0.037 Å, the resolution is just below 2 Å. While this is comparable to atomic spacings in materials, it is clear that gaining a factor of two in resolution would have tremendous benefit. The benefit in fact goes beyond just the ability to resolve smaller spacings, because if the probe becomes a factor of two smaller, the peak intensity increases fourfold, providing images of much better quality. An example of the improvement in imaging of $SrTiO_3$ with reducing probe size is shown in Figure 5.2. The sub-Ångstrom probe of the 300 kV microscope is so small that it fits easily in between the cation columns, and the image even reveals the positions of the weakly scattering oxygen columns.

The successful correction of lens aberrations means we are at a historic point in electron microscopy, with clear new views of the atomic world available for the first time. It is important to realize that the aberrations are not a defect in lens construction in the same way as with a glass lens, they arise intrinsically from the physics, the need to use electromagnetic fields to focus electrons [4]. The first suggestions for overcoming these limitations were made over 60 years ago [5], but it is only recently that aberration correctors have led to improved microscope resolution, first with the scanning electron microscope [6], then in the TEM [7–9] and the scanning TEM (STEM) [10, 11]. The reason that aberration correction has taken so long to

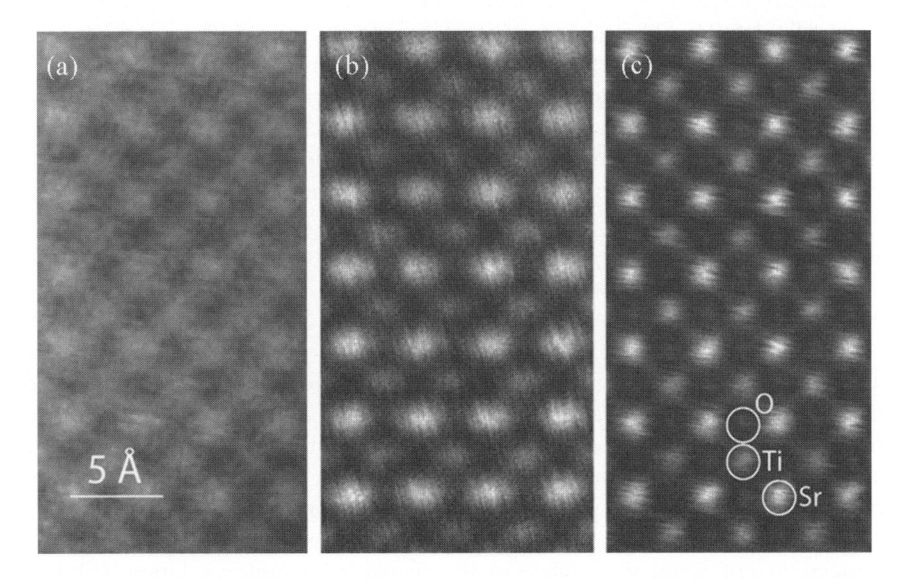

Fig. 5.2 Z-contrast images of SrTiO [100] taken on (a) an uncorrected 100 kV STEM with resolution 2.2 Å, (b) an aberration-corrected 100 kV STEM with a probe size of ~1.3 Å, and (c) an aberration-corrected 300 kV STEM with a probe size of ~0.78 Å resolving the O columns. Images courtesy M. M. McGibbon, M. Varela, and A. R. Lupini.

come to fruition is that multipole lenses are necessary for correcting aberrations, combinations of dipoles, quadrupoles, hexapoles, and octapoles which must all be controlled simultaneously to high precision: 1 part in 10^6 or 10^7 or greater, more than the perfection of the Hubble telescope. Imagine 40–60 focus controls that all need to be adjusted exactly; it is simply beyond human capability. However, with fast computers that can diagnose the aberrations and iterate to the desired solution, today this has become almost routine.

The STEM was invented by Manfred von Ardenne [12, 13] shortly after the invention of the TEM by Ernst Ruska [14] (see also Ruska's Nobel lecture [15]). Von Ardenne quickly abandoned the STEM because of the problem with noise. To produce a small scanned spot requires a bright electron source, and it was not until Albert Crewe used a cold field emission source that the STEM became a useful high resolution microscope [16, 17]. Within a few years the first spectacular observations of individual atoms were made [18, 19]. However, such microscopes were not commercially available and, furthermore, carbon support films just a few nm in thickness were necessary to avoid obscuring the atoms. As a result, STEM was not used for atomic resolution imaging of materials for many years, but instead was mostly used for microanalysis with probes of around 5 Å diameter, for example investigating grain boundary segregation and identifying defects such as precipitates [20]. Atomic resolution was developed primarily with the conventional microscope; for accounts of the history of atomic-resolution TEM see [21–24].

Eventually, atomic-sized probes became available on commercial STEMs, and atomic resolution images of materials could be obtained with scattered electrons (dark field images) as well as bright field images [25–27]. It became apparent that the nature of the atomic-resolution image obtained with high angle scattered electrons was very different from that obtained with the bright field detector [28, 29]. As we show later, the bright field image is related by time reversal symmetry to the conventional TEM image, and shows the usual characteristics of an interference or phase contrast image, for example, contrast reversals with lens focus and specimen thickness. However, the dark field image was collected over a wide range of scattering angles and showed strong atomic number contrast (as expected for high angle, Rutherford scattering from the atomic nuclei), but also showed the characteristics of an incoherent image. Just like a camera, atomic arrangements could clearly be seen on the screen without needing extensive image simulations [30–33]. Such images became known as high angle annular dark field (HAADF) or Z-contrast images.

The Z-contrast image also enabled atomic-resolution EELS [34, 35] by providing an unambiguous signal that allowed the probe to be located (by hand at that time) over a specific atomic column or plane. Atomic column-resolved detection of impurities at grain boundaries and measurement of impurity valence became feasible [36]. But data were still severely limited by signal-to-noise ratio, and the recent achievements of aberration correction have as much to do with overcoming that historic noise limitation as with improved resolution, as Figure 5.2 shows. With the first generation of aberration correctors [37] the STEM achieved sub-Ångstrom resolution imaging [38], and also the EELS sensitivity improved to allow the first spectroscopic identification of a single atom in a crystal with atomic resolution [39]. With the introduction of next-generation aberration correctors, STEM Z contrast imaging has surpassed the goal of 0.5 Å by imaging the dumbbells in ⟨114⟩ Ge spaced just 0.47 Å apart [40]. In fact, STEM has held the record for image resolution over the TEM for several years, which is in accordance with physics, since an incoherent image has higher resolution than a coherent phase contrast image, as first pointed out for light optics by Lord Rayleigh over a century ago [41,42].

Electron energy loss spectroscopy (EELS) has also benefited from the latest generation of aberration correctors. Aberration correction allows much higher probe currents to be focused into atomic scale probes, with the result that two-dimensional EELS mapping is achievable at atomic resolution [43–45], as shown in Figure 5.1. Figure 5.3 gives another example, the spectroscopic imaging of GaAs.

Light atoms scatter much less than heavy atoms and have usually been invisible in a Z-contrast image until aberration correction, especially in the presence of adjacent heavy atom columns. After correction, columns of oxygen atoms become just visible in between the heavier columns in $SrTiO_3$, see Figure 5.2(c). However, they are still easier to see in a phase contrast image, and another key advantage of the STEM is that it is possible to obtain both bright field phase contrast and Z-contrast images simultaneously. An excellent application of this capability to multiferroic $BiFeO_3$ thin film is shown in Figure 5.4, where polarization displacements are obtained directly from the Z-contrast image and O octahedral rotations are obtained from the bright field image [47]. Recently an annular bright field mode has been shown to give less Z-contrast but to retain most of the desirable incoherent characteristics of the Z-contrast image [48,49]. The reduced Z dependence means light elements are easier to see in the image.

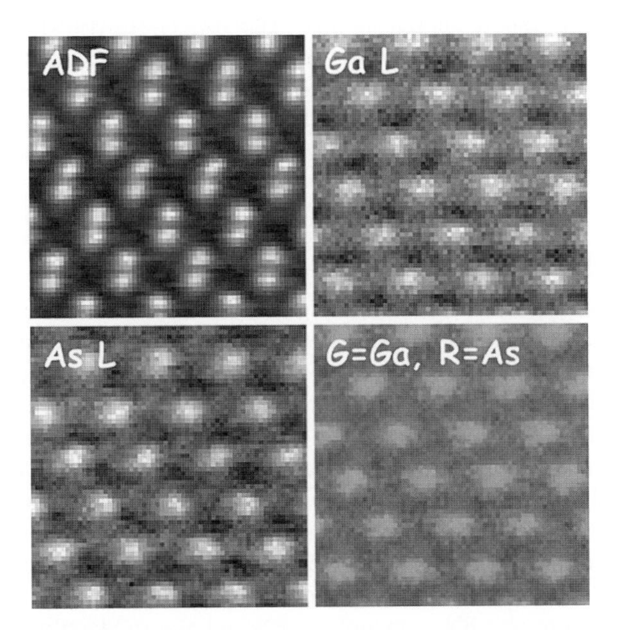

Fig. 5.3 Spectroscopic imaging of GaAs in the ⟨110⟩ projection comparing the ADF image to the Ga- and As L spectroscopic images, obtained on a Nion UltraSTEM with a 5th-order aberration corrector operating at 100 kV. Images are 64 × 64 pixels, with collection time 0.02 s/pixel and a beam current of approximately 100 pA, after noise reduction by Principle Component Analysis (PCA) [46]. This figure is reproduced in color in the color plate section.

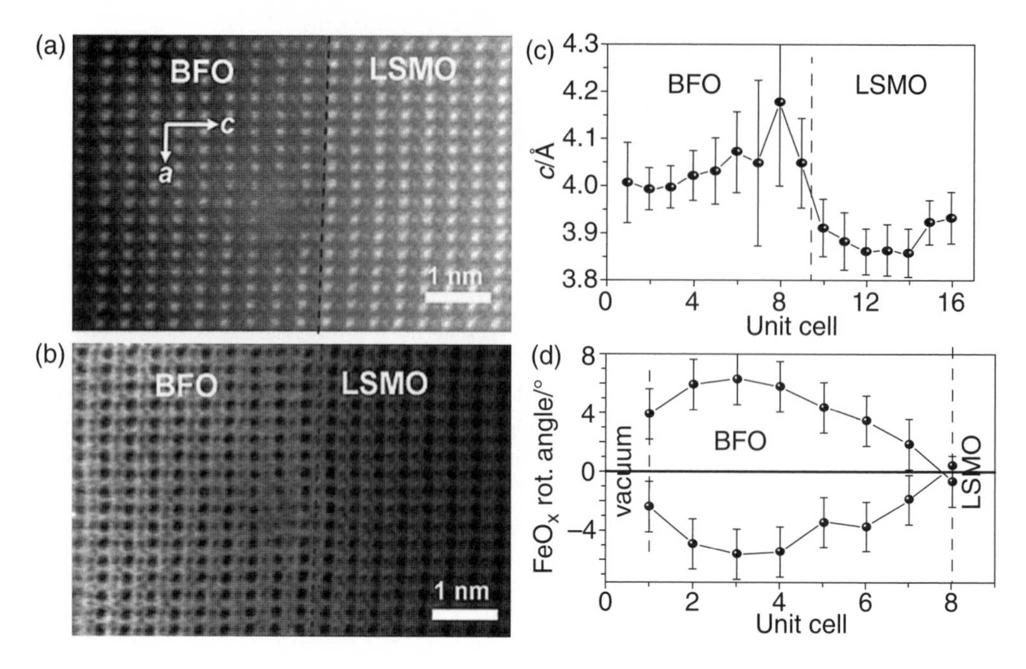

Fig. 5.4 Lattice parameters and octahedral tilts from STEM data: (a) HAADF STEM and (b) simultaneously acquired BF STEM images of the $SrTiO_3$/5 nm $La_{0.7}Sr_{0.3}MnO_3$/3.2 nm $BiFeO_3$ (BFO) ultrathin film; (c) profile of *c* lattice parameter across the interface calculated from (a); (d) profile of FeO_x octahedral tilt angles in BFO calculated from (a) and (b). Note that lattice parameter enhancement at the interface in (c) coincides with a region of suppressed octahedral tilts in (d). Adapted from [50].

5.2 STEM imaging

A schematic of the STEM is presented in Figure 5.5. The lenses before the specimen are designed to form a demagnified image of the cold field emission source on the specimen. For sufficient demagnification the probe size is limited only by the aberrations of the lenses, not by the size of the source itself. However, increasing demagnification costs current: in the limit of infinite demagnification there is zero current, and so a compromise is necessary in practice and there is always a trade-off between resolution and signal-to-noise ratio. Images are obtained serially as the probe is scanned pixel-by-pixel using a number of detectors with different geometries. The HAADF detector is normally used for Z-contrast imaging, and the detection angles are controlled by the post-specimen lenses. For higher collection efficiency a lower angle ADF detector can be used to improve signal-to-noise ratio. Such a detector has recently been used to resolve BN in monolayer BN and also to identify individual impurity atoms based solely on their image intensity [51]. There is also a (removable) bright field detector and an EELS detector, either of which can be used simultaneously with the ADF detector.

The Ronchigram camera is used for tuning the aberrations and capturing interference patterns created by the probe as a function of some parameter, e.g. probe position, depending on

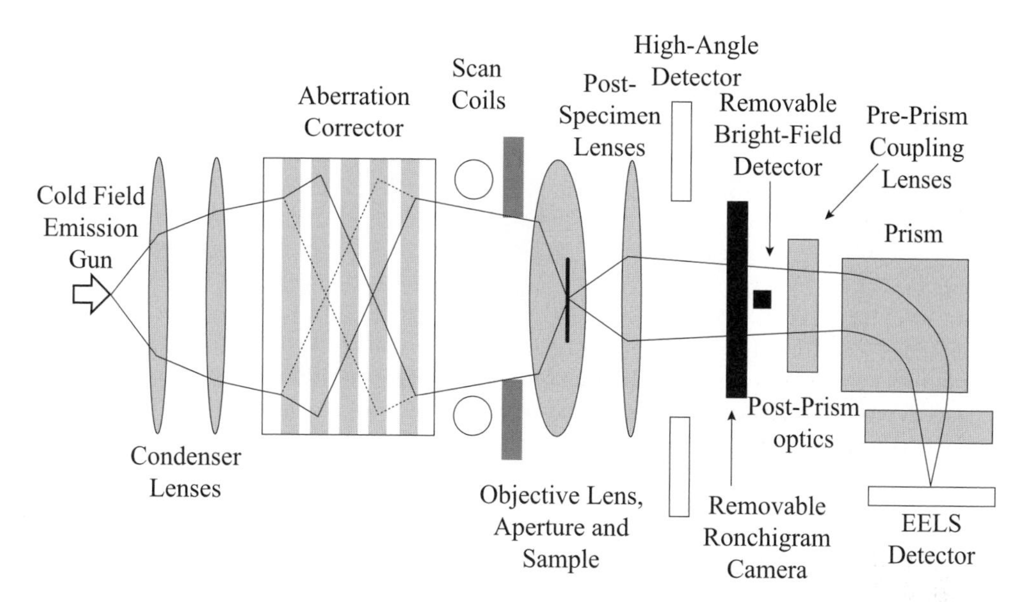

Fig. 5.5 Schematic showing the main components of a high-resolution dedicated STEM, adapted from reference [52].

the software, which are then analyzed to extract the aberration coefficients, and in turn used to adjust the aberration corrector. Note that the function of the corrector is to compensate for the intrinsic aberrations of the probe-forming optics, since they cannot be eliminated.

5.2.1 Probe formation

Let us assume for the moment that we have infinite demagnification from source to sample, in which case, in the absence of aberrations, we will have a plane wave illuminating the objective aperture in the back focal plane of the objective lens. As with light optics, the amplitude distribution in the front focal plane of the lens is given by the Fourier transform of the amplitude distribution in the back focal plane,

$$P(\mathbf{R}) = \int A(\mathbf{K}) e^{2\pi i \mathbf{K}\cdot\mathbf{R}} e^{-i\chi(\mathbf{K})} d\mathbf{K} \tag{5.1}$$

where $P(\mathbf{R})$ is the probe amplitude distribution in the specimen plane, with transverse coordinate $\mathbf{R}$, $A(\mathbf{K})$ describes the incident amplitude, which with a circular aperture is unity up to a transverse wavevector of magnitude $|\mathbf{K}| = K$, and $\chi(\mathbf{K})$ is the aberration function. By convention the aberration function is expressed as a series of coefficients in a power series, the first few rotationally symmetric terms being

$$\chi(K) = \pi\left(\Delta f \lambda K^2 + \frac{1}{2} C_3 \lambda^3 K^4 + \frac{1}{3} C_5 \lambda^5 K^6\right) \tag{5.2}$$

(a)

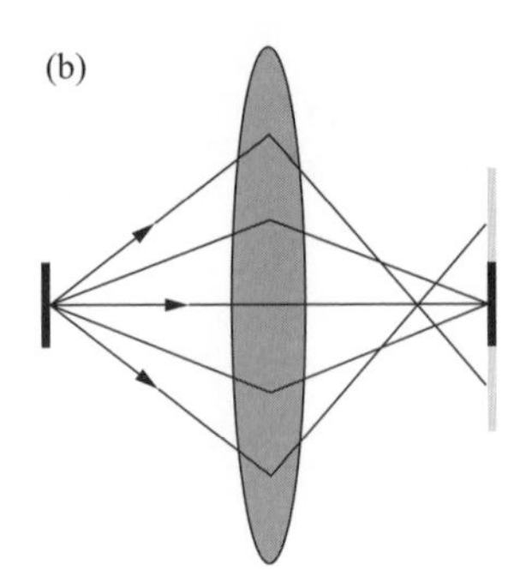

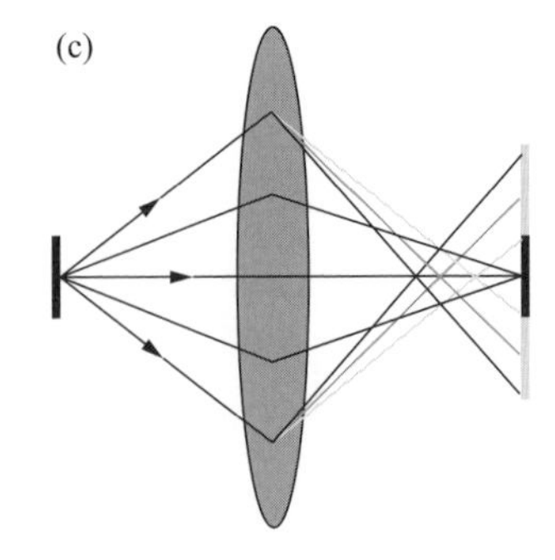

Fig. 5.6 (a) A perfect lens focuses a point source to a single image point. (b) Spherical aberration causes high angle rays to be overfocused. (c) Chromatic aberration causes rays at different energies to be focused differently. Reproduced from [52].

where Δf is lens defocus, and C_3 and C_5 are the coefficients of third- and fifth-order spherical aberration respectively. For round magnetic lenses these spherical aberration coefficients are all positive, a point first noted by [4]. Beams at high angles are focused more strongly than beams near the axis, as shown schematically in Figure 5.6. Before aberration correction it was normal to compensate for the positive C_3 by a negative defocus (weakening of the lens). In Figure 5.6 (b) this would move the focus to the right, giving less lateral spread in the specimen. Besides these geometric aberrations, chromatic aberration may also be important, especially for low accelerating voltages that are becoming more widely used with light beam sensitive materials. This causes a spread because different energies focus at different points along the optic axis (Figure 5.6 (c)). Schemes do exist for correcting chromatic aberration in TEM [53] and in STEM [54] but they are not common at present.

Figure 5.7 shows probe intensity profiles corresponding to the three situations shown in Figure 5.2, a 100 kV accelerating voltage before and after aberration correction, and a 300 kV aberration-corrected probe. All are normalized to the same total intensity through the probe-forming aperture. It is clear that along with the reduction in full width half maximum a substantial increase in peak intensity occurs, which is the reason for the improved signal-to-noise ratio. From 100 to 300 kV the wavelength decreases from 0.037 Å to 0.196 Å, with again a corresponding decrease in FWHM and an increase in peak intensity.

5.2.2 Time reversal symmetry in electron microscopy

Time reversal symmetry is a general property of elastic scattering amplitudes [55], which means that the amplitude for scattering from $\mathbf{k}$ to $\mathbf{k}'$ is equal to the amplitude for scattering in the reverse direction from $-\mathbf{k}'$ to $-\mathbf{k}$,

$$f(\mathbf{k}', \mathbf{k}) = f(-\mathbf{k}, -\mathbf{k}'). \tag{5.3}$$

Image contrast in electron microscopy is predominantly due to elastic scattering, and this fact can be used to relate the contrast in STEM and TEM. Figure 5.8 shows a schematic ray diagram for a single image point in STEM and TEM. In TEM we use a nearly parallel beam to illuminate an area of specimen and the objective lens focuses emerging waves onto a screen.

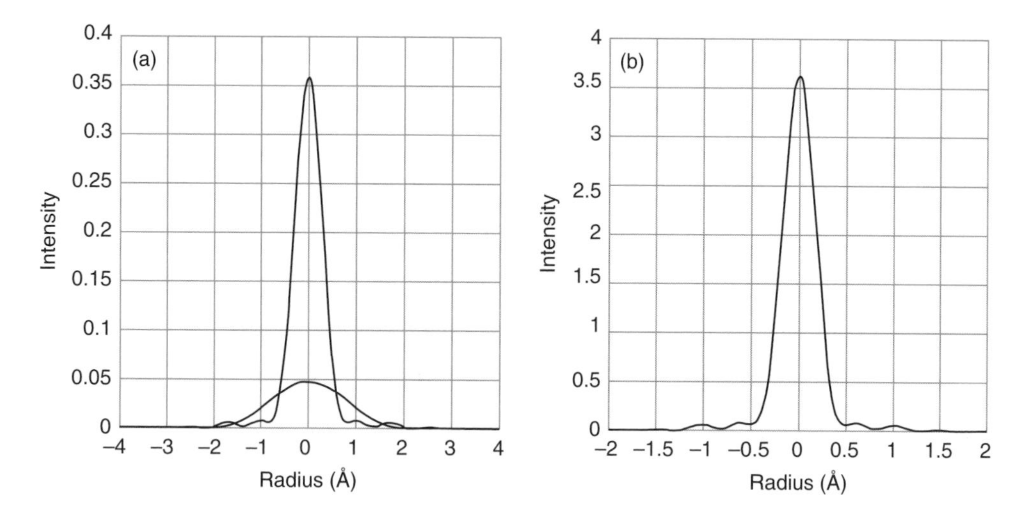

Fig. 5.7 (a) Comparison of 100 kV probes formed for typical optimum conditions before and after 3rd-order aberration correction, showing the decreased FWHM and increased peak intensity. (b) 300 kV aberration-corrected probe showing again reduced FWHM and increased peak intensity (note change of scale with respect to (a)). These probes correspond approximately to the conditions used for Figure 5.2, except that they do not include any source size broadening, chromatic aberration, or non-round aberrations.

In the STEM, the field emission source is formed into a probe focused onto the specimen, and for bright field imaging the detector is a small axial aperture. The directions of the rays are reversed from the TEM situation, and elastic scattering mechanisms will give the same image in the two cases. The STEM is a time-reversed form of the TEM. The difference is that all image points are obtained simultaneously in TEM, but sequentially in STEM through scanning the probe. This is an example of the principle of reciprocity in electron microscopy, that the amplitude scattered from a source at point A to a detector at point B is the same as for a source at B and a detector at A [56, 57]. This remains true independent of the number of optical elements involved, for example whether or not there is an aberration corrector in the two microscopes as shown in the figure, and in the presence of multiple scattering and absorptive processes.

In the case of inelastic scattering, the specimen changes from its ground state to an excited state. Reciprocity would only apply in this case if in the time-reversed situation the sample were initially to be in the same excited state, which is obviously unlikely to be the case.

5.2.3 Image simulation

The Z-contrast image has a rather deceptively simple form, in that in most circumstances the image appears to be intuitive, with heavier columns showing brighter contrast. Given the fact that electrons undergo strong multiple scattering in anything other than a very thin light sample, the apparent simplicity is quite remarkable. The origin of this convenient property is that the high angle detector does not see all the probe as it scatters through a crystal. The probe can

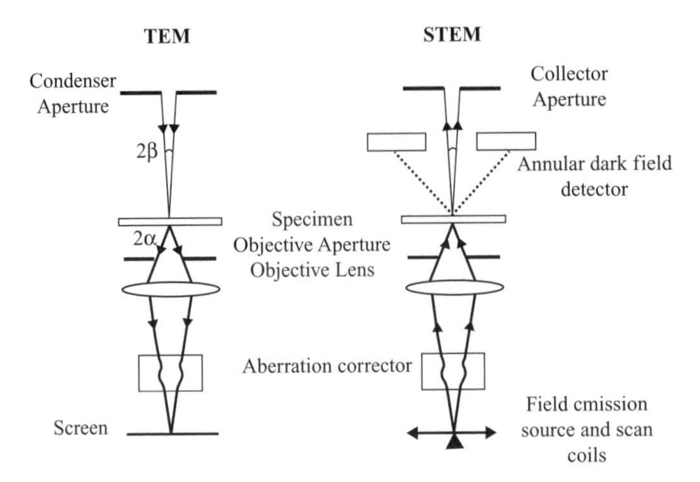

Fig. 5.8 Ray diagrams for (left) the TEM and (right) the STEM, showing the reciprocal nature of the optical pathways. The TEM image is obtained in parallel, the STEM image, pixel by pixel by scanning the probe. The STEM also provides simultaneous annular dark field (ADF) imaging. Actual microscopes have several additional lenses and the beam limiting aperture positions may differ. Reproduced from [58].

be decomposed into a set of two-dimensional Bloch states with cylindrical symmetry centered on the atomic columns. The 1s state is most highly bound, and is preferentially effective at causing high angle scattering as it is closest to the nucleus. Other more weakly bound states, such as 2p states, contribute to low angle scattering, and it is the interference between 1s and 2p Bloch states that gives the well-known thickness oscillations of bright field imaging [59]. With the 2p states ineffective at high angle scattering the Z-contrast image is effectively an image of the 1s state [28, 29].

Nevertheless, it is important to realize that precisely because it is localized in the region of strong atomic potential, it is also the most strongly absorbed, which means that the top and bottom of a column of atoms will not sample the same probe intensity. In effect only the top part of a heavy column is sampled by the beam, as shown in Figure 5.9. However, absorption is also a monotonic function of Z, and so this effect does not destroy the simple Z-dependence of the image contrast. Also, because of the depth-dependent intensity, individual dopant atoms at different depths will show different signal levels. This fact was used to locate a single La dopant atom in $CaTiO_3$ to a depth of around 100 Å, as shown in Figure 5.10.

Image simulations are more important for EELS imaging since elastic scattering can rarely be neglected. An excellent illustration of the importance of elastic scattering is seen in Figure 5.11, where O K edge EELS simulated images are shown for $LaMnO_3$ in the [010] axis (the [110] axis in pseudocubic orientation). Now there are two types of projected O sites, pure O columns and LaO columns. Because of the strong scattering when the probe is over the LaO column, the O signal is greatly depleted, in agreement with experiment.

Further complications arise with EELS imaging due to the delocalization of inelastic excitation as a result of the long-range nature of the Coulomb field. This effect tends to put a long tail on the response function, which for low energy losses can significantly reduce image

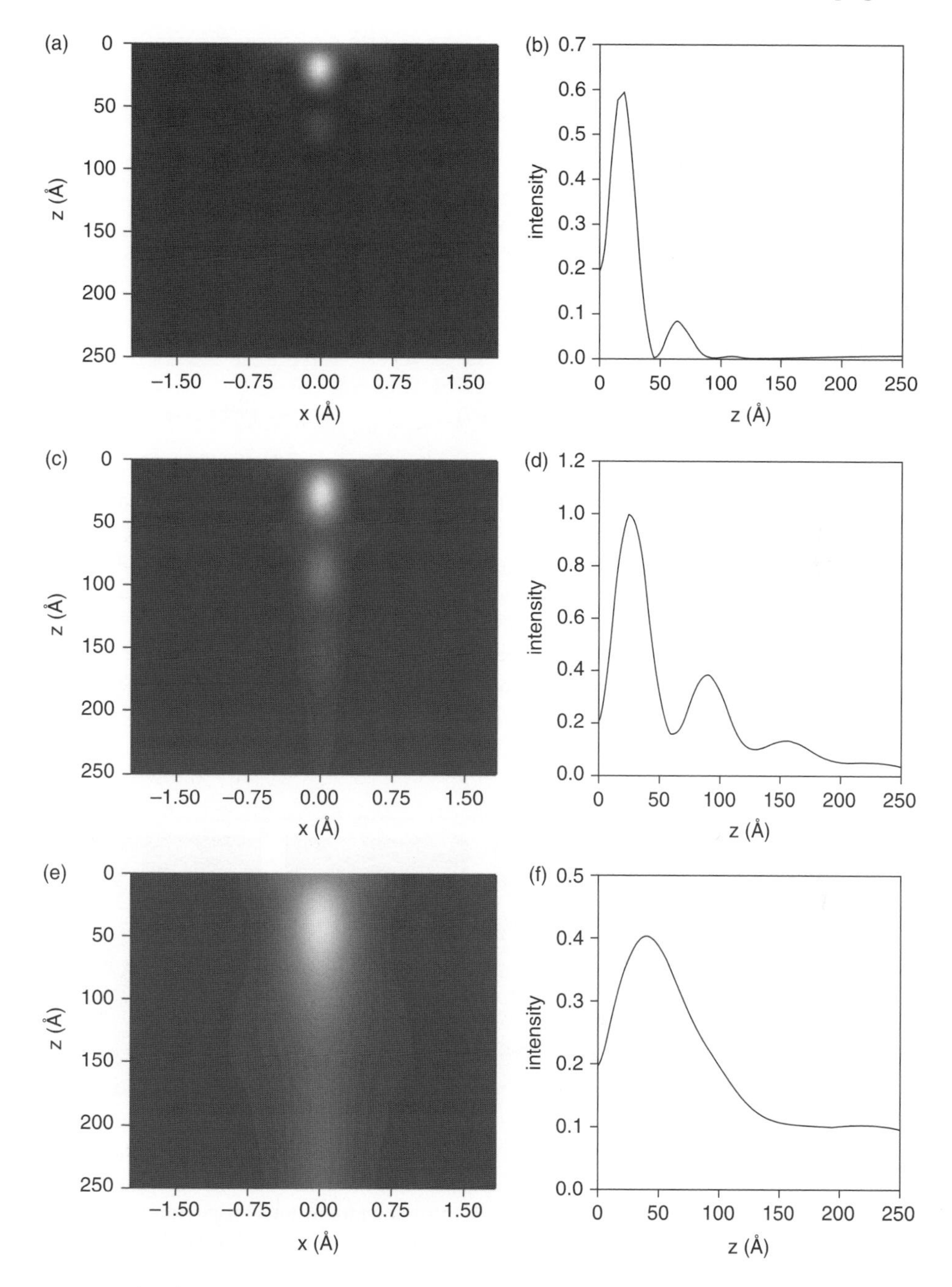

Fig. 5.9 Total probe intensity as a function of depth $LaMnO_3$ for (a) the probe located above a La column with an intensity profile along the column shown in (b). Corresponding results for Mn/O columns are shown in (c) and (d), and for O columns in (e) and (f). Different detectors sample different parts of the probe. Reproduced from [60].

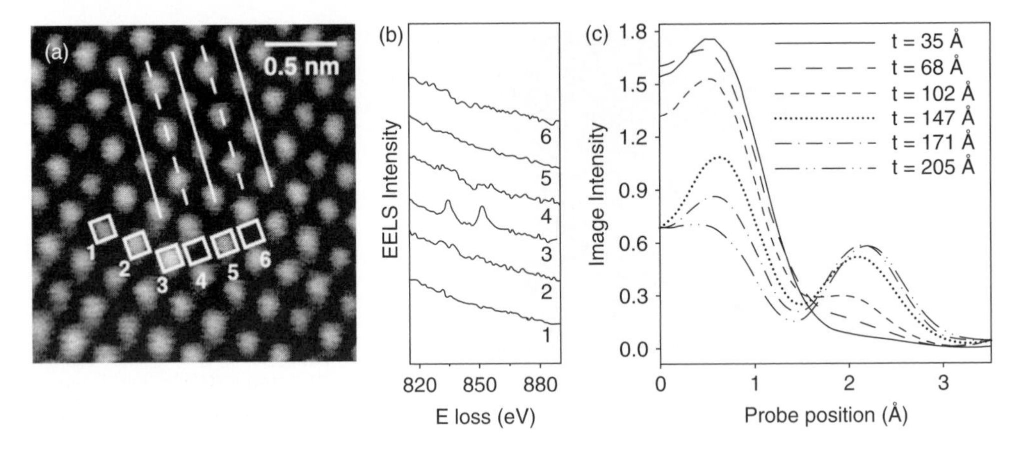

Fig. 5.10 Spectroscopic identification of an individual atom in its bulk environment by EELS. (a) Z-contrast image of $CaTiO_3$ showing traces of the CaO and TiO_2 {100} planes as solid and dashed lines respectively. A single La dopant atom in column 3 causes this column to be slightly brighter than other Ca columns, and EELS from this shows a clear La $M_{4,5}$ signal (b). Moving the probe to adjacent columns gives reduced or undetectable signals. (c) Dynamical simulation of the La $M_{4,5}$ signal, as the probe is scanned from column 3 through column 4, calculated for the La atom at different depths below the surface. The residual signal from the O column (4) is due to beam spreading and is consistent with experiment if the La atom is approximately 100 Å below the surface. Data recorded using the VG Microscope HB501UX with Nion aberration corrector, adapted from [39].

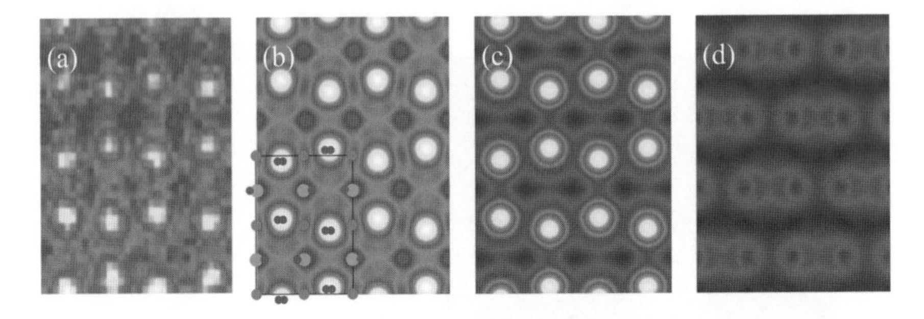

Fig. 5.11 Integrated oxygen K-shell EELS signal from $LaMnO_3$ in the [010] zone axis orientation (pseudocubic ⟨110⟩ axis). (a) Experimental image acquired on the Nion UltraSTEM operating at 60 kV. (b) Simulated image with projected structure inset. (c) Contribution to the total image from the isolated O columns. (d) Contribution to the total image from the O atoms on the La/O columns. Reproduced from [61]. This figure is reproduced in color in the color plate section.

contrast [60, 62, 63]. In addition, if there is less than 100% collection of inelastically scattered electrons nonlocal effects in the inelastic scattering potential can also lead to non-intuitive response [64, 65].

5.3 Mapping materials properties through EELS fine structure

EELS edges are a result of the excitations of inner shell electrons into unoccupied levels above the Fermi level. Therefore, the EELS fine structures ensue directly as a result of the material's unoccupied density of states and they can be used to probe optical properties (low energy excitations, less than approximately 50 eV in the spectrum) or electronic properties when core electrons are excited (energy losses above 50 eV) [66]. In oxide materials, properties such as the transition metal oxidation state can be measured from the EELS fine structure of the transition metal $L_{2,3}$ edge and the O K edge [67–70]. According to the dipole selection rule ($\Delta l = \pm 1$) the $L_{2,3}$ edge is the result of exciting the metal 2p electrons into the 3d bands, while the O K edge results from excitations of O 1s electrons into the 2p band. In perovskites, these bands are very strongly hybridized near the Fermi level, and their fine structures are strongly related to each other. For example, Figure 5.12 shows the $L_{2,3}$ spectra of a series of $La_xCa_{1-x}MnO_3$ (LCMO) compounds tilted down the pseudocubic $\langle 100 \rangle$ axis, for doping values, x, ranging from 1 down to zero. The introduction of divalent Ca atoms into the $LaMnO_3$ parent compounds provokes an increasing hole doping of the Mn 3d e_g band and a mixed Mn^{+3}/Mn^{+4} system with a valence ratio between the Mn^{+3} and the Mn^{+4} subsystems, depending on x. As x decreases, the value of the $L_{2,3}$ intensity ratio or $L_{2,3}$ ratio, which is the relative intensity ratio between the L_3 and L_2 lines (calculated after correction of the background and the continuum contribution), increases monotonically in a linear fashion. Similar behaviors have been reported in different transition metal oxides [71–75].

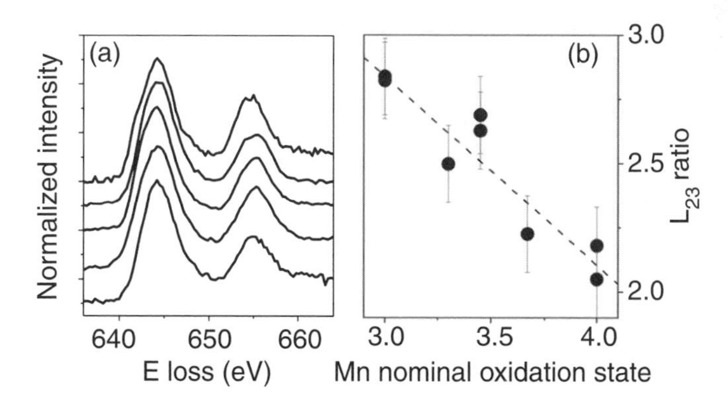

Fig. 5.12 (a) Mn $L_{2,3}$ edges (displaced vertically for clarity) for a series of $La_xCa_{1-x}MnO_3$ compounds with $x = 1, 0.7, 0.55, 0.33$ and $x = 0$ from bottom to top, respectively. The energy scale is nominal, and has been shifted and the intensity normalized so that the L_3 lines match. (b) Dependence of the L_{23} ratio with formal oxidation state for a series of LCMO compounds. The dashed line is a guide to the eye and represents a linear fit to the data. Adapted from [46].

Simultaneously, systematic changes can be observed in the O K edge [68,76–80]. Data for the LCMO series are shown in Figure 5.13, the region just above the edge onset being of special interest. It is here that the hybridization with the Mn 3d bands is more visible. The pre-peak feature exhibits very important changes when the hole doping, x, increases. As the Mn 3d e_g band becomes more and more empty due to Ca doping, the pre-peak intensity increases, and it also shifts position to lower energies when compared with the second peak of the edge. Both the pre-peak intensity and the peak separation can be quantified e.g. by fitting Gaussian functions as shown in Figure 5.13(b), and can be plotted versus the compound nominal oxidation states. Like the L_{23} ratio, they both exhibit linear behaviors (Figures 5.13(c, d)) and can also be used to quantify Mn oxidation states in these perovskite compounds [46].

The data shown in Figures 5.12 and 5.13 correspond to average illumination of LCMO specimens tilted down the pseudocubic ⟨100⟩ axis and demonstrate how the fine structures can be very useful for probing quantities such as transition metal oxidation states. However, spherical aberration correction allows probing of individual atomic columns with great detail. This capability is of major importance when it comes to understanding the properties of

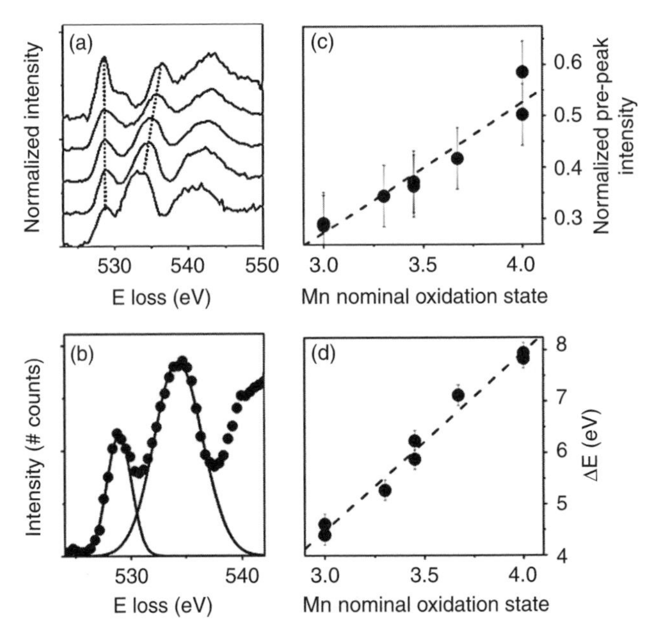

Fig. 5.13 (a) O K edges for a series of LCMO compounds with x = 1, 0.7, 0.55, 0.33 and x = 0 from bottom to top. The energy scale has been shifted so that the pre-peaks are aligned, and the intensity normalized. The spectra have been displaced vertically for clarity. (b) O K EEL spectrum showing the Gaussian curves used to extract peak intensity and position. (c) Normalized pre-peak intensity versus nominal oxidation state for the series of LCMO samples. The dashed line is a linear fit to the data. (d) Energy separation (calculated as the difference between positions of the second peak and the pre-peak) as a function of the Mn nominal oxidation state for the set of samples. Adapted from [46].

compounds such as $LaMnO_3$ (LMO), with an orthorhombic crystal structure characterized by a Jahn–Teller (JT) distortion, which lifts the degeneracy of the 3d bands (shown in Figure 5.14(a) as reported by [81]). The O octahedra in LMO are strongly distorted, and the O atoms sitting in the apex of the octahedron show a bond distance to the Mn atoms quite different from the equatorial ones. Simulations of the O K edges [80] shown in Figure 5.14(b) predict small but noticeable changes in the fine structures for both types of atoms (apical O in black, equatorial O in blue). These variations are detectable experimentally. Figure 5.14(c) shows averaged spectra obtained from different O positions on a LMO sample tilted down the pseudocubic ⟨100⟩ axis using the same color scheme. A small shift in the spectral weight of the pre-peak is detected in the blue spectrum, characteristic of a strong contribution of equatorial O atoms to the spectrum. Because of dynamical scattering, O signal contributions from individual atomic columns cannot be easily separated and the spectra in Figure 5.14 are the result of a partial averaging of different neighboring columns [46]. Nevertheless, a clear change in the pre-peak energy position is observed, well above the noise, pointing towards the possibility of distinguishing O atoms with different crystalline environments within a material.

While this possibility has yet to be examined in detail, two-dimensional maps of the peak separation parameter, ΔE, in LMO show a contrast that is commensurate with the crystal lattice, as shown in Figure 5.15 (this map derives from the EEL spectrum image used for Figure 5.11(a)). These experiments have been carried out so far at room temperature. But if, as the theoretical simulations suggest, the contrast in these ΔE maps is associated with the inequivalent O positions, these results pave the way for future experiments on the direct imaging of phenomena such as orbital ordering at low temperatures, effects of strains in JT distorted systems, interfaces, etc.

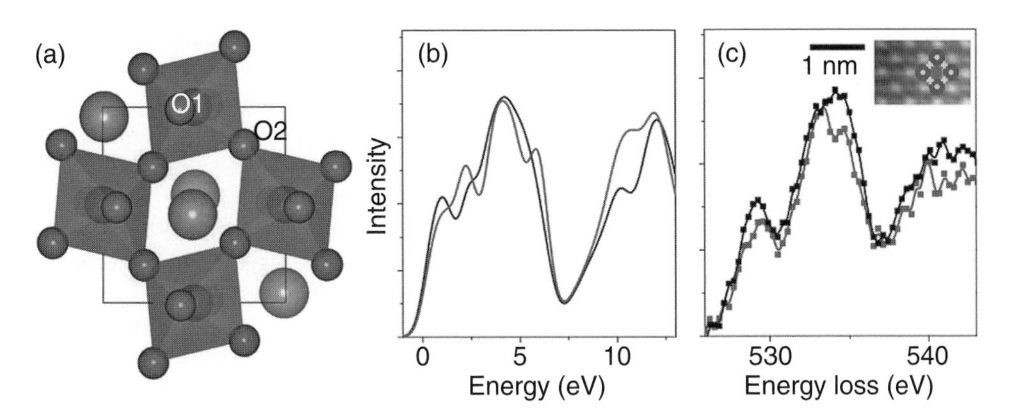

Fig. 5.14 (a) Sketch of the LMO structure, produced from the structural data reported by [81]. (b) Simulated O K EEL spectra for the O1 (black) and O2 (blue) species, adapted from [46]. (c) Experimental EEL spectra from different positions (electron beam on top of a MnO column in black and on top of a La column in blue), extracted from a linescan along the pseudocubic ⟨110⟩ direction of LMO (dataset in [46]). Principal component analysis was used to remove random noise. The LMO pseudocubic unit cell is shown on top of a Z-contrast image in the inset (La = red, Mn = blue, O = yellow). This figure is reproduced in color in the color plate section.

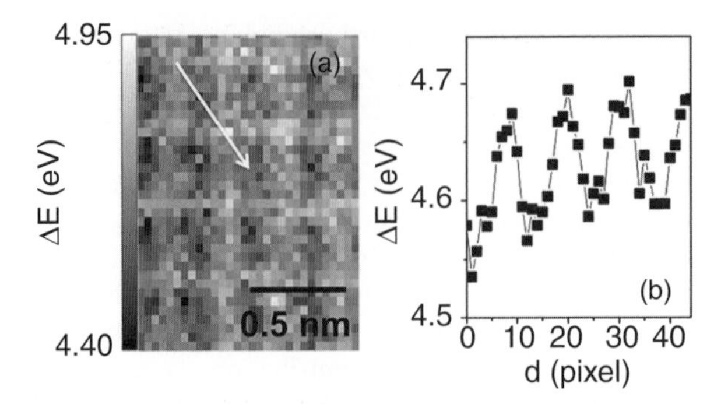

Fig. 5.15 (a) Peak separation map for the spectrum image giving rise to the O map in Figure 5.11(a), after principal component analysis was applied to the raw dataset for random noise removal. (b) Intensity profile for the image in (a), along the direction marked with a white arrow.

5.4 Applications: interfaces in manganite/cuprate heterostructures

Complex oxides with perovskite structure exhibit a plethora of different ground states including ferromagnetism, superconductivity, metallic, insulating, ferroelectric, etc. which are delicately sensitive to doping. A controlled modification of the carrier density can drive the system to the vicinity of a metal–insulator transition where the coupling between spin, charge, orbital, and lattice properties has yielded a variety of interesting behaviors in bulk samples. The ability to grow epitaxial heterostructures with highly perfect interfaces has opened the way to modification of doping by charge transfer [2, 82, 83]. The ensuing modification of the bonding at interfaces yields novel phenomena (absent in bulk samples) that challenge our understanding of interface physics. In the last decade, much interest has been focused on ferromagnetic/superconducting (F/S) heterostructures, where a high critical temperature superconductor (HTCS) and a CMR manganite are combined [84–92], such as $YBa_2Cu_3O_{7-x}/La_{0.7}Ca_{0.3}MnO_3$ (YBCO/LCMO). Superconductivity in these heterostructures is significantly depressed in the presence of the magnetic material [88–92], as shown in Figure 5.16(a), pointing to a strong F/S interplay. Early studies suggested the injection of spin polarized carriers and a proximity effect. More recent reports have suggested a transfer of charge between the ferromagnet and the superconductor [93–95]. Moreover, the reduced magnetization at the interfaces measured by polarized neutron reflectometry [96] seems to support this picture. Exploring charge transfer and its associated length scale is of primary importance to the understanding of the electronic properties of these complex interfaces with antagonistic long-range orderings. Harnessing such phenomena relies on probes capable of providing a comprehensive structural, chemical, and electronic characterization at the atomic scale. The combination of Z-contrast imaging with quantitative EELS is a very useful tool towards this aim, so we will dedicate this section to a review of this study in detail.

For a study of this sort, high quality samples are needed, since interface disorder (chemical interdiffusion or surface roughness) can directly affect the properties of interest. Specimens

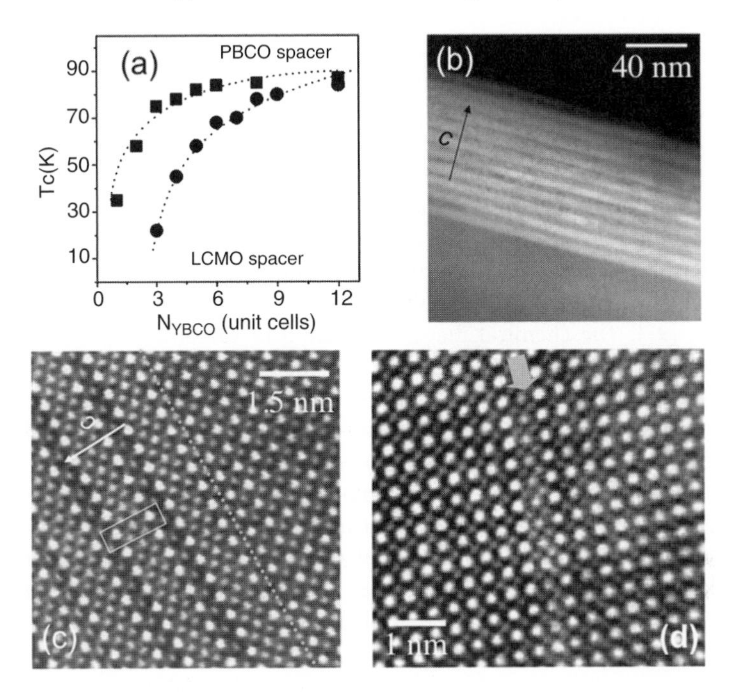

Fig. 5.16 (a) Dependence of the Tc with YBCO layer thickness for a set of superlattices with a magnetic spacer (LCMO, circles) and a non-magnetic spacer (PBCO, squares), both spacers being 6 nm thick. (b) Low magnification Z-contrast image of a nominally $[\mathrm{YBCO}_{3\mathrm{u.c.}}/\mathrm{LCMO}_{15\mathrm{u.c.}}]_{100\mathrm{nm}}$ superlattice. (c) High resolution Z-contrast image of a typical YBCO/LCMO interface. The dotted line marks the interface position. A YBCO unit cell is marked with a rectangle. (d) High resolution Z-contrast image of a $[\mathrm{YBCO}_{1\mathrm{u.c.}}/\mathrm{LCMO}_{15\mathrm{u.c.}}]_{100\mathrm{nm}}$ superlattice. The YBCO layer has been marked with an arrow. Adapted from [52, 95, 97].

grown by high oxygen pressure sputtering meet these requirements, with interdiffusion below 5% in the first unit-cell and interface step disorder below one unit-cell over micron size length scales [92,97]. Electron microscopy observations of these interfaces show layers that are flat and continuous over long lateral distances (Figure 5.16(b)) and no major secondary phases [97]. High magnification Z-contrast images reveal coherent, defect-free interfaces between YBCO and LCMO (Figures 5.16 (c) and (d)). A significant detail is observed in these images: the CuO chains of YBCO, which reveal themselves as planes with darker contrast in the YBCO structure, are not observed at the interface with LCMO in these samples. This is a very important structural feature, because the lack of CuO chains has been shown to affect hybridization between Cu and Mn orbitals through this interface [93]. While other interface terminations and major disorder have been observed in samples grown by other methods, such as pulsed laser deposition (PLD) [98], in our sputtered samples the statistically most significant termination seems to be the one in Figure 5.16. Unfortunately, the interpretation of these Z-contrast images at first sight is not unique. While the absence of an interfacial dark CuO chain plane is evident, there are two possible stacking sequences that could give rise to a contrast similar to the one observed. The first would be a LaO plane from the LCMO facing a CuO_2 plane of YBCO, while the second would be a BaO plane from the superconductor facing a MnO_2 plane from the manganite. At this point, chemically sensitive EELS is the most straightforward tool. Figure 5.17 shows a map of a LCMO/YBCO trilayer, with the simultaneously acquired ADF (Figure 5.17(a)) and an overlay of the Mn $L_{2,3}$ (red), Ba $M_{4,5}$ (blue) and La $M_{4,5}$ (green) derived atomic resolution elemental maps. The RGB image clearly demonstrates that at both the top and bottom interfaces the last plane of the LCMO is a MnO_2 plane, which faces a BaO plane from the superconductor, as shown in Figure 5.17(c). Other terminations such as a LaO termination in the LCMO are also observed, but only occasionally in these sputter grown samples [97]. The chemical maps of Figure 5.17 reveal a surprising feature of the growth of these interfaces: both interfaces of the same layer have the same termination. This is surprising, since for a ABABAB... sequence of atomic planes of binary oxides in one layer, one would expect a different termination in each interface when an integer number of unit cells are deposited. However, our analysis shows unambiguously that both interfaces of the cuprate (manganite) layers have a BaO (MnO_2) termination. This fact has been observed in other samples combining different materials grown by the same technique [99], and seems to indicate that interface chemistry and bonding and/or surface energy play an important role in determining growth properties. This suggests that the high temperature/high pressure thermalized growth is more close to thermodynamic equilibrium than other growth techniques such as pulsed laser deposition.

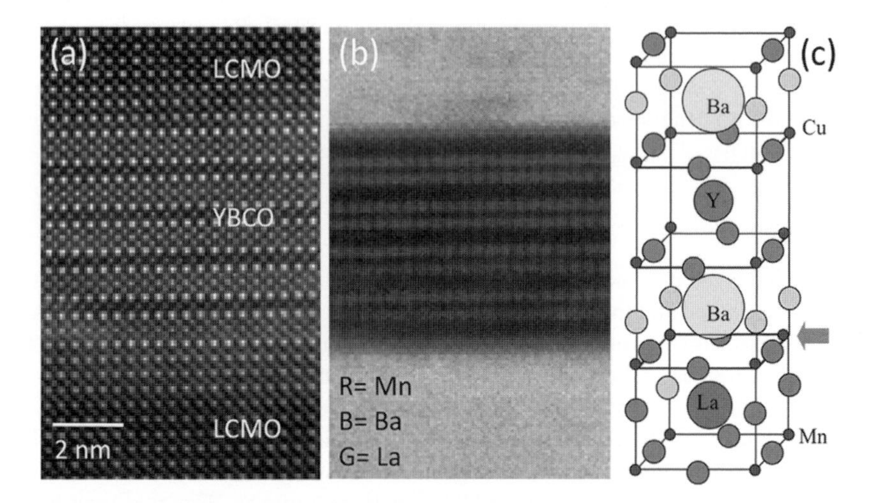

Fig. 5.17 (a) ADF signal collected simultaneously with a spectrum image in a LCMO/YBCO/LCMO trilayer. (b) False color image where three atomic resolution elemental images have been overlayed: a Mn $L_{2,3}$ image in red, a Ba $M_{4,5}$ image in blue and a La $M_{4,5}$ image in green, for the same area as the Z-contrast in (a). The EELS data were acquired in the Nion UltraSTEM column and processed with principal component analysis in order to remove random noise. (c) Sketch of the observed interface structure. An arrow marks the interface MnO_2 plane, facing a BaO plane from the superconductor. This figure is reproduced in color in the color plate section.

The implications of such interface structure, at least in a hand waving fashion, would be that the first, incomplete, YBCO unit-cell by the interface might show some degree of electron doping, just due to the lack of interface CuO chains. Their absence would prevent holes from being doped into the CuO_2 planes, and hence, superconductivity would not arise. As a matter of fact, YBCO/LCMO superlattices where the superconducting layers are nominally one YBCO unit-cell thick do not exhibit a superconducting transition at all [92]. However, a more subtle electronic mechanism develops: in YBCO/LCMO superlattices an average hole doping of the LCMO layer along with an unexpected electron doping of YBCO near the interfaces was detected in 2005 by means of EELS [95]. EELS in cuprates is also a very useful tool for measuring the number of holes in the O 2p bands. Early work by N. Browning and co-workers [100] showed how this quantity correlates directly with the intensity of the pre-peak at the O K edge onset, as shown in Figure 5.18 for a series of deoxygenated YBCO samples. Browning and co-workers measured the normalized pre-peak intensity using a three-Gaussian-window fit, as depicted in Figure 5.18. The thus measured intensity monotonically decreases when the O content per unit-cell is decreased.

The value measured for the normalized pre-peak intensity might vary slightly depending on the fitting procedure. Here, we will only use a two-Gaussian window method, as shown in Figure 5.19, for consistency with the fit within the LCMO layer described in Figure 5.13. The first fitting interval is not centered with the pre-peak itself, but shifted to slightly lower energies, thus avoiding major contribution from the spectral weight of the upper Hubbard band and hence replicating to some extent the rationale underlying Browning's three fitting window procedure (nevertheless, our procedure may somewhat overestimate the peak's area). This way, the normalized pre-peak intensity value, obtained by normalizing the area of the first Gaussian curve to the second, is 0.12 for this fully oxygenated compound. This value is consistent with the analysis of Browning and co-workers, so we will use the pre-peak intensities in Figure 5.18 (b) to compare and quantify one of our EELS linescans across a LCMO/YBCO/LCMO trilayer.

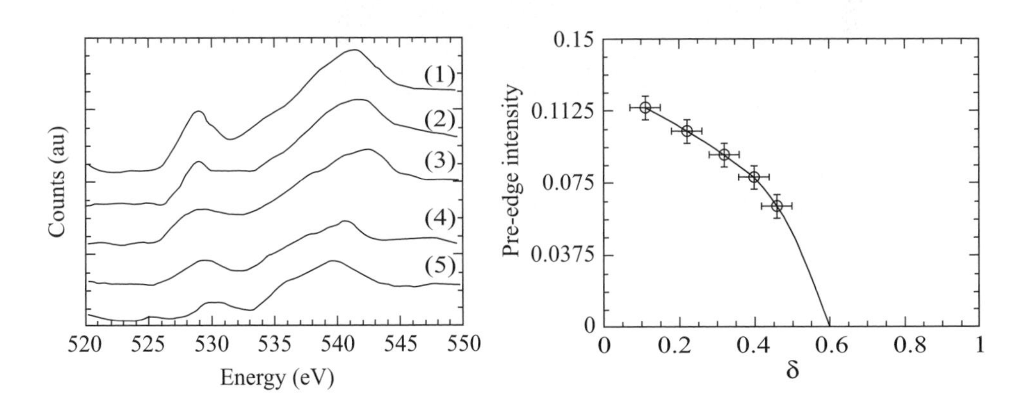

Fig. 5.18 (Left) O K edge spectra obtained from a series of YBCO samples with oxygen contents of 7-δ 6.89, 6.78, 6.67, 6.60, and 6.54 respectively, from top to bottom. (Right) Normalized pre-peak intensity using a three-Gaussian fit as a function of O concentration for the aforementioned series of YBCO samples. Reprinted from [100], copyright 1992, with permission from Elsevier.

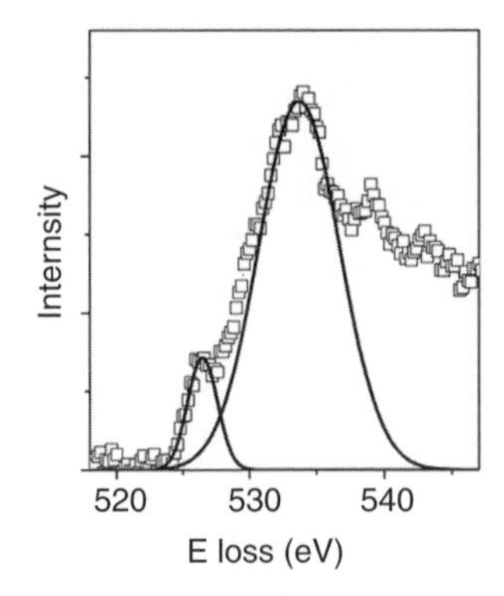

Fig. 5.19 O K edge of an optimally doped YBCO sample obtained with the electron beam perpendicular to the *c* axis. The Gaussian curves are typical fits to the pre-peak and the main peak respectively.

Note that in general the choice of suitable Gaussians depends on the system, and the Browning values would usually need to be scaled to some calibration.

Figure 5.20(a) shows an ADF image of a LCMO/YBCO/LCMO trilayer with a nominal superconducting layer thickness of five unit-cells. The CuO chains are missing at both interfaces, so the interface YBCO unit-cells are, again, incomplete. An arrow shows the direction along which the linescan in Figure 5.20(b) was acquired. In this scan, the O K, Mn $L_{2,3}$, Ba $M_{4,5}$ and La $M_{4,5}$ edges are appreciated.

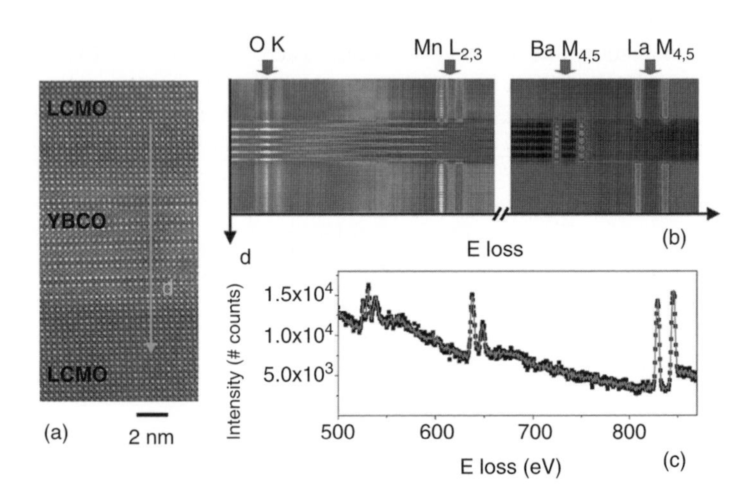

Fig. 5.20 High resolution Z-contrast image of a LCMO/YBCO/LCMO trilayer, obtained at 100 kV. (b) EELS linescan acquired along the direction marked with an arrow in (a). PCA has been used to remove random noise. (c) Sample spectrum extracted from the linescan in (b), acquisition time is 2 seconds per spectrum. The data points are a raw spectrum while the red line is the same dataset after PCA. This figure is reproduced in color in the color plate section.

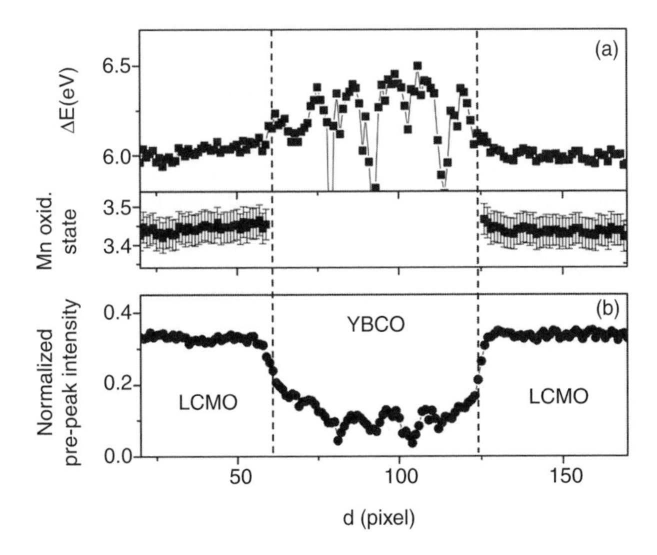

Fig. 5.21 (a) Top: For the O K edge, peak separation parameter, measured from the linescan in Figure 5.20(b). Bottom: Mn oxidation state in the LCMO layers, derived from the data in (a). (b) For the same linescan, O K edge pre-peak normalized integrated intensity.

The LCMO layers show on average a level of hole doping slightly higher than expected. Figure 5.21 shows the O K peak separation, ΔE, (a) and the normalized pre-peak intensity (b) extracted from part of the EELS linescan in Figure 5.20. Vertical dotted lines highlight the approximate positions of the interfaces. The lower panel of Figure 5.21(a) depicts, for the LCMO layers, the ensuing Mn oxidation state as a function of distance from the interface obtained using the calibration in Figure 5.13. While the nominal oxidation state for Mn in the LCMO layers should be +3.3 due to the chemical composition, a slightly higher value is obtained. Note that at the precise interface MnO_2 plane the measurement results from an average of both materials due to beam broadening (dynamical diffraction). This sets a limit to the spatial resolution that can be interpreted as real charge transfer. An indication of the spatial resolution achievable is shown by the oscillations ΔE and the normalized pre-peak intensity seen in the center of the YBCO layer. The normalized pre-peak intensity shows an oscillation with the periodicity of the unit cell: higher on the CuO_2 planes, lower on the CuO chains [52, 101], due to the fact that the holes responsible for superconductivity in YBCO arise in the planes due to transfer of electrons to the chains.

From these O K edge measurements it is therefore possible to spatially map the doping level of YBCO, using the data in Figure 5.18 as a reference. For this YBCO layer the pre-peak intensity value on top of the CuO_2 planes is, approximately, 0.12, while on the chains the values measured are around 0.06. Comparing these values to the bulk presented in Figure 5.19, the YBCO layer shows an average depletion of holes. This is especially so if we have in mind that the bulk reference value was obtained from illumination of a complete YBCO unit-cell and hence it is already averaging chains and planes. A straight comparison between the mean value of the pre-peak intensity in our YBCO layer and the Browning data [100], would yield an average doping level comparable to that of bulk $YBa_2Cu_3O_{7-x}$ with an x value around 0.20 O atoms per unit-cell. This finding evidences a noticeable level of electron doping in the YBCO layer, which is presumably due to the difference in work functions [52, 94, 95].

In YBCO/LCMO heterostructures with thicker YBCO layers, the hole density in the superconductor was found to be strongly depressed within 3 nm of the interface, while the hole density in the ferromagnet is correspondingly enhanced [52]. Figure 5.22 shows two average profiles obtained from a similar superlattice. Figure 5.22(a) shows the Mn oxidation state across a complete LCMO layer (i.e. from interface to interface), obtained from the L_{23} ratio. Note that each trace is a compilation of many individual data points, each of which was obtained by scanning a beam parallel to the interface by 8 nm [52]. Figure 5.24(b) shows the pre-peak intensity as a function of distance to the YBCO/LCMO interface. Again, every point is the result of analyzing many O K edges at every given position from an interface. This, procedure was necessary to obtain sufficient statistics to perform the analysis while avoiding oxygen depletion

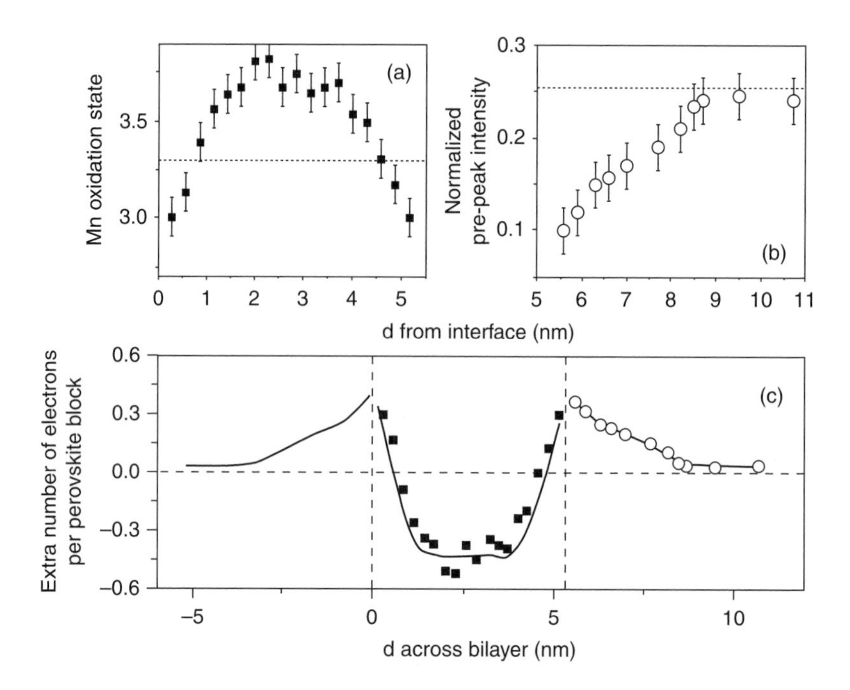

Fig. 5.22 (a) Average Mn oxidation state across a LCMO layer of a nominal $[YBCO_{12\ u.c.}/LCMO_{15\ u.c.}]_{100\ nm}$ superlattice. A horizontal dashed line marks the approximate bulk doping levels expected for the nominal compound stoichiometry. (b) O K edge pre-peak intensity as a function of distance from the interface, which is approximately at the left end of the plot. A horizontal dashed line marks the bulk doping level calculated using similar Gaussian fits. Traces adapted from [52]. (c) Schematic figure estimating the extra electrons per perovskite block (i.e. per cube volume) across the average bilayer, calculated from the average data in (a, b). Solid lines are a guide to the eye. Vertical dashed lines mark the interfaces and a horizontal dashed line shows the zero doping level (i.e. bulk nominal doping level). Data acquired in the aberration corrected VG Microscope HB501UX.

through beam damage. Each point in Figure 5.22(a, b) is therefore a representative value for that distance from the LCMO/YBCO interface, which was established by counting atomic planes in the Z-contrast image. While the middle of the LCMO layers show a Mn oxidation state higher than expected (i.e. hole doping) along with an abrupt decrease within a nm of the interfaces, the first few YBCO layers by the interface show a strong electron enrichment. It is immediately clear that the charge redistribution is extensive, much greater than any effect due to beam broadening. An attempt to convert these data into extra electrons per perovskite unit block gives the sketch of Figure 5.22(c). For the LCMO this quantity is calculated as the difference between the nominal valence, +3.3 and the measured one, in Figure 5.22(a). For the YBCO, we have used a different Gaussian fit to that shown in Figure 5.18, with appropriate scaling of the Browning data. For this dataset the fully oxygenated YBCO gives a pre-peak intensity of 0.25 in the center of the YBCO layer, and the plot can be converted into hole concentrations, as shown in Figure 5.22(c). A line reflecting the YBCO data is duplicated either side of the LCMO layer to appreciate the symmetry of the redistribution, and also the fact that the total areas under the YBCO lines (net electron doping) are consistent with the area under the LCMO data (net hole doping). It is noticeable that the two curves in the two materials, obtained with different calibrations, give a continuous carrier concentration across the interface. This gives confidence that the results are indeed indicative of a real charge transfer phenomenon.

Figure 5.23 (a,b) shows the summary of a single linescan acquired while moving the electron beam across three different bilayers in a nominal [$YBCO_{10\ u.c.}$/$LCMO_{15\ u.c.}$] superlattice.

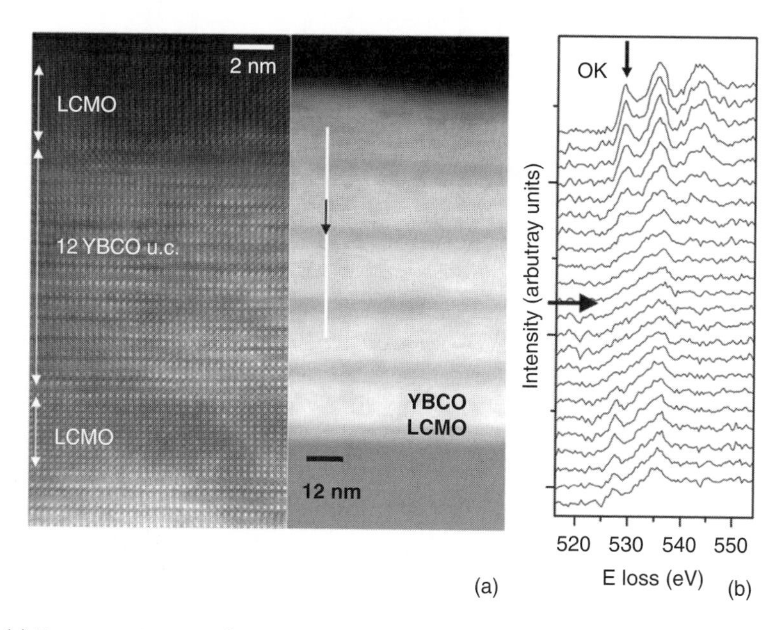

Fig. 5.23 (a) Z-contrast images of a nominal [$YBCO_{10\ u.c.}$/$LCMO_{15\ u.c.}$]$_{100\ nm}$ superlattice. (b) O K edge spectra extracted from an EELS linescan acquired while displacing the electron beam along the direction of the line in (a). Data acquired in the aberration corrected VG Microscope HB501UX.

Figure 5.23(a) shows a high resolution Z-contrast image of one of the YBCO layers (left) along with a low magnification image (right) of the sample showing the region along which the dataset was acquired. Interface steps one unit-cell high are observed giving rise to minor local fluctuations in layer thickness. These steps do not compromise the overall high crystalline quality of the sample. Figure 5.23(b) shows a fraction of the set of O K edges obtained when moving the electron beam from the center of one YBCO layer across the interface all the way up to the middle of a LCMO layer. In this case, every spectrum in the raw linescan has been averaged with its nearest neighbors to decrease random noise. The pre-peak feature (marked with an arrow) decreases dramatically near the interface (the interface approximate position has been highlighted with an arrow). Figure 5.24 shows the pre-peak intensity along the whole linescan (bottom). It is very clear that it decreases dramatically within 3 nm of the interfaces for all of the individual interfaces along the scan. The data are noisy due to the lack of lateral averaging, but the trends observed are similar to the average profile shown in Figure 5.22.

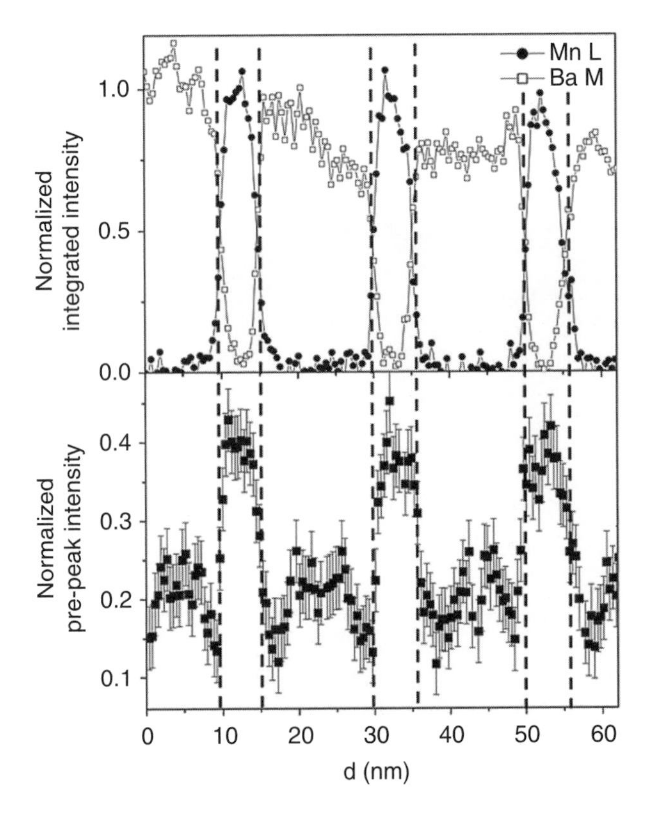

Fig. 5.24 Top panel: integrated intensities for the Ba M edge and the Mn L edge for the linescan in Figure 5.22. The approximate positions of the interfaces have been marked with vertical dashed lines. Bottom panel: Normalized pre-peak intensity for the whole linescan.

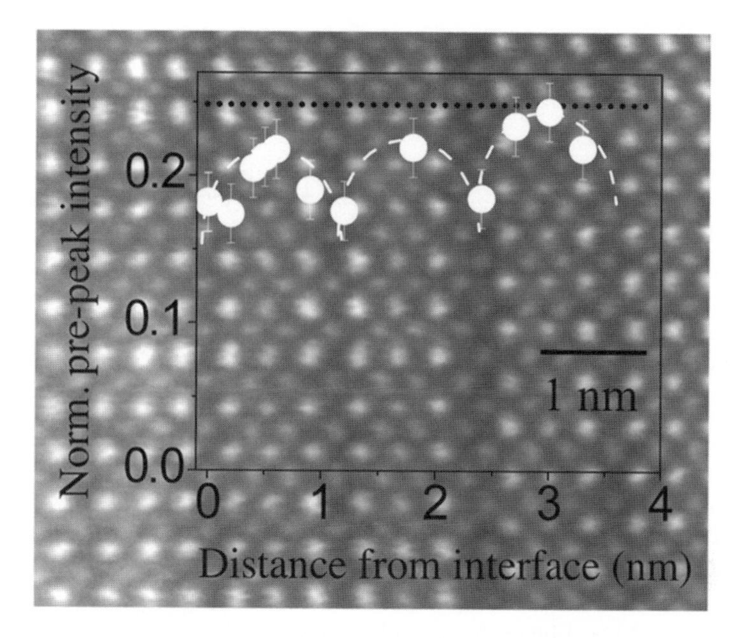

Fig. 5.25 Normalized pre-peak intensity within the YBCO in a YBCO/PBCO interface, as a function of distance to the interface. A horizontal dotted line shows the bulk doping level. The background image is a representative image to set the scale for the data points, which are average measurements, adapted from [52].

These charge transfer effects are clearly extensive, and lead directly to the altered superconducting and magnetic properties observed. Specifically, for the YBCO layers in these superlattices, Tc values are significantly below those found in YBCO/$PrBa_2Cu_3O_{7-x}$ (PBCO) multilayers, as shown in Figure 5.16(a). A plot of the normalized pre-peak intensity in the YBCO layer of a YBCO/PBCO superlattice is shown in Figure 5.25, and shows only minor oscillations between chains and planes, without the extensive hole depletion zone.

Similarly, the increased Mn oxidation state in the center of the LCMO layer correlates with the value of the saturation magnetization in these $[\mathrm{YBCO}_{n\,\mathrm{u.c.}}/\mathrm{LCMO}_{15\,\mathrm{u.c.}}]_{100\,\mathrm{nm}}$ samples, which decreases when the YBCO layer thickness increases, as shown in Figure 5.26 (saturation is reached for YBCO layer thicknesses around 5 unit-cells). This could also be an effect of the transfer of electrons into the YBCO interface layers.

Further support for the charge transfer scenario comes from the study of a set of samples with a constant superconductor thickness (of 10 nm) and a variable LCMO layer thickness. When the LCMO thickness increases, the Mn oxidation state at the center of the layer approaches its nominal value (see Figure 5.27), as does the saturation magnetization. Still, the interfaces are electronically inhomogeneous and show the fingerprints of the charge transfer effect. Interestingly, theoretical calculations where the chemical potentials in superconductor–normal–superconductor junctions are forced to match at the interfaces show qualitatively the same behavior in the dependence of the charge deviation across the layers with the

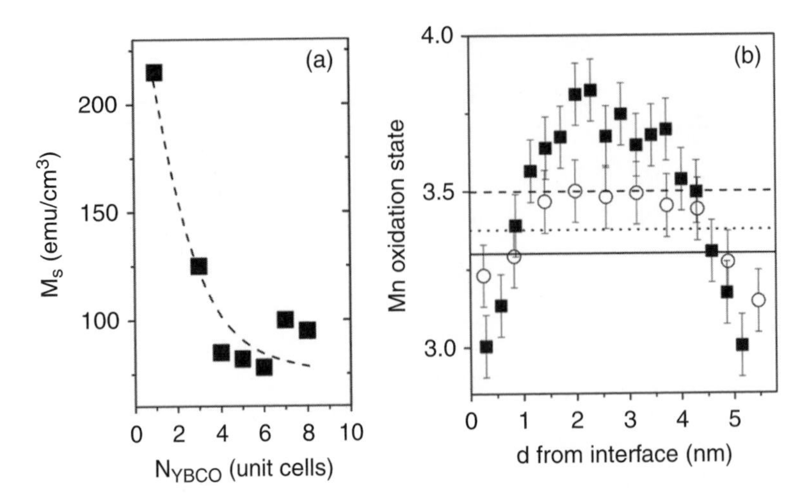

Fig. 5.26 (a) Dependence of the saturation magnetization with YBCO layer thickness for a set of $[YBCO_{n u.c.}/LCMO_{15 u.c.}]_{100 nm}$ superlattices with a fixed magnetic spacer thickness of 6 nm. (b) Average Mn oxidation state for nominal $n = 1$ (circles) and $n = 12$ (squares) superlattices. The horizontal black line marks the +3.3 oxidation state expected for the nominal stoichiometry. The dotted (dashed) line marks the average valence value of +3.37 (+3.50) measured across these LCMO layers, adapted from [52].

spacer layer thickness [102, 103]. These calculations prove that these effects originate in the mismatch of the Fermi energies across the interface, and are virtually independent of temperature.

EELS analysis of the oxidation state of Mn evidences that, besides the charge transfer, there is an extensive charge redistribution within the manganite layers. While on average there is an increased number of holes in the manganites (as expected from the transfer of electrons to the YBCO) electrons build up at the interface giving rise to a dipole moment extending over a 1–2 nm length scale. This is a puzzling result since, from elementary semiconductor junction models, one would expect the manganite interface to be depleted of electrons. This interfacial dipole may be indicative of band discontinuities or charge trapping at interface states, or may appear to cancel polar discontinuities at the interface. In summary, it is clear that the modified doping at the interface resulting from the difference in electrochemical potentials has to be taken into account to explain the superconductivity and magnetism depression at manganite/cuprate interfaces. However, experiments have shown that superconductivity is suppressed over much longer length scales than the three unit-cells over which charge transfer occurs. Moreover, in YBCO/LCMO superlattices the critical temperature is depressed over a much longer scale compared with samples having non-magnetic spacers (YBCO/PBCO superlattices) [92, 104]. This suggests that additional processes with a much longer length scale than charge transfer may be at play in the superconductivity depression [90, 105].

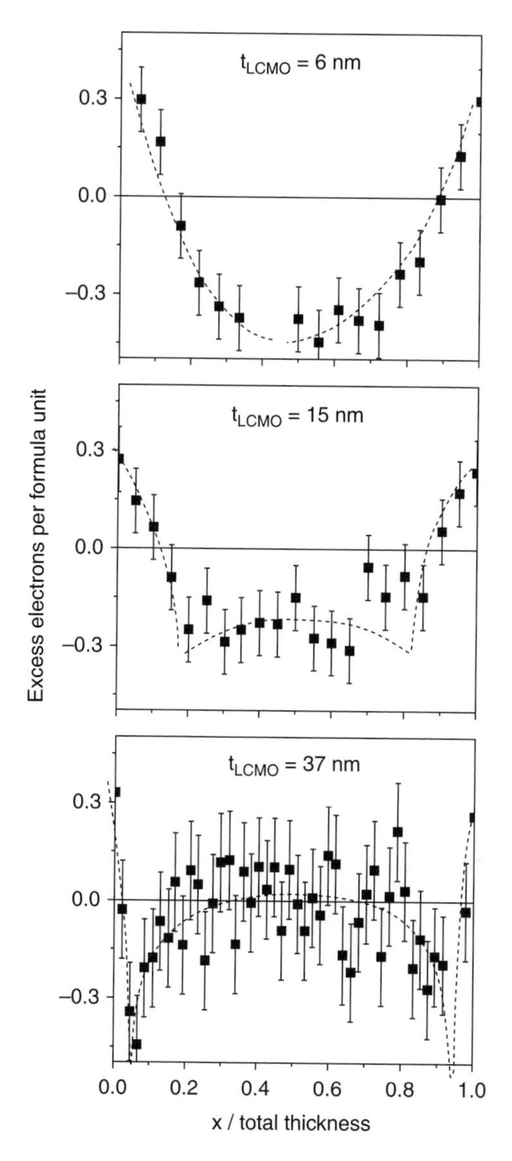

Fig. 5.27 Electronic properties of LCMO layers in $[YBCO_{10\ u.c.}/LCMO_{n\ u.c.}]_{100\ nm}$ superlattices with thick LCMO layers. (Top) Excess electrons per formula unit (difference between the nominal +3.3 and the actual Mn oxidation state) across the LCMO layer for the n = 15 sample. (Middle) n = 4. (Bottom) n = 100. The horizontal line in all cases marks the bulk doping value. Dotted lines are a guide to the eye.

5.5 Summary

The combination of STEM and EELS constitutes a very useful approach towards the study of oxide materials in real space with atomic resolution. The ultimate sensitivity, to light atoms and even to isolated impurities within solids, has been demonstrated allowing studies of materials with an unprecedented level of detail. This capability is of special importance in complex oxide materials, since very often the physical properties of these systems depend of the presence of very low densities of dopants or defects such as O vacancies. Other tasks such as the identification of inequivalent species within a given subsystem, for example O atoms within a bulk solid, have been accomplished through studies of the EELS fine structures. These studies pave the way for experiments that we can only dream of today, such as direct imaging of orbital ordering in manganites at low temperatures. For such goals we need more stable cooling holders capable of atomic resolution at, ideally, liquid helium temperatures.

Beyond the analysis of bulk-like systems, studies of nanoscale systems, interfaces, etc. benefit even more from these techniques, since it is in these systems where average macroscopic scattering techniques might not be sensitive to the quantities of interest. STEM-EELS is a very useful tool in these cases, for example in the identification of phenomena such as interface charge transfer processes in manganite/cuprate heterostructures: electron transfer from the manganite into the cuprate at LCMO/YBCO interfaces profoundly modifies their electronic properties. The excess of holes in the manganite accounts for the observed depressed magnetization at these thin LCMO layers. The depleted density of holes in the YBCO side within 3 nm of the interface must be taken into account when explaining the depressed superconductivity of these heterostructures. While this is just one example, and other interfaces such as the multiferroic system $La_{0.7}Sr_{0.3}MnO_3/BiFeO_3$ and others are of great interest, the rich variety of unexpected phenomena reported in oxide interfaces would be hard to comprehend without the help of advanced electron microscopy techniques. In the future, we need a better understanding of the physics of diffraction. In particular we need to be able to model the propagation of the coherent probe and the excitation of EELS fine structure in a unified theory, to be able to separate diffraction effects from changes in orbital occupation. Then we would be able to extract the maximum quantitative information on localized electronic structure at interfaces in a range of novel systems.

In summary, the combination of STEM and EELS with sub-Ångström resolution is a unique tool to look at oxide materials with new eyes, and will be complementary to other measurements when exploring the new properties of both bulk materials and nanoscale systems or interfaces between complex oxides with strong electron correlations.

Acknowledgments

Research supported by the U.S. Department of Energy, Office of Basic Energy Sciences, Materials Sciences and Engineering Division (SJP and MV). Work at UCM supported by MICINN CSD2009-00013 and by the European Research Council Starting Investigator Award. The authors are grateful to all the collaborators who made this work possible over the years, especially to Albina Borisevich, Nigel Browning, Rongying Jin, Julia Luck, Weidong Luo, Andy Lupini, David Mandrus, Sergio Molina, Mark Oxley, Sok Pantelides, Jing Tao, Masashi Watanabe, Zouhair Sefrioui, Cristina Visani, Vanessa Peña, Javier García-Barriocanal, Javier Tornos, and Flavio Bruno.

References

[1] Dagotto, E., Hotta, T., and Moreo, A. (2001) Colossal magnetoresistant materials: The key role of phase separation. Physics Reports **344**, 1–153.

[2] Ohtomo, A. and Hwang, H.Y. (2004) A high-mobility electron gas at the $LaAlO_3/SrTiO_3$ heterointerface. Nature **427**, 423–6.

[3] Okamoto, S. and Millis, A.J. (2004) Electronic reconstruction at an interface between a Mott insulator and a band insulator. Nature **428**, 630–3.

[4] Scherzer, O. (1936) Uber einige ehler von lektronenlinsen. Z. Physik **101**, 114–32.

[5] Scherzer, O. (1947) Sparische und chromatische Korrektur von Electronen-Linsen. Optik **2**, 114–32.

[6] Zach, J. and Haider, M. (1995) Aberration correction in a low voltage SEM by a multipole corrector. Nucl. Inst. and Meth. A **363**, 316–25.

[7] Haider, M., Rose, H., Uhlemann, S., Kabius, B. and Urban, K. (1998) Towards 0.1 nm resolution with the first spherically corrected transmission electron microscope. J. Electron Microsc. **47**, 395–405.

[8] Haider, M., Rose, H., Uhlemann, S., *et al.* (1998) A spherical-aberration-corrected 200 kV transmission electron microscope. Ultramic. **75**, 53–60.

[9] Haider, M., Uhlemann, S., Schwan, E., *et al.* (1998) Electron microscopy image enhanced. Nature 392, 768–9.

[10] Krivanek, O.L., Dellby, N., and Lupini, A.R. (1999) Towards sub-angstrom electron beams. Ultramic. **78**, 1–11.

[11] Dellby, N., Krivanek, O.L., Nellist, P.D., Batson, P.E. and Lupini, A.R. (2001) Progress in aberration-corrected scanning transmission electron microscopy. J. Electron Microsc. **50**, 177–85.

[12] von Ardenne, M. (1938) Das Elektronen-Rastermikroskop. Theoretische Grundlagen. Z. Physik **109**, 553–72.

[13] von Ardenne, M. (1938) Das Elektronen-Rastermikroskop. Praktische Ausführung. Z. Tech. Phys. **19**, 407–16.

[14] Knoll, M. and Ruska, E. (1932) Das Elektronenmikroskop. Z. Phys. A. **78**, 318–39.

[15] Ruska, E. (1987) The development of the electron-microscope and of electron-microscopy. Rev. Mod. Phys. **59**, 627–38.

[16] Crewe, A.V. (1966) Scanning electron microscopes - is high resolution possible? Science **154**, 729–38.

[17] Crewe, A.V., Wall, J., and Welter, L.M. (1968) A high-resolution scanning transmission electron microscope. J. Appl. Phys. **39**, 5861–8.

[18] Crewe, A.V. and Wall, J. (1970) A scanning microscope with 5 Å resolution. J. Mol. Biol. **48**, 375–93.

[19] Crewe, A.V., Wall, J., and Langmore, J. (1970) Visibility of single atoms. Science **168**, 1338–40.

[20] Brown, L.M. (1981) Scanning transmission electron microscopy: Microanalysis for the microelectronic age. J. Phys. **F 11**, 1–26.

[21] Herrmann, K.H. (1978) Present state of instrumentation in high-resolution electron-microscopy. J. Phys. **E 11**, 1076–91.

[22] Cowley, J.M. and Smith, D.J. (1987) The present and future of high-resolution electron-microscopy. Acta Cryst. **A 43**, 737–51.

[23] Smith, D.J. (1997) The realization of atomic resolution with the electron microscope. Rep. Prog. Phys. **60**, 1513–80.
[24] Spence, J.C.H. (1999) The future of atomic resolution electron microscopy for materials science. Mater. Sci. Eng. **R 26**, 1–49.
[25] Cowley, J.M. (1984) Scanning transmission electron microscopy and microdiffraction techniques. Bull. Mater. Sci. **6**, 477–90.
[26] Pennycook, S.J. and Boatner, L.A. (1988) Chemically sensitive structure-imaging with a scanning-transmission electron-microscope. Nature **336**, 565–7.
[27] Pennycook, S.J. (1989) Z-contrast STEM for materials science. Ultramic. **30**, 58–69.
[28] Pennycook, S.J. and Jesson, D.E. (1990) High-resolution incoherent imaging of crystals. Phys. Rev. Lett. **64**, 938–41.
[29] Pennycook, S.J. and Jesson, D.E. (1991) High-resolution Z-contrast imaging of crystals. Ultramic. **37**, 14–38.
[30] McGibbon, M.M., Browning, N.D., Chisholm, M.F. *et al.* (1994) Direct determination of grain-boundary atomic-structure in $SrTiO_3$. Science **266**, 102–4.
[31] McGibbon, A.J., Pennycook, S.J. and Angelo, J.E. (1995) Direct observation of dislocation core structures in CdTe/GaAs(001). Science **269**, 519–21.
[32] Nellist, P.D. and Pennycook, S.J. (1996) Direct imaging of the atomic configuration of ultradispersed catalysts. Science **274**, 413–5.
[33] Pennycook, S.J., Browning, N.D., McGibbon, M.M., *et al.* (1996) Direct determination of interface structure and bonding with the scanning transmission electron microscope. Philos. Trans. R. Soc. A. **354**, 2619–34.
[34] Browning, N.D., Chisholm, M.F., and Pennycook, S.J. (1993) Atomic-resolution chemical-analysis using a scanning-transmission electron-microscope. Nature **366**, 143–6.
[35] Batson, P.E. (1993) Simultaneous STEM imaging and electron energy-loss spectroscopy with atomic-column sensitivity. Nature **366**, 727–8.
[36] Duscher, G., Browning, N.D., and Pennycook, S.J. (1998) Atomic column resolved electron energy-loss spectroscopy. Phys. Stat. Solidi A **166**, 327–42.
[37] Batson, P.E., Dellby, N., and Krivanek, O.L. (2002) Sub-angstrom resolution using aberration corrected electron optics. Nature **418**, 617–20.
[38] Nellist, P.D., Chisholm, M.F., Dellby, N. *et al.* (2004) Direct sub-angstrom imaging of a crystal lattice. Science **305**, 1741.
[39] Varela, M., Findlay, S.D., Lupini, A.R. *et al.* (2004) Spectroscopic imaging of single atoms within a bulk solid. Phys. Rev. Lett. **92**, 095502.
[40] Erni, R., Rossell, M.D., Kisielowski, C., and Dahmen, U. (2009) Atomic-resolution imaging with a sub-50-pm electron probe. Phys. Rev. Lett. **102**, 096101.
[41] Rayleigh, Lord. (1896) On the theory of optical images with special reference to the microscope. Philos. Mag. (5) **42**, 167–95.
[42] Scherzer, O. (1949) The theoretical resolution limit of the electron microscope. J. Appl. Phys. **20**, 20–9.
[43] Bosman, M., Keast, V.J., Garcia-Munoz, J.L., *et al.* (2007) Two-dimensional mapping of chemical information at atomic resolution. Phys. Rev. Lett. **99**, 086102.
[44] Kimoto, K., Asaka, T., Nagai, T., *et al.* (2007) Element-selective imaging of atomic columns in a crystal using STEM and EELS. Nature **450**, 702–4.

[45] Muller, D.A., Kourkoutis, L.F., Murfitt, M. *et al.* (2008) Atomic-scale chemical imaging of composition and bonding by aberration-corrected microscopy. Science **319**, 1073–6.

[46] Varela, M., Oxley, M.P., Luo, W. *et al.* (2009) Atomic-resolution imaging of oxidation states in manganites. Phys. Rev. B **79**, Art. No. 085117.

[47] Jia, C., Nagarajan, V., He, J. *et al.* (2006) Unit-cell scale mapping of ferroelectricity and tetragonality in epitaxial ultrathin ferroelectric films. Nature Mat. **6**, 64–9.

[48] Okunishi, E., Ishikawa, I., Sawada, H., *et al.* (2009) Visualization of light elements at ultrahigh resolution by STEM annular bright field microscopy. Microsc. Microanal. **15**, 164–5.

[49] Findlay, S.D., Shibata, N., Sawada, H. *et al.* (2009) Robust atomic resolution imaging of light elements using scanning transmission electron microscopy. Appl. Phys. Lett. **95**, 191913.

[50] Borisevich, A.Y., Chang, H.J., Huijben, M. *et al.* (2010) Suppression of octahedral tilts and associated changes in electronic properties at epitaxial oxide heterostructure interfaces. Phys. Rev. Lett. **105**, 087204.

[51] Krivanek, O.L., Chisholm, M.F., Nicolosi, V. *et al.* (2010) Atom-by-atom structural and chemical analysis by annular dark-field electron microscopy. Nature **464**, 571–4.

[52] Varela, M., Lupini, A.R., van Benthem, K. *et al.* (2005) Materials characterization in the aberration-corrected scanning transmission electron microscope. Annu. Rev. Mater. Res. **35**, 539–69.

[53] Kabius, B., Hartel, P., Haider, M. *et al.* (2009) First application of Cc-corrected imaging for high-resolution and energy-filtered TEM. J. Electron Microsc. **58**, 147–55.

[54] Krivanek, O.L., Ursin, J.P., Bacon, N.J. *et al.* (2009) High-energy-resolution monochromator for aberration-corrected scanning transmission electron microscopy/electron energy-loss spectroscopy. Phil. Trans. R. Soc. A **367**, 3683–97.

[55] Glauber, R. and Schomaker, V. (1953) The theory of electron diffraction. Phys. Rev. **89**, 667–71.

[56] Cowley, J.M. (1969) Image contrast in a transmission scanning electron microscope. Appl. Phys. Lett. **15**, 58–9.

[57] Pogany, A.P. and Turner, P.S. (1968) Reciprocity in electron diffraction and microscopy. Acta Cryst. A **24**, 103–9.

[58] Pennycook, S.J. (2006) Microscopy: Transmission electron microscopy. In: Bassani, F Liedl, J, and Wyder, P, eds. *Encyclopedia of Condensed Matter Physics*, pp. 240–7, Elsevier Science Ltd., Kidlington, Oxford.

[59] Kambe, K. (1982) Visualization of Bloch waves of high-energy electrons in high-resolution electron-microscopy. Ultramic. **10**, 223–7.

[60] Oxley, M.P., Varela, M., Pennycook, T.J. *et al.* (2007) Interpreting atomic-resolution spectroscopic images. Phys. Rev. B **76**, 064303.

[61] Oxley, M., Chang, H., Borisevich, A., Varela, M., and Pennycook, S. (2010) Imaging of light atoms in the presence of heavy atomic columns. Microsc. Microanal. **16**, 92–3.

[62] Oxley, M.P. and Pennycook, S.J. (2008) Image simulation for electron energy loss spectroscopy. Micron **39**, 676–84.

[63] Cosgriff, E.C., Oxley, M.P., Allen, L.J., and Pennycook, S.J. (2005) The spatial resolution of imaging using core-loss spectroscopy in the scanning transmission electron microscope. Ultramic. **102**, 317–26.

[64] D'Alfonso, A.J., Findlay, S.D., Oxley, M.P., and Allen, L.J. (2008) Volcano structure in atomic resolution core-loss images. Ultramicrosc. **108**, 677–87.
[65] Oxley, M.P., Cosgriff, E.C., and Allen, L.J. (2005) Nonlocality in imaging. Phys. Rev. Lett. **94**, 203906.
[66] Egerton, R.F. (1996) *Electron Energy-loss Spectroscopy in the Electron Microscope*, Plenum Press, New York.
[67] Krivanek, O.L. and Paterson, J.H. (1990) Elnes of 3d transition-metal oxides: I. Variations across the periodic table. Ultramicrosc. **32**, 313–8.
[68] Kurata, H. and Colliex, C. (1993) Electron-energy-loss core-edge structures in manganese oxides. Phys. Rev. B **48**, 2102–8.
[69] Paterson, J.H. and Krivanek, O.L. (1990) ELNES of 3d transition-metal oxides2. Variations with oxidation-state and crystal-structure. Ultramicrosc. **32**, 319–25.
[70] Rask, J.H., Miner, B.A., and Buseck, P.R. (1987) Determination of manganese oxidation states in solids by electron energy-loss spectroscopy. Ultramicrosc. **21**, 321–6.
[71] Waddington, W.G., Rez, P., Grant, I.P., and Humphreys, C.J. (1986) White lines in the $L_{2,3}$ electron-energy-loss and x-ray absorption spectra of 3d transition metals. Phys. Rev. B **34**, 1467.
[72] Wang, Z.L., Yin, J.S., and Jiang, Y.D. (2000) EELS analysis of cation valence states and oxygen vacancies in magnetic oxides. Micron **31**, 571–80.
[73] Leapman, R.D., Grunes, L.A., and Fejes, P.L. (1982) Study of the L_{23} edges in the 3d transition-metals and their oxides by electron-energy-loss spectroscopy with comparisons to theory. Phys. Rev. B **26**, 614–35.
[74] Pearson, D.H., Ahn, C.C., and Fultz, B. (1993) White lines and d-electron occupancies for the 3d and 4d transition-metals. Phys. Rev. B **47**, 8471–8.
[75] Pearson, D.H., Fultz, B., and Ahn, C.C. (1988) Measurements of 3d state occupancy in transition metals using electron energy loss spectrometry. Appl. Phys. Lett. **53**, 1405–7.
[76] de Groot, F.M.F., Grioni, M., Fuggle, J.C., *et al.* (1989) Oxygen 1s x-ray-absorption edges of transition-metal oxides. Phys. Rev. B **40**, 5715.
[77] Grunes, L.A., Leapman, R.D., Wilker, C.N., Hoffmann, R., and Kunz, A.B. (1982) Oxygen- near-edge fine-structure - an electron-energy-loss investigation with comparisons to new theory for selected 3d transition-metal oxides. Phys. Rev. B **25**, 7157–73.
[78] Thole, B.T. and van der Laan, G. (1988) Branching ratio in x-ray absorption spectroscopy. Phys. Rev. B **38**, 3158.
[79] Luo, W., Franceschetti, A., Varela, M., *et al.* (2007) Orbital-occupancy versus charge ordering and the strength of electron correlations in electron-doped $CaMnO_3$. Phys. Rev. Lett. **99**, 036402.
[80] Luo, W., Varela, M., Tao, J., Pennycook, S.J., and Pantelides, S.T. (2009) Electronic and crystal-field effects in the fine structure of electron energy-loss spectra of manganites. Phys. Rev. **B 79**, 052405.
[81] Rodriguez-Carvajal, J., Hennion, M., Moussa, F., *et al.* (1998) Neutron-diffraction study of the Jahn-Teller transition in stoichiometric $LaMnO_3$. Phys. Rev. B **57**, R3189.
[82] Ohtomo, A., Muller, D.A., Grazul, J.L., and Hwang, H.Y. (2002) Artificial charge-modulation in atomic-scale perovskite titanate superlattices. Nature **419**, 378–80.
[83] Yamada, H., Ogawa, Y., Ishii, Y. *et al.* (2004) Engineered interface of magnetic oxides. Science **305**, 646–8.

[84] Przyslupski, P., Komissarov, I., Paszkowicz, W., *et al.* (2004) Magnetic properties of $La_{0.67}Sr_{0.33}MnO_3/YBa_2Cu_3O_7$ superlattices. Phys. Rev. B **69**, 134428.
[85] Sa de Melo, C.A.R. (1997) Magnetic exchange coupling in ferromagnet/superconductor/ferromagnet multilayers. Phys. Rev. Lett. **79**, 1933–6.
[86] Sa de Melo, C.A.R. (2000) Magnetic exchange coupling through superconductors: A trilayer study. Phys. Rev. B **62**, 12303.
[87] Vasko, V.A., Larkin, V.A., Kraus, P.A. *et al.* (1997) Critical current suppression in a superconductor by injection of spin-polarized carriers from a ferromagnet. Phys. Rev. Lett. **78**, 1134–7.
[88] Holden, T., Habermeier, H.U., Cristiani, G. *et al.* (2004) Proximity induced metal-insulator transition in $YBa_2Cu_3O_7/La_{2/3}Ca_{1/3}MnO_3$ superlattices. Phys. Rev. B **69**, 064505.
[89] Peña, V., Sefrioui, Z., Arias, D., *et al.* (2004) Long length scale interaction between magnetism and superconductivity in $La_{0.3}Ca_{0.7}MnO_3/YBa_2Cu_3O_7$ superlattices. European Physical Journal B **40**, 479–82.
[90] Peña , V., Sefrioui, Z., Arias, D. *et al.* (2004) Coupling of superconductors through a half-metallic ferromagnet: Evidence for a long-range proximity effect. Phys. Rev. B **69**, 224502.
[91] Sefrioui, Z., Arias, D., Peña, V. *et al.* (2003) Ferromagnetic/superconducting proximity effect in $La_{0.7}Ca_{0.3}MnO_3/YBa_2Cu_3O_{7-}$ superlattices. Phys. Rev. B **67**, 214511.
[92] Sefrioui, Z., Varela, M., Pena, V. *et al.* (2002) Superconductivity depression in ultrathin $YBa_2Cu_3O_{7-}$ layers in $La_{0.7}Ca_{0.3}MnO_3/YBa_2Cu_3O_{7-}$ superlattices. Appl. Phys. Lett. **81**, 4568–70.
[93] Chakhalian, J., Freeland, J.W., Habermeier, H.U. *et al.* (2007) Orbital reconstruction and covalent bonding at an oxide interface. Science **318**, 1114–7.
[94] Yunoki, S., Moreo, A., Dagotto, E., *et al.* (2007) Electron doping of cuprates via interfaces with manganites. Phys. Rev. B **76**, 064532.
[95] Varela, M., Lupini, A.R., Peña, V. *et al.* (2005) Direct measurement of charge transfer phenomena at ferromagnetic/superconducting oxide interfaces. http://arxiv.org/pdf/cond-mat/0508564v2.
[96] Hoffmann, A., te Velthuis, S.G.E., Sefrioui, Z. *et al.* (2005) Suppressed magnetization in $La_{0.7}Ca_{0.3}MnO_3/YBa_2Cu_3O_{7-\delta}$ superlattices. Phys. Rev. B **72**, 140407.
[97] Varela, M., Lupini, A.R., Pennycook, S.J., Sefrioui, Z., and Santamaria, J. (2003) Nanoscale analysis of $YBa_2Cu_3O_{7-x}/La_{0.67}Ca_{0.33}MnO_3$ interfaces. Solid State Electron. **47**, 2245–8.
[98] Zhang, Z.L., Kaiser, U., Soltan, S., Habermeier, H.U., and Keimer, B. (2009) Magnetic properties and atomic structure of $La_{2/3}Ca_{1/3}MnO_3$–$YBa_2Cu_3O_7$heterointerfaces. Appl. Phys. Lett. **95**, 242505.
[99] Garcia-Barriocanal, J., Bruno, F.Y., Rivera-Calzada, A. *et al.* (2010) "Charge leakage" $LaMnO_3/SrTiO_3$ interfaces. Adv. Mater. **22**, 627–32.
[100] Browning, N.D., Yuan, J., and Brown, L.M. (1992) Determination of the local oxygen stoichiometry in $YBa_2Cu_3O_{7-}$ by electron-energy loss spectroscopy in the scanning-transmission electron-microscope. Physica C **202**, 12–18.
[101] Mizoguchi, T., Varela, M., Buban, J.P., Yamamoto, T., and Ikuhara, Y. (2008) Site dependence and peak assignment of $YBa_2Cu_3O_{7-x}$O K-edge electron energy loss near-edge fine structure. Phys. Rev. B **77**.

[102] Nikolic, B.K., Freericks, J.K., and Miller, P. (2002) Equilibrium properties of double-screened dipole-barrier SINIS osephson junctions. Phys. Rev. B **65**, 064529.

[103] Nikoliv, B.K., Freericks, J.K., and Miller, P. (2001) Intrinsic reduction of osephson critical current in short ballistic SNS weak links. Phys. Rev. B **64**, 212507.

[104] Varela, M., Arias, D., Sefrioui, Z., *et al.* (2000) Epitaxial mismatch strain in $YBa_2Cu_3O_{7-\delta}/PrBa_2Cu_3O_7$ superlattices. Phys. Rev. B **62**, 12509.

[105] Soltan, S., Albrecht, J., and Habermeier, H.U. (2004) Ferromagnetic/superconducting bilayer structure: A model system for spin diffusion length estimation. Phys. Rev. B **70**, 144517.

6

Advanced modes of piezoresponse force microscopy for ferroelectric nanostructures

A. Gruverman

6.1 Introduction

Over the last several years the field of ferroelectric and multiferroic oxides has been experiencing a significant revival. A clear manifestation of this trend is inclusion of new developments in this field into the "breakthroughs of the year" by the Science magazine [1]. This is partly due to recent experimental advances allowing synthesis of complex oxides with atomic scale precision. Additionally, understanding of the physics of nanoscale ferroic phenomena has increased dramatically in recent years due to the advent of advanced first-principles theoretical methods. These developments have lead to the discovery of entirely new classes of compounds exhibiting unusual physical properties which, besides being of fundamental interest, hold much promise for future generations of novel electronic devices. As the quality of ferroelectric and multiferroic interfaces has proven to be essential to their functionality, a necessity to characterize their properties down to the atomic scale has lead to the wide application of electron and scanning probe microscopy (SPM) methods in this field. Specifically, Piezoresponse Force Microscopy (PFM), along with other SPM-based methods, has proved to be an indispensable tool for high-resolution characterization of ferroelectrics. The versatility of PFM, which allows measurements of local electronic properties along with nanoscale polarization control [2, 3], makes it a method of choice for addressing the problems relevant to the scaling behavior of ferroelectrics. Standard implementation of this technique, where an electrically biased probe scans the sample surface to modify or visualize the domain structure, has been around for over 15 years. However, recent years have witnessed the development of advanced modes of PFM such as resonance-enhanced PFM, stroboscopic PFM, switching spectroscopy PFM, and so on.

This chapter considers several aspects related to the application of the advanced PFM modes to investigation of the static and dynamic properties of ferroelectric structures.

6.2 Ferroelectric structures and size effects

Distinction of ferroelectrics from other polar dielectrics (pyro- and piezoelectrics) lies in the presence of electrically reversible spontaneous polarization [4]. Electric field dependence of polarization is characterized by a hysterestic behavior indicative of the existence of two stable polar states (Figure 6.1). This feature along with other unique properties which include high pyroelectric coefficients, strong electromechanical coupling, and high dielectric constant, make ferroelectrics key materials for modern electronic devices, such as nonvolatile memories, microwave phase filters, infrared sensors, and microactuators [5, 6, 7, 8]. Along with the technological importance, experimental and theoretical studies of the switching processes in ferroelectrics represent one of the most interesting problems in the theory of structural phase transitions.

Although ferroelectric materials can have unit cells of various degrees of complexity, most ferroelectrics of practical importance have a perovskite structure with chemical formula ABO_3 (where A and B represent metal cations), such as $BaTiO_3$ or $Pb(Zr,Ti)O_3$ (Figure 6.1). A microscopic mechanism of structural phase transitions in ferroelectrics involves asymmetric ionic displacement (oxygen relative to B ions) resulting in a spontaneous dipole moment. In the framework of lattice dynamics, this transition is associated with freezing of the low-frequency lattice vibration at the Brillouin zone center—the soft-mode concept [9, 10]. The softening of a given phonon mode is explained in terms of a competition between short-range and long-range Coulomb interactions. In the thermodynamic approach, phase transitions are

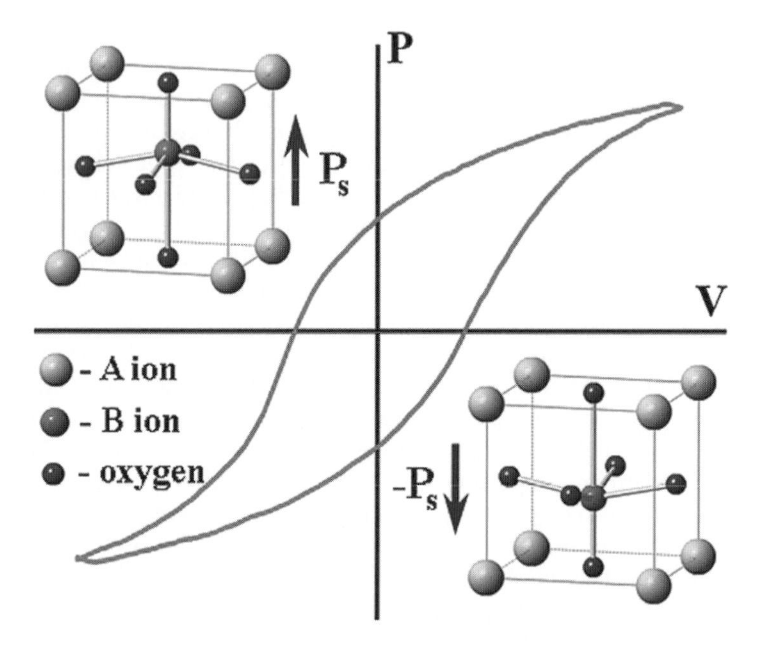

Fig. 6.1 Ferroelectric polarization hysteresis loop and unit-cells of the ABO_3 perovskite ferroelectric, showing two opposite polarization directions.

described by using an expansion of free energy with respect to polarization and strain as order parameters (the Landau–Ginzburg–Devonshire theory) [11, 12]. This approach provides a very good description of the temperature-dependent behavior of the dielectric, piezoelectric, and elastic coefficients as well as polarization configurations in ferroelectric thin films.

Ferroelectric materials are characterized by the presence of regions with uniform electrical polarization, called domains. Domain existence is one of the fundamental properties of ferroelectrics. Domains form during phase transition to relieve mechanical stress associated with transformation strains, and to minimize the depolarization energy related to spontaneous polarization. Domain arrangements reflect the effects of sample conductivity, mechanical strains, and imperfections such as dislocations, vacancies, and impurities.

In most device applications, switching of the polarization direction is the most essential aspect of the functional behavior of ferroelectrics. For decades, because of its fundamental and technological importance, switching behavior in ferroelectrics has been a focal point of experimental and theoretical studies [13]. The thermodynamic consideration of polarization reversal yields coercive field values that are orders of magnitude higher than those registered experimentally. This implies that electrically induced ferroelectric switching occurs via inhomogeneous domain nucleation and domain wall motion. That these are indeed two basic mechanisms involved in polarization reversal has been confirmed by numerous experiments in a variety of ferroelectric materials, and described in a number of monographs [5, 13, 14].

There is a large amount of experimental evidence that both switching mechanisms are strongly affected by the defect structure of the ferroelectric medium, which leads to different functional field dependencies [15, 16]. The exponential field dependence of the domain wall velocity has been reported in systems with long-range variations in pinning potential (so-called random-field disorder) [17, 18]. Random-field disorder is typically associated with the directional internal fields due to oxygen vacancies that create conditions favoring a certain polarization state [19]. The short-range (random-bond) disorder which is characterized by symmetric variations in the depth of the ferroelectric double-well potential generally leads to a power law dependence of the wall velocity. In terms of the hysteresis switching behavior, random-bond disorder leads to a symmetric increase in the coercive field, while random-field defects result in the polarization imprint manifesting as a horizontal shift of the hysteresis loop (Figure 6.2).

The role of microstructural defects has been emphasized in the domain nucleation problem called "Landauer paradox" [20]. In the late 1950s, it was shown that formation of a domain nucleus due to thermal fluctuations would lead to such a change in free energy that "would permit only implausibly large domains to expand in the presence of typical coercive field" [20]. For a typical range of external electric fields ($\sim$100 kV/cm) the energy barrier height that needs to be overcome is $10^3 k_B T$, which makes the probability of domain nucleation negligibly small. It has been suggested that, in reality, nucleation is assisted by local or extended defects (such as passive layers or dislocations [21]), which effectively reduce the coercive fields to the experimentally observed values.

Microstructural imperfections are considered to be the root cause of the qualitatively different switching kinetics observed in epitaxial and polycrystalline thin films. The statistical theory of Kolmogorov–Avrami–Ishibashi (KAI) [22, 23, 24] describes the switching process as the phase transformation in infinite media with a constant nucleation rate and domain wall velocity being a rate-limiting parameter [8, 25]. However, the KAI model is inapplicable to

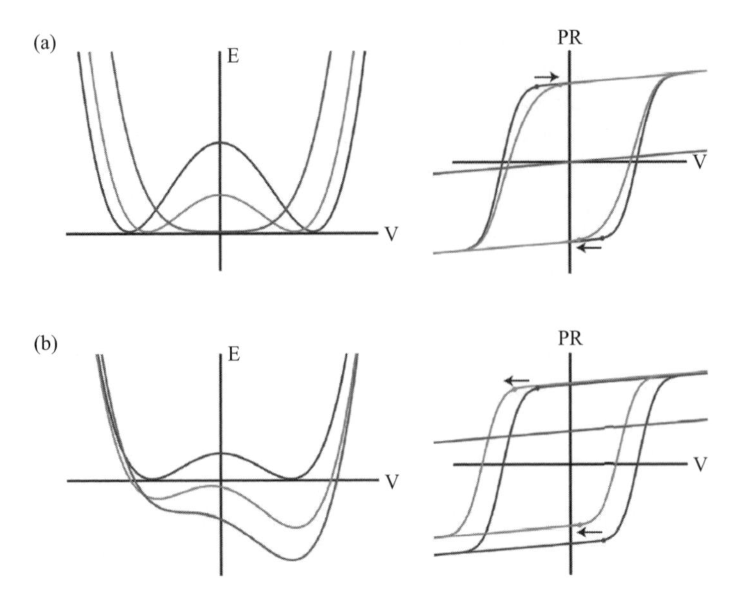

Fig. 6.2 The role of defects on the energetics of polarization switching. (a) Random-bond (symmetric) and (b) random-field (asymmetric) disorder in a ferroelectric material and their effects on local hysteresis loop shape. The blue curves correspond to a defect-free material, green curves to the presence of a weak random-bond (a) and random-field (b) defect, and red curves to the limiting cases of a nonpolar phase and polar non-ferroelectric phases, respectively. Reprinted by permission from Macmillan Publishers Ltd: Nature Materials [16], copyright 2008. This figure is reproduced in color in the color plate section.

the situation where the number of nuclei is small [26]. In addition, experiments show that in the high-field regime the nucleation rate is not constant, and polarization reversal occurs as a non-steady process [27]. In polycrystalline films, domain nucleation and wall motion are strongly affected by grain boundaries and other defects, which results in an exponentially wide distribution of local nucleation times [28]. In this case, nucleation becomes a rate-limiting parameter and so the nucleation-limited model (NLS) has been developed to provide a better description of the switching kinetics [29, 30, 31].

A number of classical monographs have served as a comprehensive introduction into the field for several generations of scientists. However, recent advances in the synthesis and fabrication of micro- and nanoscale ferroelectric structures have brought to life new physical phenomena that need to be studied and understood at this size range. As structure dimensions are getting smaller, ferroelectric materials exhibit a pronounced size effect, manifesting itself in significant deviation of the properties of low-dimensional structures from the bulk properties. In this sense, ferroelectrics are similar to magnetic materials, since surface energy cannot be neglected in small volumes and long-range dipole interaction is significantly modified in confined geometries.

With ferroelectric film thickness reaching the range of the Thomas–Fermi screening length, polarization instability increases due to the uncompensated depolarizing field. Formation of periodic domain structures with an alternating sign of spontaneous polarization (Figure 6.3)

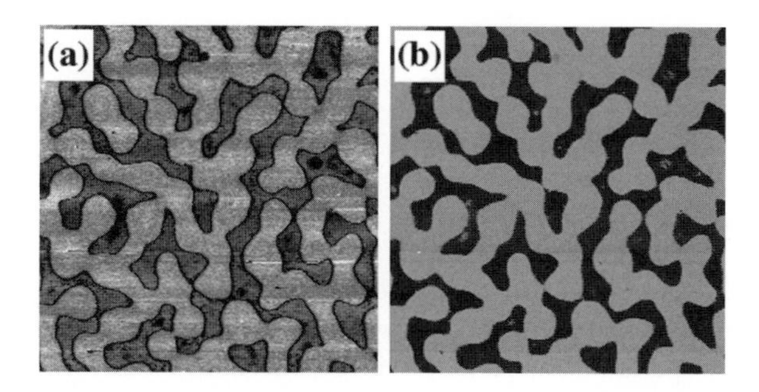

Fig. 6.3 PFM amplitude (a) and phase (b) images of a 50-nm-thick $PbTiO_3/SrTiO_3$ thin film, illustrating a maze structure of antiparallel 180° domains. Image size is $2 \times 2\mu m^2$. Sample courtesy H. Funakubo.

is one of the mechanisms that reduce the electrostatic energy associated with the depolarizing field. It has been shown that a balance between domain wall energy and electrostatic energy leads to a square root relationship between the domain periodicity w and film thickness d[32]. The validity of this relationship has been confirmed experimentally for a thickness range of several orders of magnitude down to several nanometers, most recently for $PbTiO_3$ films, where a stripe domain period of 12 nm for 10-nm-thick films has been reported [33]. However, in general, the type of equilibrium polarization state strongly depends on the electrical and mechanical boundary conditions. For example, antiparallel domain formation can be suppressed in the presence of ionic adsorbates, with polarization direction being dependent on the chemical nature of the adsorbate. Enhanced ferroelectricity has been observed in ferroelectric films coherently strained by the substrate [34].

Size effects in ferroelectric thin films also include smearing of the dielectric anomaly due to stress gradient near the film–substrate interface [35]. Because of the flexoelectric effect, the stress gradient is equivalent to the external electric field and thus can shift and smear out the paraelectric–ferroelectric phase transition. As was recently noted by Bratkovsky and Levanyuk [36], the surface may be considered as a defect of a "field" type and an internal field might arise simply due to the difference in work functions of both metal electrodes or compositional inhomogeneity. This could also cause smearing of the ferroelectric anomaly in thin films or the abrupt appearance of domain structure [37].

The failure of continuous theories to describe adequately the physical properties of systems with the dimensions of several unit cells, necessitates first-principles calculations. Quite a few review papers and book chapters containing detailed descriptions of the latest advances in this field have been published recently [38, 39]. In one of the first seminal papers on the scaling effect, Junquera and Ghosez [40] performed first-principles calculations for a fully strained $SrRuO_3/BaTiO_3/SrRuO_3$ heterostructure grown on a $SrTiO_3$ substrate. By taking into account the interface chemistry, strain relaxation, and compensation of the depolarizing field in $SrRuO_3$, they found that the "intrinsic" size effect for such heterostructures was below 2.4 nm, i.e. six unit cells. Thus, domain formation was not actually needed to maintain a stable ferroelectric phase in such ultrathin films. Another interesting conclusion was made by Fu and Bellaiche [41], who predicted a novel "vortex"-type polarization state in $BaTiO_3$ quantum dots. The polarization could still be found in dots with lateral dimensions of about 2.5 nm, where the polarization is oriented parallel to the surface, and electromechanical properties greatly reduced. Indeed, ferroelectricity was observed in very thin $BaTiO_3$ films of about 5 nm thickness [42].

6.3 Advanced modes of PFM

PFM has been developed as a viable alternative to electron microscopy to allow non-destructive high-resolution visualization and modification of domain structures in ferroelectric thin films. The principle of PFM operation has been described in detail in a number of recent reviews and book chapters [43]. Briefly, domain imaging in PFM is based on two things: (1) piezoelectric activity of ferroelectrics and its linear coupling with polarization in the low-field range; (2) the ability of SPM to detect sub-Angstrom displacements of the cantilever. Because of the converse piezoelectric effect, the ferroelectric sample expands or contracts in response to electrical excitation. Domain mapping in PFM is carried out by scanning the sample surface with the probe in the repulsive force (contact) regime and monitoring the piezoelectric strain. In PFM, the role of the probe is two-fold: it is used both as an actuator, which allows electric field application through the nanoscale contact with the sample, and as a sensor, which measures the mechanical response of the sample. The response magnitude is a measure of the effective piezoelectric coefficient d_{33} value, which can be related to the polarization value, while the polarization direction can be determined from the sign of the induced piezoelectric strain. Lateral resolution of PFM is in the nanoscale range, which is primarily determined by the size of the tip-sample contact area, which in turn depends on the tip curvature and elastic and dielectric constants of the sample material, indentation force, etc. [44]. Resolution in the sub-10 nm range has been routinely demonstrated in a variety of ferroelectric materials. This feature of PFM, along with the possibility of effective nanoscale domain control, has made PFM one of the mainstream methods for nanoscale ferroelectric studies. However, in spite of the significant advantages it offers, until recently PFM was still mainly considered as an imaging technique. Recent progress in the development of more advanced modes of PFM has allowed acquisition of quantitative data related to domain wall dynamics, nucleation, disorder potential, energy dissipation, etc. [18, 28, 45, 46, 47]. We now consider several of the most important modes of advanced PFM.

6.3.1 Resonance-enhanced PFM: static domain imaging

Extension of PFM application to a broader range of materials, such as piezoelectrics semiconductors and biomaterials characterized by weak electromechanical behavior (signal, response), as well as to ultrathin ferroelectric films and superlattices, has brought about a necessity to drastically increase the sensitivity of PFM measurements.

Although PFM domain imaging, in principle, can be performed in the static deflection mode of AFM, conventional implementation of PFM is based on detection of the dynamic electromechanical response of the sample. A small ac (imaging) voltage, $V_{tip} = V_0 \cos \omega t$, is applied to a conductive probe in contact with a bare surface of the sample, i.e. without a deposited top electrode. Domain imaging is performed by monitoring the piezoelectric strain using the same probe: the surface vibration, $d = d_0 \cos(\omega t + \varphi)$, is transferred to the oscillation of the cantilever, which is detected using the lock-in technique. This allows detection of a piezoelectric displacement as small as tens of picometers over a relatively rough topographic landscape (of the order of tens of nanometers). Typically, the frequency of the modulation voltage is kept well below the cantilever resonant frequency to avoid topographic cross-talk. This modulation approach provides around three orders of magnitude improvement in PFM sensitivity in comparison with static detection [48]. However, even in this case, reliable measurements of the piezoelectric response can only be realized mainly for materials with strong piezoelectric properties, such as $Pb(Zr,Ti)O_3$, with a d_{33} coefficient above 100 pm/V [49]. In the case of ferroelectrics, an increase in PFM sensitivity by using a higher imaging voltage is limited by the requirement that it should not exceed the threshold voltage, in order to avoid polarization switching and to ensure the non-destructive character of domain detection. A typical amplitude range for PFM imaging voltage is from 1 V to 10 V-far above the coercive voltage in ultrathin ferroelectric structures [50].

A significant enhancement of the PFM sensitivity can be achieved by employing a piezoelectric excitation at frequencies close to a resonance [51]. Resonance imaging is a standard procedure in a number of SPM techniques, such as electrostatic force microscopy (EFM) and atomic force acoustic microscopy (AFAM) [52]. Typically, contact resonance frequencies are in the range of several hundred kHz. In resonance-enhanced PFM, the signal-to-noise ratio is increased by a quality factor, Q, of the cantilever, allowing significant signal amplification and detection of a weak piezoelectric response (Figure 6.4).

However, there are certain reasons why the resonance approach cannot be used straightforwardly in PFM. First, the contact resonance frequency is strongly dependent on surface roughness and variations in local elastic properties. This instability of the contact resonance results in topographic cross-talk with the amplified contribution from topographic features. The associated changes in the phase shift will mask the phase contrast due to the presence of domain structure. Second, even for a perfectly smooth ferroelectric surface, tracking of the resonance frequency by a phase-locked feedback loop, which maintains a constant phase shift between the driving and response signals will be inherently problematic given that the response signal phase changes by 180° with crossing the boundary between antiparallel domains.

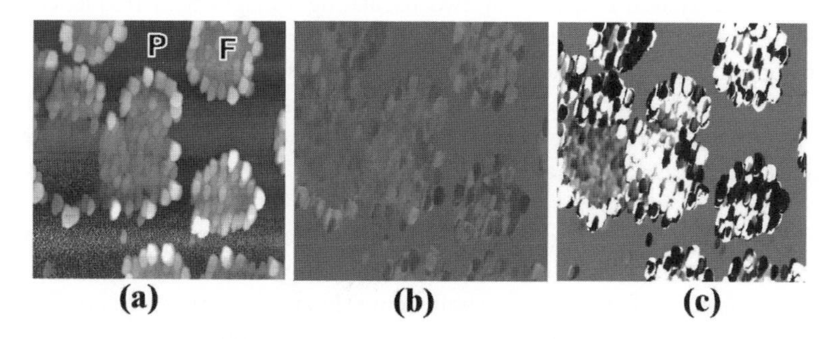

Fig. 6.4 Topography (a) and PFM images obtained of resonance (b) and near contact resonance (c) $Pb(Zr_{0.3}Ti_{0.7})O_3$ films showing both ferroelectric and non-polar phases. Scan size is 5 mm, z-scale of (a) is 20 nm, the z-scale of (b) and (c) is the same, 10 mV full scale. Image courtesy C. Harnagea. This figure is reproduced in color in the color plate section.

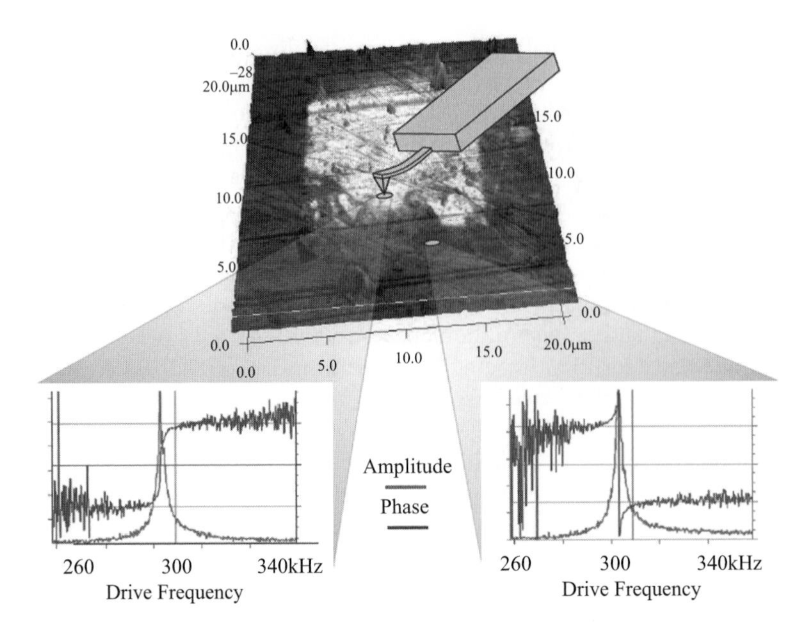

Fig. 6.5 Schematic illustration of amplitude and phase cantilever response, showing a 180° phase offset over antiparallel domains, demonstrating that it cannot reliably be used as a feedback signal for a phase-locked loop. Reprinted with permission from [53], courtesy of the Institute of Physics. This figure is reproduced in color in the color plate section.

Figure 6.5 illustrates this effect showing that a constant phase shift cannot be used as a feedback signal [53].

This problem has been elegantly solved by introducing a dual-frequency excitation method that allows amplitude-based resonant frequency tracking in PFM [53]. In contrast to conventional PFM with single-frequency excitation far from the resonance, in this approach, known currently as dual ac resonant tracking—DART (implemented in the SPMs by Asylum Research), the system is driven by two oscillating voltages with frequencies near the resonance—one below and one above the resonance, as shown in Figure 6.6. The response signals for both driving frequencies are detected by two separate lock-in amplifiers and the difference between the amplitudes A_1 and A_2 is used as a feedback loop input signal. A change in the resonance frequency alters the A_1-A_2 value and the feedback system adjusts the driving frequencies so as to maintain A_1-A_2 constant, thereby allowing effective tracking of the resonance.

Figure 6.7 shows an example of single-frequency PFM imaging in comparison with the DART-PFM method in ultrathin (2.4 nm) $BaTiO_3$ film. Application of single-frequency PFM to ultrathin ferroelectric films is quite challenging. First, as mentioned above, an imaging voltage high enough to induce a detectable surface vibration could easily exceed the coercive voltage, rendering non-destructive domain imaging impossible. Second, PFM testing of these films may suffer from increased conductivity and possible leakage. Third, in the case of modification of the domain structure by a tip-generated electric field, the electrostatic contribution from the deposited surface charge (not necessarily related to the switched polarization) can

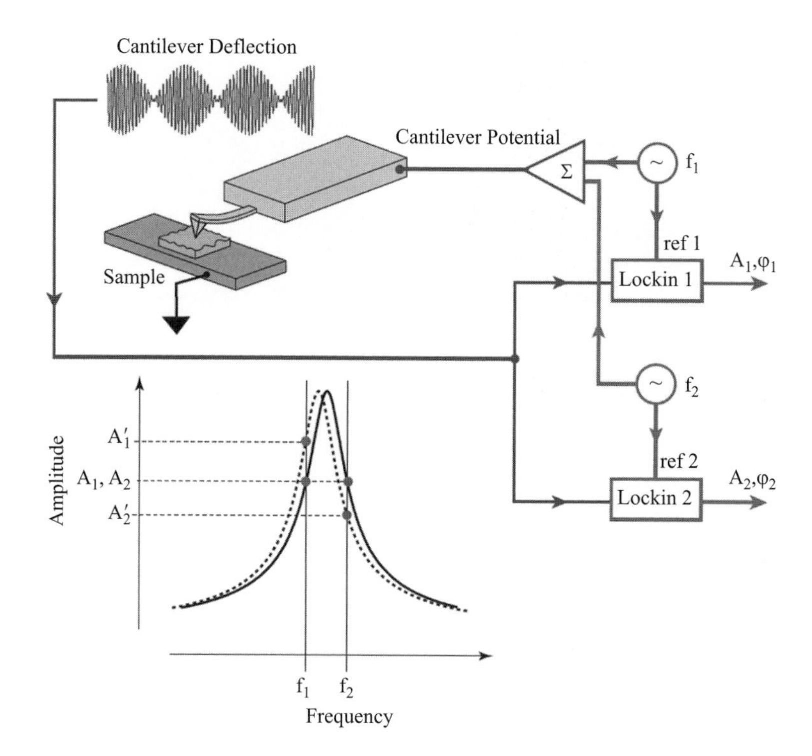

Fig. 6.6 Schematic diagrams of the experimental set-up and the dual-frequency excitation principle based on resonant-amplitude tracking. Reprinted with permission from [53], courtesy of the Institute of Physics.

be misinterpreted as a weak piezoresponse signal. In Figures 6.7(a, b), single-frequency PFM amplitude and phase images of the electrically written bi-domain polarization pattern a exhibit a rather fazed contrast, which cannot be reliably attributed to true piezoresponse. On the other hand, the DART method produces "textbook examples" of the amplitude and phase images of the 180° domain patterns: (a) opposite domains appear as regions with the same level of PFM amplitude separated by black lines corresponding to domain walls; (b) phase difference between antiparallel domains is close to 180°. Note that the DART images were obtained at 200 mVpp, while single-frequency PFM were acquired at 500 mVpp.

6.3.2 Stroboscopic PFM: domain switching dynamics

Considering the switching studies in ferroelectric heterostructures, PFM allows two different approaches. An *imaging* approach involves visualization of domain structure resulting from application of a poling voltage, locally or to an extended area. Measurements of the voltage–time dependence of the switched domains provide information on the mechanism of domain growth, local activation energies, wall interactions with defects and pinning centers, and the role of disorder potential in wall motion [54].

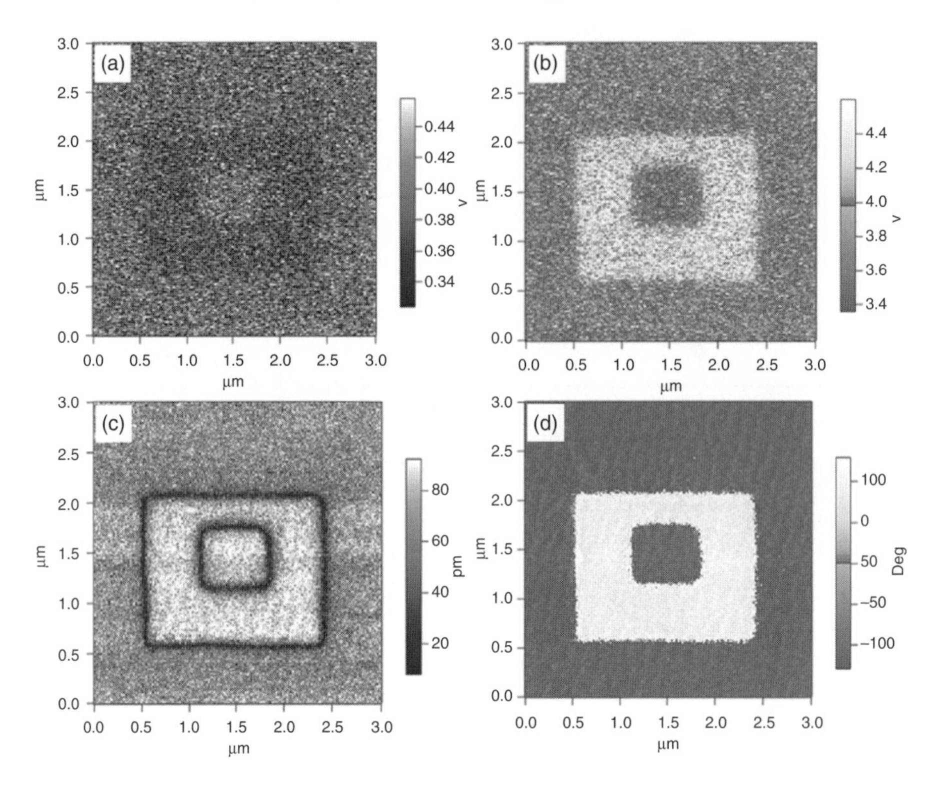

Fig. 6.7 (a, b) PFM amplitude (a) and phase (b) images of poled domains in *ultra-thin* (2.4-nm-thick) $BaTiO_3$ film acquired by a conventional PFM method; (c, d) PFM amplitude (c) and phase (d) images of poled domains in the same film by resonant-enhanced DART method. Note a significantly improved resolution and contrast in the DART images. Sample courtesy C.B. Eom.

A *spectroscopic* approach is based on measurement of the local electromechanical hysteresis loops which represent the dc bias dependence of the local piezoelectric response directly related to the process of domain formation specifically under the probing tip [54]. This method can be used to relate the global switching characteristics of ferroelectric capacitors, such as remanent polarization and coercive field, to the local switching parameters.

One of the key features of the PFM imaging method is that it allows application of the input voltage and detection of the electromechanical response, both through the conductive probing tip and deposited electrodes (Figure 6.8), which facilitates not only nanoscale visualization of static domain configurations in capacitor structures, but also their transformations under an applied electric field. In PFM imaging of ferroelectric capacitors, the probing tip is in contact with the deposited top electrode. Although in this case during imaging the whole volume underneath the electrode is electrically excited, the electromechanical response is still probed locally. Scanning the electrode surface while measuring local piezoresponse provides spatially resolved information on domain structure underneath the electrode. Lateral resolution, determined as the width of the domain wall image profile, linearly scales with the thickness of the

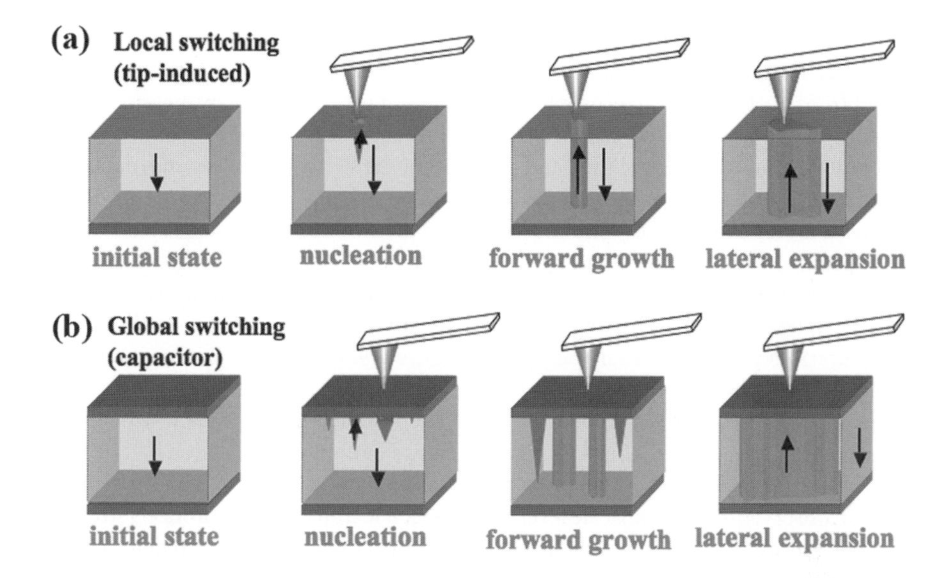

Fig. 6.8 Schematic diagram of domain switching for different PFM experimental geometry: (a) local switching, and (b) global switching.

ferroelectric layer H and top electrode L. For typical material parameters and thin top electrode ($L << H$), the resolution is expected to be $w \sim 0.2 \cdot H$, which represents the ultimate limit for PFM resolution in capacitors [55].

An external bias applied to the top electrode generates a uniform electrical field within the capacitor so that the PFM tip senses the response, from the whole thickness of the ferroelectric layer. Thus, this approach allows one to get around the problem of an inhomogeneous field distribution generated by the probing tip in a film without a top electrode [43]. Additionally, domain imaging in capacitors provides a possibility for direct studies of polarization reversal dynamics, which on a microscopic level occurs via the nucleation and growth of a large number of domains. This PFM-assisted insight into domain kinetics has a significant advantage over conventional electrical testing of the switching behavior, as it allows direct assessment of the relative contribution of nucleation and domain wall motion into the polarization reversal process, field-dependent motion of domain walls, and capacitor size effect on its switching behavior.

Generally, application of PFM to investigation of dynamic processes is limited by its low time resolution, determined by the acquisition time for a single frame (of the order of several minutes). A high-speed version of PFM (HSPFM) has been developed by Huey's group to allow complete image acquisition in several seconds, thus increasing time resolution by two orders of magnitude over conventional PFM imaging [56]. This approach, which involves high-speed scanning of a bare ferroelectric surface with a tip under a superposition of a switching and imaging bias, allows effective study of the dynamics of domain nucleation and growth, but it requires relatively smooth surfaces.

Development of a stroboscopic PFM (S-PFM) method has extended the time resolution of PFM imaging into the sub-100 ns range [57]. This method is based on visualization of domain configurations developing in ferroelectric capacitors during step-by-step polarization reversal. Switching characteristics such as nucleation rate and domain wall velocity can be calculated from a set of PFM snapshots taken at different time intervals, by measuring the time dependence of the number and size of growing domains. The time resolution of the S-PFM method is determined by the rise time and duration of the switching pulses and, depending on the capacitor size and time constant of the external circuitry, can be in the order of 10 ns.

Two main modifications of the stroboscopic PFM method have been suggested (Figure 6.9). In one approach, the domain switching behavior is visualized by applying a series of short input pulses with fixed amplitude and incrementally increasing duration ($\tau_1 < \tau_2 < \ldots < \tau_n < t_s \tau < t_s$, where t_s is a switching time for a given voltage). PFM imaging of the resulting domain pattern is performed after each pulse (Figure 6.9(a)) [58]. At the beginning of each switching cycle, the capacitor is reset into the initial polarization state.

Applicability of this approach depends upon the reproducibility of domain switching kinetics from cycle to cycle. Indeed, the deterministic nature of domain nucleation had been shown earlier in bulk crystals of lead germanate by means of optical stroboscopy [27]. The same behavior has been observed at the nanoscale level in ferroelectric thin film capacitors via stroboscopic PFM studies: in each switching cycle, domain nucleation occurs in the predetermined sites most likely to correspond to the local defects at the film–electrode interface [59]. In epitaxial $Pb(Zr,Ti)O_3$ (PZT) capacitors, nucleation probability is above 90% while in polycrystalline capacitors it is close to 100%.

In another approach, proposed to reduce the detrimental effect of stochastic nucleation events in epitaxial structures, after a capacitor is being set, the switching pulses of the same duration are applied to the capacitor ($\tau_1 = \tau_2 = \ldots = \tau_n < t_s$) with PFM imaging between the pulses (Figure 6.9(b)) [60]. In this case, it is assumed that the PFM image obtained after

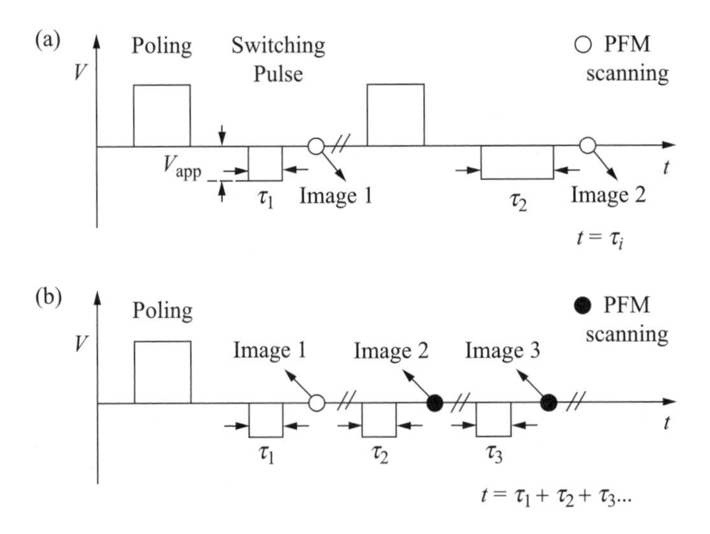

Fig. 6.9 Schematic of pulse sequences for IS-PFM imaging of domain switching kinetics: (a) step-by-step switching approach, (b) successive switching approach. Reprinted with permission from [60]. Copyright 2008, American Institute of Physics.

the n-th pulse is the same as that after a single pulse with duration of $t = \tau_1 + \tau_2 + \ldots + \tau_n$. Then, all the PFM images taken before the $(n + 1)$-th pulse reveal the successive domain wall evolution during the time period of t.

Both variations of the S-PFM method rely on stability of instantaneous domain patterns between pulse applications. Whether polarization relaxation takes place or not can be checked by comparing the PFM switching with transient current measurements. Little or no discrepancy between the two sets of data obtained in most reports is solid proof of the reliability of the stroboscopic PFM approach [61].

An example of stroboscopic PFM imaging is illustrated in Figure 6.10. It shows snapshots of domain structure evolution during switching in epitaxial (001) PZT capacitors [62]. Switching as a whole occurs via dynamics of 180° domains and no 90° wall formation has been observed in agreement with the earlier conclusion based on transient current measurements [63]. For $E = 700$ kV/cm, the nucleation density was estimated to be 7.1×10^8 cm^{-2} (Figure 6.10(a)), which is well below the nucleation density of 3.2×10^{12} m^{-2} measured in polycrystalline PZT capacitors [57, 58]. Given a higher concentration of point defects associated with grain boundaries and relatively rough interfaces in polycrystalline films, this difference is reasonable. The time-dependent evolution of domain structure reveals that the wall velocity is isotropic and independent of domain size in the range at least up to 400 nm in diameter. A considerable reduction in the wall velocity has been observed for domains in close proximity ($<$ 100 nm) to other growing domains.

Significant anisotropy of the wall velocity has been detected during switching in the lower field range ($<$500 kV/cm). This anisotropy is manifested by the formation of characteristically elongated domains (Figure 6.10(b)). Measurement of domain size as a function of time shows that in the y-axis direction the wall velocity v_y is almost three times higher than velocity v_x in the orthogonal x-axis direction (0.39 m/s and 0.12 m/s, respectively). Overall, the switching kinetics in the low fields is characterized by a long-range (up to 1 μm) lateral growth of just a few nucleated domains, in stark contrast to the switching in the high fields.

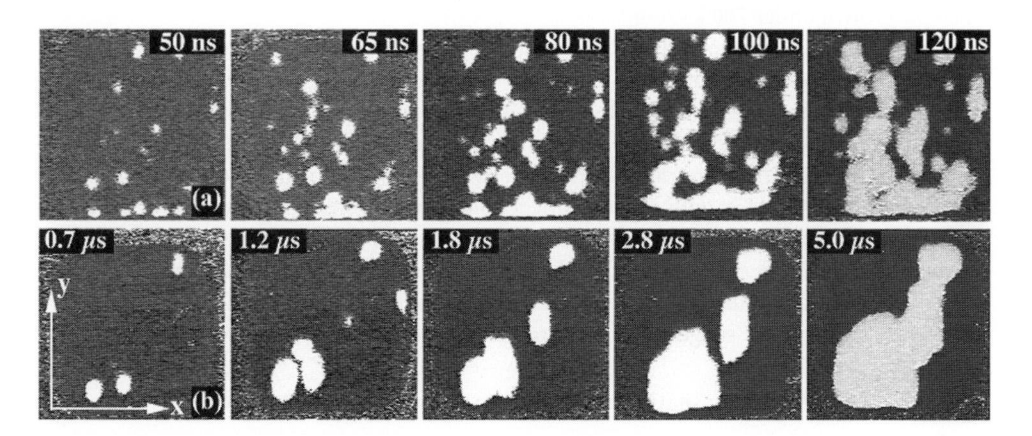

Fig. 6.10 PFM phase images of domain configurations developing in epitaxial (001) PZT capacitors (a) E = 700 kV/cm; (b) E = 500 kV/cm. The scan size is $2.5 \times 2.5 \mu m^2$. Reprinted with permission from [62]. Copyright 2010, American Institute of Physics.

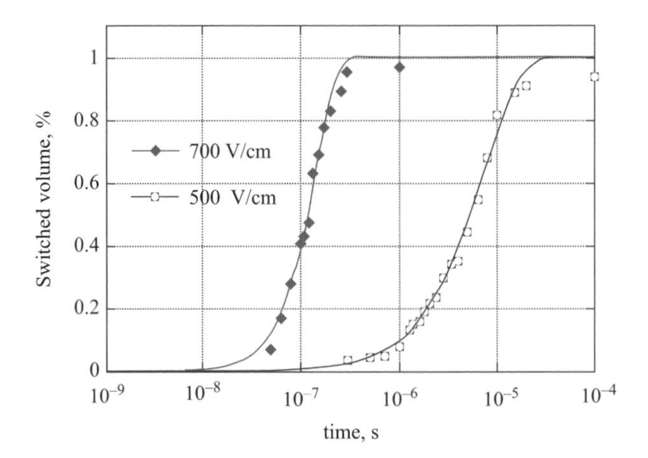

Fig. 6.11 Fitting of the polarization switching kinetics by the KAI model. Domain dimensionality n is 1.96 for E=700 kV/cm and 1.12 for E=500 kV/cm. Reprinted with permission from [62]. Copyright 2010, American Institute of Physics.

For epitaxial capacitors, the KAI model provides excellent description of the time-dependent switching behavior in epitaxial capacitors, contrary to the case of polycrystalline PZT capacitors (Figure 6.11). It has been found that n values are different for low and high fields and very close to integer values: 1.12 for 500 kV/cm and 1.96 for 700 kV/cm, indicating a change in the switching mechanism. In the KAI model, $n = 2$ corresponds to a growth of cylindrical domains (2D growth) and $n = 1$ indicates a lateral expansion of lamellar domains (1D growth). An increase in n with an applied field, also reported by So *et al.* [64], is consistent with the change in domain growth dimensionality revealed by stroboscopic PFM studies—transition from one-dimensional anisotropic growth under 500 kV/cm to two-dimensional isotropic growth under 700 kV/cm.

A general approach adopted by Noh *et al.* [65] treats domain wall motion as a nonlinear dynamic process, resulting from competition between elastic and pinning forces. Investigations of domain growth in a wide range of temperatures and electric fields showed that the wall dynamics could be classified into creep, depinning, and flow regimes. Wall velocity exhibits high temperature dependence in low fields, while in high fields a crossover to the flow regime makes it temperature independent. In the creep regime at low fields, the wall dynamics can be described as thermally activated propagation between pinning sites: $v \sim \exp[-(U/k_B T)(E_c/E)^{\mu}]$, where U is an energy barrier and μ is a dynamic exponent, reflecting the nature of the pinning potential. In epitaxial PZT capacitors $\mu = 0.9 \pm 0.1$ over a wide temperature range, suggesting a long-range pinning potential.

6.3.3 PFM Spectroscopy: spatial variability of switching parameters

Obtaining information on spatially resolved local switching parameters is important both for the fundamental understanding of the scaling effect in ferroelectrics and for development of ferroelectric-based devices, such as ferroelectric random access memory (FeRAM) [5]. FeRAM devices have been in commercial production for more than 10 years by several semiconductor companies, such as Ramtron, Toshiba, Matsushita, and Fujitsu, and can be found in RFID cards, smart meters, data controllers, and game consoles to name just a few. However, development of high-density FeRAM devices requires capacitors with sub-μm dimensions, thus necessitating new levels of understanding of their physical properties. Among the most important questions is the effect of capacitor scaling on its switching parameters, the minimum size of operational FeRAM cell, capacitor-to-capacitor variations in remanent polarization, reliability, and retention.

Scaling of device dimensions to the range where ferroelectric materials start to show a pronounced size effect has emphasized the importance of nanoscale studies of ferroelectric switching at the appropriate spatial and time scales. Measurement of polarization hysteresis loops and transient currents are typically the ways to test the switching behavior of ferroelectric capacitors. These methods, however, provide little insight in the actual mechanism of switching, which involves domain nucleation and growth via motion of domain walls of different types. Particularly, interpretation of the hysteresis loops is quite difficult and needs integral calculation methods.

Because of the extremely small size of the ferroelectric volume excited during PFM measurements ($\sim 10^3$ nm^3), this approach offers the unique possibility of studying the local switching parameters at a level commensurate with structural defects [16, 66]. By establishing correlation between global switching behavior and local switching parameters, PFM can predict how the average coercive bias, imprint, and remanent polarization will scale with capacitor size.

The PFM spectroscopic approach is based on monitoring the electromechanical response at a fixed tip position as a function of a dc poling voltage swept in the cyclic manner [54]. The measured behavior is referred to as a local piezoelectric, or PFM, hysteresis loop [67, 68, 69, 70, 71]. If performed on a macroscopic scale, this loop corresponds to a weak-field piezoelectric coefficient tuned by a continuously varying bias field. According to a linearized electrostriction equation, the piezoelectric coefficient can often be expressed as $d_{33} = 2Q\varepsilon_{33}P_3$, where Q is the longitudinal electrostriction coefficient and ε_{33} and P_3 are the corresponding dielectric constant and spontaneous polarization values. Therefore, the d_{33} variation reflects polarization switching with corresponding tuning of the dielectric permittivity $\varepsilon(\mathrm{E})$ and polarization $P(\mathrm{E})$, which both affect the piezoelectric coefficient. This method was further developed by Kalinin's group into the switching-spectroscopy PFM method (SS-PFM) [72, 73, 74], which involves point-by-point acquisition of the local hysteresis loops over a grid of points with an average spacing of approximately several tens of nanometers. Subsequent numerical analysis of the obtained loops allows extraction of the local switching parameters, such as threshold (nucleation) bias, imprint, and remanent polarization, which can be represented in the form of 2D maps.

Structural imperfections of the ferroelectric layers (point defects, dislocations, etc.) result in spatial inhomogeneity of the local switching parameters, reflected in strong variations in the local hysteresis loops. This effect is illustrated in Figure 6.12(a), which shows local PFM hysteresis loops obtained in three different locations on the top electrode of a polycrystalline PZT capacitor [75]. All three locations are characterized by different switching parameters, such as imprint and coercive bias, which are also quite different from the corresponding macroscopic parameters of the typical polarization P-V hysteresis loops measured by the Sawyer–Tower method (Figure 6.12(b)). Further insight into the spatial variability of properties is provided by a two-dimensional map of the local imprint bias obtained by SS-PFM (Figure 6.12(c)). Spatial

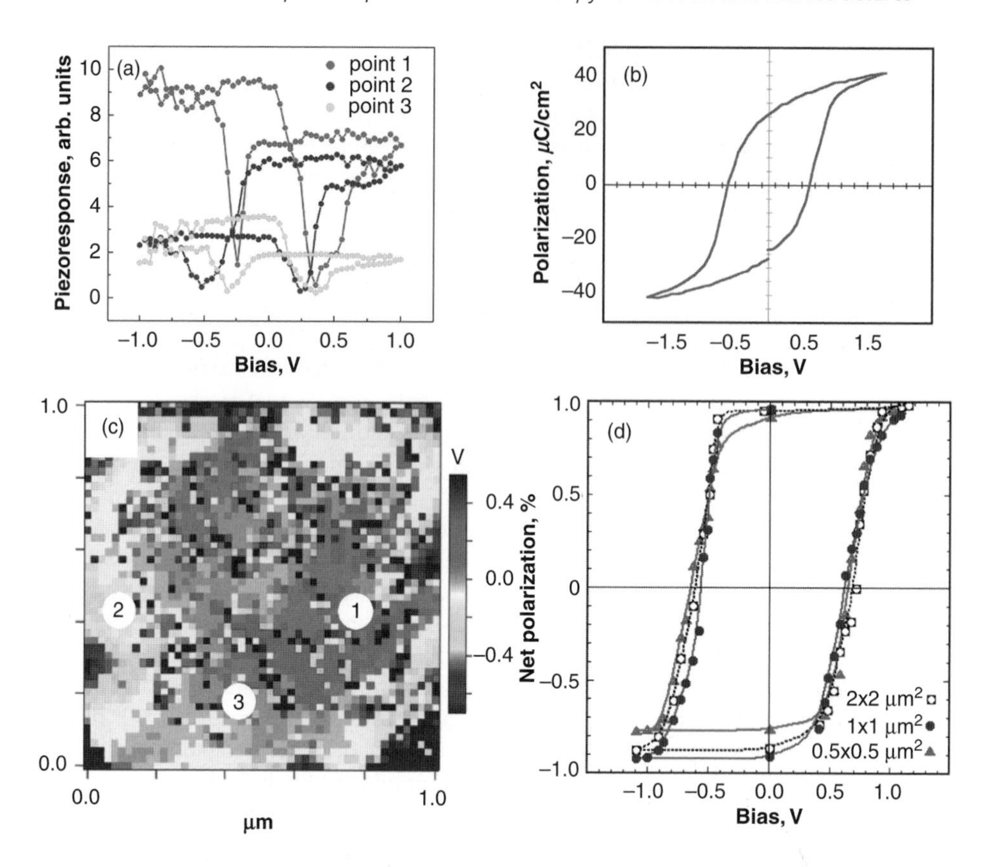

Fig. 6.12 (a) Local PFM hysteresis loops measured in different locations on the surface of the $1.0 \times 1.0 \mu m^2$ PZT capacitor. Loop numbers correspond to location numbers shown in (c). (b) Conventional P-E hysteresis loop measured in the same type of PZT capacitor, but larger dimensions ($65 \times 65 \mu m^2$). (c) 2D map of an imprint bias in the $1.0 \times 1.0 \mu m^2$ PZT capacitor generated by SS-PFM. (d). Global PFM hysteresis loops measured in capacitors of different dimensions. Reprinted with permission from [75]. Copyright 2009, American Institute of Physics. This figure is reproduced in color in the color plate section.

distribution of the imprint and coercive bias gives a sense of the expected domain dynamics in the capacitor during switching.

The relationship between macroscopic polarization hysteresis and local switching parameters can be established by employing a bias-dependent PFM (BD-PFM) approach to visualize domain structure evolution as a function of the dc poling voltage. Figures 6.13 and 6.14 show profiles of domain patterns developing during switching in $0.5 \times 0.5 \mu m^2$ and $2 \times 2 \mu m^2$ PZT capacitors, respectively, under an external dc bias, incrementally increasing in the positive and negative directions. It can be seen that larger capacitors exhibit a number of nucleation sites. For smaller capacitors, the switching is mainly dominated by domain wall motion. An

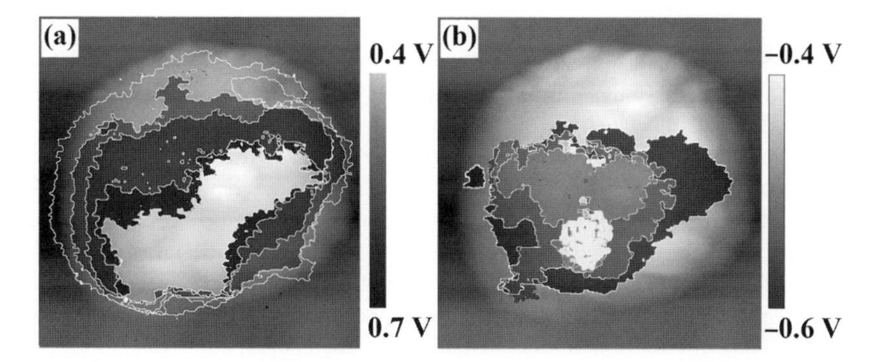

Fig. 6.13 Domain structure evolution delineated by means of BD-PFM in the 0.5 × 05μm² capacitor (scan size is 0.65 × 0.65μm²); (a) Switching to a positive polarization state. Domain profiles are shown for a dc bias of 0.40 V, 0.55 V, 0.60 V, 0.70 V. (b) Switching to a negative polarization state. Domain profiles are shown for a dc bias of -0.40 V, −0.45 V, −0.50 V, −0.60 V. Reprinted with permission from [75]. Copyright 2009, American Institute of Physics.

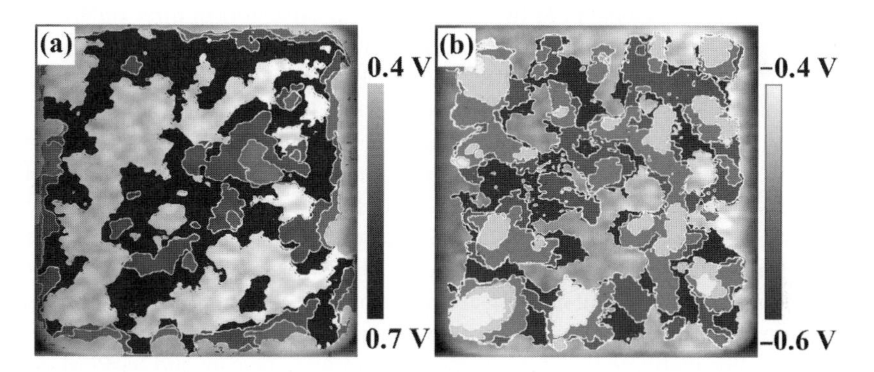

Fig. 6.14 Domain structure evolution delineated by means of BD-PFM in the 2.0 × 2.0μm² capacitor. (a) Switching to a positive polarization state. Domain profiles are shown for a dc bias of 0.4 V, 0.5 V, 0.55 V, 0.7 V. (b) Switching to a negative polarization state. Domain profiles are shown for a dc bias of −0.45 V, −0.50 V, −0.55 V, −0.60 V, −0.65 V. Reprinted with permission from [75]. Copyright 2009, American Institute of Physics.

interesting feature here is asymmetry of switching in the positive and negative directions for the smaller capacitor (Figure 6.13). While for the negative direction switching proceeds via lateral expansion of a singe domain nucleating in the capacitor center, for the positive direction it starts from the electrode edge via movement of the residual domain wall. A similar effect of the electrode edge, although to a much smaller extent, can be observed in the larger capacitor. This observation is consistent with the imprint behavior in the electrode edge region detected by hysteresis loop measurements (Figure 6.12(a)). This effect can be explained by a higher defect concentration at the perimeter region generated during top electrode fabrication [76]. Another feature is the roughening of domain wall topology, appearing as a number of nanoscale protrusions with a curvature radius of 10–30 nm (Figure 6.14). This observation implies that the domain walls do not propagate in a continuous manner, but rather by triggering switching in nanoscale regions in front of the wall. As the average grain size in the PZT layer is about 50 nm, it can be also concluded that these nanoscale switched regions correspond to individual grains. This effect can be attributed to a short-range electrostatic or mechanical coupling between the adjacent grains, reported elsewhere [77, 78]. Given that the coercive voltage of a grain depends upon its crystallographic orientation [66], this type of domain wall propagation is consistent with the nucleation-limited switching mechanism reported in polycrystalline PZT capacitors, which infers a wide distribution of local switching parameters [79, 80].

Using commercial software (WS × M v.2.2, Nanotec Electronica), the PFM phase and amplitude images in Figures 6.13 and 6.14 have been converted to the switched capacitor volume for a given value of the applied bias, and subsequently to global PFM hysteresis loops (Figure 6.12(d)), which are analogous to the conventional polarization P-V hysteresis loop. What is important is that the global PFM loops can be measured for capacitors with dimensions well into the submicron range, thereby overcoming the limitations of the P-E measurements and allowing one to address the capacitor scaling effect on the global switching parameters. Figure 6.12(d) shows the global PFM hysteresis loops for capacitors of three different dimensions. Surprisingly, hysteresis loop width, coercive voltage, and imprint are very similar for all three capacitor sizes. These values are also close to the parameters of the P-V loops measured in capacitors of much larger dimensions (Figure 6.12(b)). The capacitor scaling effect is manifested only in the decrease of the remanent negative polarization for $0.5 \times 0.5 \mu m^2$ capacitors (by about 20%) due to the pinned positive polarization along the capacitor perimeter. Indeed, local hysteresis loop measurements (Figure 6.12(a)), SS-PFM mapping (Figure 6.12(c)), and BD-PFM imaging data (Figures 6.13 and 6.14) all indicate that the perimeter area exhibits positive imprint, which precludes its complete switching.

Thus, combined application of imaging and spectroscopic PFM approaches allows bridging between the macroscopic and local switching behavior in submicron ferroelectric capacitors by numerically analyzing the bias-dependent PFM imaging data and SS-PFM switching parameters.

6.4 Summary

Rapid development of ferroelectric-based devices with reduced dimensions generated a strong need for extensive investigation of the size effects in these materials. Application of the advanced modes of PFM provides a unique opportunity to study the physical mechanisms underlying the static and dynamic properties of ferroelectric structures at the nanoscale level. In future, state-of-the-art SPM methods, in conjunction with first-principles modeling and atomically controlled growth, will play an increasingly important role in investigating the fundamental issues related to the critical behavior and switching dynamics in ultrathin ferroelectric and multiferroic heterostructures, the interplay between ferroelectric and ferromagnetic order parameters, the role of structural defects and interfacial properties in electronic transport behavior, and the scalability of these heterostructures to micro-, meso-, and nano-dimensions.

Acknowledgments

This work was supported by the National Science Foundation (Grant No. MWN DMR-1007943) and by the U.S. Department of Energy, Office of Basic Energy Sciences, Division of Materials Sciences and Engineering under Award DE-SC0004876. The author would like to thank Dr. S. V. Kalinin, Dr. B. J. Rodriguez, Dr. R. Proksch, Dr. C. Harnagea and Professor T.W. Noh for their kind permission to use their published data and Professor H. Funakubo and Professor C.-B. Eom for use of their samples.

References

[1] http://www.scienceonline.org/cgi/content/full/318/5858/1848.
[2] *Ferroelectrics at Nanoscale: Scanning Probe Microscopy Approach*, M. Alexe and A. Gruverman (eds), (Springer, 2004).
[3] A. Gruverman and A. Kholkin, Nanoscale Ferroelectrics: Processing, Characterization and Future Trends Rep. Prog. Phys. **69**, 2443–2474 (2006).
[4] M. E. Lines and A. M. Glass, *Principles and Applications of Ferroelectric and Related Materials* (Clarendon, Oxford, 1977).
[5] J. F. Scott, *Ferroelectric Memories* (Springer, Berlin, 2000).
[6] *Nanoelectronics and Information Technology: Advanced Electronic Materials and Novel Devices*, R. Waser (ed), (Wiley-VCH, Berlin, 2005).
[7] K. Uchino, *Ferroelectric Devices* (Marcel Decker, Inc, New York, 2000).
[8] M. Dawber, K. M. Rabe, and J. F. Scott, Physics of thin-film ferroelectric oxides, Rev. Mod. Phys. **77**, 1083 (2005).
[9] V. L. Ginzburg, Polarization and Piezoelectric Effect in $BaTiO_3$ Near the Ferroelectric Transition Point, Zh. Eksp. Teor. Fiz. **19**, 36 (1949).
[10] W. Cochran, Crystal stability and the theory of ferroelectricity, Adv. Phys. **9**, 387 (1960).
[11] V. L. Ginzburg, The Dielectric Properties of Crystals of Seignettoelectric Substances and of Barium Titanate, Zh. Eksp. Teor. Fiz. **15**, 739 (1945).
[12] A. F. Devonshire, Theory of $BaTiO_3$ (I), Philos. Mag. **40**, 1040 (1949).
[13] A. K. Tagantsev, L. E. Cross and J. Fousek, *Domains in Ferroic Crystals and Thin Films* (Springer , 2010).
[14] E. Fatuzzo and W. J. Merz, *Ferroelectricity* (North-Holland Publishing Company: Amsterdam, John Wiley & Sons Inc.: New York, 1967).
[15] Scott, J. F., Kammerdiner, L., Parris, M., *et al.* Switching Kinetics of Lead Zirconate Titanate Submicron Thin Film Memories, J. Appl. Phys., **64**, 787 (1988).
[16] Jesse, S., Rodriguez, B. J., Choudhury, S., *et al.* Direct imaging of the spatial and energy distribution of nucleation centres in ferroelectric materials, Nature Mater. **7**, 209 (2008).
[17] Jo, J. Y., Yang, S. M., Kim, T. H., *et al.* Nonlinear Dynamics of Domain-Wall Propagation in Epitaxial Ferroelectric Thin Films, Phys. Rev. Lett. **102**, 045701 (2009).
[18] Tybell, T., Paruch, P., Giamarchi, T., and Triscone, J. M., Domain Wall Creep in Epitaxial Ferroelectric $Pb(Zr_{0.2}Ti_{0.8})O_3$ Thin Films, Phys. Rev. Lett. 89, 097601 (2002).
[19] V. Gopalan and M. C. Gupta, Origin of internal field and visualization of 180° domains in congruent $LiTaO_3$ crystals, J. Appl. Phys. **80**, 6099 (1996).
[20] Landauer, R., J. Electrostatic Considerations in $BaTiO_3$ Domain Formation during Polarization Reversal, Appl. Phys. **28**, 227 (1957).

[21] A. M. Bratkovsky, and A. P. Levanyuk, Easy Collective Polarization Switching in Ferroelectrics Phys. Rev. Lett. 85, 4614 (2000).

[22] Y. Ishibashi and Y. Takagi, Note on Ferroelectric Domain Switching, J. Phys. Soc. Jap. 31, 506 (1971).

[23] M. Avrami, Kinetics of Phase Change. I General Theory, J. Chem. Phys. **7**, 1003 (1939).

[24] A.N. Kolmogorov, On the Statistical Theory of Metal Crystallization, Izv. Akad. Nauk USSR; Ser. Math. **3**, 355 (1937).

[25] A. Gruverman and Y.Ikeda, Characterization and Control of Domain Structure in $SrBi_2Ta_2O_9$ Thin Films by Scanning Force Microscopy, Jpn. J. Appl. Phys. 37, Part 2, L939 (1998).

[26] N. W. Dalton, J. T. Jacobs, and B. D. Silverman, Phys. Rev. **133 A**, 1034 (1964).

[27] A. Gruverman, N. Ponomarev, and K. Takahashi, Domain Nucleation During Polarization Reversal in Lead Germanate Jpn. J. Appl. Phys. 33, Part 1, 5536 (1994).

[28] J. Y. Jo, H. S. Han, J.-G. Yoon, *et al.* Domain Switching Kinetics in Disordered Ferroelectric Thin Films, Phys. Rev. Lett. 99, 267602 (2007).

[29] X.F. Du, and I. W. Chen, Model experiments on fatigue of $Pb(Zr_{0.53}Ti_{0.47})O_3$ ferroelectric thin films, Appl. Phys. Lett. **72**, 1923 (1998).

[30] Tagantsev, A. K., Stolichnov, I., Setter, N., Cross, J. S., and Tsukada, M., Non-Kolmogorov-Avrami switching kinetics in ferroelectric thin films, Phys. Rev. B 66, 214109 (2002).

[31] Grain-size effects in ferroelectric switching, H. M. Duiker and P. D. Beale, Phys. Rev. B **41**, 490 (1990).

[32] T. Mitsui and J. Furuichi, Domain Structure of Rochelle Salt and KH_2PO_4, Phys. Rev. 90, 193 (1953).

[33] C. Thompson, D.D. Fong, R.V. Wang, *et al.* Imaging and alignment of nanoscale 180° stripe domains in ferroelectric thin films, Appl. Phys. Lett. **93**, 182901 (2008).

[34] K. J. Choi, M. Biegalski, Y. L. Li, *et al.* Enhancement of Ferroelectricity in Strained $BaTiO_3$ Thin Films, Science **306**, 1005 (2004).

[35] G. Catalan, B. Noheda, J. McAneney, L. J. Sinnamon, and J. M. Gregg, Strain gradients in epitaxial ferroelectrics, Phys. Rev. **B 72**, 020102 (2005).

[36] A. M. Bratkovsky and A. P. Levanyuk, Smearing of Phase Transition due to a Surface Effect or a Bulk Inhomogeneity in Ferroelectric Nanostructures, Phys. Rev. Lett. **94**, 107601 (2005).

[37] A. M. Bratkovsky and A. P. Levanyuk, Abrupt Appearance of the Domain Pattern and Fatigue of Thin Ferroelectric Films, Phys. Rev. Lett. **84**, 3177 (2000).

[38] K.M.Rabe, Theoretical investigations of epitaxial strain effects in ferroelectric oxide thin films and superlattices, Curr. Opin. Solid State Mat. Sci. **9**, 122 (2005).

[39] J. Junquera and P. Ghosez, First-Principles Study of Ferroelectric Oxide Epitaxial Thin Films and Superlattices: Role of the Mechanical and Electrical Boundary Conditions, J. Comput. Theor. Nanosci. **5**, 2071 (2008).

[40] J. Junquera and P. Ghosez, Critical thickness for ferroelectricity in perovskite ultrathin films, Nature **422**, 506 (2003).

[41] H. Fu and L. Bellaiche, Ferroelectricity in Barium Titanate Quantum Dots and Wires, Phys. Rev. Lett. **91**, 257601 (2003).

[42] Y. S. Kim, D. H. Kim, J. D. Kim *et al.*, Critical thickness of ultrathin ferroelectric $BaTiO_3$ films, Appl. Phys. Lett. **86**, 102907 (2005).

[43] A. Gruverman and S.V. Kalinin, Piezoresponse force microscopy and recent advances in nanoscale studies of ferroelectrics, J. Mat. Sci. **41**, 107 (2006).

[44] L. Tian, A. Vasudevarao, A. N. Morozovska, *et al.* Nanoscale polarization profile across a 180° ferroelectric domain wall extracted by quantitative piezoelectric force microscopy, J. Appl. Phys. **104**, 074110 (2008).

[45] R. Nath, Y.-H. Chu, N. A. Polomoff, R. Ramesh, and B. D. Huey, High speed piezoresponse force microscopy: <1 frame per second nanoscale imaging, App. Phys. Lett. **93**, 072905 (2008).

[46] S. V. Kalinin, B. J. Rodriguez, S. Jesse, *et al.* Intrinsic single-domain switching in ferroelectric materials on a nearly ideal surface, Proc. Nat. Acad. Sci. **104**, No 51, 20204–20209 (2007).

[47] C. Dehoff, B. J. Rodriguez, A. I. Kingon, *et al.* Atomic force microscopy-based experimental setup for studying domain switching dynamics in ferroelectric capacitors, Rev. Sci. Instrum. **76**, 023708 (2005).

[48] Wang YG, Kleemann W, Woike T, and Pankrath R., Atomic force microscopy of domains and volume holograms in $Sr_{0.61}Ba_{0.39}Nb_2O_6{:}Ce^{3+}$, Phys. Rev. **B 61**, 3333 (2000).

[49] A. Gruverman, B.J. Rodriguez, R.J. Nemanich, and A.I. Kingon, Nanoscale observation of photoinduced domain pinning and investigation of imprint behavior in ferroelectric thin films, J. Appl. Phys. **92**, 2734 (2002).

[50] J. F. Scott, Nanoferroelectrics: statics and dynamics, J. Phys.: Condens. Matter **18,** R361 (2006).

[51] C. Harnagea, A. Pignolet, M. Alexe, and D. Hesse, Higher-order electromechanical response of thin films by contact resonance piezoresponse force microscopy, IEEE Transactions on Ultrasonics, Ferroelectrics, and Frequency Control, **53**, No. 12, 2309–2322 (2006).

[52] U. Rabe, M. Kopycinska, S. Hirsekorn, *et al.* High-resolution characterization of piezoelectric ceramics by ultrasonic scanning force microscopy techniques, J. Phys. D, Appl. Phys. **35**, 2621 (2002).

[53] B.J. Rodriguez, C. Callahan, S. Kalinin, and R. Proksch, Dual-frequency resonance-tracking atomic force microscopy, Nanotechnology **18**, 475504 (2007).

[54] *Scanning Probe Microscopy of Electrical and Electromechanical Phenomena at the Nanoscale*, edited by S.V. Kalinin and A. Gruverman (Springer, 2006).

[55] S. V. Kalinin, B. J. Rodriguez, S.-H. Kim, *et al.* Imaging mechanism of piezoresponse force microscopy in capacitor structures, Appl. Phys. Lett. **92**, 152906 (2008).

[56] N. A. Polomoff, R. Nath, J. L. Bosse, and B. D. Huey, Single ferroelectric domain nucleation and growth monitored by high speed piezoforce microscopy, J. Vac. Sci. & Technol. B **27**, 1011 (2009).

[57] A. Gruverman, B. J. Rodriguez, C. Dehoff, *et al.* Direct studies of domain switching dynamics in thin film ferroelectric capacitors, Appl. Phys. Lett. **87**, 082902 (2005).

[58] A. Gruverman, D. Wu, and J.F. Scott, Piezoresponse Force Microscopy Studies of Switching Behavior of Ferroelectric Capacitors on a 100-ns Time Scale, Phys. Rev. Lett. **100**, 097601 (2008).

[59] D. J. Kim, J. Y. Jo, T. H. Kim, *et al.* Observation of inhomogeneous domain nucleation in epitaxial $Pb(Zr,Ti)O_3$ capacitors, Appl. Phys. Lett. **91**, 132903 (2007).

[60] S. M. Yang, J. Y. Jo, D. J. Kim, *et al.* Domain wall motion in epitaxial Pb(Zr,Ti)O_3 capacitors investigated by modified piezoresponse force microscopy, Appl. Phys. Lett. **92**, 252901 (2008).

[61] D.A. Bonnell, S.V. Kalinin, A.Kholkin, and A. Gruverman, Piezoresponse Force Microscopy: A Window into Electromechanical Behavior at the Nanoscale, MRS Bulletin **34**, 648 (2009).

[62] D. Wu, I. Vrejoiu, M. Alexe, and A. Gruverman, Anisotropy of domain growth in epitaxial ferroelectric capacitors, Appl. Phys. Lett. **96**, 112903 (2010).

[63] W. Li, and M. Alexe, Investigation on switching kinetics in epitaxial Pb($Zr_{0.2}Ti_{0.8}$)O_3 ferroelectric thin films: Role of the 90° domain walls, Appl. Phys. Lett. **91**, 262903 (2007).

[64] Y. W. So, D. J. Kim, T. W. Noh, J.-G. Yoon, and T. K. Song, Polarization switching kinetics of epitaxial Pb ($Zr_{0.4}Ti_{0.6}$)O_3 thin films, Appl. Phys. Lett. **86**, 092905 (2005).

[65] J. Y. Jo, S. M. Yang, T. H. Kim, *et al.* Phys. Nonlinear Dynamics of Domain-Wall Propagation in Epitaxial Ferroelectric Thin Films, Rev. Lett. **102,** 045701 (2009).

[66] S. V. Kalinin, A. Gruverman, and D. A. Bonnell, Quantitative analysis of nanoscale switching in $SrBi_2Ta_2O_9$ thin films by piezoresponse force microscopy, Appl. Phys. Lett. **85**, 795 (2004).

[67] J. Hong, H. W. Song, S. Hong, H. Shin, and K. No, Fabrication and investigation of ultrathin, and smooth Pb(Zr,Ti)O_3 films for miniaturization of microelectronic devices, J. Appl. Phys. **92**, 7434 (2002).

[68] H. Y. Guo, J. B. Xu, I. H. Wilson, *et al.*, Study of domain stability on ($Pb_{0.76}Ca_{0.24}$)TiO_3 thin films using piezoresponse microscopy, Appl. Phys. Lett. **81**, 715 (2002).

[69] M. Abplanalp, J. Fousek, and P. Gunter, Higher Order Ferroic Switching Induced by Scanning Force Microscopy, Phys. Rev. Lett. **86,** 5799 (2001).

[70] C. Harnagea, A. Pignolet, M. Alexe, *et al.*, Quantitative ferroelectric characterization of single submicron grains in Bi-layered perovskite thin films, Appl. Phys. A - Mater. **70**, 261 (2000).

[71] C. Harnagea, A. Pignolet, M. Alexe, *et al.*, Nanoscale Switching and Domain Structure of Ferroelectric $BaBi_4Ti_4O_{15}$ Thin Films, Jpn. J. Appl. Phys. **38**, Part 2, L1255 (1999).

[72] S. Jesse, A. P. Baddorf, and S. V. Kalinin, Switching spectroscopy piezoresponse force microscopy of ferroelectric materials, Appl. Phys. Lett. **88,** 062908 (2006).

[73] B. J. Rodriguez, S. Jesse, M. Alexe, and S. V. Kalinin, Spatially Resolved Mapping of Polarization Switching Behavior in Nanoscale Ferroelectrics, Adv. Mat. **20**, 109 (2008).

[74] P. Bintachitt, S. Trolier-McKinstry, K. Seal, S. Jesse, and S. V. Kalinin, Switching spectroscopy piezoresponse force microscopy of polycrystalline capacitor structures, Appl. Phys. Lett. **94**, 042906 (2009).

[75] D. Wu,I. Kunishima, S. Roberts, and A. Gruverman, Spatial variations in local switching parameters of ferroelectric random access memory capacitors, Appl. Phys Lett. **95**, 092901 (2009).

[76] M. Grossmann, O. Lohse, D. Bolte, *et al.* Imprint in Ferroelectric $SrBi_2Ta_2O_9$ Capacitors for Non-Volatile Memory Applications, Integrated Ferroelectrics **22**, 95 (1998).

[77] I. Stolichnov, L. Malin, E. Colla, A. K. Tagantsev, and N. Setter, Microscopic aspects of the region-by-region polarization reversal kinetics of polycrystalline ferroelectric Pb(Zr,Ti)O_3 films, Appl. Phys. Lett. **86**, 012902 (2005).

[78] B. J. Rodriguez, A. Gruverman, A. I. Kingon, R. J. Nemanich, and J. S. Cross, Investigation of the mechanism of polarization switching in ferroelectric capacitors by three-dimensional piezoresponse force microscopy, Appl. Phys. **A80**, 99 (2005).

[79] X. F. Du and I. W. Chen, Model experiments on fatigue of $Pb(Zr_{0.53}Ti_{0.47})O_3$ ferroelectric thin films , Appl. Phys. Lett. **72,** 1923 (1998).

[80] A. K. Tagantsev, I. Stolichnov, N. Setter, J. S. Cross, and M. Tsukada, Non-Kolmogorov-Avrami switching kinetics in ferroelectric thin films, Phys. Rev. B **66**, 214109 (2002).

Part III

Oxide films and interfaces: functional properties

7 General considerations of the electrostatic boundary conditions in oxide heterostructures

Takuya Higuchi and Harold Y. Hwang

7.1 Introduction

When the size of materials is comparable to the characteristic length scale of their physical properties, novel functionalities can emerge. For semiconductors, this is exemplified by the "superlattice" concept of Esaki and Tsu, where the width of the repeated stacking of different semiconductors is comparable to the "size" of the electrons, resulting in novel confined states now used routinely in opto-electronics [1]. For metals, a good example is magnetic/non-magnetic multilayer films that are thinner than the spin-scattering length, from which giant magnetoresistance (GMR) emerged [2, 3], used in the read heads of hard disk drives. For transition metal oxides, a similar research program is currently underway, broadly motivated by the vast array of physical properties that they host. This long-standing notion has recently been invigorated by the development of atomic-scale growth and probe techniques, which enable the study of complex oxide heterostructures approaching the precision idealized in Figure 7.1(a). Taking the subset of oxides derived from the perovskite crystal structure, the close lattice match across many transition metal oxides presents the opportunity, in principle, to develop a "universal" heteroepitaxial materials system.

Hand-in-hand with the continual improvements in materials control, an increasingly relevant challenge is to understand the consequences of the electrostatic boundary conditions which arise in these structures. The essence of this issue can be seen in Figure 7.1(b), where the charge sequence of the sublayer "stacks" for various representative perovskites is shown in the ionic limit, in the (001) direction. To truly "universally" incorporate different properties using different materials components, be it magnetism, ferroelectricity, superconductivity, etc., it is necessary to access and join different charge sequences, labeled here in analogy to the designations "group IV, III-V, II-VI" for semiconductors. As we will review, interfaces between different families create a host of electrostatic issues. This can be avoided somewhat if, as in many semiconductor heterostructures, only one family is used, with small perturbations (such

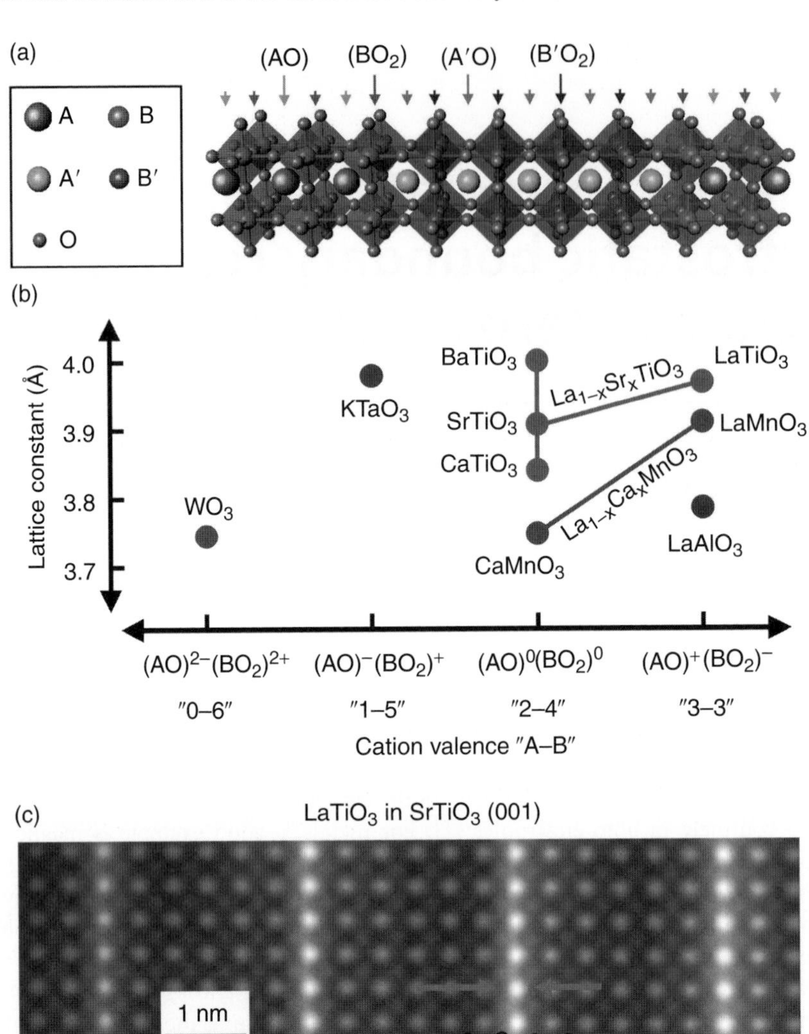

Fig. 7.1 (a) Schematic illustration of ideal heterointerfaces of two perovskites ABO_3 and $A'B'O_3$ stacked in the [001] direction. (b) Charge sequences of the AO and BO_2 planes of perovskites plotted together with their pseudocubic lattice parameters. (c) Scanning transmission electron microscopy image of a $LaTiO_3/SrTiO_3$ (001) superlattice (Ohtomo *et al.* [4]). This figure is reproduced in color in the color plate section.

as n-type or p-type doping) around them.[1] However, for most transition metal oxides, this is greatly restrictive. For example, $LaMnO_3$ and $SrMnO_3$ are both insulators in part due to strong electron correlations, and only in their solid solution does "colossal magnetoresistance" emerge in bulk [6]. Similarly, the metallic superlattice shown in Figure 7.1(c) can be considered a nanoscale deconstruction of $(La,Sr)TiO_3$ to the insulating parent compounds. Therefore the aspiration to arbitrarily mix and match perovskite components requires a basic understanding of, and ultimately control over, these issues.

In this context, we present here basic electrostatic features that arise in oxide heterostructures which vary the ionic charge stacking sequence. In close relation to the analysis of the stability of polar surfaces and semiconductor heterointerfaces, variation of the dipole moment across a heterointerface plays a key role in determining its stability. Different self-consistent assignments of the unit-cell are presented, allowing the *polar discontinuity* picture to be recast in terms of an equivalent *local charge neutrality* picture. The latter is helpful in providing a common framework with which to discuss electronic reconstructions, local-bonding considerations, crystalline defects, and lattice polarization on an equal footing, all of which are the subject of extensive current investigation.

7.2 The *polar discontinuity* picture

7.2.1 Stability of ionic crystal surfaces

The surface of crystals determines many of their physical, mechanical, and chemical properties. Because of the lack of translational symmetry in the perpendicular direction, the stable charge distribution at the surface can be completely different from that of the bulk, and the surface may reconstruct in a manner different from the bulk states. Imagine an ideal ionic crystal which consists of charged ions bound together by their attractive interactions, and all the ions are taken as fixed point charges. Since the charges are locally preassigned to the ions in this model, the ideal ionic surface apparently requires no reassignment of the charges from that of the bulk. However, the electrostatic potential in an ionic crystal diverges when there is a dipole moment in the unit-cell perpendicular to the surface. The potential[2] ϕ should be constant in vacuum in the absence of external fields, and the potential can be obtained by integrating the electric field caused by the charged sheets, as shown in Figure 7.2. A finite shift in the potential emerges due to the dipole moment of each unit-cell, and as the unit-cells are stacked, so the potential grows, and diverges into the crystal. Because of this effectively infinite surface energy, such surfaces cannot exist without reconstructions, and the stability of an ionic surface randomly cut from the bulk is not trivial without knowing the stacking sequence of the charged sheets precisely. This surface instability and the associated reconstructions have indeed been observed by means of low-energy electron diffraction (LEED) and ion scattering, where absorption of foreign atoms, surface roughening, or changes in surface stoichiometry were found [7–9].

[1] These effects can in principle also be reduced by choosing a (110) growth orientation, but other aspects of stability may be limiting [5].

[2] Note that literature on this topic uses both the electrostatic potential (for a positive test charge) and the electron potential energy (as in band diagrams)—we use the former here.

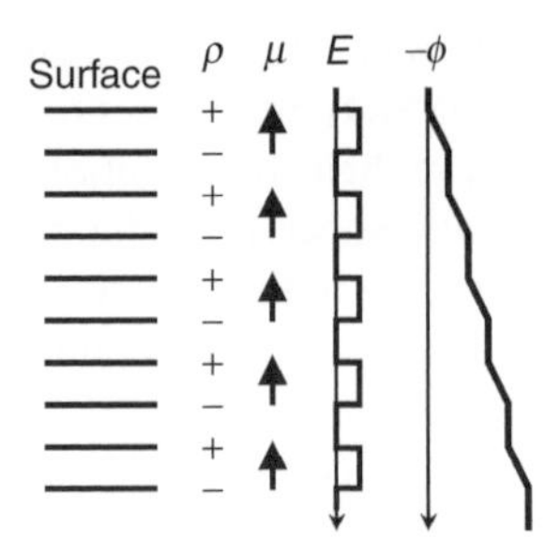

Fig. 7.2 Schematic illustration of the charge density ρ, the dipole μ in the unit-cell starting from the top-most layer, the electric field E induced by the dipoles, and the electrostatic potential ϕ at the surface of an ionic crystal with dipole moment in each unit cell.

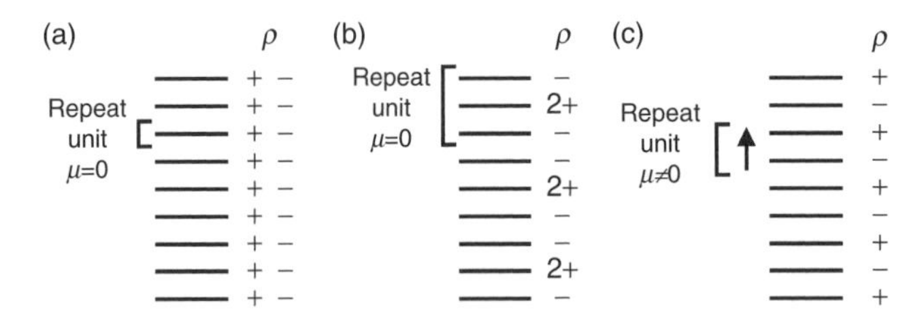

Fig. 7.3 Distribution of charges ρ on planes for three stacking sequences parallel to the surface. (a) Type 1, (b) type 2, and (c) type 3. Reprinted with permission from [10], courtesy of the Institute of Physics.

In order to survey the stability of such surfaces, Tasker introduced a classification of the surfaces from the viewpoint of the charge of the atomic sheets and the dipole in a unit-cell which *starts from the top-most layer* [10]. Note, he discussed the stability of *bulk frozen* surfaces,[3] where the top-most layer is one of the constituent atomic sheets of the bulk crystal and has no reconstruction. Tasker described three types of the surfaces, as shown in Figure 7.3:

- Type 1 has equal numbers of anions and cations on each plane, and therefore the unit-cell has no dipole moment. For example, the (001) and (110) surfaces of the rocksalt structure MX (e.g. NaCl, MgO, NiO) are classified as this type.
- Type 2 has charged planes, but no net dipole moment perpendicular to the surface. The anion X terminated (001) surface of the fluorite structure MX_2 (e.g. UO_2, ThO_2) is an example.
- Type 3 has charged planes and a net dipole moment normal to the surface. Examples include the (111) surface of the rocksalt structure, and (001) or (111) surfaces of the zincblende structure MX (e.g. GaAs, ZnS).

[3] This definition of "*bulk frozen*" follows the description of Goniakowski *et al.* [11].

In Tasker's classification, type 1 and type 2 surfaces are stable while type 3 is not, since the instability of the surface comes from the stacking of the dipole in each unit-cell. Macroscopically, the instability of the type 3 surface arises from the change in the potential slope when crossing the surface. The term "polar surface" can be defined following this classification, namely we call a surface "non-polar" when it is type 1 or type 2, and "polar" when it is type 3.

These definitions are analogous to the definition of polar crystals, although polar surfaces and surfaces of polar crystals are not equivalent. Dielectric polarization is observed when an electric field is applied to a material, but even in the absence of the field, some crystals retain a "spontaneous" polarization [12]. Only 10 out of 32 point groups show this behavior, and their members are called polar crystals, while the others are non-polar. Twenty-one out of 32 point groups do not have inversion centers, and the polar crystals are included among them. When a material shows macroscopic spontaneous polarization, the electrostatic potential of one end is different from the other end, as a result of the stacking of the dipole in each unit-cell, and the surface usually has "compensating" charge to reconcile this potential difference. Since Tasker took unit-cells from the top-most layer, even crystals with inversion symmetry can show dipoles in the unit-cells in his model. For example, although NaCl is cubic and has inversion symmetry, its (111) surface is classified into type 3. Therefore, the word "polar" should be used with some care since it has different meanings in different contexts.

7.2.2 Stability of covalent surfaces

At the surface of covalent crystals, lacking full coordination, the top-most atoms have valence electrons which are not used for bond formation. These non-bonding electrons are called dangling bonds, and have higher energy than the bonding electrons, which causes movement of the atom positions to decrease their number [13].

In a covalent crystal, since the bonds are formed as a hybridization of the valence electrons of charge-neutral atoms, one can describe the charge distribution starting from a bulk unit-cell which is charge neutral and dipole-free. However, when the electronegativities of the atoms are different e.g. Ga and As in GaAs, charge transfer between anions and cations occurs, similarly to the ionic case. This charge transfer is realized by the difference in contribution of each bond, namely $1+\alpha$ electrons to the anions and $1-\alpha$ electrons to the cations. Here α is a parameter to describe the ionicity of the bond, determined by the electronegativity of the two bonding atoms. As a consequence, the unit-cell can have a dipole moment normal to the surface, which causes the same instability as that in the ionic crystal case. Therefore, even in a covalent crystal, a surface instability emerges from the dipole in the unit-cell, independent of the surface instability naturally arising from dangling bond formation. Even though Tasker's classification was introduced to describe the stability of ionic surfaces, it is also relevant for covalent surfaces in the presence of finite ionicity.

Both dangling bond formation and the instability of polar surfaces are at play, and they are reconciled simultaneously at the surface of covalent crystals. Therefore, direct observation of the instability of polar surfaces in covalent systems has been difficult. When we consider an epitaxial interface which has a similar charge structure to the polar surface, the instability from the dangling bonds disappears, and we can solely discuss the stability in the same manner as that for the ideal ionic surfaces, as discussed in the next section.

7.2.3 Polar semiconductor interfaces

Similarly to the instability of polar surfaces, dipoles in the unit-cells stacking from the interface can cause potential divergence and instability, and require reconstruction. This point was first proposed at the heteroepitaxial interface of GaAs and Ge in the [001] direction by three groups from different starting points for treating ionicity. Based on their considerations, we define polar and non-polar interfaces, in analogy to polar and non-polar surfaces.

Charge transfer based on electronegativity

Frensley and Kroemer calculated the energy band diagram at abrupt semiconductor heterojunctions [14]. Their starting point was to describe the alignment of atoms around the interface without considering the ionicity, and then calculate the charge transfer based on the ionicity using the electronegativity of the atoms, under the assumption that the charge transfer only occurs between nearest neighbors. The Phillips electronegativity values X_{Ph} were used ($X_{\mathrm{Ph}}(\mathrm{Ga}) = 1.13, X_{\mathrm{Ph}}(\mathrm{Ge}) = 1.35$, and $X_{\mathrm{Ph}}(\mathrm{As}) = 1.57$ [15]). In the bulk zincblende structure AB, the A site is tetrahedrally coordinated by four B atoms and vice versa, and based on their calculation [16], the ionic charges e^* of the atoms are given by

$$e^*(A) = -e^*(B) = 0.76q_0 \times [X_{\mathrm{Ph}}(B) - X_{\mathrm{Ph}}(A)], \tag{7.1}$$

where q_0 is the elementary charge. This is equivalent to assuming a charge transfer of $0.76q_0 \times \frac{1}{4}[X_{\mathrm{Ph}}(B) - X_{\mathrm{Ph}}(A)]$ between any pair of nearest neighbors.

Consider the case of Ge/GaAs interfaces, as shown in Figure 7.4(a), where the charge $e^*(\mathrm{Ga_{int}})$ on the Ga ions adjacent to the interface is

$$e^*(\mathrm{Ga_{int}}) = 0.76q_0 \times \left[\frac{1}{2}X_{\mathrm{Ph}}(\mathrm{Ge}) + \frac{1}{2}X_{\mathrm{Ph}}(\mathrm{As}) - X_{\mathrm{Ph}}(\mathrm{Ga})\right] = 0.25q_0. \tag{7.2}$$

Similarly, the charges $e^*(\mathrm{Ge_{int}})$ on the Ge ions at the interface and $e^*(\mathrm{As})$ at the As sites are $e^*(\mathrm{Ge_{int}}) = -0.08q_0$ and $e^*(\mathrm{As}) = -0.33q_0$.

Without ionicity, the electrostatic potential is constant, and even with ionicity, the potential does not diverge. This can easily be understood by tracking the virtual charge transfer processes from the starting alignment of the charge neutral atoms. The charge transfer is equivalent to the situation that each neutral atom loses $0.76q_0 \times X_{\mathrm{Ph}}$ charges, and half of them are transferred to the left atoms and the other half to the other side, as shown in Figure 7.4(b). Therefore, charges are always transferred symmetrically, and each modulation creates no dipole, resulting in no potential shift. Here, a change of the number of electrons in the ions compared to the bulk state is assumed, which is equivalent to changing the valence assignments.

Self-consistent calculation and counting electrons of bonds

Baraff *et al.* performed a self-consistent calculation of the potential, charge density, and interface states for the abrupt interface between Ge and GaAs, terminated on a (001) Ga plane [17]. As shown in Figure 7.5(a), their calculation showed fractional occupancy of electronic states only at the interface, which cannot exist in bulk.

These interface states can be discussed from a local-bond counting point of view as well. The number of the valence electrons is 4 for Ge, 3 for Ga, and 5 for As. When we assume all of

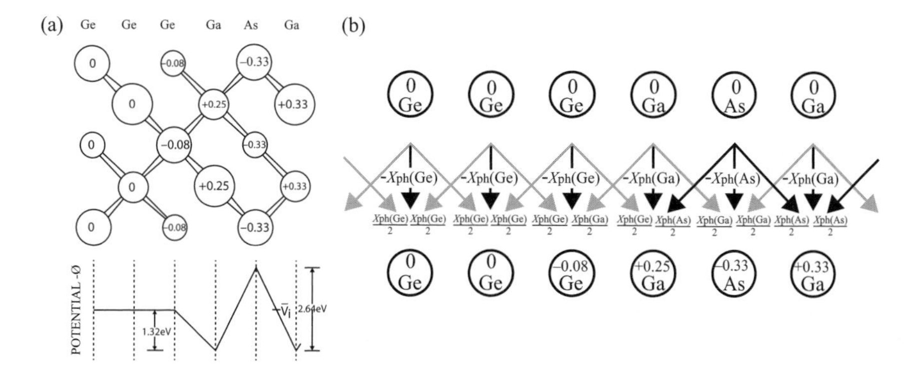

Fig. 7.4 (a) Model for a (001) Ge/GaAs heterojunction considering the ionic character of the bonds. The atomic positions and effective ionic charges are shown above. Below is a diagram of the plane-averaged potential. (b) Schematic diagram of the charge transfer from the neutral atoms with respect to the electronegativity. Reprinted with permission from [14]. Copyright 1977 by the American Physical Society.

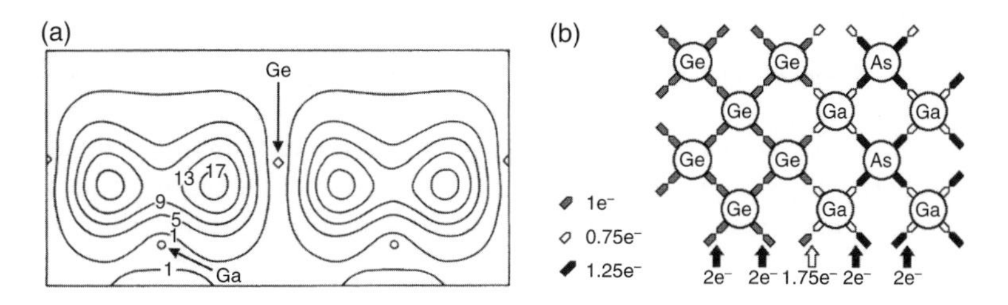

Fig. 7.5 (a) Calculated contour plot of charge density for the partially occupied interface band around Ga-Ge bonds. (b) Schematic model for counting electrons from the local-bond point of view. Reprinted with permission from [17]. Copyright 1977 by the American Physical Society.

the valence electrons of an atom are equally distributed to the four covalent bonds around it, Ge, Ga, and As atoms supply 1, 0.75, and 1.25 electrons to each bond, respectively. As shown in Figure 7.5(b), the Ge–Ga bonds at the interface have 1.75 electrons per bond, while other bonds have 2 electrons in each. These partially occupied bonds are considered to form the interface states.

When the number of electrons is smaller than that in the bulk, the attractive interaction between the bonded atoms should be weaker. According to their calculation, the energy is minimized when the Ge–Ga bond length is 4% larger than that of the bulk. Without the change of bonding length, the system requires long-range disturbance of the lattice, which is unlikely to be realized.

Solving the potential divergence from the ionic picture

Both the charge transfer model by Frensley and Kroemer and the electron counting model by Baraff *et al.* predict (indeed require) that there are interface states at a Ge/GaAs (001) interface, even if it is perfectly abrupt with no crystalline defects. The central point raised by these studies is that, despite having the same crystal structure, and having very closely matching lattice constants, this interface must accomodate charge arising from interface boundary conditions. However, experimentally no considerable density of interface states was observed [18], and a model was required to treat this interface without changing the number of charges at the interface.

Harrison *et al.* constructed a model for the Ge/GaAs (001) heterojunctions by stacking the fully ionized atoms, and they calculated the electrostatic potential based on the fixed assignment of the charges [19]. As shown in Figure 7.6(a), the potential is very similar to the case of polar surfaces since the unit-cells which *start from the interface* have dipoles in GaAs, while the unit-cells in Ge are always charge neutral. Therefore, the stacking of the dipoles causes potential divergence in this case as well.

Since the charge of each ion was fixed, the solution to the instability of this interface requires compensation by changing the stoichiometry at the interface. They proposed a simple model where 1/4 of the Ge atoms are replaced by As atoms at the interface, while 1/4 of the Ga atoms adjacent to the interface are replaced by Ge. In this reconstructed model with two transition layers, the electrostatic potential does not diverge, as shown in Figure 7.6(b).

It might be surprising that from two completely different starting points, namely one from covalent (Frensley and Kroemer) and the other from ionic (Harrison *et al.*) pictures, exactly the same potential diagrams were obtained. However, rearranging the number of charges at the perfectly abrupt interface or changing the interface composition while maintaining the ionic charges of the atoms can give the same net charge distribution. The experimental absence of localized states suggests the atomic reconstruction based on the ionic picture. This is equivalent to saying that the electronic state at this semiconductor heterointerface cannot

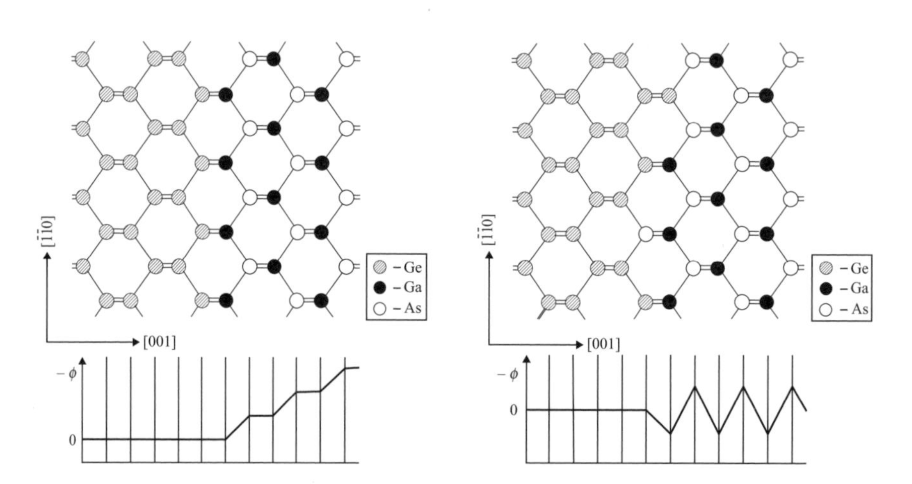

Fig. 7.6 Schematic crystal structure and electrostatic potential ϕ in the heterojunctions of Ge and GaAs in the [001] direction. (a) An atomically abrupt interface. (b) A Ge/GaAs heterojunction with two off-stoichiometric transition layers. Reprinted with permission from [19]. Copyright 1978 by the American Physical Society.

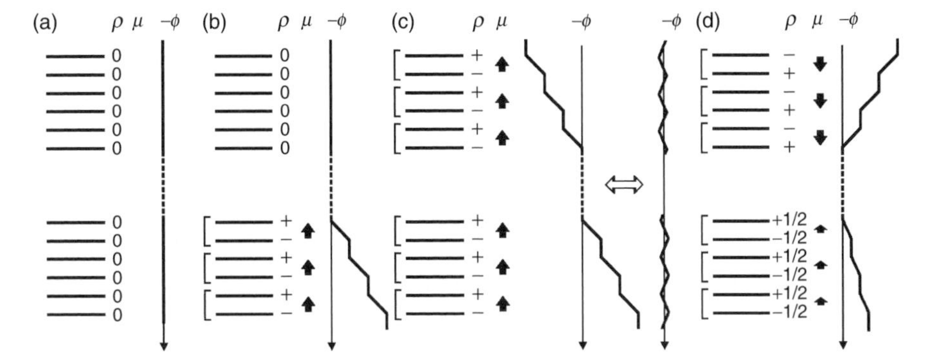

Fig. 7.7 Schematic distribution of charges ρ on planes, the dipole moment μ in unit-cells starting from the interface, and the electrostatic potential for the four stacking sequences parallel to the interface, (a) type I, (b) type II, (c) type III, and (d) type IV. In order to treat a *bulk frozen* interface as a set of two *bulk frozen* surfaces, vacuum is inserted between them (dashed lines). The electrostatic potential ϕ was calculated taking the vacuum as the potential reference except for the right plot in (c), where a constant electric field was added to show the absence of macroscopic band bending at the interface.

deviate so strongly from that of the bulk constituents—it is energetically inaccessible. This is the fundamental aspect which can be quite different in complex oxide heterointerfaces, and is the subject of much current excitement. Namely, there is a possibility that the charge transfer picture (Frensley and Kroemer) and the electron counting picture (Baraff *et al.*), which require large deviations of electron numbers from the bulk values, can be energetically accessed in transition metal oxides with multi-valency, as described in Section 7.3.

Definition of the polar interface

Following the model by Harrison *et al.*, we can define the polar nature of a *bulk frozen* interface between two materials, where the interface consists of two *bulk frozen* surfaces connected together.[4] First let us classify the interfaces by the polar nature of the two constituent surfaces, as shown in Figure 7.7.

- Type I is formed between two non-polar surfaces, and the potential is flat.
- Type II is formed between polar and non-polar surfaces, and the potential diverges from the interface in the material with the polar surface.
- Type III is formed between two polar surfaces, and the direction and the size of the dipoles in the unit cells which start from the interface is the same. Although ϕ looks to diverge from the interface in both materials, there is no macroscopic difference in the potential slope across the interface. We can cancel the potential slope on both sides by adding a constant electric field, as shown in Figure 7.7(c). Therefore, this interface is stable as constructed.

[4] Here, we consider only interfaces between two semi-infinite materials—we ignore the coupling to other interfaces or surfaces, which is discussed in Sections 7.4.3 and 7.4.4.

- Type IV is formed between two polar surfaces, where the dipoles in the two different unit-cells are not identical, which results in a macroscopic difference in the potential slope at the interface. It is impossible to find any constant electric field to cancel out the potential divergence in both media, due to this difference.[5]

Type I and type III are stable due to the absence of a macroscopic difference in the potential slopes, while type II and type IV are not stable. This difference arises from the continuity/discontinuity of the dipoles in the unit-cells which start from the interface. A *bulk frozen* interface is non-polar, when the dipoles in the unit-cells starting from the interface are identical across the interface, and thus no change in the macroscopic potential slope exists. On the other hand, it is polar if a *bulk frozen* interface has a discontinuity in the dipole moment in each unit-cell. For example, from a purely ionic viewpoint (Harrison *et al.*) the abrupt Ge/GaAs (001) interface is classified as type II, and thus polar.

This definition of the polar nature of interfaces is consistent with that of surfaces. When the vacuum is treated as a charge neutral medium, the non-polar surface is type I, and the polar surface is type II in this classification of the interfaces, and the polar nature is maintained following the definitions for the interface. We can treat surfaces and interfaces in one framework, which is the polar nature of discontinuities at materials boundaries. In summary, polar discontinuities, which consist of polar surfaces and interfaces, are unstable due to the macroscopic potential folding arising from the discontinuity of the stacking of dipoles in the unit-cells, and require reconstructions to stabilize them. Interfaces with continuous polarity, on the other hand, are stable without any reconstructions.

7.3 Metallic conductivity between two insulators

As noted in the introduction, the modern ability to approach atomic control in complex oxide heterostructures has recently enabled the experimental investigation of their polar discontinuities. As illustrated in Figure 7.1(a), an immediate question arises regarding the choice of the termination layer at the interface, and the consequences of this degree of freedom. This issue has been most explicitly raised, and hotly debated, for the electron gas observed at the interface of two perovskite insulators, $LaAlO_3$ and $SrTiO_3$ [20]. Specifically, the (001) heterointerface was found to be insulating when grown using a $SrTiO_3$ substrate which was SrO-terminated, and conducting when TiO_2-terminated. Given the rapid evolution of the field, and the numerous reviews of this heterostructure in the literature, we do not attempt a comprehensive review here. Instead the $LaAlO_3/SrTiO_3$ interface will be used to illustrate the various mechanisms suggested to explain the interface electronic structure, and the electrostatic boundary conditions which arise. It should be stressed, however, that all oxide heterointerfaces should be considered type IV to varying degrees, and thus these issues are quite general—even underlying the interface charge in the superlattice shown in Figure 7.1(c).

[5] Note, type IV is the most common and general case in reality, since the electronegativity can never perfectly match between different materials.

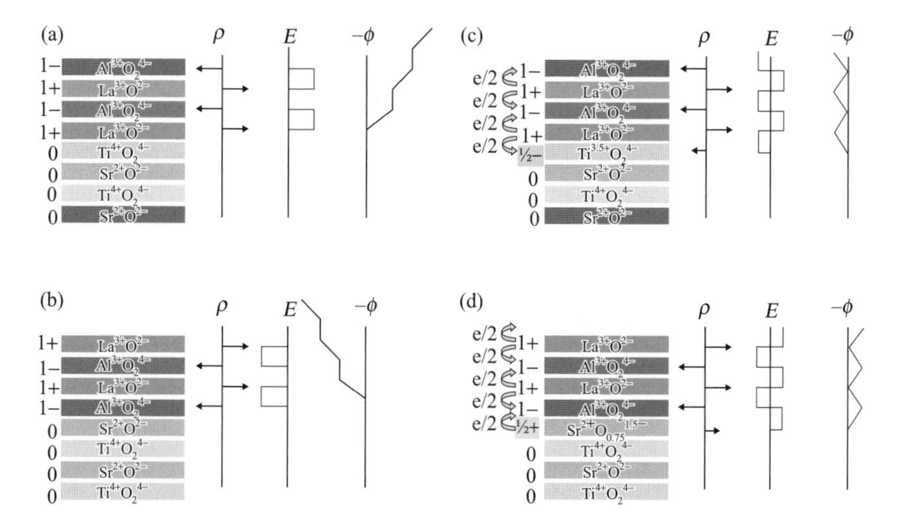

Fig. 7.8 Polar reconstructions at the $LaAlO_3/SrTiO_3$ interfaces. The unreconstructed (a) LaO/TiO_2 terminated interface, (b) AlO_2-SrO terminated interface, (c) and (d) the corresponding reconstructed interfaces, respectively (Nakagawa *et al.* [24]).

7.3.1 The *polar discontinuity* scenario

Assuming pure ionicity, the charge sequence of the (001) perovskite plane is different in these two materials, namely the planes of $LaAlO_3$ are $(La^{3+}O^{2-})^+$ and $(Al^{3+}O_2^{2-})^-$, while those of $SrTiO_3$ are $(Sr^{2+}O^{2-})^0$ and $(Ti^{4+}O_2^{2-})^0$. Therefore, the abrupt interface between $LaAlO_3$ and $SrTiO_3$ is type II polar,[6] and requires reconstruction as shown in Figure 7.8, just as for the GaAs/Ge (001) interface.

Unlike polar semiconductor interfaces where only atomic reconstructions are available due to the fixed ionic charge of each element,[7] we have another possibility to reconcile the instability of polar interfaces, through electronic reconstructions [23]. At the $LaAlO_3/SrTiO_3$ interface, when the interface termination is LaO/TiO_2, it requires a net half negative charge per 2D unit cell to reconcile the potential divergence (n-type). Accessing Ti^{3+} can source this negative charge by accommodating electrons at the Ti 3*d* level, as was spectroscopically observed [24]. On the other hand, when the interface is terminated by AlO_2/SrO, the sign of the required charges is opposite (p-type). Because of the difficulty of accommodating holes in this structure (such as Ti^{5+}), the positive charges are realized by the formation of oxygen vacancies close to the interface, as inferred from measurements of the O-K edge fine structure. Oxygen vacancies are known as electron donors, but in this case they are formed to provide positive

[6] Allowing for covalency, SrO and TiO_2 planes in $SrTiO_3$ are no longer charge neutral, and thus the (001) surface is weakly polar, but still the dipole size of the unit-cells which start from the interface is different from that of $LaAlO_3$.

[7] While here we discuss large scale charge modifications, small polar discontinuities can induce free carriers in semiconductors, such as in AlGaN/GaN heterostructures [21, 22].

charges, and thus no electrons are supplied from these vacancies compensating the polar discontinuity. Therefore the system does not have itinerant electrons and remains insulating [20, 25].

One of the key corollaries of this scenario is the $LaAlO_3$ thickness dependence. This is because the size of the potential difference arising from the stacking of the dipoles in the unreconstructed structure is finite in thin films, and if it is small enough, the system may be stable without any reconstruction. Indeed, a critical thickness of $LaAlO_3$ was observed [26], where the n-type $LaAlO_3/SrTiO_3$ interface is insulating if the thickness of $LaAlO_3$ is up to three unit-cells, and metallic above that thickness. A similar tendency was also observed in $SrTiO_3/LaAlO_3/SrTiO_3$ trilayer structures, where the distance between the two polar interfaces was varied [27].

7.3.2 Oxygen vacancy formation during growth

$SrTiO_3$ is known to be a material which readily accommodates oxygen vacancies that act as donors to provide itinerant electrons [28, 29]. Either kinetic bombardment of the $SrTiO_3$ substrate by the ablated species (early studies of this interface all used pulsed laser deposition), or gettering by a reduced film, can induce oxygen vacancies, and they were suggested to be the dominant origin for the observed conductivity by several groups [30–32]. Indeed, the first report found a strong dependence of the Hall density for n-type interfaces on the oxygen partial pressure (P_{O_2}) during growth, while the p-type interface was robustly insulating [20]. For n-type samples with similar variations in the transport properties, Kalabukhov *et al.* found when grown at $P_{O_2} = 10^{-6}$ Torr, the samples exhibited blue cathode- and photo-luminescence at room temperature [30], similar to that of reduced $SrTiO_3$ by Ar bombardment [33]. In addition to transport studies, Siemons *et al.* demonstrated that the photoemission spectra from these interfaces showed a larger amount of Ti^{3+} in samples grown at low pressures without oxygen annealing [31]. Herranz *et al.* observed Shubnikov–de Haas oscillations in reduced $LaAlO_3/SrTiO_3$ samples, which were quite similar to bulk doped $SrTiO_3$, and rotation studies indicated a three-dimensional Fermi surface [32].

A strong P_{O_2} dependence of the conducting channels in $LaAlO_3/SrTiO_3$ was observed and mapped by means of conducting-tip atomic force microscopy on cross-sections of the interface, which revealed a conducting region extending >1 μm into the substrate for samples grown at low pressure [34], as shown in Figure 7.9. After annealing, the width of the conductive layer decreased to $\sim$7 nm, as limited by the radius of the probing tip. The consensus of these and other studies was that the free carriers in samples grown at low P_{O_2} were dominated by oxygen vacancies, since the density far exceeded that needed to stabilize the polar discontinuity. For high P_{O_2}, or after post-annealing, the origin was less clear.

7.3.3 Intermixing and local bonding at the interface

Willmott *et al.* studied a five-unit-cell film of $LaAlO_3$ on $SrTiO_3$ (001) by means of surface X-ray scattering [35], using coherent Bragg rod analysis (COBRA) [36] and further structural refinement. Their analysis revealed intermixing of the cations (La, Sr, Al, and Ti) at the interface [Figure 7.10(a)], as well as significant local displacement of the atomic position, both in the film, and in $SrTiO_3$ close to the interface [Figure 7.10(b)]. The distribution of Ti valence was also inferred by minimizing the electrostatic potential [Figure 7.10(c)] following the obtained

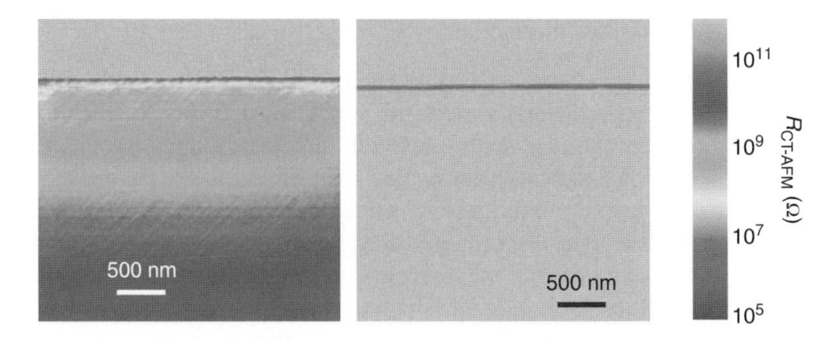

Fig. 7.9 Conducting-tip atomic force microscopy mapping around the $LaAlO_3/SrTiO_3$ interface of (a) the as grown sample ($P_{O_2} = 10^{-5}$ Torr) and (b) the postannealed sample. Reprinted by permission from Macmillan Publishers Ltd: Nature Materials [34], copyright 2008. This figure is reproduced in color in the color plate section.

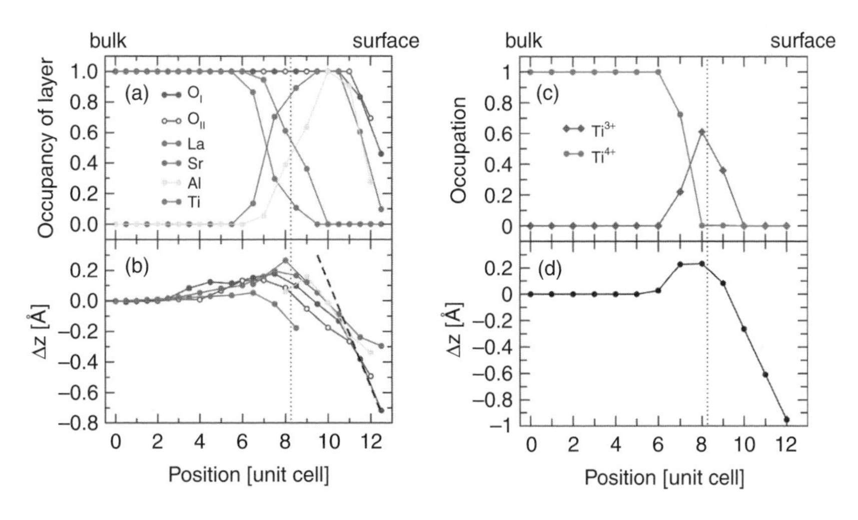

Fig. 7.10 (a) Occupancies and (b) cumulative displacements Δ_z of the atoms at the $LaAlO_3/SrTiO_3$ interface. (c) Concentration of Ti^{3+} determined by a minimization of the electrostatic potential. (d) Predicted cumulative unit-cell displacements from bulk positions, based on the component ionic radii. Reprinted with permission from [35]. Copyright 2007 by the American Physical Society. This figure is reproduced in color in the color plate section.

atomic positions, and the atomic displacements were explained by the larger ionic radii of Ti^{3+} compared to Ti^{4+} [Figure 7.10(d)]. Based on these observations, the origin of the interface conductivity was suggested to be the formation of the bulk-like solid solution $La_{1-x}Sr_xTiO_3$ in a region of approximately three unit-cells. This explanation can be considered a diffused version of local bonding arguments—i.e. that even in the abrupt limit, the Ti at the interface has La on one side, and Sr on the other.

7.3.4 Reconciling the various mechanisms

At present, it can be fairly stated (we believe) that no one scenario can completely explain the vast and growing body of experimental work on this system. Even theoretical calculations show an extreme sensitivity to the choice of boundary conditions and assumptions of site-occupancy [37–41]. To give examples for each perspective: The polar discontinuity picture should lead to a significant internal field in ultrathin $LaAlO_3$, while experiments [42] put an upper bound far below that expected theoretically, even allowing for significant lattice polarization [43]. Oxygen vacancies induced by growth are difficult to reconcile with the notion that a single layer of SrO can prevent their formation. Local bonding and interdiffusion considerations do not address the constraints of global charge neutrality. Furthermore, discriminating between these mechanisms is often difficult, since the change of Ti valence shows similar transport, spectroscopic, and optical properties, independent of its origin.

It is extremely likely that multiple contributions exist to varying degrees, dependent on the growth details of a given sample. While this is a matter for further experimental investigation and refinement, an equally difficult issue appears to be one of semantics. For example, one of the conceptual difficulties of the *polar discontinuity* picture has been the question: Where do the electrons come from? In many presentations [24, 26, 44, 45], the charges at the interface are depicted to originate from the surface of the $LaAlO_3$ film, but it is not so obvious that they travel through an insulating film independent of its thickness [26]. Fundamentally, the "non-locality" of these electrostatic descriptions has sometimes been considered less intuitive and compelling than "local chemistry" mechanisms such as vacancies or interdiffusion [30, 31, 34, 35, 46].

A related difficulty is how to treat the charge density to describe the macroscopic electric field. Based on the *polar discontinuity* picture, the stacking of the dipoles in the unit-cells creates a macroscopic electric field, resulting in a change in the potential slope at the discontinuity. In this picture, the unit-cells start from the discontinuity, and thus the composition is always the same as that of the bulk, which is charge neutral. It might be strange that the potential starts to bend at the discontinuity, although all the unit-cells start out charge neutral, because the source of an electric field is charge. In fact, the source is the macroscopic bound charge density at the interface, as discussed in Section 7.5, but the existence of such implicit charge density has caused a fair bit of confusion. To address these concerns, it is useful to treat the electrostatics in a purely local description, as well as the boundary charges explicitly. Furthermore, the effects of defects and diffusion can be discussed more simply by re-framing the polar discontinuity picture in local form. Therefore, this *local charge neutrality* picture based on dipole-free unit-cells is developed in the next section first for idealized models, followed by discussion of incorporation of chemical defects.

7.4 The *local charge neutrality* picture

7.4.1 Unit-cells in ionic crystals

One of the origins of confusion regarding the stability and reconstruction of polar discontinuities arises because of the various choices of a unit-cell in a crystal, which determine the size and direction of the dipole with in it. For example, when we take the unit-cell of $LaAlO_3$, this stacking can be treated as a dipole of $[(AlO_2)^- \ (LaO)^+]$, as shown in Figure 7.11(a). However, it is also possible to take $[(LaO)^+ \ (AlO_2)^-]$ as a unit-cell, and the sign of the dipole

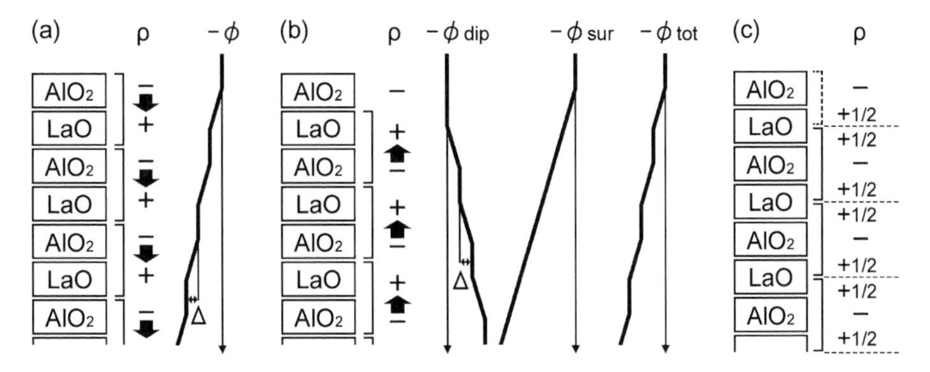

Fig. 7.11 Schematic illustrations of the various choices of unit-cells in $LaAlO_3$, charge distribution ρ, and associated potential ϕ. The filled arrows indicate the orientation of the dipoles with size Δ in the unit-cells. (a) Taking $[(AlO_2)^- \ (LaO)^+]$ as a unit-cell. (b) Taking $[(LaO)^+ \ (AlO_2)^-]$ as a unit-cell, and the total potential ϕ_{tot} is the sum of the potential ϕ_{dip} arising from the stacking of the dipoles and the potential ϕ_{sur} from the surface charge. (c) Taking a dipole-free unit-cell $[\frac{1}{2}(LaO)$ - (AlO_2) - $\frac{1}{2}(LaO)]$.

is the opposite of the previous case, as shown in Figure 7.11(b). For simplicity, we started our discussion from a point-charge model of the ion, where the distribution of valence charges is neglected. More generally, the charge distribution of the formal ionic charge can be described using contributions from ion cores and valence Wannier functions, and the ambiguity of the choice of unit-cells appears in this extended case as well [47, 48].

The choice of unit-cells should not change the electrostatic potential in the crystal. Indeed, the difference between the two choices is compensated for by the potential ϕ_{sur} arising from the surface layer. If the surface is terminated by the $(AlO_2)^-$ layer, it remains as an extra negatively charged layer, when we take $[(LaO)^+ \ (AlO_2)^-]$ as the unit-cell, as shown in Figure 7.11(b), and the total electrostatic potential remains the same as in the $[(AlO_2)^- \ (LaO)^+]$ unit-cell case.

Therefore, it is impossible to fix the direction of the dipoles without knowing the surface termination. In other words, it is the surface that determines the stability of the system. So the problem can be simplified by considering the surface locally, and not by counting the number of dipoles in the material. In order to avoid the dipoles arising from the stacking of charged layers, the simplest approach is to take dipole-free unit-cells. This can be achieved in any crystal [11], and in the $LaAlO_3$ case, this is done by taking $[\frac{1}{2}(LaO)$ - (AlO_2) - $\frac{1}{2}(LaO)]$ (or $[\frac{1}{2}(AlO_2)$ - (LaO) - $\frac{1}{2}(AlO_2)]$) as a unit-cell, as shown in Figure 7.11(c). This is analogous to the unit-cell in a type 2 model in Tasker's classification. From group theory, it is known that spontaneous polarization can be observed only in a direction where the crystal does not have mirror symmetry. Since cubic perovskites do have mirror symmetry in the [001] direction (we neglect surface-or interface-induced lattice polarization for now), it is useful to take dipole-free unit-cells to reflect the lack of polarization in the bulk.

When we take this unit-cell, the stability of a polar surface or interface can be discussed by only considering the charge neutrality of each unit-cell, since now there is no net dipole created by the stacking of charged layers. For example, the instability of the (AlO_2) terminated surface of $LaAlO_3$ is naturally derived because the top-most unit-cell is $[(AlO_2)^- \ \frac{1}{2}(LaO)^+]^{-0.5}$, which violates charge neutrality, as shown in Figure 7.11(c). Thus the (001) surface of $LaAlO_3$ cannot keep the bulk termination, either by an AlO_2 or LaO layer, and must reconstruct to compensate this charge [49, 50]. Once charge neutrality of all the unit-cells is achieved, the system is free to undergo electron/hole modulation or interdiffusion/displacement of the atoms, which creates only finite dipoles and does not violate neutrality.[8] These perturbations are important because they determine the real charge structure, for example via lattice distortion close to the surface of $LaAlO_3$ [49, 50].

Following these arguments, the polar nature of given *bulk frozen* surfaces and interfaces is clearly defined, considering local charge neutrality using dipole-free unit-cells: a *bulk frozen* surface or interface is polar if the unit-cell at the surface or interface cannot keep charge neutrality when we take dipole-free unit-cells in the bulk.

7.4.2 $LaAlO_3/SrTiO_3$ in the *local charge neutrality* picture

Taking dipole-free unit-cells of perovskites ABO_3 in the [001] direction, namely $[\frac{1}{2}(AO)$ - (BO_2) - $\frac{1}{2}(AO)]$, the interface unit-cell can be treated as a δ-dopant at the interface. As shown in Figure 7.12, the interface unit-cell does not keep the stoichiometry of either of the bulk materials, not even a simple mixture of them. This issue is actually one of the central opportunities of the interface science of heterostructures. Namely, the LaO/TiO_2 terminated $LaAlO_3/SrTiO_3$ interface has $La_{0.5}Sr_{0.5}TiO_3$ as the interface unit-cell. Considering the formal electronic charges of La^{3+}, Sr^{2+}, and O^{2-}, the Ti ion in this unit-cell should take $Ti^{3.5+}$ as a formal valence, which is the same reconstructed state as that for the previous discussion by the *polar discontinuity* picture.

Similarly, the AlO_2/SrO terminated interface has $La_{0.5}Sr_{0.5}AlO_3$ as the interface unit-cell, and due to the fixed ionic charges of the elements, such a unit-cell is not charge neutral and

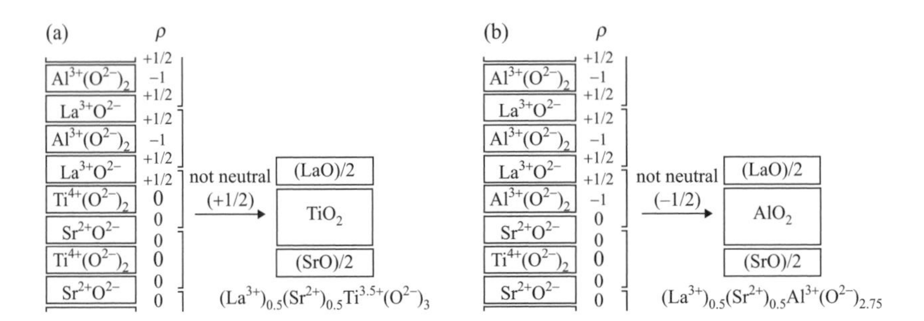

Fig. 7.12 Schematic illustrations of the charge structure across the two types of *bulk-frozen* $LaAlO_3/SrTiO_3$ interfaces, assuming the ionic charges based on the valence states in bulk $LaAlO_3$ and $SrTiO_3$. Taking dipole-free unit-cells, the interface unit-cell cannot keep charge neutrality, and the simplest neutral interface stoichiometry is written on the right. (a) LaO/TiO_2 terminated $LaAlO_3/SrTiO_3$ interface and (b) AlO_2/SrO terminated $LaAlO_3/SrTiO_3$ interface.

[8] The effect of diffusion is further discussed in Section 7.6.1 as an example of sources of such finite dipoles. Another source of interface dipoles, the *quadrupolar discontinuity*, is discussed in Section 7.6.3.

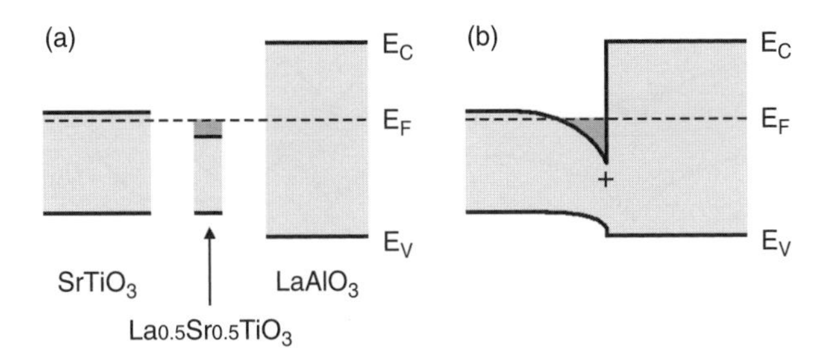

Fig. 7.13 (a) Schematic band structures of $SrTiO_3$, $LaAlO_3$ and $La_{0.5}Sr_{0.5}TiO_3$, assuming that $SrTiO_3$ and $LaAlO_3$ are intrinsic semiconductors. (b) Schematic energy band diagram of the $LaAlO_3/SrTiO_3$ interface with the δ-dopant $La_{0.5}Sr_{0.5}TiO_3$ at the interface.

thus unstable: $(La^{3+}_{0.5}Sr^{2+}_{0.5}Al^{3+}O^{2-}_3)^{0.5-}$. Instead, allowing a change of the oxygen number in the interface unit-cell, $La_{0.5}Sr_{0.5}AlO_{2.75}$ is charge neutral,[9] and the decrease of the oxygen content at the interface to stabilize a p-type $LaAlO_3/SrTiO_3$ interface is naturally derived. Note in this case, these oxygen vacancies are introduced to keep charge neutrality, and thus do not provide any free electrons. Thus, the electrons/oxygen vacancies to reconcile the polar instability of the *bulk frozen* $LaAlO_3/SrTiO_3$ interfaces are supplied by the interface unit-cell itself.

In semiconductors, δ-doping is usually achieved in a symmetric geometry [51], that is the dopant layer is sandwiched between the same host material. In the $LaAlO_3/SrTiO_3$ case, by contrast, the two sandwiching materials have different band structures, with $SrTiO_3$ having the narrower bandgap. Therefore, broadening of the charge distribution from the δ-dopant occurs only in the $SrTiO_3$ side [46]. Figure 7.13 shows how the $LaAlO_3/SrTiO_3$ interface can be depicted in a semiconductor energy band diagram.

7.4.3 Coupling of polar discontinuities

When two polar discontinuities are brought in proximity to one another, coupling of the charges can occur to minimize the total energy of the system. This is just like the coupling of dopant layers in δ-doped semiconductor heterostructures, which can be understood in terms of depleted and undepleted structures. Figure 7.14(a) shows the band diagram of an undepleted semiconductor with one layer of δ-dopant. Since the dopant is positively ionized, an equal number of free electrons are left, and they screen out the potential created by the δ-dopant. As a consequence, the structure is neutral, and has zero electric field sufficiently far away from the δ-dopant layer.

On the other hand, Figure 7.14(b) shows the band diagram of a depleted δ-doped structure, where the number of donors is equal to that of acceptors, and they are sufficiently close in space. As a result, all the free carriers recombine and the structure is depleted. The critical parameters to treat electronic coupling of two δ-doping layers are the distance between them

[9] This composition is not stable in bulk perovskite form, but can be considered as a mixture of the bulk compounds, $LaAlO_3$, $Sr_3Al_2O_6$, and $SrAl_2O_4$, stabilized epitaxially at the interface.

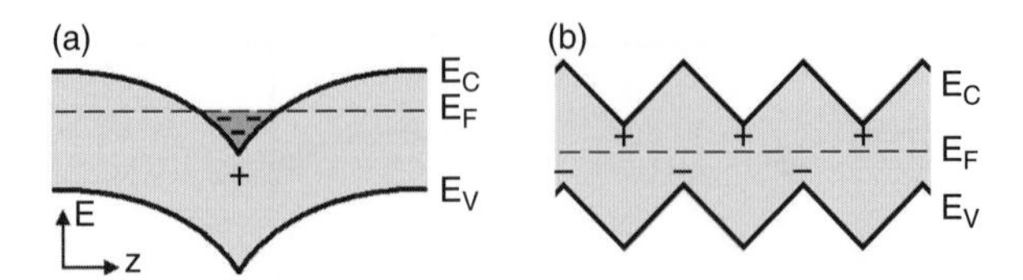

Fig. 7.14 Schematic band diagram of an (a) undepleted and (b) depleted δ-doped semiconductor. The undepleted structure contains free carriers as well as ionized impurities. The depleted structure contains the same amount of donor and acceptor impurities (derived from Gossmann and Schubert [51]).

and the dielectric constant of the medium. When the distance between them is smaller than the length scale of band bending of the undepleted structure, they couple and the system goes to a depleted state. Note in this case, charge neutrality in the neighborhood of one dopant layer is not necessarily maintained. Superlattice calculations using density functional theory show that the above threshold is observed by changing the thickness of each layer in $LaAlO_3/SrTiO_3$ superlattices, which can be captured in terms of a simple capacitor model [40, 41].

In δ-doped semiconductor heterostructures, only considering modulation of free carriers is sufficient to describe the coupling of the δ-dopant layers. However, when the system allows possibilities of excess charges other than the free carriers, which come from outside of the constructed crystal—e.g. anion or cation vacancies or foreign atoms absorbed to the surface, the electrostatic potential can be minimized via these. For example, the instability of the polar AlO_2-terminated $LaAlO_3$ surface can be solved by introducing positively charged surface oxygen vacancies [50]. The $LaAlO_3$ thickness dependence of LaO-TiO_2 terminated $LaAlO_3/SrTiO_3$ can be explained by a simple assumption, where we only consider coupling of the free electrons provided by the $LaAlO_3/SrTiO_3$ interface and the surface oxygen vacancies, to keep the local charge neutrality of the polar $LaAlO_3$ surface. Figure 7.15 shows a schematic structure of the $LaAlO_3/SrTiO_3$ heterojunction, where two polar discontinuities exist at the $LaAlO_3/SrTiO_3$ interface and the $LaAlO_3$ surface. When the $LaAlO_3$ film is sufficiently thick [Figure 7.15(a)], these polar discontinuities are decoupled and charge neutrality is preserved locally by introducing oxygen vacancies at the surface and taking the Ti valence of 3.5+ at the interface unit-cell. Fractionally filled Ti valence provides itinerant electrons, and therefore the system is metallic in the thick limit.

If the two polar discontinuities are brought closer [Figure 7.15(b)], the external charges and the free carriers (surface oxygen vacancies and extra electrons in Ti valence) can recombine in an environment with oxygen gas, and the system is depleted and the conductivity disappears.[10] Here, the word "deplete" is used to mean "reduce the amount of charge other than charge bound to the crystal", and the full depletion of the $LaAlO_3$ surface means extinguishing the positively charged oxygen deficiency at the surface—i.e. the surface turns to an unreconstructed *bulk frozen* state. Manipulating this transition appears to roughly capture the essence of writing nanoscale features [44], which corresponds to the "writing" of surface charge [53].

[10] This is similar to the recombination of free electrons and holes, and the actual recombination can be written by following Kröger–Vink notation [52] as: $\frac{1}{2}O_2 + 2e' + V_O^{\cdot\cdot} \rightarrow 0$.

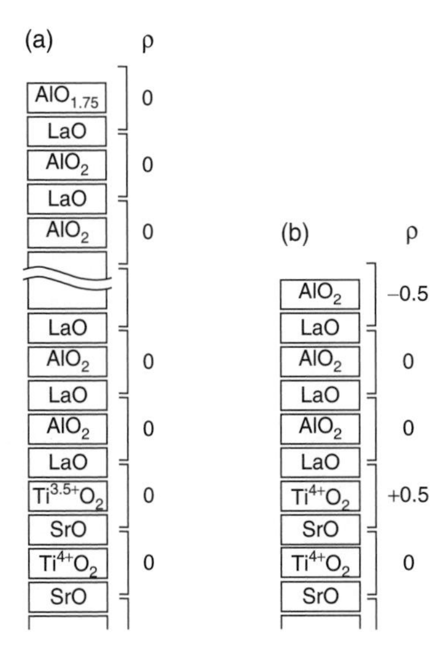

Fig. 7.15 Schematic illustration of the charges (ρ) of the dipole-free unit-cells of atomically abrupt $LaAlO_3/SrTiO_3$ interfaces in (a) the thick $LaAlO_3$ limit with the polar discontinuities locally neutralized, and (b) the thin $LaAlO_3$ limit with depleted polar discontinuities.

7.4.4 Modulation doping by a proximate polar discontinuity

Even in systems containing only one polar discontinuity, coupling between the polar discontinuity and the other layers can occur in analogy to modulation doping by a δ-doping layer [54], as shown in Figure 7.16. At the semiconductor interface, lineup of the conduction and valence

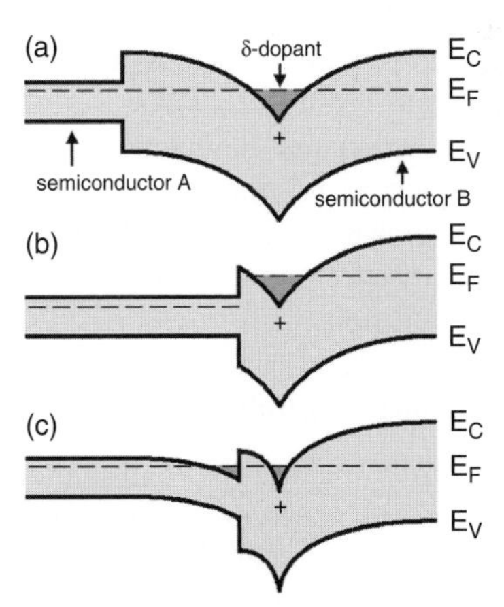

Fig. 7.16 Schematic band diagram of the interface between two intrinsic semiconductors A and B, with a layer of δ-dopant in B. (a) A is sufficiently far from the δ-dopant in B. (b) Model with no charge modulation although the distance between A and the δ-dopant is close, resulting in the mismatch of the chemical potential. (c) Charges transfer from B to A, so as to match the chemical potential.

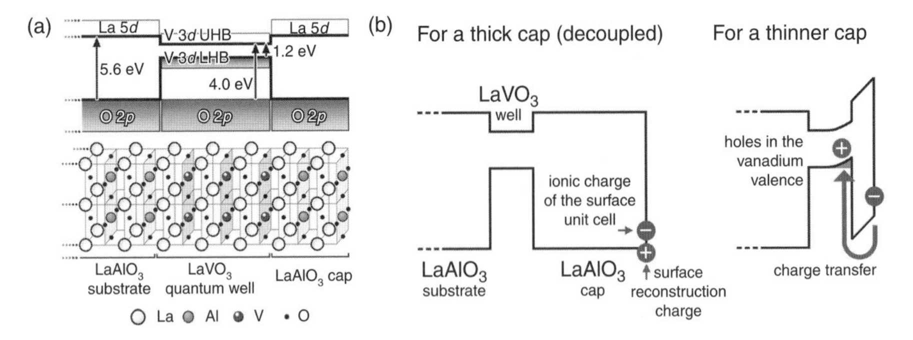

Fig. 7.17 (a) Schematic band diagram and crystal structure of a $LaVO_3$ quantum well embedded close to an AlO_2-terminated $LaAlO_3$ (001) surface. (b) Illustrations showing how reconstruction charge at the $LaAlO_3$ surface is transferred to the buried $LaVO_3$ quantum well. In order to solve the instability caused by the polar nature of the AlO_2 terminated surface, positive charge is required. When the $LaAlO_3$ cap is sufficiently thick (left), the $LaAlO_3$ surface and the $LaVO_3$ well layer are separated, and the positive compensating charge remains at the surface. For a thinner spacing (right), the $LaVO_3$ well layer accommodates this positive charge, which is energetically more favored [45, 55].

bands should be maintained, as well as a fixed chemical potential, which causes the modulation of carriers resulting in band bending.

Assume an interface between two intrinsic semiconductors A (narrow bandgap) and B (wide bandgap). If the δ-dopant in B is sufficiently far from A, the flat band condition at the interface is maintained [Figure 7.16(a)]. When A is brought in proximity to the δ-dopant in B, in order to keep the conduction band lineup, the conduction band minimum of A lies at lower energy than the chemical potential in B [Figure 7.16(b)]. Since the free carriers around the δ-dopant in B have higher energy than the conduction band minimum of A, they transfer to A and band bending occurs in A [Figure 7.16(c)]. As a result, the δ-dopant in B is depleted, and A is doped close to the interface.

An experimental example of such charge modulation was observed from a type 3 polar surface of $LaAlO_3$ (001), which acts as a δ-dopant, to a narrow layer of the Mott insulator $LaVO_3$ with smaller bandgap, as shown in Figure 7.17 [45, 55]. This system is noteworthy in the discussion of polar discontinuity effects, in that the observed hole-doping can neither arise from oxygen vacancies nor interdiffusion.

7.4.5 Advantages of the *local charge neutrality* picture

As discussed, the *local charge neutrality* picture gives a clear and self-consistent explanation for the various phenomena at surfaces and interfaces. The stability of given surfaces and interfaces can be simply judged by looking at the local composition. When the interface composition differs from that of the bulk or a simple superposition of them, a different electronic and/or atomic state can be expected. Enforcing local charge neutrality at the interface unit-cell naturally explains the source of the charges which can change the stoichiometry or the electron number of the constituent atoms from those of the bulk. For example, in a LaO/TiO_2 terminated $LaAlO_3/SrTiO_3$ *bulk frozen* interface, the interface unit cell is $La_{0.5}Sr_{0.5}TiO_3$ and the Ti is 3.5+

to achieve charge neutrality. Therefore, the difference between the Ti^{4+} in the bulk $SrTiO_3$ and the interface $Ti^{3.5+}$ comes from the interface unit-cell itself, and not from anywhere else.

Another advantage of taking the *local charge neutrality* picture becomes clear when we discuss the coupling of polar discontinuities. The stability of the system can be discussed through the distance between the polar discontinuities and the screening length of the host material. In other words, the stability is determined by the balance of the activation energy of the dopant and the electrostatic energy allowing polarization of the media between the dopant layers. Therefore, the total polarizability of the media can be considered in the calculation, and is connected to the bulk permittivity in the thick limit. Note when the media is thin, the local effective permittivity is nontrivial, since the local atomic displacements can be different from that in the bulk, and the local dielectric approximation breaks down [56]. Recently, this macroscopic to microscopic connection has been theoretically studied in detail for the $LaAlO_3/SrTiO_3$ interface, where the use of calculated bulk $SrTiO_3$ properties was found to correspond well to a full first-principles treatment of the interface [57].

7.5 Equivalence of the two pictures

Thus far we have discussed boundary conditions based on two different choices for the unit-cell. This was implicitly taking a microscopic viewpoint, since the electric field **E** and the total charge density ρ_{tot} were connected by Gauss' law: $\varepsilon_0 \nabla \cdot \mathbf{E} = \rho_{tot}$, where ε_0 is the vacuum permittivity. Since ρ_{tot} is used (hence the atomic-scale stepped or sawtooth potentials), different choices for the unit-cell were irrelevant as long as global charge neutrality was considered. Here we confirm the equivalence of the two pictures from the macroscopic electrostatic viewpoint.

7.5.1 Gauss' law for infinite crystals

In media, treating ρ_{tot} is often quite complicated, which can be simplified by using the electric displacement **D**:

$$\mathbf{D} = \varepsilon_0 \mathbf{E} + \mathbf{P}, \tag{7.3}$$

where **P** is the polarization. Then Gauss' law is given by

$$\nabla \cdot \mathbf{D} = \rho_{free}, \tag{7.4}$$

where ρ_{free} is the free charge—the part of the macroscopic charge density due to excess charge not intrinsic to the medium, which is designated as bound charge ρ_{bound}. These definitions do not depend on whether the charges are localized or itinerant, and are just introduced for practical convenience to treat the displacement of the bound charges as the dielectric response of the media to the electric field. Since the total charge is conserved ($\rho_{tot} = \rho_{free} + \rho_{bound}$), by taking the divergence of (7.3),

$$\rho_{bound} = -\nabla \cdot \mathbf{P}. \tag{7.5}$$

In infinite crystals, only the divergence of **E**, **P**, and **D** has physical meaning, and the polarization arising from the density of unit-cell dipole moments $\mathbf{P}_{dipole}$ can be neglected, since it merely adds a constant value to **P**.

7.5.2 Gauss' law for finite crystals

When the crystal is finite, $\mathbf{P}_{\text{dipole}}$ drops to zero at the surface. Thus the choice of unit-cell is important, since it changes the nature of the discontinuity at the surface. According to (7.5) the magnitude of the polarization discontinuity at the surface is determined by ρ_{bound} there. Since ρ_{tot} at the surface does not depend on the choice of unit-cell, uncertainty in the dipole moment of the unit-cell is compensated by whether the charges at the surface are defined to be free or bound—the surface charge is bound when it belongs to a bulk unit-cell, and free when not, and this definition does not depend on the origin of the charges [12].

For simplicity, let us adopt the bulk dielectric constant ε to connect $\mathbf{E}$ and the induced polarization $\mathbf{P}_{\text{ind}}$ by the field, namely

$$\mathbf{P}_{\text{ind}} = (\varepsilon - \varepsilon_0)\mathbf{E}, \tag{7.6}$$

hence

$$\mathbf{P} = \mathbf{P}_{\text{ind}} + \mathbf{P}_{\text{dipole}}. \tag{7.7}$$

Substituting them in (7.3) we obtain

$$\mathbf{D} = \varepsilon\mathbf{E} + \mathbf{P}_{\text{dipole}}. \tag{7.8}$$

Now, let us calculate the value of $\mathbf{E}$, $\mathbf{P}_{\text{dipole}}$, $\mathbf{D}$, and the charge density, considering the unreconstructed AlO_2 terminated (001) surface of $LaAlO_3$ as an example, for the two different unit-cell assignments previously discussed.

Taking $[(AlO_2)^-\ (LaO)^+]$ as a unit cell—the *polar discontinuity* picture

The z-component of the vectors $\mathbf{D}$, $\mathbf{E}$, and $\mathbf{P}$ are denoted as D, E, and P. As shown in Figure 7.18(a), all the ionic charges are included in the unit-cells, and thus bound. To fix the constant in E, the vacuum can be taken as the reference for $E = 0$. Because of the absence of free charges, $D = 0$ from (7.4). In this case, the $[(AlO_2)^-\ (LaO)^+]$ unit-cell has a dipole moment of $\frac{q_0 a}{2}$, where q_0 is the elementary charge and a is the pseudocubic lattice constant of $LaAlO_3$. Considering the unit-cell volume a^3, P_{dipole} is given by $P_{\text{dipole}} = \frac{q_0}{2a^2}\theta(z)$, where $\theta(z)$ is the step function. Equation (7.8) immediately provides E as a function of the position, namely $E = -\frac{q_0}{2\varepsilon a^2}\theta(z)$. The total polarization is given by $P = \left((\varepsilon - \varepsilon_0) \cdot \frac{-q_0}{2\varepsilon a^2} + \frac{q_0}{2a^2}\right)\theta(z) = \frac{\varepsilon_0}{\varepsilon}\frac{q_0}{2a^2}\theta(z)$, following (7.6) and (7.7), and from (7.5), $\rho_{\text{bound}} = -\frac{\mathrm{d}}{\mathrm{d}z}P = -\frac{\varepsilon_0}{\varepsilon}\frac{q_0}{2a^2}\delta(z)$, where $\delta(z)$ is the Dirac δ function. This indicates that the system has bound charges of $-\frac{\varepsilon_0}{\varepsilon}\frac{q_0}{2a^2}$ at the surface.

Taking $[\frac{1}{2}(LaO)$ - (AlO_2) - $\frac{1}{2}(LaO)]$ as a unit cell—the *local charge neutrality* picture

When taking dipole-free unit-cells [Figure 7.18(b)], the topmost unit-cell is $[(AlO_2)$ - $\frac{1}{2}(LaO)]$, which is not a bulk unit-cell. Therefore, the half negative charge which belongs to this

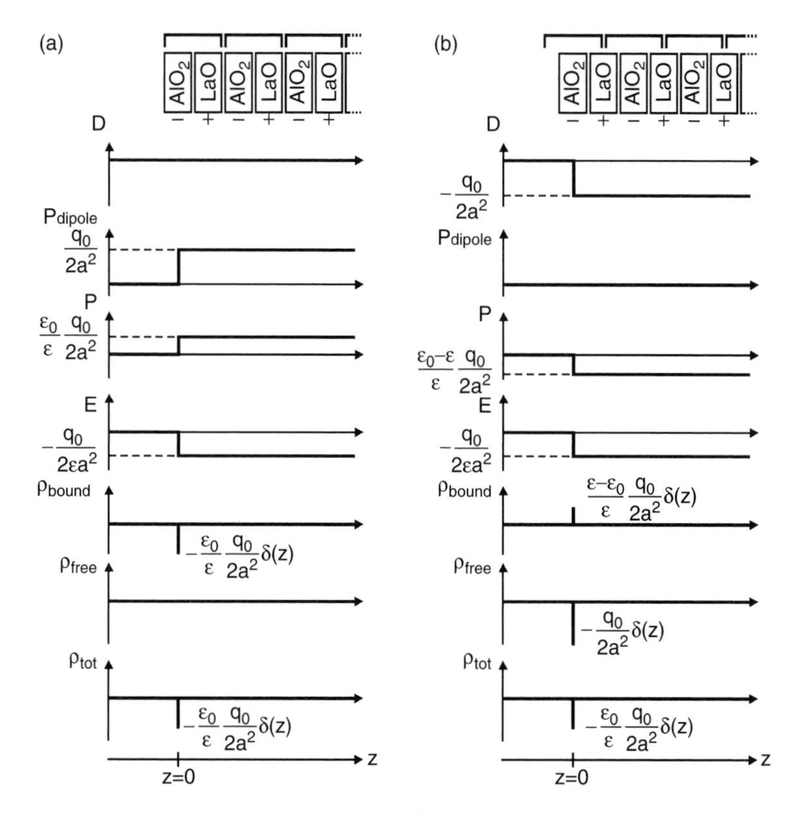

Fig. 7.18 Schematic illustration of macroscopic D, P_{dipole}, P, E, ρ_{bound}, ρ_{free}, and ρ_{tot} close to the (001) surface of $LaAlO_3$. (a) Taking $[(AlO_2)^- \ (LaO)^+]$ as a unit-cell and (b) taking $[\frac{1}{2}(LaO)$ - (AlO_2) - $\frac{1}{2}(LaO)]$ as a unit-cell.

unit-cell is free, and $\rho_{\text{free}} = -\dfrac{q_0}{2a^2}\delta(z)$. By integrating (7.4) and using the boundary condition that $D = 0$ in the vacuum, we obtain $D = -\dfrac{q_0}{2a^2}\theta(z)$. Since $P_{\text{dipole}} = 0$, E is obtained as $E = \dfrac{D - P_{\text{dipole}}}{\varepsilon} = -\dfrac{q_0}{2\varepsilon a^2}\theta(z)$. According to (7.3), the total polarization is $P = D - \varepsilon_0 E = \dfrac{\varepsilon - \varepsilon_0}{\varepsilon}\dfrac{q_0}{2a^2}\theta(z)$, and ρ_{bound} is obtained by (7.5), namely $\rho_{\text{bound}} = \dfrac{\varepsilon - \varepsilon_0}{\varepsilon}\dfrac{q_0}{2a^2}\delta(z)$. This bound charge density appeared as a response to the electric field E arising from ρ_{free}. Thus we confirm that ρ_{total} is independent of the choice of unit-cell.

7.6 Further discussions

7.6.1 Effect of interdiffusion

In real systems, interdiffusion of atoms across the interface is inevitable, and we should note its effect on the electrostatic stability of the system. Interdiffusion is a process where atoms are exchanged locally, and does not change the charge neutrality conditions around the interface, except for the finite dipoles induced by the modulation of charges. Since the instability of polar interfaces can be derived from the lack of charge neutrality around them, simple interdiffusion in a finite region cannot compensate this. That is, local stoichiometric interdiffusion can neither create, nor remove, a potential divergence. Here, this point is emphasized by considering a simple model.

Figure 7.19(a) shows the charge structure ρ and the calculated electrostatic potential ϕ of an abrupt polar interface. By taking dipole-free unit-cells, the interface unit-cell has a half positive charge per 2D unit-cell, and the system does not keep charge neutrality as it is, and is unstable. Figure 7.19(b) shows an example of interdiffusion, where half of the negatively charged layer and half of the charge neutral layer close to the interface are swapped, compared to the model in Figure 7.19(a). In this case, the extra positive charge appears at a different position, but the amount of the charge is conserved, and a similar instability in ϕ arises as in the abrupt case.

In order to show that interdiffusion does not change the stability or instability of a polar interface, it is useful to consider charge neutrality in a cluster consisting of a sufficiently large number of dipole-free unit-cells covering the interface region. By taking dipole-free unit-cells, the electrostatic structure in the bulk can be neglected to determine the stability of the system. As an example, let us take a cluster shown in Figures 7.19(a) and (b). Since the modulation of the charge to achieve the interdiffused model occurs inside the cluster, the total amount of the charge in the cluster is conserved, and thus the same instability appears in both cases.

For comparison of these models, let us focus on the difference of the charge distributions $\rho(\mathrm{b}) - \rho(\mathrm{a})$ in the two models, as shown in Figure 7.19(c), where a dipole with a finite size appears. Since ϕ is calculated by spatially integrating the charge distribution twice, the difference of ϕ in Figures 7.19(a) and (b) should be equal to the potential shift created by the dipole in Figure 7.19(c). Indeed, ϕ in Figure 7.19(b) has a shift from that in Figure 7.19(a), which is the same amount as the dipole shift in Figure 7.19(c). It should be highlighted that this dipole shift

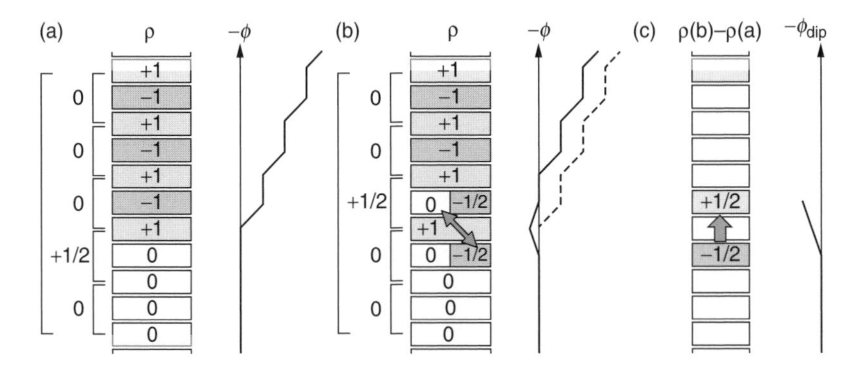

Fig. 7.19 Schematic charge structure ρ and electrostatic potential ϕ of (a) an abrupt interface and (b) an interface with interdiffusion. Five dipole-free unit-cells are taken as an example of a cluster covering the interface region. The dashed line in (b) is the duplication of ϕ in (a). (c) Difference $\rho(\mathrm{b}) - \rho(\mathrm{a})$ of the charge distributions in (a) and (b), and created potential shift ϕ_{dip} by the dipole indicated by the shaded arrow.

can indeed be an interface-specific additional driving force for interdiffusion, and has been suggested to fundamentally limit the abruptness of some interfaces [24]. The energy associated with a band offset, for example, can be reduced by forming this dipole.

Finally, we should note the difference between simple interdiffusion and change of interface composition. In the interdiffusion process, only exchanging atoms in the finite interface region is allowed, and therefore the total number of atoms of each element in the region is conserved. On the other hand, interface composition can be changed, for example by inserting other materials, segregation of atoms, or creating vacancies. For example, the reconstruction model in Figure 7.6(b) to compensate the instability of a polar Ge/GaAs interface cannot be achieved by a simple roughening at the interface: the numbers of Ge, Ga, and As atoms are different compared to those in the abrupt model [Figure 7.6(a)].

In summary, the effect of the interdiffusion appears as an extra interface dipole moment at the interface from the viewpoint of electrostatics. However, it does not change the total number of charges at the vicinity of the interface, and thus cannot screen the charge imbalance at polar interfaces. In order to compensate the instability of a polar interface, therefore, introduction of a compositional change or other compensating charge is required.

7.6.2 Role of correlation effects

So far we have tried simply to describe the perovskite polar discontinuity using semiconductor language by taking dipole-free unit-cells and treating the interface unit-cell as a δ-dopant. It should be mentioned that in the presence of strong electron–electron correlations, commonly found in transition metal oxides, it is formally impossible to draw semiconductor energy band diagrams based on the single-particle picture of independent electrons [59, 60].[11] Also, in order to draw band diagrams, we should know which part of the charges are to be assigned as free carriers, which is nontrivial in correlated systems, such as Mott insulators. For example, a perovskite with 1 electron per unit-cell can give rise to an effective carrier density ranging from $\sim 1 \times 10^{22}$ cm^{-3} to zero, depnding on the correlation strength.

However, the *local charge neutrality* picture can still provide important information on the interface charge structure of transition metal oxides, since correlations cannot change the amount of total charge. While the distribution of this charge may be significantly modified by correlation features in the electronic compressibility [61], the charge can still be determined in a cluster consisting of a sufficiently large number of dipole-free unit-cells. This is because the material outside of the charge modulation region consists of dipole-free unit-cells of bulk, which are charge neutral and thus stable.

7.6.3 Quadrupolar discontinuity

Thus far, we have considered the stability of polar discontinuities, and showed that the amount of charge needed at the interface can be determined by taking dipole-free unit-cells and considering local charge neutrality in each unit-cell. Of course, local charge modulation is allowed

[11] Owing to the relatively small (still non-negligible) correlation effects in $SrTiO_3$ with the Ti $3d^0$ configuration and weak $2p$-$3d$ hybridization in the coherent state of doped $SrTiO_3$ [58], the schematic band diagram shown in Figure 7.13 is still reasonably valid, approximating $SrTiO_3$ to be a band semiconductor.

once neutrality is obtained, since it does not violate global charge neutrality. This dipole energy arising from the modulation of charges determines the real charge distribution in the system.

Here, we note that there is an intrinsic dipole shift at the interface arising from the charge stacking sequence, which should be distinguished from this charge modulation. For example, consider a $LaAlO_3/SrTiO_3$ (001) interface with $Ti^{3.5+}$ at the interface to solve the instability of the polar interface, as shown in Figure 7.20(a). Although there is no potential divergence, a finite shift Δ remains between the two materials when considering the averaged electrostatic potential on both sides.

The origin of this potential shift can be understood based on the discontinuity of the quadrupole moment of the unit-cells. In one dimension, the quadrupole moment density Q is defined as $\frac{d}{dx}Q = P$, where P is the dipole moment density. It has the same form as that of Gauss' law connecting the dipole moment and the charge—i.e. the source of the quadrupole moment is the dipole moment. Therefore, following the same argument as in Section 7.4.1, a finite interface dipole moment appears at a quadrupolar discontinuity, when the quadrupole moment is different on the two sides.

Moreover, this quadrupole moment is proportional to the potential shift at that point in the absence of free charge, since $D = 0$ and thus $\varepsilon_0 E = -P$. Therefore if the unit-cell has a finite quadrupole moment, that means that the averaged potential in the unit-cell is shifted by the corresponding value. Note the quadrupole moment does not induce a potential shift outside of the unit-cell, while the dipole moment does change.

For example, $[(\frac{1}{2}\mathrm{LaO})^{+1/2}\text{-}\mathrm{AlO}_2^{-1}\text{-}(\frac{1}{2}\mathrm{LaO})^{+1/2}]$, a dipole-free unit-cell of $LaAlO_3$ (001) [Figure 7.20(b)], has a finite quadrupole moment, while the charge neutral stacking of $SrTiO_3$ creates no quadrupole moment in the unit-cells.[12] This difference of quadrupole moment

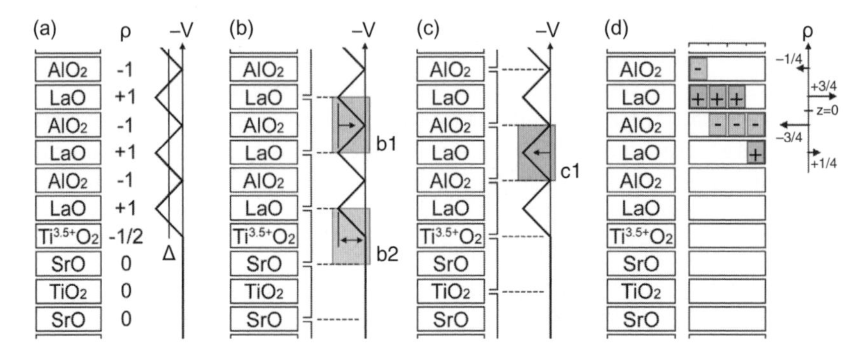

Fig. 7.20 (a) Schematic structure, charge density ρ, and electrostatic potential ϕ of a $LaAlO_3/SrTiO_3$ interface with a $Ti^{3.5+}O_2$ interface layer, showing a finite shift Δ. Taking (b) $[(\frac{1}{2}\mathrm{LaO})^{+1/2}\text{-}\mathrm{AlO}_2^{-1}\text{-}(\frac{1}{2}\mathrm{LaO})^{+1/2}]$ and (c) $[(\frac{1}{2}\mathrm{AlO}_2)^{-1/2}\text{-}\mathrm{LaO}^{+1}\text{-}(\frac{1}{2}\mathrm{AlO}_2)^{-1/2}]$ as a unit-cell. The shaded areas (1) and (1) show the unit cells with opposite signs of quadrupole moment, respectively, and the green shaded area in (b) shows the unit-cell with a finite dipole moment. (d) An example of dipole- and quadrupole-free unit-cells.

[12] Absence of covalency is assumed.

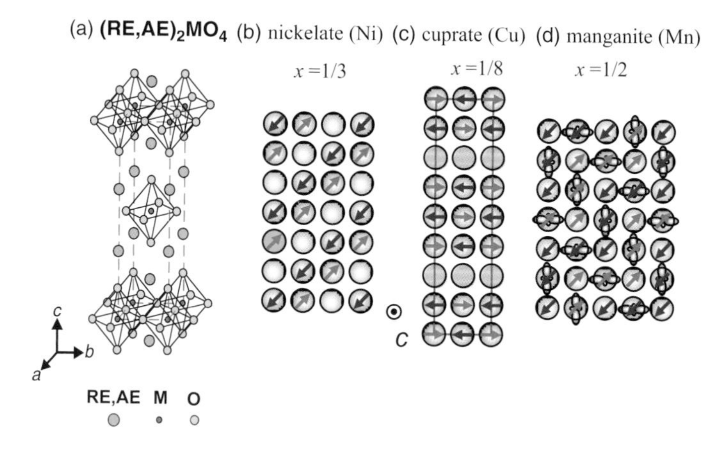

Fig. 1.7 Charge order in the form of "stripes" in hole-doped two-dimensional metal oxide sheets of layered perovskite materials with (a) K_2NiF_4-type structure; (b) $La_{2-x}Sr_xNiO_4$ ($x \approx 1/3$) [22], (c) $La_{2-x}Ba_xCuO_4$ ($x \approx 1/8$) [23, 24], and (d) $La_{1-x}Sr_{1+x}MnO_4$ ($x \approx 1/2$) [25].

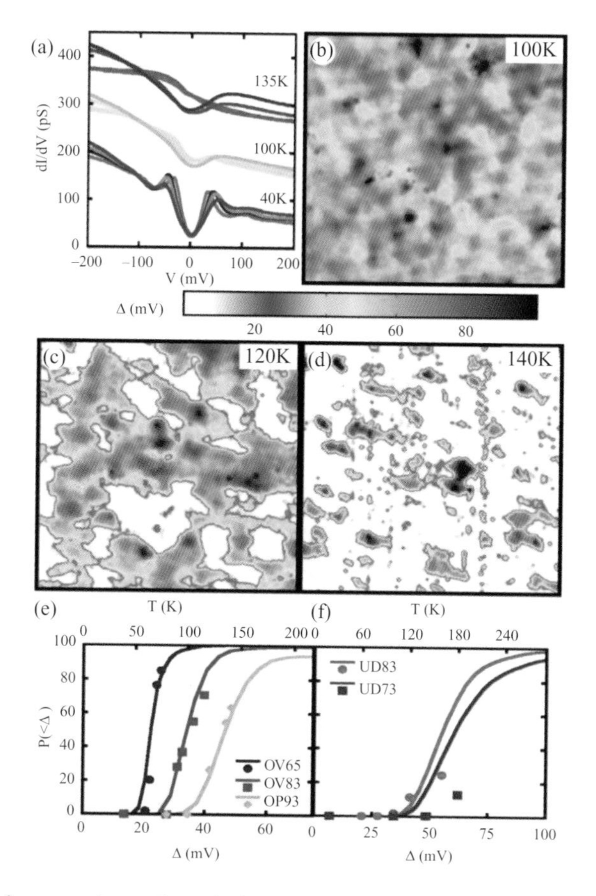

Fig. 1.11 Example of an experimental result that illustrates the presence of inhomogeneous states in a transition-metal oxide. Shown are results reported in [29], containing scanning tunneling microscopy (STM) data gathered for a high-temperature superconductor material with a critical (optimal) temperature of 93 K. The colored regions have a *d*-wave STM spectrum, and the remarkable result is that these superconducting areas are observed even at temperatures as high as 120 K, well above the critical temperature, contrary to the behavior of typical BCS low-temperature superconductors. Reprinted by permission from Macmillan Publishers Ltd [29], copyright 2007.

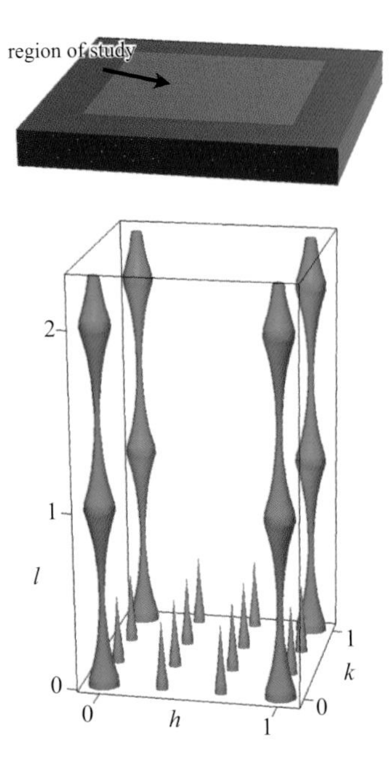

Fig. 4.1 Schematic depiction of surface X-ray diffraction. The real space image of the sample lies above its reciprocal space counterpart. CTRs are in green while SRs are blue. Courtesy of S. O. Hruszkewycz.

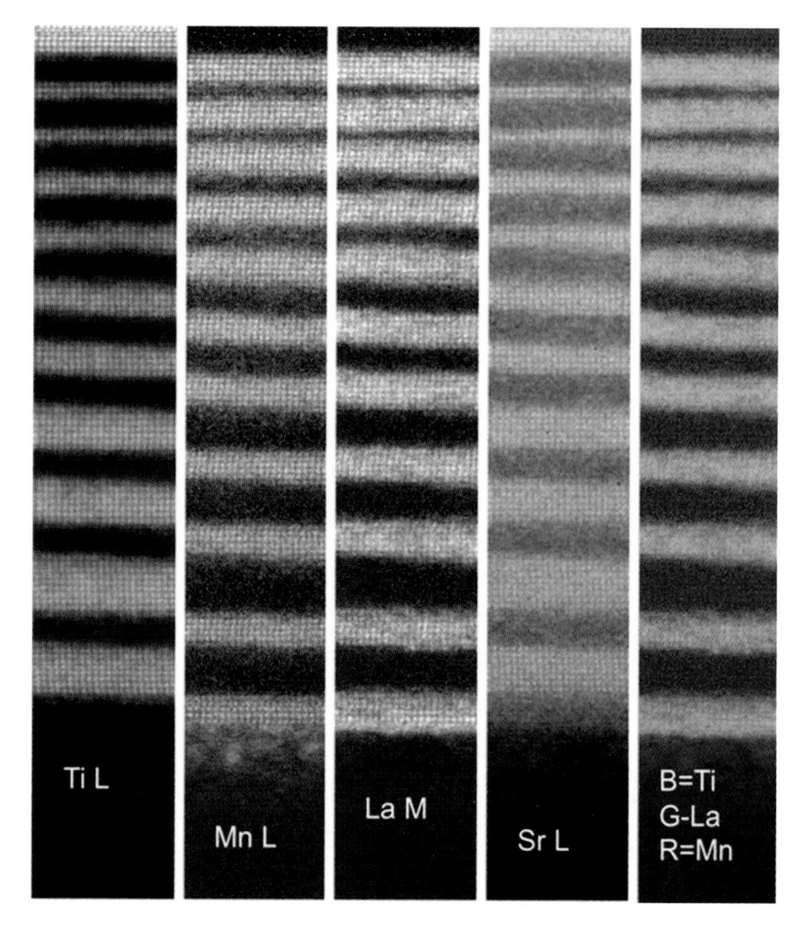

Fig. 5.1 Atomic resolution elemental maps of a $La_{0.7}Sr_{0.3}MnO_3/SrTiO_3$ superlattice with changing $SrTiO_3$ layer thicknesses, obtained in the aberration corrected Nion UltraSTEM 100. From left to right: Ti $L_{2,3}$ map, Mn $L_{2,3}$ map, La $M_{4,5}$ map, and Sr $L_{2,3}$ maps. The RGB is an overlay of the Ti, La, and Mn images.

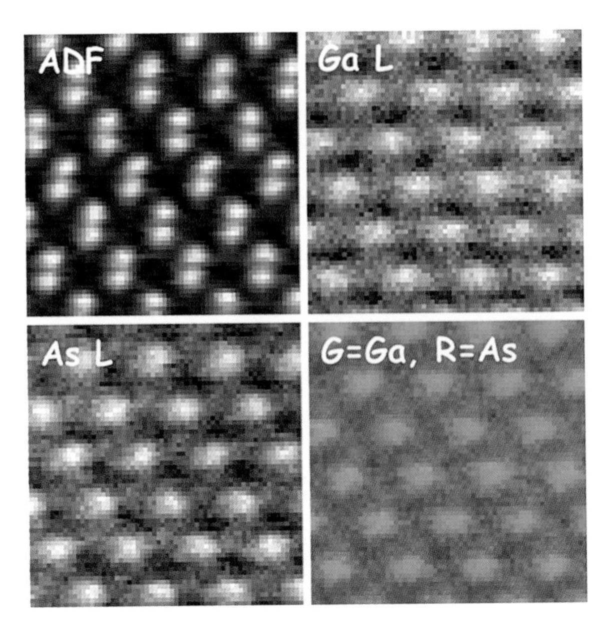

Fig. 5.3 Spectroscopic imaging of GaAs in the ⟨110⟩ projection comparing the ADF image to the Ga- and As L spectroscopic images, obtained on a Nion UltraSTEM with a 5th-order aberration corrector operating at 100 kV. Images are 64 × 64 pixels, with collection time 0.02 s/pixel and a beam current of approximately 100 pA, after noise reduction by Principle Component Analysis (PCA) [46].

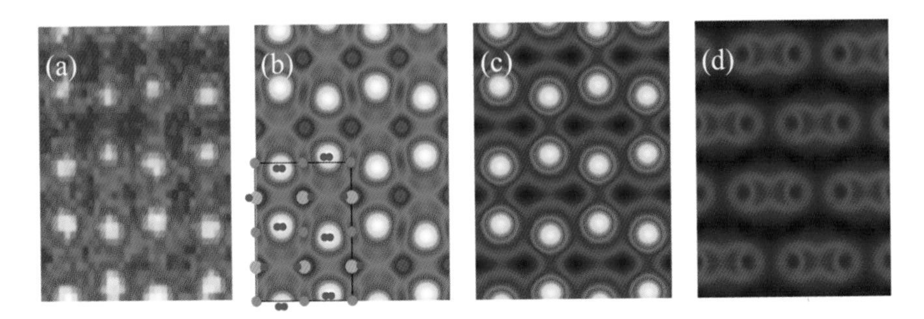

Fig. 5.11 Integrated oxygen K-shell EELS signal from $LaMnO_3$ in the [010] zone axis orientation (pseudocubic ⟨110⟩ axis). (a) Experimental image acquired on the Nion UltraSTEM operating at 60 kV. (b) Simulated image with projected structure inset. (c) Contribution to the total image from the isolated O columns. (d) Contribution to the total image from the O atoms on the La/O columns. Reproduced from [61].

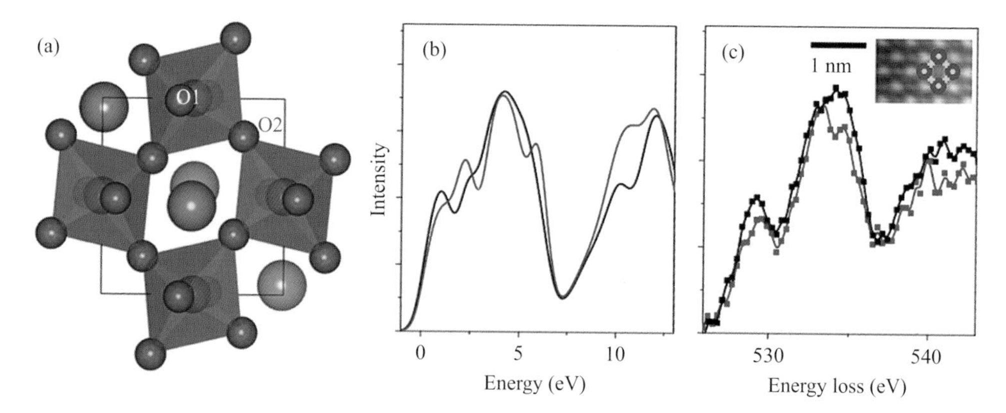

Fig. 5.14 (a) Sketch of the LMO structure, produced from the structural data reported by [81]. (b) Simulated O K EEL spectra for the O1 (black) and O2 (blue) species, adapted from [46]. (c) Experimental EEL spectra from different positions (electron beam on top of a MnO column in black and on top of a La column in blue), extracted from a linescan along the pseudocubic ⟨110⟩ direction of LMO (dataset in [46]). Principal component analysis was used to remove random noise. The LMO pseudocubic unit cell is shown on top of a Z-contrast image in the inset (La = red, Mn = blue, O = yellow).

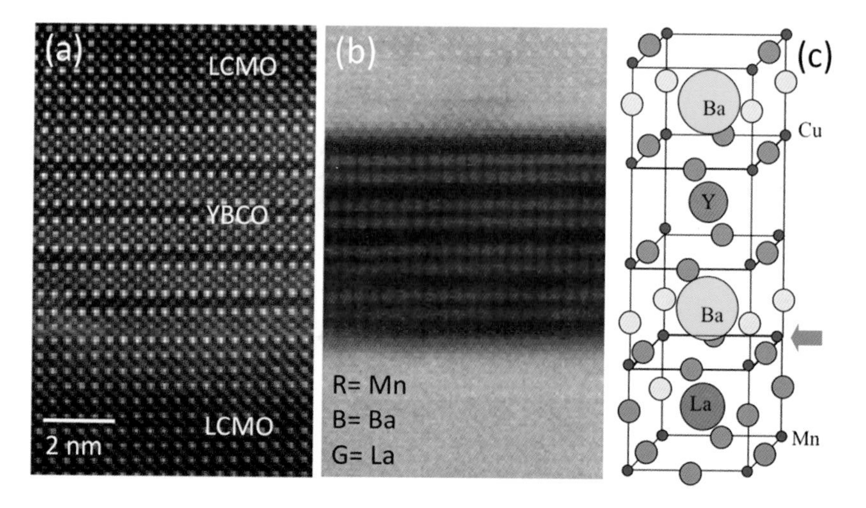

Fig. 5.17 (a) ADF signal collected simultaneously with a spectrum image in a LCMO/YBCO/LCMO trilayer. (b) False color image where three atomic resolution elemental images have been overlayed: a Mn $L_{2,3}$ image in red, a Ba $M_{4,5}$ image in blue and a La $M_{4,5}$ image in green, for the same area as the Z-contrast in (a). The EELS data were acquired in the Nion UltraSTEM column and processed with principal component analysis in order to remove random noise. (c) Sketch of the observed interface structure. An arrow marks the interface MnO_2 plane, facing a BaO plane from the superconductor.

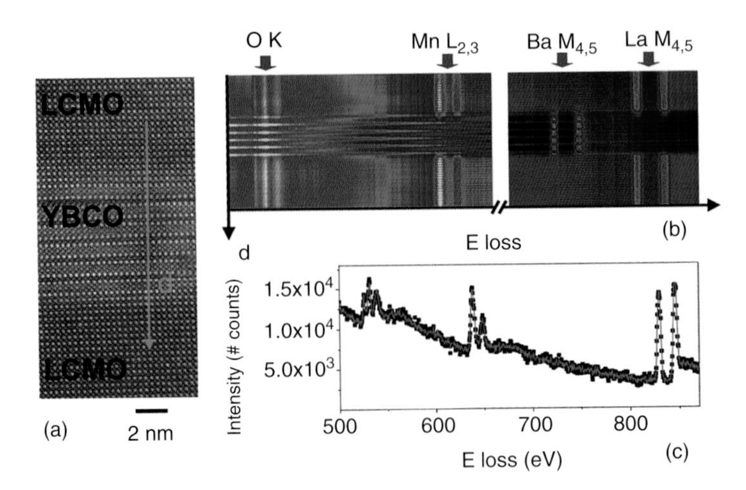

Fig. 5.20 High resolution Z-contrast image of a LCMO/YBCO/LCMO trilayer, obtained at 100 kV. (b) EELS linescan acquired along the direction marked with an arrow in (a). PCA has been used to remove random noise. (c) Sample spectrum extracted from the linescan in (b), acquisition time is 2 seconds per spectrum. The data points are a raw spectrum while the red line is the same dataset after PCA.

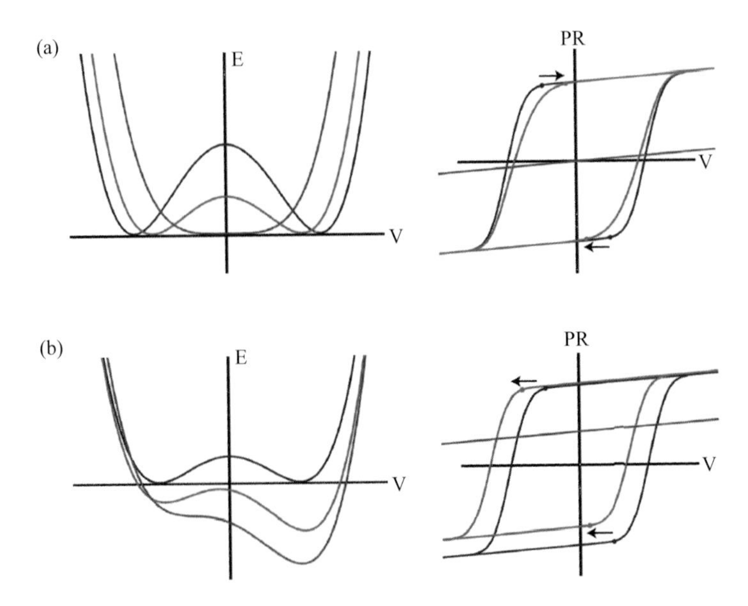

Fig. 6.2 The role of defects on the energetics of polarization switching.(a) Random-bond (symmetric) and (b) random-field (asymmetric) disorder in a ferroelectric material and their effects on local hysteresis loop shape. The blue curves correspond to a defect-free material, green curves to the presence of a weak random-bond (a) and random-field (b) defect, and red curves to the limiting cases of a nonpolar phase and polar non-ferroelectric phases, respectively. Reprinted by permission from Macmillan Publishers Ltd: Nature Materials [16], copyright 2008.

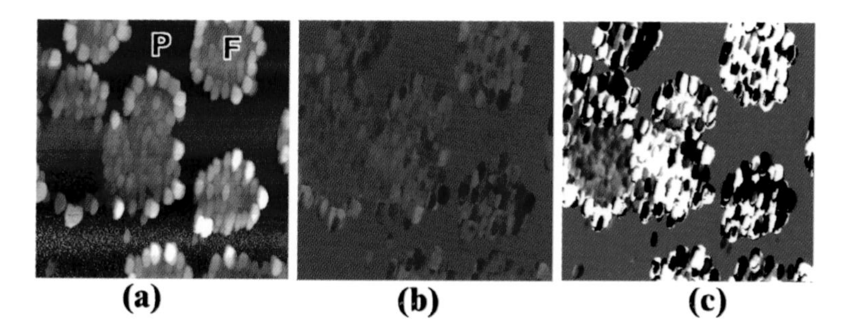

Fig. 6.4 Topography (a) and PFM images obtained of resonance (b) and near contact resonance (c) $Pb(Zr_{0.3}Ti_{0.7})O_3$ films showing both ferroelectric and non-polar phases. Scan size is 5 mm, z-scale of (a) is 20 nm, the z-scale of (b) and (c) is the same, 10 mV full scale. Image courtesy C. Harnagea.

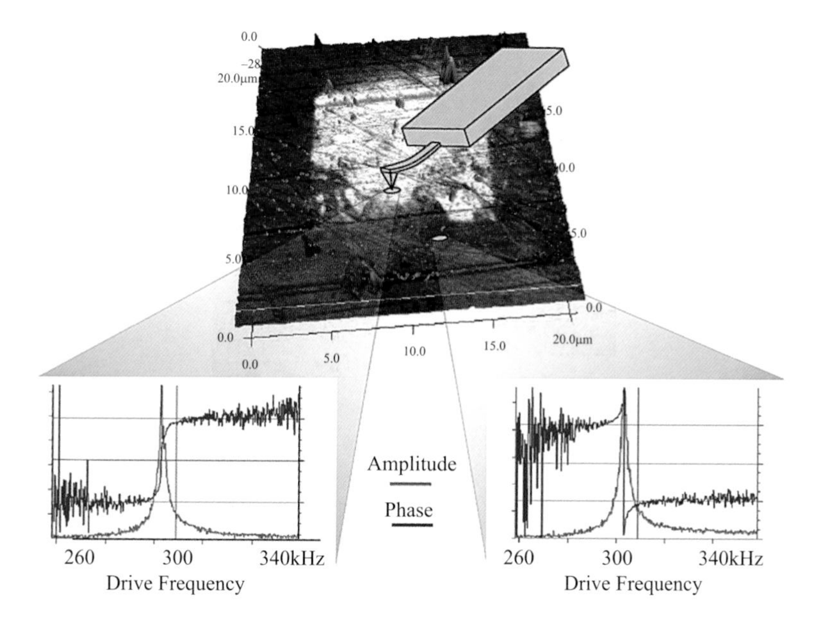

Fig. 6.5 Schematic illustration of amplitude and phase cantilever response, showing a 180° phase offset over antiparallel domains, demonstrating that it cannot reliably be used as a feedback signal for a phase-locked loop. Reprinted with permission from [53], courtesy of the Institute of Physics.

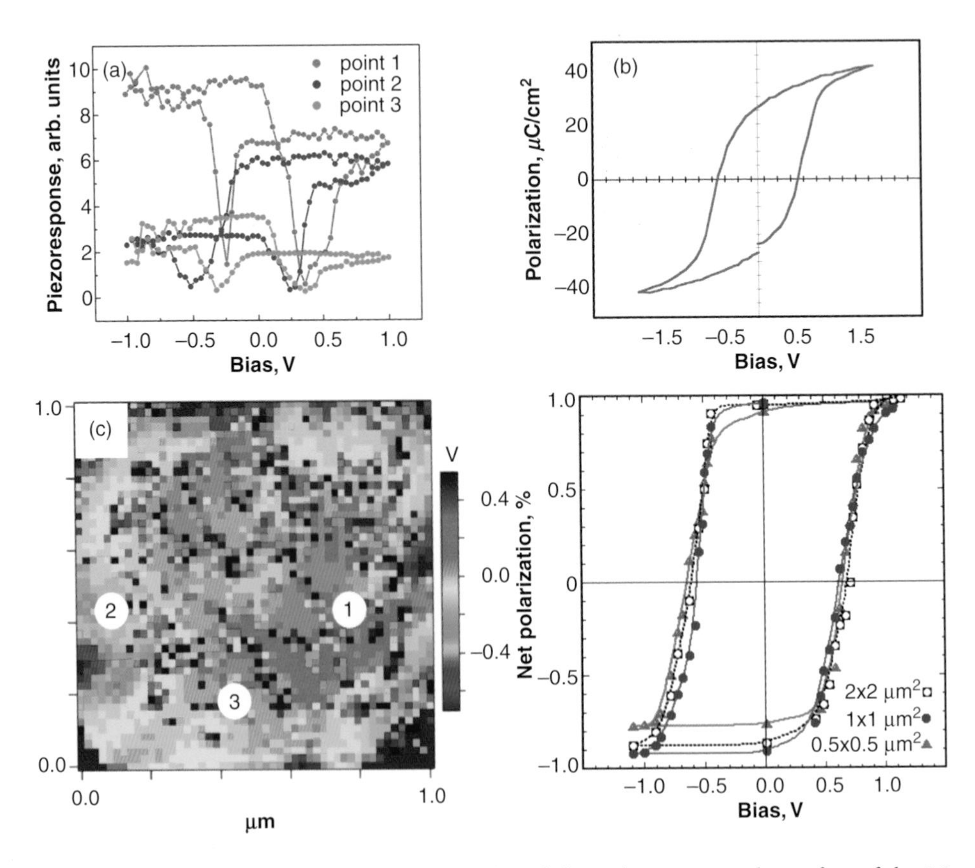

Fig. 6.12 (a) Local PFM hysteresis loops measured in different locations on the surface of the $1.0 \times 1.0\mu m^2$ PZT capacitor. Loop numbers correspond to location numbers shown in (c). (b) Conventional P-E hysteresis loop measured in the same type of PZT capacitor, but larger dimensions ($65 \times 65\mu m^2$). (c) 2D map of an imprint bias in the $1.0 \times 1.0\mu m^2$ PZT capacitor generated by SS-PFM. (d). Global PFM hysteresis loops measured in capacitors of different dimensions. Reprinted with permission from [75]. Copyright 2009, American Institute of Physics.

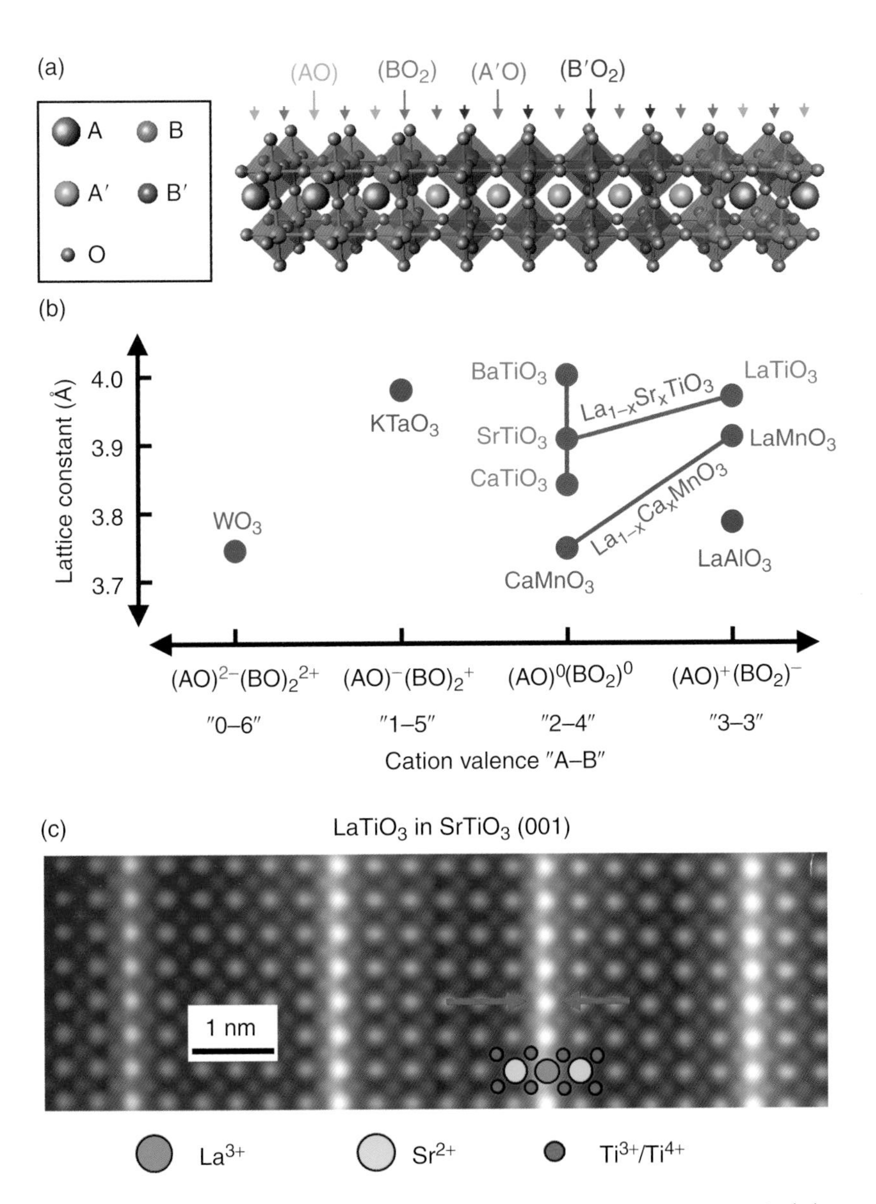

Fig. 7.1 (a) Schematic illustration of ideal heterointerfaces of two perovskites ABO_3 and $A'B'O_3$ stacked in the [001] direction. (b) Charge sequences of the AO and BO_2 planes of perovskites plotted together with their pseudocubic lattice parameters. (c) Scanning transmission electron microscopy image of a $LaTiO_3/SrTiO_3$ (001) superlattice (Ohtomo *et al.* [4]).

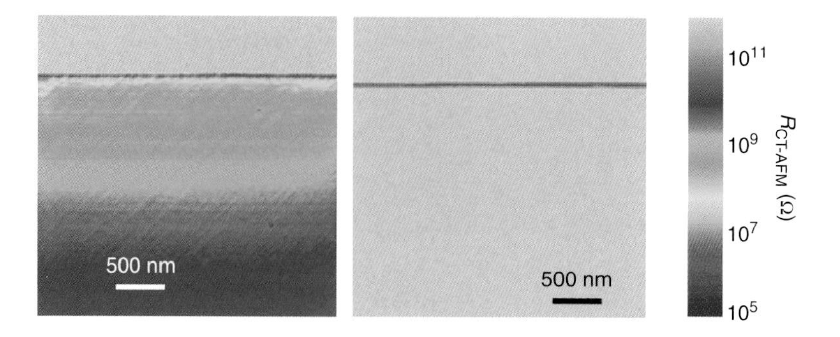

Fig. 7.9 Conducting-tip atomic force microscopy mapping around the $LaAlO_3/SrTiO_3$ interface of (a) the as grown sample ($P_{O_2} = 10^{-5}$ Torr) and (b) the postannealed sample. Reprinted by permission from Macmillan Publishers Ltd: Nature Materials [34], copyright 2008.

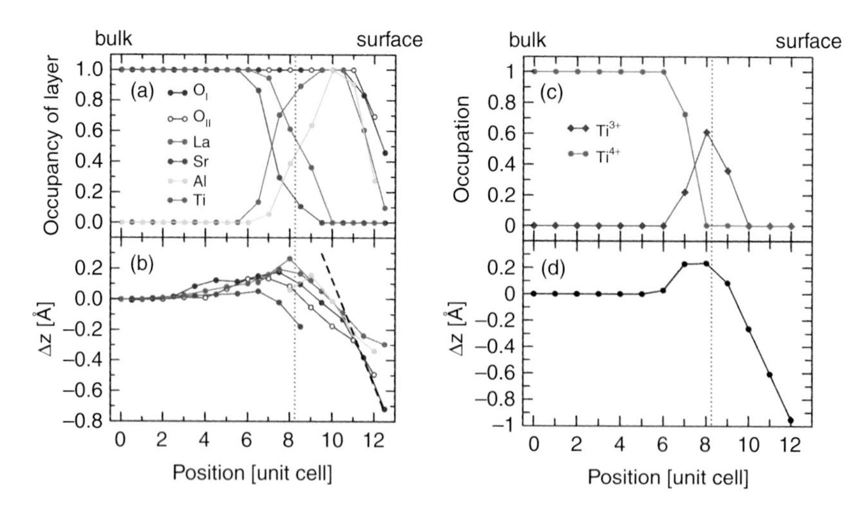

Fig. 7.10 (a) Occupancies and (b) cumulative displacements Δ_z of the atoms at the $LaAlO_3/SrTiO_3$ interface. (c) Concentration of Ti^{3+} determined by a minimization of the electrostatic potential. (d) Predicted cumulative unit-cell displacements from bulk positions, based on the component ionic radii. Reprinted with permission from [35]. Copyright 2007 by the American Physical Society.

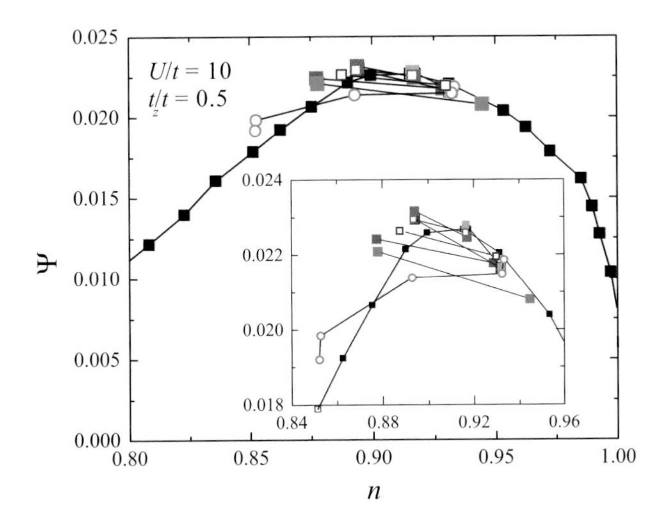

Fig. 8.8 The superconducting order parameter Ψ as a function of carrier density n for the quasi-two-dimensional Hubbard model computed by CDMFT with the ED impurity solver at $T = 0$. Black symbols are the results for the uniform systems. Shaded symbols connected by a thin line denote the results for the SL. The inset shows the magnification around the optimally doped region. Figure reproduced from [75].

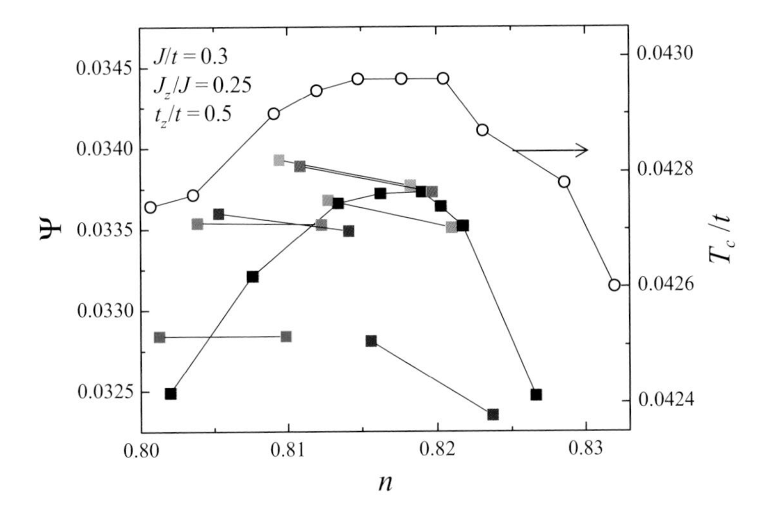

Fig. 8.10 The superconducting order parameter Ψ as a function of carrier density n for the quasi-two-dimensional t-J model computed by the DCA-NCA at a temperature $T = 0.041t$ close to T_C. The results for the uniform systems are indicated by the black symbols, while the shaded symbols connected by a thin line denote the result for the SL. Also shown is the doping dependence of T_C in the uniform system (open circles). Figure reproduced from [75].

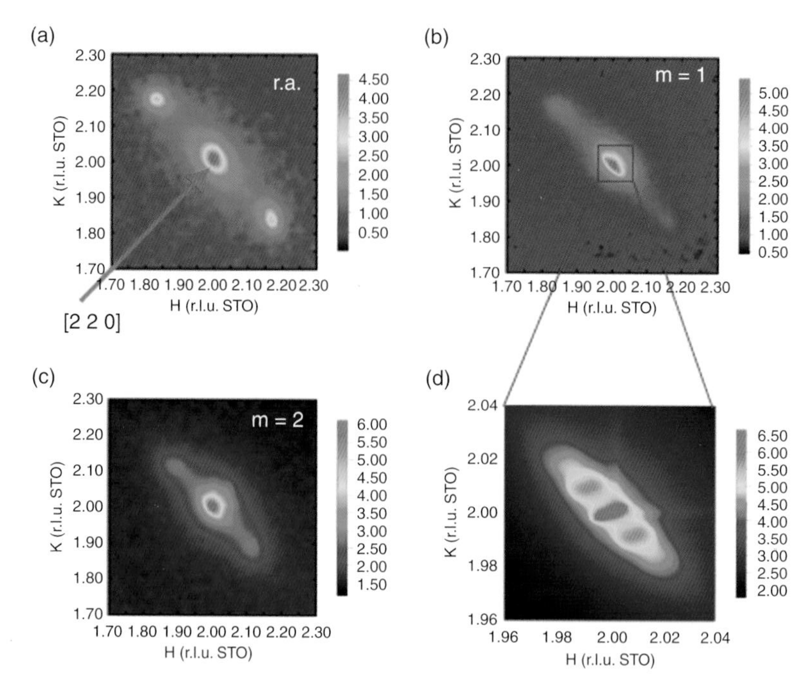

Fig. 9.10 Reciprocal space maps around the $(2\ 2\ \delta)$ reflection in grazing incidence geometry for the (a) random alloy (r.a.) (b) $m = 1$ and (c) $m = 2$ superlattices. The red arrow in (a) shows the scattering vector q along (2 2 0). Satellite peaks are seen with Δq orthogonal to q, indicating a transverse modulation that is strongest for *r.a.* and weakest for $m = 1$. The region near the 220 peak in the $m = 1$ sample is magnified in (d) to reveal a transverse modulation peak at much longer wavelength. The transverse modulation wavelength for *r.a.* is $\sim$ 1.6 nm, for $m = 2$, is $\sim$ 2.5 nm, and for $m = 1$, $\sim$ 32 nm. Reproduced from [92]. Copyright 2009, Macmillan Publishers Limited.

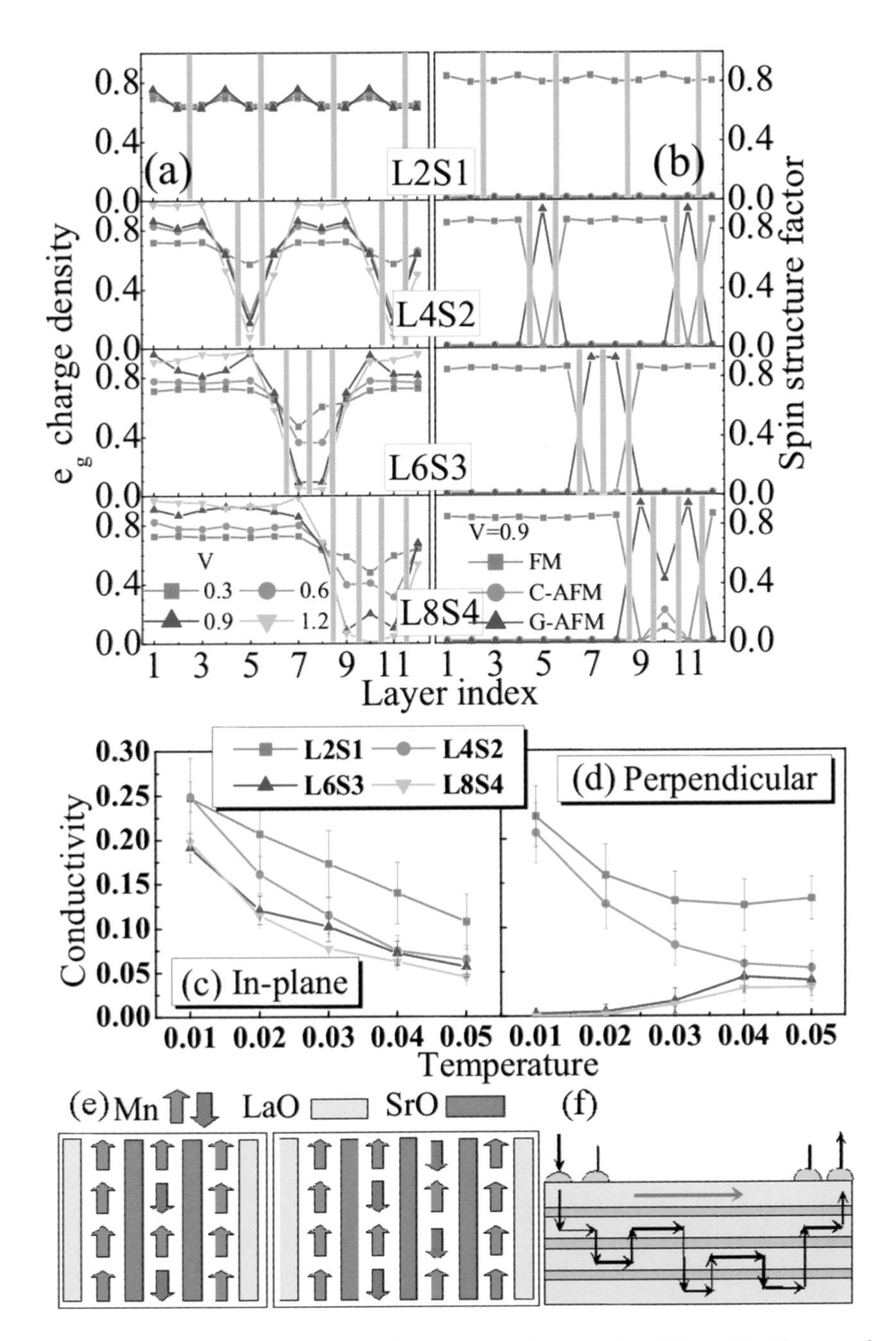

Fig. 9.18 (a) Layer-averaged e_g electron density in the superlattices with different Vs. The case of $n = 1$ with $V = 1.2$ cannot be obtained due to phase separation. Two reference values 0.5 and 2/3 are marked as the horizontal lines. (b) In-plane spin structure factor at $V = 0.9$. In (a) and (b), pink bars denote SrO sheets in the superlattice, while LaO sheets are not shown. $L_{2n}S_n$ denotes $(LMO)_{2n}/(SMO)_n$. Conductivities as a function of T: (c) in-plane one (d) out-of-plane (perpendicular) one. (e) Sketch of the spin order at interfaces for the $n = 2$ (left) and $n = 3$ (right) cases. (f) Sketch of experimental setup for resistance measurements. Pink bars are SMO layers. Typical conducting paths via the double-exchange process (black curves) connect NN interfaces, which are broken when $n \geq 3$. Reproduced from [118]. Copyright 2008, the American Physical Society.

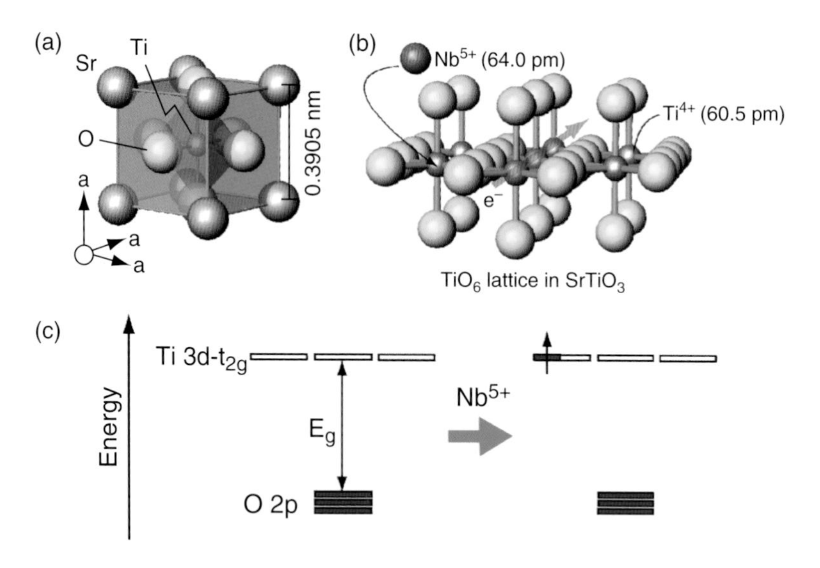

Fig. 10.6 (a),(b) Schematic crystal structure and (c) Nb doping mechanism of $SrTiO_3$.

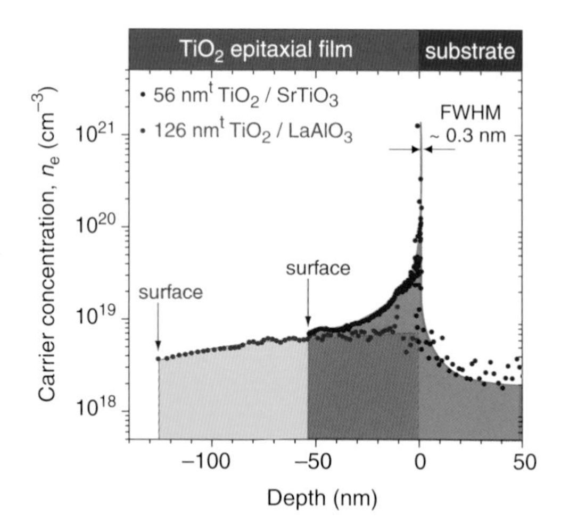

Fig. 10.10 (left) Depth profile around the interface between a 56-nm-thick TiO_2 epitaxial layer and the $SrTiO_3$ substrate. The profile for the interface between a 126-nm-thick epitaxial layer and insulating $LaAlO_3$ is also plotted for comparison (red). An intense carrier concentration peak ($n_e \sim 1.4 \times 10^{21}$ cm^{-3}) with a full-width at half maximum of $\sim$0.3 nm is seen at the TiO_2/$SrTiO_3$ interface. [H. Ohta *et al.*, *Nature Mater.* **6**, 129 (2007)].

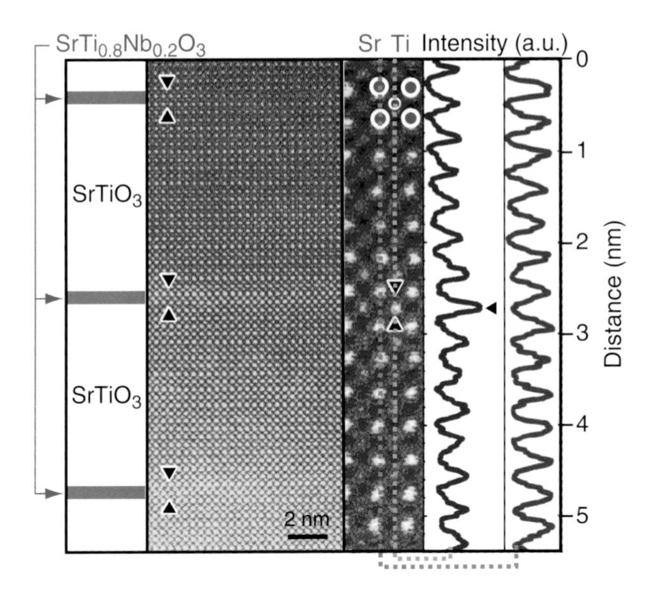

Fig. 10.11 HAADF-STEM image of $[(SrTiO_3)_{24}/(SrTi_{0.8}Nb_{0.2}O_3)_1]_{20}$ superlattice and intensity profiles across Ti and Sr atomic columns. Stripe-shaped contrast is clearly seen. Also, the intensity of the Ti column in one-unit-cell's thickness of $SrTi_{0.8}Nb_{0.2}O_3$ shows higher intensity than that of the $SrTiO_3$ barrier layer, confirming that the doped Nb^{5+} ions are exclusively confined in the $SrTi_{0.8}Nb_{0.2}O_3$ layer. [H. Ohta *et al.*, *Nature Mater*. **6**, 129 (2007)].

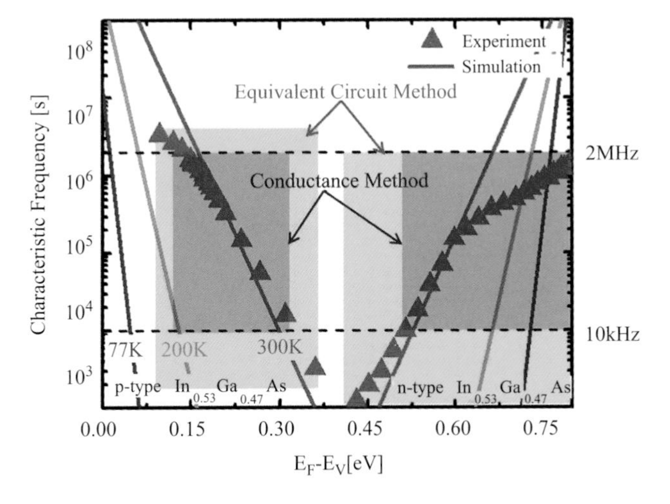

Fig. 11.8 Full conductance versus equivalent circuit methods for extracting the interface state density, Dit, and the trap response time at InGaAs/high-K dielectric interface.

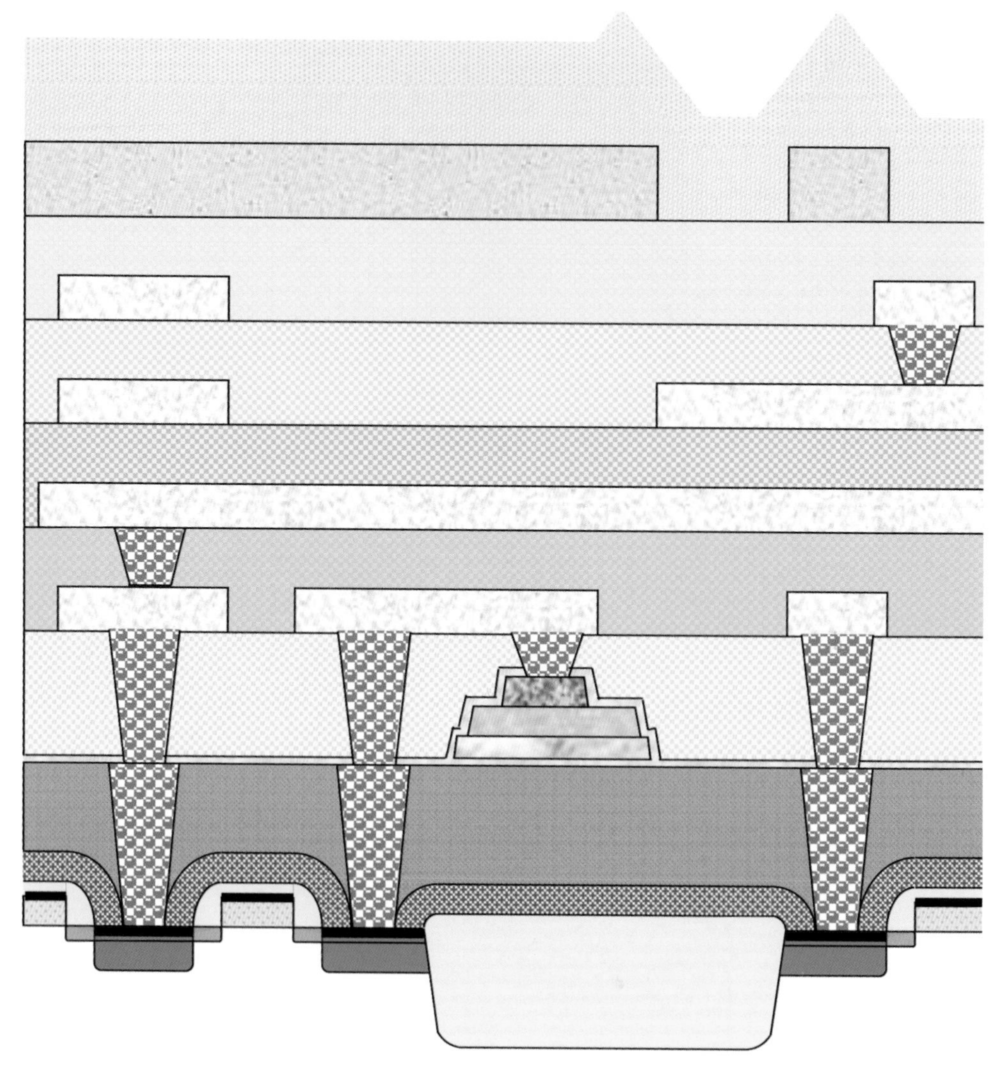

Fig. 12.11 A schematic drawing of a cross-section of actual FeRAM cells with planar ferroelectric capacitors. Courtesy of Fujitsu Semiconductor Ltd.

between the two sides adds a finite potential shift to the averaged potential in the unit-cell. Figures 7.20(b) and (c) show two ways of taking dipole-free unit-cells in cubic perovskites, where the sign of the quadrupole moment is opposite. The choice of unit-cells cannot change the electrostatic potential, and indeed this difference of the quadrupole moment of the unit-cells is compensated by the dipole moment of the interface unit-cell in Figure 7.20(b).

The ambiguity of the quadrupole moment in the unit-cells is reminiscent of the uncertainty of the dipole moment, as discussed in Section 7.4.1, and we can take the same prescription as we did previously—taking quadrupole-free unit-cells. It is more complicated since these quadrupole-free unit-cells should not have a dipole moment at the same time, in order to avoid the instability of the polar discontinuity, and Figure 7.20(d) shows one way to take such unit-cells. Note it is always possible to take unit-cells which do not have either dipole or quadrupole moments in one direction in any bulk crystal.[13] Once we take this dipole- and quadrupole-free unit-cell, the size of the interface dipoles can be determined locally by considering the dipole moment in the interface unit-cell(s), since there is no shift in the averaged potential in the bulk unit-cells due to the absence of quadrupole moment.

7.7 Summary

As pointed out by Tasker, the order of the ionic charge stacking plays an important role for the stability of the surfaces and interfaces, and sometimes they show a diverging potential when following bulk compositions and electronic states. In semiconductors, where the number of ionic charges is fixed, they are usually stabilized by atomic reconstructions, as shown by Harrison *et al.* However, in transition metal oxides, electronic reconstructions provide another possibility to reconcile the instability. Exploiting this degree of freedom has been a central topic of recent research in oxide heterostructures, but with much debate over the relative contribution of this effect, compared to those common to all interfaces, such as defects and diffusion.

We hope the reader will agree that the attribution of various experimental measurements in real materials to purely the polar discontinuity, as opposed to chemical defects, is not really a valid separation. It is rather governed by the response of the total system to the electro-chemical potential (in the chemistry sense), subject to specific thermodynamic, kinetic, and geometric boundary conditions. Given the freedom to define the unit-cell, taking dipole-free units provides a convenient format for this perspective. While not explicitly discussed here, interface screening by local lattice polarization can be directly incorporated, including subtleties that arise at very short length scales. Thus these topics connect to current research in surface and interface phenomena for ferroelectrics and multi-ferroics. Furthermore, recent discussion of the role of film stoichiometry can be understood together with these other aspects [62]. Ultimately, the requirement to stabilize polar discontinuities is independent of whether the charge is free or bound, itinerant or localized, and electronic or ionic. Perhaps the central message of this simple review chapter is that one cannot claim atomically pre-cise oxide heterostuctures without understanding how the electrostatic boundary conditions

[13] This can easily be proved from the periodicity of the charge structure and the macroscopic charge neutrality of the lattice by using the intermediate-value theorem.

are accomodated. This is both a problem to be solved, as well as a fascinating synthetic opportunity.

Acknowledgments

We thank our many colleagues and collaborators in this field who have helped to develop these topics, and T. Yajima for pointing out the quadrupole–dipole connection. We acknowledge support from the Japan Science and Technology Agency, the Japan Society for the Promotion of Science, and the Department of Energy, Office of Basic Energy Sciences, under contract DE-AC02-76SF00515 (H.Y.H.).

References

[1] Esaki, L. and Tsu, R. (1970). Superlattice and negative differential conductivity in semiconductors. *IBM J. Res. Devel.* **14**, 61–65.

[2] Baibich, M.N., Broto, J.M., Fert, A., *et al.* (1988). Giant magnetoresistance of (001)Fe/(001)Cr magnetic superlattices. *Phys. Rev. Lett.* **61**, 2472–2475.

[3] Binasch, G., Grünberg, P., Saurenbach, F., and Zinn, W. (1989). Enhanced magnetoresistance in layered magnetic structures with antiferromagnetic interlayer exchange. *Phys. Rev. B* **39**, 4828–4830.

[4] Ohtomo, A., Muller, D.A., Grazul, J.L., and Hwang, H.Y. (2002). Artificial charge-modulation in atomic-scale perovskite titanate superlattices. *Nature (London)* **419**, 378–380.

[5] Mukunoki, Y., Nakagawa, N., Susaki, T., and Hwang, H.Y. (2005). Atomically flat (110) $SrTiO_3$ and heteroepitaxy. *Applied Physics Letters* **86**, 171908 1–3.

[6] Tokura, Y., Urushibara, A., Moritomo, Y., *et al.* (1994). Giant magnetotransport phenomena in filling-controlled Kondo lattice system: $La_{1-x}Sr_xMnO_3$. *J. Phys. Soc. Jpn.* **63**, 3931–3935.

[7] Kinniburgh, C.G. (1975). A LEED study of MgO(100). II. Theory at normal incidence. *J. Phys. C: Solid State Phys.* **8**, 2382–2394.

[8] Netzer, F.P. and Prutton, M. (1975). LEED and electron spectroscopic observations on NiO (100). *J. Phys. C: Solid State Phys.* **8**, 2401–2412.

[9] Benson, G.C. and Yun, K.S. (1967). Surface energy and surface tension of crystalline solids. In: *The Solid-Gas Interface*, E.A. Flood, ed., vol. 1, pp. 203–269. Marcel Dekker Inc., New York.

[10] Tasker, P.W. (1979). The stability of ionic crystal surfaces. *J. Phys. C: Solid State Phys.* **12**, 4977–4984.

[11] Goniakowski, J., Finocchi, F., and Noguera, C. (2008). Polarity of oxide surfaces and nanostructures. *Rep. Prog. Phys.* **71**, 016501 1–55.

[12] Ashcroft, N.W. and Mermin, N.D. (1976). *Solid State Physics*, pp. 533–559. Thomson Learning, Tampa.

[13] Harrison, W.A. (1976). Surface reconstruction on semiconductors. *Surf. Sci.* **55**, 1–19.

[14] Frensley, W.R. and Kroemer, H. (1977). Theory of the energy-band lineup at an abrupt semiconductor heterojunction. *Phys. Rev. B* **16**, 2642–2652.

[15] Phillips, J.C. (1973). *Bonds and Bands in Semiconductors*, pp. 26–56. Academic, New York.

[16] Frensley, W.R. and Kroemer, H. (1977). Interstitial potential differences, electronegativity differences, and effective ionic charges in zinc-blende-type semiconductors. *Applied Physics Letters* **31**, 48–50.

[17] Baraff, G.A., Appelbaum, J.A., and Hamann, D.R. (1977). Self-consistent calculation of the electronic structure at an abrupt GaAs-Ge interface. *Phys. Rev. Lett.* **38**, 237–240.

[18] Grant, R.W., Waldrop, J.R., and Kraut, E.A. (1978). Observation of the orientation dependence of interface dipole energies in Ge-GaAs. *Phys. Rev. Lett.* **40**, 656–659.

[19] Harrison, W.A., Kraut, E.A., Waldrop, J.R., and Grant, R.W. (1978). Polar heterojunction interfaces. *Phys. Rev. B* **18**, 4402–4410.

[20] Ohtomo, A. and Hwang, H.Y. (2004). A high-mobility electron gas at the $LaAlO_3/SrTiO_3$ heterointerface. *Nature (London)* **427**, 423–426.

[21] Ambacher, O., Smart, J., Shealy, J.R., *et al.* (1999). Two-dimensional electron gases induced by spontaneous and piezoelectric polarization charges in N- and Ga-face AlGaN/GaN heterostructures. *J. Appl. Phys.* **85**, 3222–3233.

[22] Ibbetson, J.P., Fini, P.T., Ness, K.D., *et al.* (2000). Polarization effects, surface states, and the source of electrons in AlGaN/GaN heterostructure field effect transistors. *Appl. Phys. Lett.* **77**, 250–252.

[23] Hesper, R., Tjeng, L.H., Heeres, A., and Sawatzky, G.A. (2000). Photoemission evidence of electronic stabilization of polar surfaces in K_3C_{60}. *Phys. Rev. B* **62**, 16046–16055.

[24] Nakagawa, N., Hwang, H.Y., and Muller, D.A. (2006). Why some interfaces cannot be sharp. *Nature Mater.* **5**, 204–209.

[25] Nishimura, J., Ohtomo, A., Ohkubo, A., Murakami, Y., and Kawasaki, M. (2004). Controlled carrier generation at a polarity-discontinued perovskite heterointerface. *Jpn. J. Appl. Phys.* **43**, L1032–L1034.

[26] Thiel, S., Hammerl, G., Schmehl, A., Schneider, C.W., and Mannhart, J. (2006). Tunable quasi-two-dimensional electron gases in oxide heterostructures. *Science* **313**, 1942–1945.

[27] Huijben, M., Rijnders, G., Blank, D.H.A., *et al.* (2006). Electronically coupled complementary interfaces between perovskite band insulators. *Nature Mater.* **5**, 556–560.

[28] Tufte, O.N. and Chapman, P.W. (1967). Electron mobility in semiconducting strontium titanate. *Phys. Rev.* **155**, 796–802.

[29] Frederikse, H.P.R. and Hosler, W.R. (1967). Hall Mobility in $SrTiO_3$. *Phys. Rev.* **161**, 822–827.

[30] Kalabukhov, A., Gunnarsson, R., Börjesson, *et al.* (2007). Effect of oxygen vacancies in the $SrTiO_3$ substrate on the electrical properties of the $LaAlO_3/SrTiO_3$ interface. *Phys. Rev. B* **75**, 121404(R) 1–4.

[31] Siemons, W., Koster, G., Yamamoto, H., *et al.* (2007). Origin of charge density at $LaAlO_3$ on $SrTiO_3$ heterointerfaces: possibility of intrinsic doping. *Phys. Rev. Lett.* **98**, 196802 1–4.

[32] Herranz, G., Basletić, M., Bibes, M., *et al.* (2007). High mobility in $LaAlO_3/SrTiO_3$ heterostructures: origin, dimensionality, and perspectives. *Phys. Rev. Lett.* **98**, 216803 1–4.

[33] Kan, D., Terashima, T., Kanda, R., *et al.* (2005). Blue-light emission at room temperature from Ar^+-irradiated $SrTiO_3$. *Nature Mater.* **4**, 816–819.

[34] Basletic, M., Maurice, J.L., Carrétéro, C., *et al.* (2008). Mapping the spatial distribution of charge carriers in $LaAlO_3/SrTiO_3$. *Nature Mater.* **7**, 621–625.

[35] Willmott, P.R., Pauli, S.A., Herger, R., *et al.* (2007). Structural basis for the conducting interface between $LaAlO_3$ and $SrTiO_3$. *Phys. Rev. Lett.* **99**, 155502 1–4.
[36] Yacoby, Y., Sowwan, M., Stern, E., *et al.* (2003). Direct determination of epitaxial film and interface structure: Gd_2O_3 on GaAs (1 0 0). *Physica B: Condensed Matter* **336**, 39–45.
[37] Park, M.S., Rhim, S.H., and Freeman, A.J. (2006). Charge compensation and mixed valency in $LaAlO_3/SrTiO_3$ heterointerfaces studied by the FLAPW method. *Phys. Rev. B* **74**, 205416 1–6.
[38] Popović, Z.S., Satpathy, S., and Martin, R.M. (2008). Origin of the two-dimensional electron gas carrier density at the $LaAlO_3$ on $SrTiO_3$ Interface. *Phys. Rev. Lett.* **101**, 256801 1–4.
[39] Lee, J. and Demkov, A.A. (2008). Charge origin and localization at the *n*-type $SrTiO_3/LaAlO_3$ interface. *Phys. Rev. B* **78**, 193104 1–4.
[40] Ishibashi, S. and Terakura, K. (2008). Analysis of screening mechanisms for polar discontinuity for $LaAlO_3/SrTiO_3$ thin films based on *ab initio* calculations. *J. Phys. Soc. Jpn.* **77**, 104706 1–10.
[41] Bristowe, N.C., Artacho, E., and Littlewood, P.B. (2009). Oxide superlattices with alternating *p* and *n* interfaces. *Phys. Rev. B* **80**, 045425 1–5.
[42] Segal, Y., Ngai, J.H., Reiner, J.W., Walker, F.J., and Ahn, C.H. (2009). X-ray photoemission studies of the metal-insulator transition in $LaAlO_3/SrTiO_3$ structures grown by molecular beam epitaxy. *Phys. Rev. B* **80**, 241107 1–4.
[43] Pentcheva, R. and Pickett, W.E. (2009). Avoiding the polarization catastrophe in $LaAlO_3$ overlayers on $SrTiO_3(001)$ through polar distortion. *Phys. Rev. Lett.* **102**, 107602 1–4.
[44] Cen, C., Thiel, S., Hammerl, G., *et al.* (2008). Nanoscale control of an interfacial metal-insulator transition at room temperature. *Nature Mater.* **7**, 298–302.
[45] Takizawa, M., Hotta, Y., Susaki, T., *et al.* (2009). Spectroscopic evidence for competing reconstructions in polar multilayers $LaAlO_3/LaVO_3/LaAlO_3$. *Phys. Rev. Lett.* **102**, 236401 1–4.
[46] Yoshimatsu, K., Yasuhara, R., Kumigashira, H., and Oshima, M. (2008). Origin of metallic states at the heterointerface between the band insulators $LaAlO_3$ and $SrTiO_3$. *Phys. Rev. Lett.* **101**, 026802 1–4.
[47] Vanderbilt, D. and King-Smith, R.D. (1993). Electric polarization as a bulk quantity and its relation to surface charge. *Phys. Rev. B* **48**, 4442–4455.
[48] Stengel, M. and Vanderbilt, D. (2009). Berry-phase theory of polar discontinuities at oxide-oxide interfaces. *Phys. Rev. B* **80**, 241103(R) 1–4.
[49] Francis, R.J., Moss, S.C., and Jacobson, A.J. (2001). X-ray truncation rod analysis of the reversible temperature-dependent [001] surface structure of $LaAlO_3$. *Phys. Rev. B* **64**, 235425 1–9.
[50] Lanier, C.H., Rondinelli, J.M., Deng, B., *et al.* (2007). Surface reconstruction with a fractional hole: $(\sqrt{5} \times \sqrt{5})R26.6°$ $LaAlO_3$ (001). *Phys. Rev. Lett.* **98**, 086102 1–4.
[51] Gossmann, H.J. and Schubert, E.F. (1993). Delta doping in silicon. *Crit. Rev. Solid State Mater. Sci.* **18**, 1–67.
[52] Kosuge, K. (1993). *Chemistry of Non-Stoichiometric Compounds*, pp. 18–44. Oxford University Press, Oxford.
[53] Xie, Y., Bell, C., Yajima, T., Hikita, Y., and Hwang, H.Y. (2010). Charge writing at the $LaAlO_3/SrTiO_3$ surface. *Nano Letters* **10**, 2588–2591.

[54] Dingle, R., Störmer, H.L., Gossard, A.C., and Wiegmann, W. (1978). Electron mobilities in modulation-doped semiconductor heterojunction superlattices. *Appl. Phys. Lett.* **33**, 665–667.

[55] Higuchi, T., Hotta, Y., Susaki, T., Fujimori, A., and Hwang, H.Y. (2009). Modulation doping of a Mott quantum well by a proximate polar discontinuity. *Phys. Rev. B* **79**, 075415 1–6.

[56] Stengel, M. and Spaldin, N.A. (2006). Origin of the dielectric dead layer in nanoscale capacitors. *Nature* **443**, 679–682.

[57] Stengel, M. (2011). First-Principles Modeling of Electrostatically Doped Perovskite Systems. *Phys. Rev. Lett.* **106**, 136803.

[58] Ishida, Y., Eguchi, R., Matsunami, M., *et al.* (2008). Coherent and incoherent excitations of electron-doped $SrTiO_3$. *Phys. Rev. Lett.* **100**, 056401 1–4.

[59] Eskes, H., Meinders, M.B.J., and Sawatzky, G.A. (1991). Anomalous transfer of spectral weight in doped strongly correlated systems. *Phys. Rev. Lett.* **67**, 1035–1038.

[60] Fujimori, A., Hase, I., Namatame, H., *et al.* (1992). Evolution of the spectral function in Mott-Hubbard systems with d^1 configuration. *Phys. Rev. Lett.* **69**, 1796–1799.

[61] Oka, T. and Nagaosa, N. (2005). Interfaces of correlated electron systems: proposed mechanism for colossal electroresistance. *Phys. Rev. Lett.* **95**, 266403 1–4.

[62] Chambers, S., Engelhard, M., Shutthanandan, V., *et al.* (2010). Instability, intermixing and electronic structure at the epitaxial heterojunction. *Surf. Sci. Rep.* **65**, 317–352.

8
Strongly correlated heterostructures

SATOSHI OKAMOTO

8.1 Introduction

"Correlated-electron systems" such as transition-metal oxides have been a focus of materials science for decades because of their exotic behavior including high-T_c superconductivity and colossal magnetoresistance effect. These phenomena originate from strong electron–electron or electron–lattice interactions rendering them incompatible with the standard "density functional" theory (DFT) which describes most compounds used for present electronic devices. In the past decade, we have seen tremendous progress in the physics and materials science of correlated-electron systems. Improvements in crystal growth and measurement techniques have led to the discovery of a variety of novel phases (superconducting, magnetic, charge and orbitally ordered) [1, 2]. In addition, recent rapid developments in theoretical techniques have led to a much improved understanding of the bulk properties of these materials.

As discussed in detail in this book, recent progress in synthesizing atomically controlled "digital" heterostructures opened a new area of materials science. The study of oxide heterostructures is thus rapidly developing into one of the most promising areas of research in strongly correlated electronic systems. In addition to its basic importance as a fundamental question in materials science, exploring new quantum phenomena by means of carrier doping and structural control, correlated-electron surface/interface science should provide the necessary scientific background for the study of potential electronic devices exploiting correlated-electron properties. This is because essentially any device must be coupled to the rest of the world via motion of electrons through interfaces.

The fundamental interest of bulk correlated-electron materials is the novel phases they exhibit. Therefore the fundamental issue for the field of "correlated-electron interface" is "how does the electronic phase at the interface differ from that in the bulk?" In other words, "what is the electronic reconstruction?" This question has begun to attract experimental attention. For example, Hesper and co-workers have shown that the [111] surface of K_3C_{60} differs from the bulk because of charge transfer caused by a polar surface [3]. "Electronic reconstruction" was originally introduced to describe this behavior. Currently, electronic reconstruction is often used for the redistribution of electronic charge at interfaces. But, we propose that it is a more general concept referring to any changes in the electronic phase behavior at interfaces.

As in ordinary surface or interface science, many physics and material science issues arise in considering the behavior of correlated electrons near surfaces and interfaces. *Atomic reconstruction* may occur and may change the underlying electronic structure. For example, Moore *et al.* argued that, in two-dimensional ruthenates $Sr_{1-x}Ca_xRuO_4$, a change in tilt angle of the surface RuO_6 octahedra increases the electronic hopping, thereby allowing the metallic phase to persist to lower temperature than in the bulk [4]. *Reduced coordination number* at surfaces is supposed to enhance correlation effects, as discussed by Potthoff and Nolting [5, 6], Schwieger *et al.* [7], and Liebsch [8]. Also, Hesper and co-workers noted that the structural change will lead to changes in Madelung potentials and many-body interaction parameters [3]. *Leakage of charge* across an interface may change densities away from the commensurate values required for insulating behavior.

Perovskite-type transition-metal oxides (TMOs) offer an attractive starting point for these problems. In these systems, the near lattice match (the in-plane lattice parameter or the distance between the neighboring transition-metal ions $\sim$4Å) enables us to combine a wide range of compounds in an "artificial crystal" using various techniques such as (laser) molecular beam epitaxy (MBE). Ohtomo, Muller, Grazul, and Hwang have demonstrated the fabrication of atomically precise digital heterostructures involving a controllable number n of planes of $LaTiO_3$ (a correlated-electron Mott-insulating material) separated by a controllable number m of planes of $SrTiO_3$ (a more conventional band-insulating material) and have measured both the variation of electron density transverse to the planes and metallic conductivity [9]. These systems have attracted great interest and have been analyzed by photoemission spectroscopy [10] and infrared spectroscopic ellipsometry [11]. Furthermore, the near Fermi surface states are derived mainly from transition-metal d orbitals and oxygen p orbitals forming relatively narrow bands and, therefore, it is possible to model such a system using a multi-orbital tight-binding model with various interactions. Of course, DFT calculations can provide important information specific to the system under consideration. Therefore, model calculations and DFT calculations could play complementary roles.

On the experimental side, a variety of heterostructures have been fabricated and studied. Heterostructures involving manganites have been intensively studied since the discovery of the colossal magnetoresistance effect [12, 13]. First, the effect of strain on the electronic properties was investigated in detail [14, 15]. Then, people started to investigate the properties of multilayers involving two end-members of manganites, $LaMnO_3$ and $SrMnO_3$ [16–22]. Heterostructures involving manganites and other TMO's were also studied, including, for example, $LaAlO_3/La_{1-x}Sr_xMnO_3/SrTiO_3$ trilayers which break inversion symmetry [23, 24], $La_{1-x}Sr_xFeO_3$ [25], and $CaRuO_3$ [26, 27]. From a technological viewpoint, perovskite manganites may be ideal systems as ferromagnetic metallic leads of a tunneling magnetoresistance (TMR) junction[28, 29] because of their large spin polarization and high Curie temperature T_C, especially $La_{1-x}Sr_xMnO_3$ with $0.2 < x < 0.5$, where T_C reaches 350K [30]. It was first reported that the TMR disappears far below the bulk T_C [31]. A short time later, the TMR effect was confirmed above room temperature [32]. However, the magnetoresistance ratio[1] (MR) at

[1] Recent convention of the magnetoresistance ratio MR is given by $MR = (G_P - G_{AP})/G_{AP}$ with $G_{P(AP)}$ the tunneling conductance with parallel (antiparallel) alignment of the magnetization of the electrodes. This quantity becomes divergingly large when G_{AP} is small.

room temperature remains significantly small. Increasing MR at room temperature is the key to realizing a manganite-based TMR device for practical application.

Thin films of high-T_c cuprates[33] have also been studied. Early work is reported in [34]. It has been pointed out that the strain coming from the substrate affects T_c, and especially, the in-plane compressive strain increases T_c [35–37]. Multilayers of high-T_c cuprates were fabricated to investigate the proximity effect of superconductivity [38, 39]. More recently, various heterostructures involving cuprates were fabricated and the effect of "proximity" between different doping regimes or between cuprate and other material on T_c were investigated [40–43].

Heterostructures involving cuprates and manganites have also been attracting great interest [44–49]. In addition to the interfacial magnetism [46–48], interfacial orbital states are also investigated [49]. More recently, Nemes and co-workers reported an inverse spin–switch behavior in trilayer systems involving $La_{1-x}Ca_xMnO_3$ and $Y_2Ba_3CuO_{7-\delta}$ [50]. Interestingly, superconductivity was found to favor parallel alignment of magnetization in the ferromagnetic layers under an applied magnetic field. Such an effect may become useful for the spintronic application. Recently, Salafranca and the author developed a phenomenological model focusing on antiferromagnetic coupling between the cuprate and the manganite layers [51]. Based on this model, the inverse spin–switch effect was qualitatively explained as a partial cancelation between the internal field, which is opposite to the spin polarization in the manganite layers, and the applied magnetic field.

Among other strongly correlated materials, heterostructures involving $LaFeO_3$ and $LaCrO_3$ were fabricated and their magnetic properties were investigated [52]. Hotta and co-workers considered $LaVO_3/SrTiO_3$ heterostructures. They have demonstrated carrier doping to a Mott insulator $LaVO_3$ in such heterostructures by utilizing the polar discontinuity [53]. The ability to control magnetism by an electric field was investigated at an interface between ferroelectric (multiferroic) $BiFeO_3$ and ferromagnetic $SrRuO_3$ [54]. More recently, the robust exchange bias effect was observed in heterostructures involving $BiFeO_3$ and $La_{1-x}Sr_xMnO_3$ accompanying the interfacial orbital reconstruction [55]. Such an effect may provide another route for applying perovskite TMOs to electronics devices.

Many physics and material science issues influence the behavior of correlated electrons near interfaces and surfaces. In addition to the strain effects studied in manganites [14, 15] and cuprates [35–37], atomic reconstruction may occur, and may change the underlying electronic structure. For example, Matzdorf *et al.* discussed the possibility of surface ferromagnetism stabilized by the larger rotation angle of RuO_6 octahedra at a surface of Sr_2RuO_4 [56]. More recently, Wakabayashi and co-workers reported that surface orbital ordering in the single-layered manganite $La_{0.5}Sr_{1.5}MnO_4$ is much rougher than the crystallographic surface [57].

Finally, Ohtomo and Hwang found that an n-type interface between two band insulators, $LaAlO_3$ and $SrTiO_3$, becomes metallic [58]. Although, these may not be considered as "strongly correlated systems," metallic behavior at an interface between $LaAlO_3$ and $SrTiO_3$ is technologically important, and an intensive study of such systems has been carried out [59–63].

On the theoretical side, there have also been some significant developments. The enhanced correlations near the surface due to reduced coordination number could induce surface magnetic ordering. This had been discussed in a mean field treatment of the Hubbard model by Potthoff and Nolting [64]. Going beyond the study of static properties, Freericks applied the dynamical-mean-field method [65] to correlated [001] heterostructures comprised of non-correlated and strongly-correlated regions, and computed the conductance perpendicular to the plane [66]. In his model, the correlated region is described by the Falicov–Kimball model. Helmes *et al.* considered the proximity effect between a metal and a Mott insulator [67], which is described by the single-band Hubbard model, within single-site dynamical-mean-field theory (DMFT). In both cases, the computations are limited to the particle–hole symmetric case for simplicity. Extension to the general model, and in particular, to the situation where the charge density is spatially modulated are important future directions. The author and co-workers have carried out a series of analyses on $LaTiO_3/SrTiO_3$ heterostructures [68–71]. As discussed in detail later, we performed the DFT calculations to investigate the lattice effects in the $LaTiO_3/SrTiO_3$ heterostructure. Similar analysis has also been carried out by Hamman and co-workers [72]. We also performed DMFT calculations for simplified model heterostructures involving a Mott insulator and a band insulator. The layer-extension of the DMFT has also been applied to "polar" interfaces involving a Mott insulator and a band insulator by Lee and MacDonald [73, 74]. Further, the author and Maier applied the cluster-extension of the DMFT to model superlattices (SLs) involving high-T_c cuprates with different carrier concentrations to investigate the proximity effect of superconductivity [75]. In addition, the author has tried to combine the layer DMFT and the Keldysh Green's function technique to study the non-equilibrium steady-state properties of strongly correlated heterostructures [76].

For manganite-based systems, there have appeared several theoretical studies. Calderon *et al.* discussed possible surface antiferromagnetism in manganites arising from competition between reduced kinetic energy and antiferromagnetic superexchange interaction between the nearest-neighboring local spins [78]. Brey has investigated the possibility of phase separation in interfaces between manganite and an insulator [79]. Motivated by experimental work, $LaMnO_3/SrMnO_3$-type SLs have been studied by using double-exchange-based models with layer DMFT [80] and the classical Monte-Carlo [81], and by the DFT [82, 83].

There have appeared a number of DFT studies on TMOs. The effect of bulk strain on magnetic ordering in perovskite manganites was discussed by Fang *et al.* [84] Aforementioned $LaTiO_3/SrTiO_3$ SLs have been studied by the author, Millis and Spaldin [70], Hamann and co-workers [72], Popovic and Satpathy [85], and Pentcheva and Pickett [86]. Using the generalized gradient approximation (GGA) with on-site U to the DFT, it has been shown in [86] that the interface between $LaTiO_3$ and $SrTiO_3$ can become ferromagnetic and insulating accompanying the Ti^{3+}/Ti^{4+}-type charge ordering when U is sufficiently large [86]. Nanda and Satpathy have studied $CaRuO_3/CaMnO_3$ SLs and have shown that leaked electrons from $CaRuO_3$ to $CaMnO_3$ induce canted ferromagnetism in the latter [87]. After the discovery of metallicity at the n-type $LaAlO_3/SrTiO_3$ interface in [58], $LaAlO_3/SrTiO_3$-type SLs have been intensively studied using the DFT, one of the early works was reported in [88]. Among other TMO-based heterostructures, Pardo and Pickett studied TiO_2/VO_2 nanostructures. They have proposed that, by confining conductive VO_2 layers between insulating TiO_2 layers, there would appear half-metallic semi-Dirac dispersions, in which the dispersion relation is linear in momentum (mass-less) along one direction and quadratic (massive) along the other [89]. It was predicted that such dispersion relations give rise to the Landau level quantized as $(n + \frac{1}{2})^{2/3}$, which differs from $(n + \frac{1}{2})$ for the conventional dispersion and $\sqrt{n}$ the Dirac dispersion [90]. Exploring such exotic behavior in other systems using the DFT and going beyond the mean-field level to see the effect of many-body interaction on electronic properties would be very interesting future directions.

The field of strongly correlated heterostructures is thus developing extremely rapidly, and it is far beyond the author's capability to cover all the important subjects. Therefore, in this article, we review our theoretical analysis of lattice-matched digital heterostructures in which different systems are grown along the [001] direction. We focus on the effect of strong correlations on the electronic phase behavior in such heterostructures. Specifically, we present the theoretical results on $LaTiO_3/SrTiO_3$ (Mott-insulator/band-insulator) heterostructures and model SLs involving high-T_c cuprates with different doping concentrations.

The rest of this article is organized as follows: in Section 8.2, we discuss how to model strongly correlated heterostructures. We also discuss how to analyze such a model. Section 8.3 presents our results on $LaTiO_3/SrTiO_3$ heterostructures, and section 8.4 presents our results on cuprate multilayers. In Section 8.5, we discuss future directions by presenting two of our recent theoretical results. Finally, we summarize in Section 8.6. Work presented in Sections 8.3, 8.4, and 8.5 have been published in [68–71], [75], and [76, 77], respectively.

8.2 Theoretical description

In this section, we discuss how to deal with strongly correlated systems with spatial inhomogeneity, i.e. correlated heterostructures. First, we introduce a general model for correlated electron systems. Then, we consider how to model heterostructures. Because of many-body interactions, one has to apply some kind of approximate method to solve a many-body problem unless the exact solution becomes available. For this purpose, we introduce the layer-extension of dynamical-mean-field theory (DMFT) as a theoretical technique to deal with correlated heterostructures. Finally, we present some other techniques that are used to solve model correlated heterostructures.

8.2.1 Model

In general, the Hamiltonian describing the electronic system in a solid is given as a sum of several terms as

$$H = H_t + H_C + H_\lambda + H_L, \tag{8.1}$$

where H_t is the single-particle part, H_C the Coulomb interaction, H_λ the relativistic spin–orbit coupling, and H_L the electron–lattice coupling. For a multi-orbital system, H_t becomes

$$H_t = \sum_{i\alpha\sigma} \varepsilon_{i\alpha} c^\dagger_{i\alpha\sigma} c_{i\alpha\sigma} + \sum_{ij\alpha\beta\sigma} t^{\alpha\beta}_{ij} c^\dagger_{i\alpha\sigma} c_{j\beta\sigma}, \tag{8.2}$$

where $c_{i\alpha\sigma}$ is an annihilation operator of an electron at position $\vec{r}_i = (x_i, y_i, z_i)$ orbital α with spin σ, $\varepsilon_{i\alpha}$ single-particle level at site i and orbital α, and $t^{\alpha\beta}_{ij}$ hopping integral between orbital α on site i and orbital β on site j. In a solid, the Coulomb interaction between two electrons is screened by other charges, and one of the strongest interactions appears at $i = j$, on-site interaction given by

$$H_U = \frac{1}{2} \sum_{i\alpha\beta\alpha'\beta'\sigma\sigma'} U_i^{\alpha\beta\alpha'\beta'} c_{i\alpha\sigma}^{\dagger} c_{i\beta\sigma'}^{\dagger} c_{i\beta'\sigma'} c_{i\alpha'\sigma}. \tag{8.3}$$

Here, we have presented the most general form with the coefficient given by

$$U_i^{\alpha\beta\alpha'\beta'} = \int dr_1^3 dr_2^3 \varphi_{i\alpha}^*(\vec{r}_1) \varphi_{i\beta}^*(\vec{r}_2) \frac{e^2}{r_{12}} \varphi_{i\beta'}(\vec{r}_2) \varphi_{i\alpha'}(\vec{r}_1), \tag{8.4}$$

where $e(> 0)$ is the unit charge, and $r_{12} = |\vec{r}_1 - \vec{r}_2|$. $\varphi_{i\alpha}(\vec{r})$ is a wave function centered at site i with the symmetry or orbital index α. The form of $U_i^{\alpha\beta\alpha'\beta'}$ is in general very complicated. Therefore, simplified forms are often used by assuming the symmetry of the atomic orbitals. For transition-metal oxides (TMOs), interactions between d electrons are important. When either t_{2g} or e_g orbitals are considered, we have the well-known relation $U = U' + 2J$, where $U = U^{\alpha\alpha\alpha\alpha}$ (intraorbital Coulomb), and $U' = U^{\alpha\beta\alpha\beta}$ (interorbital Coulomb) and $J = U^{\alpha\beta\beta\alpha}$ (interorbital exchange) $= U^{\alpha\alpha\beta\beta}$ (interorbital pair transfer) for $\alpha \neq \beta$, and no other components. But these relations do not hold when both t_{2g} and e_g orbitals or more general orbitals are considered. Detailed structure for Coulomb interaction of d orbitals can be found in, for example, [91].

Sometimes, parts of the d electrons are localized having spins and interact with other itinerant electrons via interorbital exchange interactions. Such a situation may be more easily described by a model in which the localized electrons on a site are replaced by a local spin with length S, rather than including full matrix elements of $U_i^{\alpha\beta\alpha'\beta'}$. The double-exchange (DE) model is one of those effective models in which t_{2g} electrons form a localized spin $S = 3/2$ and e_g electrons are itinerant. This model has often been adopted to discuss the properties of CMR manganites [92–96]. An identical theoretical model has been derived for the s-d interaction by Kasuya [97].

As non-local interactions, important terms are the long-range Coulomb repulsion given by

$$H_{long} = \frac{1}{2} \sum_{ij\alpha\beta\sigma\sigma'} \frac{e^2 n_{i\alpha\sigma} n_{j\beta\sigma'}}{\varepsilon|\vec{r}_i - \vec{r}_j|}, \tag{8.5}$$

and the direct exchange

$$H_{ex} = -\sum_{ij\alpha\beta} J_{ij}^{\alpha\beta} \vec{S}_{i\alpha} \cdot \vec{S}_{j\beta}. \tag{8.6}$$

Here, $n_{i\alpha\sigma} = c_{i\alpha\sigma}^{\dagger} c_{i\alpha\sigma}$ and $\vec{S}_{i\alpha} = \frac{1}{2} \sum_{\sigma\sigma'} c_{i\alpha\sigma}^{\dagger} \vec{\tau}_{\sigma\sigma'} c_{i\alpha\sigma'}$ with $\vec{\tau}$ the Pauli matrix. $J_{ij}^{\alpha\beta}$ is given by

$$J_{ij}^{\alpha\beta} = \int dr_1^3 dr_2^3 \varphi_{i\alpha}^*(\vec{r}_1) \varphi_{j\beta}^*(\vec{r}_2) \frac{e^2}{r_{12}} \varphi_{i\alpha}(\vec{r}_2) \varphi_{j\beta}(\vec{r}_1). \tag{8.7}$$

The negative sign in the right-hand side of (8.6) indicates that H_{ex} is normally ferromagnetic. This interaction becomes important when the neighboring magnetic ions have large overlap. In the perovskite-type TMOs, H_{ex} is usually very small because transition-metal ions are separated by $\sim 4\text{Å}$ or more. In (8.5), the dielectric constant $\varepsilon > 1$ is included explicitly representing the screening effect, and the overlap between $\varphi_{i\alpha}$ and $\varphi_{j\beta}$ is neglected for simplicity. In fact, both on-site and long-range Coulomb interactions and direct exchange are screened in a solid, and several ways to evaluate the screening effect have been proposed (for example, see [98]). Long-range Coulomb interaction, H_{long}, plays an important role determining the charge distribution, in particular, when the system is spatially inhomogeneous, as in heterostructures.

The forms of H_λ and H_L depend on the detail of the system under consideration. Therefore, we will not present their explicit forms here. But, this does not mean that these interactions are not important. In fact, these terms play leading roles for some systems. For example, it was suggested that H_L, especially the Jahn–Teller–type electron–lattice coupling, is one of the main ingredients giving rise to the colossal magnetoresistance effect in manganites [99]. In 5d systems, the strong spin–orbit coupling may dominate low-energy behaviors. For example, oxides involving iridium Ir are attracting interest both experimentally and theoretically [100, 101].

When the characteristic energy scale of H_U (H_L) becomes larger than that of H_t, Hartree–Fock-type single-particle (Born–Oppenheimer-type adiabatic) approximations become less accurate. Systems in such a situation are called "strongly correlated electron systems."

Next, we consider constructing heterostructures. When different systems are combined, essentially all physical parameters are spatially modulated and can even differ from their bulk values. Evaluating these parameters using a first-principles technique is highly desired for performing detailed model calculations considering many-body effects. However, most theoretical studies of oxide heterostructures appearing so far have employed either DFT methods or model Hamiltonians with less-elaborate parameterizations.

When a heterostructure is formed by chemically similar systems, the spatial variation of parameters may not be so large. Even in this case, when the electron densities of the constituent systems are different, one has to consider the change in the on-site potential. For example, $[ABO_3]_m[A'BO_3]_n$-type heterostructure can be modeled by adding the Coulomb interactions between the conduction electrons and ionic charges to the on-site level as

$$\varepsilon_{i\alpha} = -e \sum_{l_A} \frac{\rho_{l_A}}{\varepsilon |\vec{r}_i - \vec{r}_{l_A}|}, \tag{8.8}$$

where l_A labels the A-site ion in the perovskite ABO_3 structure, $\vec{r}_{l_A}$ its position, and ρ_{l_A} its ionic charge. One can then proceed to solve a set of equations with applying the Hartree approximation to H_{long}. This procedure has been adopted for $[LaTiO_3]_m/[SrTiO_3]_n$-type heterostructures by the author and co-workers. In the previous work, only Ti d orbitals were considered, the theoretical model shown schematically in Figure 8.1 (a). As discussed later, this model was used to extract model parameters such as hopping integrals and dielectric constant ε due mainly to lattice relaxation by fitting to the DFT results. When the contribution of oxygen ions becomes stronger, one may extend the theoretical model by including oxygen ions as ionic charges and/or electrically active p levels as shown in Figure 8.1 (b). Using such a model, it may be possible to explicitly include the screening effect due to the oxygen displacements. But such a study has not appeared yet because the numerical complexity is multiplied.

When considering chemically distinct materials, on-site levels for an effective model are no longer given as a sum of Coulomb terms coming from charged ions and other carriers.

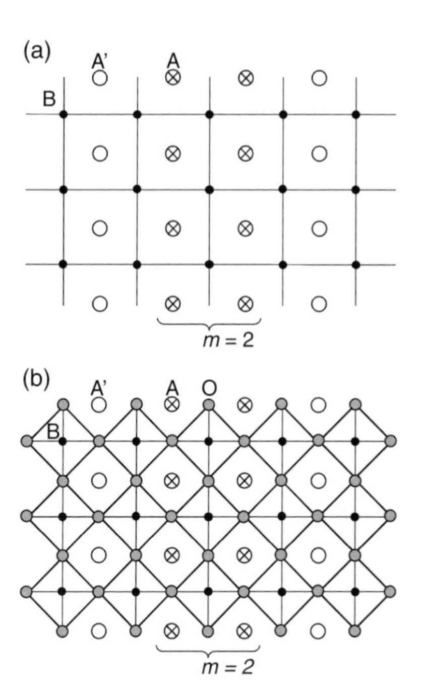

Fig. 8.1 Schematic figures of a model [001] heterostructure $[ABO_3]_m/[A'BO_3]_n$ with $m = 2$ projected on the [010] plane. (a) *d* model in which only transition-metal *d* orbitals are considered to be electronically active and (b) *d-p* model in which both transition-metal *d* and oxygen *p* orbitals are considered. Crossed and open circles show the positions of A and A' ions located on the "A" site of the perovskite ABO_3 lattice, and dots show the positions of the "B" site. For example, A and A' correspond to La and Sr ions, respectively, and B corresponds to Ti ion in the $[LaTiO_3]_m/[SrTiO_3]_n$ heterostructure. Gray circles in (b) represent oxygen ions.

In bulk, such chemical effects influence the work function, the level difference between the highest occupied state and the vacuum, and in heterostructures the difference in the work function, i.e. band alignment, dominates the direction of charge redistribution. To extract such information from first-principles methods may still be challenging. Instead, Yunoki, the author, and co-workers used experimental data to construct a band diagram including cuprates, manganites, and titanates as shown in Figure 8.2. [102] They further considered interfaces between manganites and cuprates, and predicted electron doping to the cuprate region from the manganite region.

Quite often, an interface becomes "polar" creating a potential slope inside a system. As a result, the amount of charge redistribution depends on whether or not the interface is polar, and, if it is polar, the thickness of the region with the finite potential slope. This effect has been proposed for understanding the different behaviors of "LaO"-terminated and "SrO"-terminated $LaAlO_3/SrTiO_3$ interfaces [103]. For correlated materials, Hotta *et al.* demonstrated carrier doping to $LaVO_3$ (a bulk Mott insulator) by utilizing the polar discontinuity at interfaces between $LaVO_3$ and $SrTiO_3$ [53]. By carefully considering the work functions, the

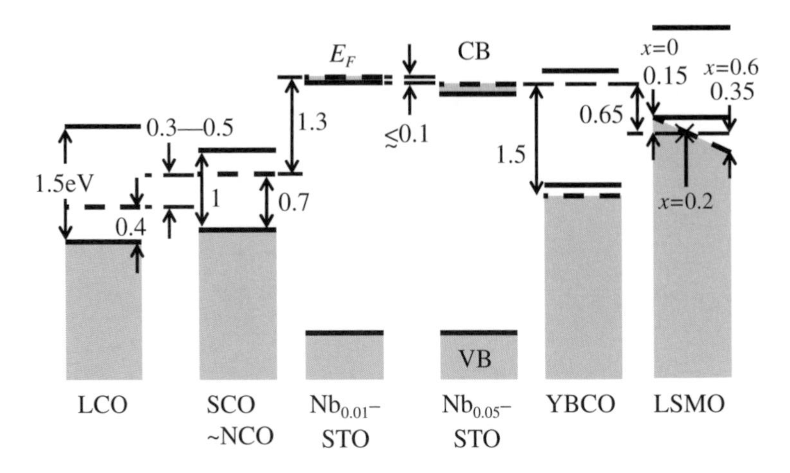

Fig. 8.2 Schematic band diagrams of La_2CuO_4 (LCO), Sm_2CuO_4 (SCO)[Nd_2CuO_4 (NCO)], 0.01wt% Nb-doped $SrTiO_3$ ($Nb_{0.01}$-STO), $Nb_{0.05}$-STO, $Y_2Ba_3CuO_{7-\delta}$ (YBCO), and $La_{1-x}Sr_xMnO_3$ (LSMO) based on diffusion voltage measurements and photoemission spectroscopy. Tops of valence bands (VB) and bottoms of conduction bands (CB) are indicated by solid lines, while chemical potentials are indicated by dashed lines. For details, see [102] from which the figure is reproduced.

polar discontinuity can be used as an alternative method for carrier doping without introducing "chemical disorder" to explore a variety of phases that the strongly correlated systems exhibit.

8.2.2 Layer-extension of dynamical-mean-field theory

The model Hamiltonian, (8.1), involves many-body interactions, and the exact solution is not available except for some special cases, one-dimensional or infinite-dimensional systems. Therefore, one has to apply some approximation depending on the problem of interest. The dynamical-mean-field theory (DMFT) has been developed to deal with strongly correlated systems [65]. This method becomes exact in the infinite dimension, but has been successful to describe a variety of systems in finite dimensions. Here, we consider the layer extension of the DMFT.

The basic object of study is the electron Green's function, which is given by

$$\hat{G}_\sigma(i\omega_n) = \left[i\omega_n + \mu - \hat{H}_t - \hat{\Sigma}_\sigma(i\omega_n)\right]^{-1}, \tag{8.9}$$

where $\omega_n = (2n+1)\pi T$ is the Matsubara frequency with temperature T. In the DMFT for a spatially uniform system, we approximate the self-energy as $\Sigma_\sigma(\vec{r}_i, \vec{r}_j; i\omega_n) = \Sigma_\sigma(i\omega_n)$, i.e. the local approximation. Since the self-energy is local, the lattice problem is mapped to the Anderson impurity model, whose effective conduction band (energy dispersion ε_k and hybridization V_k with the impurity site) is fixed self-consistently. Although the Anderson impurity model is a single-site problem, it has many-body interactions. Therefore, a numerical calculation is necessary to solve the problem and compute physical quantities accurately.

As the simplest extension of the dynamical-mean-field method for a layered system, Potthoff and Nolting introduced self-energy which is site-diagonal but depends on the layer index z_i [5]. In this case, the self-energy is written as

$$\Sigma_\sigma(\vec{r}_i, \vec{r}_j; i\omega_n) = \Sigma_\sigma(z_i; i\omega_n)\delta_{ij}. \tag{8.10}$$

For a system with a number N of inequivalent layers with inplane translational invariance there appear N components of self-energy. The self-energy $\Sigma_\sigma(z_i; \omega_n)$ is obtained by solving a number N of Anderson impurity models whose effective conduction bands are determined self-consistently as

$$\begin{aligned}\Delta_\sigma(z_i, i\omega_n) &= \sum_k \frac{|V_{z_i k\sigma}|^2}{i\omega_n - \varepsilon_{z_i k\sigma}} \\ &= i\omega_n + \mu - \varepsilon_i - \Sigma_\sigma(z_i; i\omega_n) - G_\sigma^{-1}(z_i; i\omega_n).\end{aligned} \tag{8.11}$$

Here, $G_\sigma(z_i; i\omega_n)$ is the local Green's function given by

$$G_\sigma(z_i; i\omega_n) = \int \frac{dk_\parallel^2}{(2\pi)^2} G_\sigma(z_i, z_i, \vec{k}_\parallel; i\omega_n), \tag{8.12}$$

with

$$G_\sigma(z_i, z_j, \vec{k}_\parallel; i\omega_n) = \left[\left\{i\omega_n + \mu - \hat{H}_t(\vec{k}_\parallel) - \hat{\Sigma}_\sigma(i\omega_n)\right\}^{-1}\right]_{ij}. \tag{8.13}$$

We have used the mixed representation with the z-axis coordinate and the inplane momentum $\vec{k}_\parallel = (k_x, k_y)$ because of translational invariance in the xy plane. Each impurity model is isolated, and therefore one may think that the interlayer coupling is totally ignored. However, the effect of the coupling with the other impurities is included through the effective conduction band which is fixed by the self-consistent condition given by (8.12) and (8.13). $\Delta_\sigma(z_i, i\omega_n)$ is called the hybridization function. This corresponds to the Weiss field in a classical mean field theory. It is equivalent to using the non-interacting Green's function $\mathcal{G}_{0\sigma}^{-1}(z_i, i\omega_n) = i\omega_n + \mu - \varepsilon_i - \Delta_\sigma(z_i, i\omega_n)$ as the Weiss mean field. The procedure of the layer DMFT is shown schematically in Figure 8.3. So far, we have not considered any symmetry breaking in the xy plane. When there is such symmetry breaking, the number of impurity models is multiplied.

Similar layer-extension of the DMFT was used to investigate the Josephson effect in which the insulating barrier is described by a Falicov–Kimball model in [104]. In fact, the procedure described above is equivalent to the one adopted to study the stripe phase in the two-dimensional Hubbard model by Fleck and co-workers [105]. All real-space extensions of the DMFT follow essentially the same self-consistent loop shown in Figure 8.3 with different mapping between the lattice problem and the impurity problems.

The layer DMFT is used to discuss the physical properties of Mott-insulator/band-insulator heterostructures in the next section. As another model for strongly correlated systems, the DE model in which the localized spins are treated as classical spins has been analyzed using the DMFT[93–96] to discuss the physical properties of CMR manganites. The layer DMFT presented here was also applied to investigate the magnetic properties of multilayers involving the DE interaction with classical spins in [106].

For a system in which the interlayer coupling is relatively weak compared with the intralayer coupling, such as high-T_c cuprates, the single-site DMFT is not a good approximation because the intralayer short-range correlations become as important as the local correlations. Therefore,

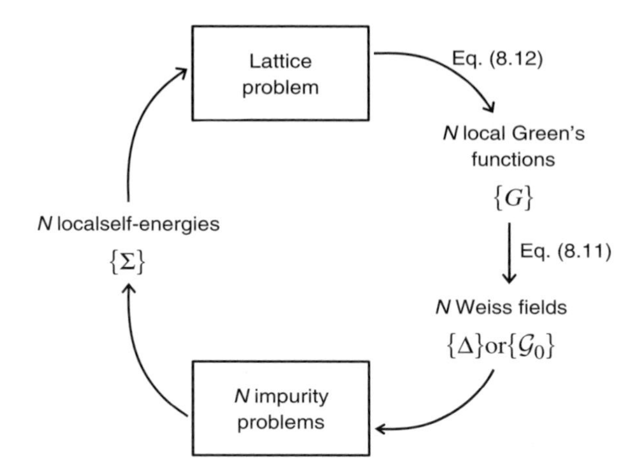

Fig. 8.3 Self-consistent loop of the layer DMFT with layer number N. Starting from N hybridization functions $\{\Delta\}$ or Weiss fields $\{\mathcal{G}_0\}$, N local interacting Green's functions and then N local self-energies $\{\Sigma\}$ are obtained by solving N impurity models. The N local self-energies are used for the lattice Green's function which is a $N \times N$ matrix. Using Eq. (8.12), N local Green's functions $\{G\}$ are computed, and using Eq. (8.11), the Weiss fields are updated. This loop is iterated until the convergence is reached between old Weiss functions and updated functions.

several extensions of DMFT have been proposed. One of the extensions of the DMFT is including finite-ranged correlations in a real space. This is called "cellular DMFT (CDMFT)" [107–109]. For example, spatial correlations up to a second-neighbor site in a square lattice can be included by considering a 2×2 plaquette as a unit of the DMFT. Extending this CDMFT to the multilayer geometry, the electron self-energy may be expressed as

$$\Sigma_\sigma(\vec{r}_i, \vec{r}_j; i\omega_n) = \Sigma_\sigma(\vec{r}_i, \vec{r}_j; i\omega_n)\delta_{z_i z_j}\Box_{ij}. \tag{8.14}$$

Here, $\Box_{ij}$ indicates that sites i and j are on the same plaquette. The CDMFT necessarily breaks the translational invariance. Another cluster extension of the DMFT, called the dynamic cluster approximation (DCA) [110, 111], maps lattice problem to a finite-size cluster problem with a periodic boundary condition. In this case, the translational invariance is not broken, and therefore, using this symmetry, a variety of numerical techniques can be used to deal with larger clusters than the CDMFT.

Both the CDMFT[109, 112] and the DCA[113] have been used to discuss the possibility of d-wave superconductivity in the two-dimensional Hubbard model and tJ model for high-T_c cuprates. In Section 8.4, we discuss the proximity effect of d-wave superconductivity in multilayers consisting of over-doped and under-doped cuprates using the layer extensions of these techniques.

The analytic continuation $i\omega_n \to \omega + i\delta$ (δ is infinitesimal) in (8.12) gives the self-consistent condition for the retarded Green's function $G^R_\sigma(z_i;\omega)$. By imposing the same form for the Keldysh Green's function $G^K_\sigma(z_i;\omega)$, the layer DMFT may be used to investigate the nonequilibrium steady-state properties of strongly-correlated heterostructures. We present an example applying such a method to study the transport properties of two-terminal correlated heterostructures in Section 8.5.

8.2.3 Auxiliary particle methods

The DMFT and its cluster extensions are very powerful techniques to deal with strongly correlated electron systems. However, sometimes it is difficult to understand their results. Therefore, it is useful to apply physically transparent techniques such as auxiliary particle methods. For strongly correlated electron systems such as Hubbard and t-J models, there are roughly two approaches, slave-boson and slave-fermion methods. Let us here consider a single-band model for simplicity. In the slave-boson method, an electron operator is described by products of a fermionic spinon representing the spin degree of freedom at a singly occupied site, a bosonic holon representing the charge degree of freedom, a doped hole or an empty site, and a doublon representing the double occupancy [114, 115]. For the infinite U case, doublons are simply neglected [116, 117]. In the slave-fermion method, statistics of spinon and boson (and doublon) are interchanged [118, 119]. Because of the spinon bose-condensation, this technique is suitable for studying magnetically ordered states in correlated models. On the other hand, the slave-boson method is suitable for describing the spin liquid and superconducting states.

In these formalisms, an electron operator $c_{i\sigma}$ is represented as products of spinon $s_{i\sigma}$, holon h_i, and doublon d_i operators as

$$c_{i\sigma} = h_i^\dagger s_{i\sigma} + \sigma s_{i-\sigma}^\dagger d_i, \tag{8.15}$$

with the local constraint

$$h_i^\dagger h_i + d_i^\dagger d_i + \sum_\sigma s_{i\sigma}^\dagger s_{i\sigma} = 1. \tag{8.16}$$

This local constraint is normally treated on a static mean-field level, and fluctuations around the mean-field solution are considered using a random-phase approximation. For a spatially uniform system, solving the self-consistent equations is straightforward. However, when these methods are applied to a spatially inhomogeneous situation, such as heterostructures, self-consistent equations for different regions are coupled non-linearly. In general, when there are more than two inequivalent regions, these self-consistent equations cannot be solved by a simple bisection method, and instead one has to employ, for example, a conjugate gradient algorithm.

Alternative formalism of the slave-boson method developed by Kotliar and Ruckenstein is more tractable [120]. Rüegg and co-workers have applied this method to a model Mott-insulator/band-insulator heterostructure [121] and obtained qualitatively similar results by the layer DMFT [71].

To study the bulk DE model for manganites, Sarker has applied the Schwinger-boson method [122]. This method is similar in spirit to the slave-fermion method since spin degrees of freedom are treated by bosonic auxiliary particles, i.e. Schwinger bosons. Applying this technique to multilayers or surfaces of a DE system is also possible with some effort to solve the self-consistent equations. As an example, a study of the surface magnetism of the DE model for manganites will be presented in Section 8.5.

There are many other theoretical techniques. For example, a conventional Monte-Carlo technique has been used by a number of authors to study models for both bulk systems and heterostructures in which electrons are interacting with classical degrees of freedom such as classical spins and lattice distortions [81, 123].

8.3 Mott-insulator/band-insulator heterostructures

In this section, we consider heterostructures in which $LaTiO_3$ (a Mott insulator in bulk) and $SrTiO_3$ (a band insulator) are grown along the [001] (z) direction as in the experiments of Ohtomo *et al* [9]. An interesting issue is that $SrTiO_3$ is a nearly ferroelectric material and the static dielectric constant becomes larger than 2000 at long wavelength and low temperatures [124, 125]. However, it is much smaller at high frequencies, room temperature, or short length scales. In order to estimate the strong screening effect mainly originating from the lattice distortion, we first present the results of the density functional theory (DFT) calculations. The DFT calculations do not address the Mott physics associated with the strong electron–electron interactions. We will then present the results of the layer DMFT calculation for model Mott-insulator/band-insulator heterostructures to discuss the effect of the strong correlation on spectral transfer and the metallicity of such heterostructures.

8.3.1 Lattice relaxation and charge redistribution

To investigate the lattice effect, we used the projector augmented wave (PAW) approach [126] as implemented in the *Vienna Ab initio* Simulation Package [127, 128], with the rotationally invariant LDA+U method of Liechtenstein *et al.* [129]. During the DFT calculations, the lattice constants in the x and y directions were fixed to the experimental value for cubic $SrTiO_3$ (3.91 Å), the substrate used in the experiments [9]. The c axis lattice constant and atomic z coordinates were optimized retaining the tetragonal symmetry of the crystal. Numerical details are presented in [70].

Figure 8.4 shows the calculated relaxed-lattice structures for two representative cases: $[LaTiO_3]_1[SrTiO_3]_8$ [(1–8) in short] and $[LaTiO_3]_2[SrTiO_3]_7$ [(2–7) in short] heterostructures. The largest structural relaxations occur in the TiO_2 layer at the $LaTiO_3$–$SrTiO_3$ interface ($Ti_{0.5}$ in the upper panel and Ti_1 in the lower panel) and involve a ferroelectric-like distortion in which the negatively charged O and positively charged Ti ions are displaced relative to each other by 0.15Å in the (1–8) case and 0.18 Å in the (2–7) structure. This distortion produces a local ionic dipole moment which screens the Coulomb field created by the substitution of Sr^{2+} by La^{3+} ions, and also leads to an increase in the z direction Ti–Ti distance by about 0.08Å. Moving further away from the interface, the magnitude of the ferroelectric-like distortion decays rapidly, while the Ti–Ti distance reverts to a constant value very close to bulk. The sizes of the lattice distortions are comparable for the one- and two-La-layer cases.

An important quantity for physical insight and further theoretical analysis is the conduction band charge density: loosely speaking, this is the Ti d occupancy, which is not unambiguously defined with density functional calculations. To obtain this we make use of the fact that within LDA+U the ground state is a highly polarized ferromagnet in which the magnetization density can be ascribed to the conduction bands. Following the procedure used for total charge density in [130] we compute the smoothed magnetization density $\overline{m}(z)$ by averaging in the xy plane

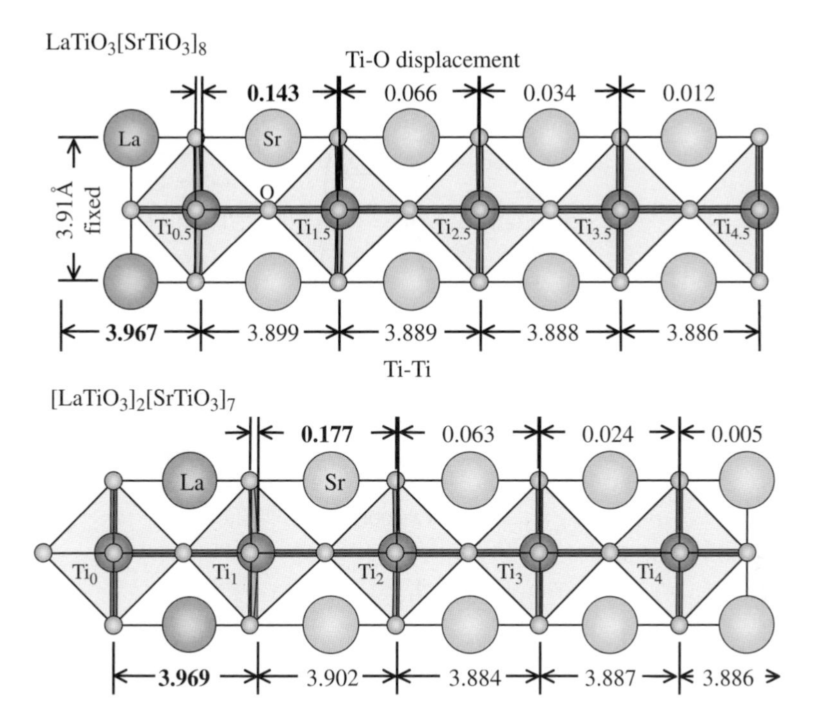

Fig. 8.4 Calculated optimized lattice structures of superlattices $[LaTiO_3]_1[SrTiO_3]_8$ (upper figure, z-axis lattice constant $c = 35.09$ Å) and $[LaTiO_3]_2[SrTiO_3]_7$ (lower figure, z-axis lattice constant 35.17 Å); half of the unit-cell is shown in each case. The inter-titanium distances (lower lines) and displacements of the Ti ions relative to the O_2 planes (upper lines) are also indicated. The center of the $LaTiO_3$ region is taken as the zero of the z coordinate in each case and the Ti ions are labeled by their relative z positions. Figure reproduced from [70].

and smoothing in z over the range $\pm a/2$. We identify the integral of $\overline{m}(z)$ over a unit-cell with the conduction-band charge density within that unit-cell. When the spin polarization is weak, we renormalize the density appropriately.

Figure 8.5 compares our calculated Ti d charge densities for relaxed (black squares) and unrelaxed (white squares) superlattices (SLs) for $[LaTiO_3]_1[SrTiO_3]_8$ (upper panel) and $[LaTiO_3]_2[SrTiO_3]_7$ (lower panel). In both SLs, the screening provided by lattice relaxation reduces the charge density on the central Ti layer and produces a long "tail" in the charge distribution, extending far away from the interface. The effect is particularly large in the two-layer structure, reducing the middle-layer density by almost a factor of 2. We note that in each case the interface layer ($Ti_{0.5}$ or Ti_1) remains electronically well defined, with the density dropping by approximately 0.3 electrons between the Ti at the interface, and its neighbor surrounded by two Sr-O layers. The relaxed-lattice charge densities agree within experimental uncertainties with the Ti^{3+} values reported by Ohtomo *et al.* [9] both in terms of peak values (0.3–0.4 in the central region) and the slow decay away from the central region.

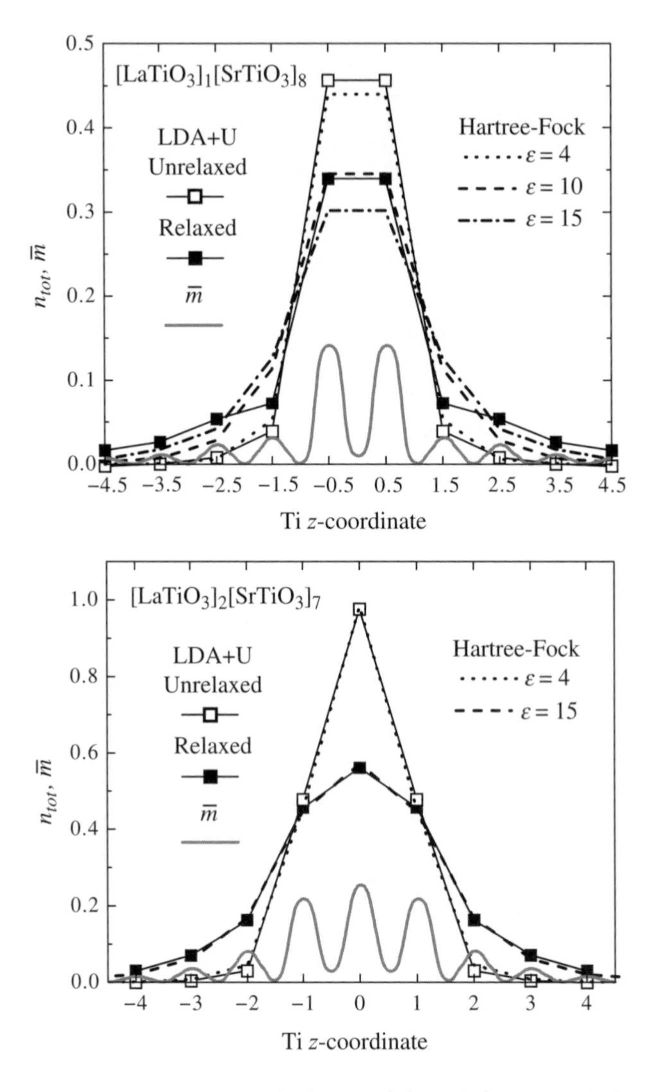

Fig. 8.5 Charge and magnetization densities for $[LaTiO_3]_1[SrTiO_3]_8$ (1–8) and $[LaTiO_3]_2[SrTiO_3]_7$ (2–7) heterostructures. Open and filled squares: conduction-band charge densities per unit-cell obtained as described in the text for unrelaxed and relaxed superlattices, respectively. Light lines: smoothed magnetization densities $\overline{m}(z)$ of relaxed heterostructures. Dotted, dashed, and dash-dotted lines: results of model Hartree–Fock calculation with hopping parameters taken from fits to LDA+U calculations, $U = 5$ and $J = 0.64$ eV, and dielectric constant ε indicated. Figure reproduced from [70].

We next simulate the screening due to the lattice relaxation and the filled electronic bands by a fixed dielectric constant ε. Here, we perform the Hartree–Fock (HF) approximation calculations using a three-band Hubbard model for t_{2g} orbitals with long-ranged Coulomb interaction between conduction electrons and between a conduction electron and positive $+e$ charge located on the La site of the $[LaTiO_3]_m[SrTiO_3]_n$ heterostructure. One-particle parameters for the three-band model were extracted from the band structures obtained by the LDA with the optimized structures described above. We compute the charge density to fit the LDA results of the density profile in the near-La region as shown in Figure 8.5. The charge density for the unrelaxed lattices is reproduced by $\varepsilon = 4$ fairly well. The additional screening from lattice relaxations can be simulated reasonably well by increasing ε to 15. The main deficiency is that the model does not capture the highly enhanced long wavelength dielectric response of $SrTiO_3$ and hence overestimates the rate at which the charge density decays far from the $LaTiO_3$ layer ($n_{tot} \lesssim 0.05$ region) for the (1–8) heterostructure.

The lattice relaxation in $[LaTiO_3]_m[SrTiO_3]_n$ heterostructures was also studied by Hamman and co-workers. They used the GGA approximation to DFT and reported lattice distortions similar to those found in our study [72].

Within a HF approximation, we can get a rough picture of the metallicity appearing at an interface between a band insulator and a Mott insulator, as in $[LaTiO_3]_m[SrTiO_3]_n$-type heterostructures as follows: a set of contiguous $LaTiO_3$ layers form a quantum well whose electronic states are organized into subbands. The substantial charge spreading means that neither the binding energy nor the inter-subband energy splitting is large, so that in the near-interface region there are several partially occupied subbands, which lead to metallic behavior [68, 69].

Using the HF approximation, we have also computed complicated spin–orbital phase diagrams of $LaTiO_3/SrTiO_3$ heterostructures with and without the lattice relaxation (Fig. 4 of [70]). Since the HF approximation is a single-particle approximation, it is known to overestimate the tendency associated with spin and/or orbital symmetry breaking. Therefore, it is worth investigating such phase diagrams using beyond-HF methods such as the DMFT. In fact, in the next subsection, we will see that the picture of metallicity is dramatically modified.

8.3.2 Mott physics

Both the density functional theory and model Hartree–Fock calculations discussed in the previous subsection are based on a single-particle approximation, which may not necessarily be a good representation for the strongly correlated electron systems. Here, we investigate the electronic properties of $LaTiO_3/SrTiO_3$-type heterostructures using the layer DMFT and show that, even going beyond the single-particle approximation, metallic interfaces survive.

Our model heterostructure is described by a single-band Hubbard model with the long-range Coulomb repulsion between electrons. Difference between $LaTiO_3$ and $SrTiO_3$ regions is introduced by placing $+e$ point charges (corresponding to the difference between the ionic charges of La and Sr in $ATiO_3$ compounds) at the crystallographic La positions. As the dimensionless parameter for the long-ranged Coulomb interaction, we took $E_c = e^2/\varepsilon ta = 0.8$, where t is the nearest-neighbor electron transfer, and a the lattice constant taken to be unity. Taking $t \sim 0.3$ eV and $a \sim 4$ Å, $E_c = 0.8$ corresponds to $\varepsilon \sim 15$. We solve the self-consistent equations for the layer DMFT described in Section 8.2 with the two-site DMFT impurity solver [131].

One of the most useful observables to see the dynamical property of the correlated electron is the single particle spectral function. The spectral functions are in principle measurable in photoemission or scanning tunneling microscopy. Numerical results for the layer-resolved spectral function $A(z;\omega) = -\frac{1}{\pi}\int \frac{d^2k_\parallel}{(2\pi)^2}\mathrm{Im}G(z,z,\vec{k}_\parallel;\omega+i\delta)$ for a 10-layer heterostructure consisting of 10 $LaTiO_3$ layers with different values of U are presented in Figure 8.6. The left panel shows results for the weak coupling ($U/t = 10$), and the right panel for the strong coupling ($U/t = 16$ about 10% greater than the critical value which drives the Mott transition in a bulk system described by H with $n = \infty$). The critical value for the bulk Mott transition is estimated to be $U_c/t \approx 14.7$ by the two-site DMFT. Outside the $LaTiO_3$ region ($|z| \gg 6$), the spectral function is essentially identical in form to that of the free tight-binding model H_{band} for both the weak coupling and strong coupling results. The electron density is negligible, as can be seen from the fact that almost all of the spectral function lies above the chemical potential. As one approaches the heterostructure ($|z| = 6$), the spectral function begins to broaden. For the weak coupling case, spectral functions at $|z| < 6$ are also quite similar to the results of the HF analysis except for tiny peaks outside of the central quasiparticle band corresponding to the upper- and lower-Hubbard bands. On the other hand, for the strong coupling case, spectral weight around $\omega = 0$ begins to decrease rapidly and the characteristic strong correlation structure of lower and upper Hubbard bands with a central quasiparticle peak begins to form. The sharp separation between these features is an artifact of the two-site DMFT.

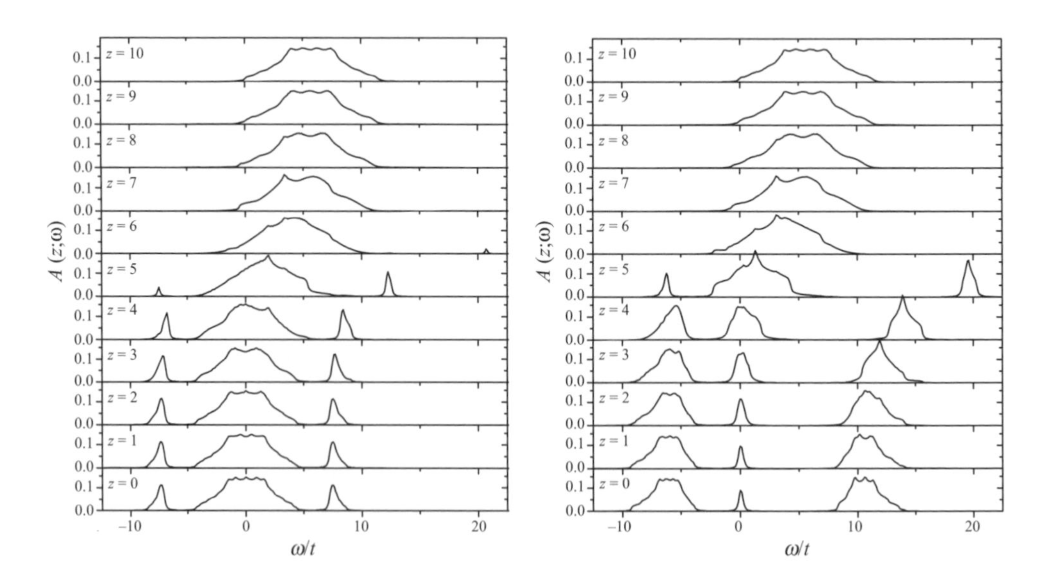

Fig. 8.6 Layer-resolved spectral function calculated for 10-layer heterostructure with $E_c = 0.8$. Left panel: $U/t = 10$, right panel: $U/t = 16$. Charge $+e$ counterions defining the $LaTiO_3$ region are placed at $z = \pm 0.5, \ldots, \pm 4.5$ so the electronic (B) sites are at integer values of z. Left figure reproduced from [132], and right figure reproduced from [71].

Experience with bulk calculations suggests that the existence of three features and the weight in the quasiparticle region are reliable. Towards the center of the heterostructure, the weight in the quasiparticle band becomes very small, indicating nearly insulating behavior. For very thick heterostructures, we find the weight approaches 0 exponentially.

The behavior shown in Figure 8.6 is driven by the variation in density caused by leakage of electrons out of the $LaTiO_3$ region. Figure 8.7 shows as open squares the numerical results for the charge–density distribution $n_{tot}(z)$ for the heterostructure whose photoemission spectra are shown in Figure 8.6. One sees that in the center of the heterostructure ($z = 0$) the charge density is approximately 1 per site, and that there exists an edge region, of about three-unit-cell width, over which the density drops from ~ 1 to ~ 0. The overall charge profile is determined mainly by the self-consistent screening of the Coulomb fields which define the heterostructure, and is only very weakly affected by the details of the strong on-site correlations (although the fact that the correlations constrain $n_{tot} < 1$ is obviously important). To show this, we have used the HF approximation to recalculate the charge profile: the results are shown as filled circles in Figure 8.7 and are seen to be almost identical to the DMFT results.

In order to study the metallic behavior associated with the quasiparticle subband, we computed the charge density from the quasiparticle bands n_{coh} by integrating $A(z;\omega)$ from $\omega = 0$ down to the first point at which $A(z;\omega) = 0$. Results are shown as open circles in Figure 8.7. It is obvious that these near-Fermi-surface states contain a small but non-negligible fraction of the total density, suggesting that edges should display relatively robust metallic behavior. The results represent a significant correction to the HF calculation, which leads, in the edge region,

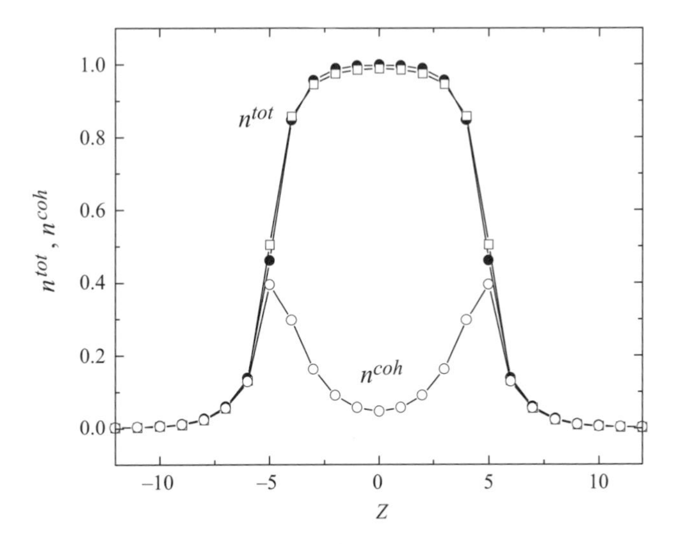

Fig. 8.7 Total charge density (open squares) and charge density from the coherent part near the Fermi level (open circles). For comparison, the total charge density calculated by applying the Hartree–Fock approximation to the Hamiltonian is shown as filled symbols. The parameters are the same as in the right panel of Figure 8.6. Figure reproduced from [71].

to a metallic quasiparticle density essentially equal to the total density. This is because the HF approximation does not give mass renormalization.

There have appeared experimental results supporting the "metallic interface" originating from "electronic reconstruction." Takizawa and co-workers performed photoemission spectroscopy measurements of $[LaTiO_3]_m[SrTiO_3]_n$ SLs varying the thickness of the capping $SrTiO_3$ layer [10]. They observed a relatively sharp quasiparticle spectrum at the Fermi level E_F, and found that its dependence on the thickness of capping $SrTiO_3$ layer (controlling the degree of electron transfer from $LaTiO_3$ region to $SrTiO_3$ region) and the roughness of the interface were consistent with the numerical results by layer DMFT. Further, Seo and co-workers used infrared spectroscopic ellipsometry and estimated the carrier density of $[LaTiO_3]_m[SrTiO_3]_n$ SLs [11]. They found that the carrier density is about $-0.5e$ per interface between $LaTiO_3$ and $SrTiO_3$ irrespective of the periodicity of SLs and the individual layer thicknesses, suggesting that the metallic carriers are located at interfaces. On the theory side, Ishida and Liebsch performed the DMFT calculation for the bulk $LaTiO_3$ with tetragonal distortion mimicking the crystal grown on the $SrTiO_3$ substrate. They claim that the Ti t_{2g} levels are strongly influenced by tetragonal distortion, and the critical interaction U_c to realize a Mott insulating state becomes larger than the realistic value [133]. Instead, they proposed that the metallic behavior of $[LaTiO_3]_m[SrTiO_3]_n$ SLs comes from whole $LaTiO_3$ regions. In this case, the two-dimensional carrier density must be proportional to the thickness of $LaTiO_3$. The apparent conflict between the experimental results and the theoretical one has not yet been reconciled.

As another system of Mott-insulator/band-insulator heterostructures, Smadici and co-workers studied the electronic properties of $[LaMnO_3]_m[SrMnO_3]_n$ [18]. They used resonant soft-X-ray scattering to demonstrate that the interfacial electronic state is distinct from those in $LaMnO_3$ and $SrMnO_3$ regions. In particular, the interface density of states was found to exhibit a pronounced peak at the Fermi level, and its intensity correlated with the conductivity and magnetization. This suggests that the "electronic reconstruction" at interfaces between $LaMnO_3$ and $SrMnO_3$ is responsible for the metallic behavior of such heterostructures.

8.4 Superlattices of under-doped-cuprate/over-doped-cuprate

In this section, we consider superlattices (SLs) made up of high-T_c cuprate layers with different carrier concentrations. High-T_c cuprates have been one of the central issues of condensed matter physics since the discovery of superconductivity in $La_{2-x}Ba_xCuO_4$ systems in 1986 [33]. Intensive experimental studies revealed that there are two distinct doping regimes in high-T_c cuprates, under-doped (UD) regime and over-doped (OD) regime with the optimal-doping regime, where the critical temperature is maximized, in between. Such a global phase behavior has been theoretically reproduced, at least qualitatively, based on the Hubbard model or t-J model. Furthermore, in the UD regime, a high pairing scale is observed both experimentally [134] and theoretically [135], while in the OD regime, stronger carrier coherence exists, although T_c is lower than at optimal doping in both regions. This behavior may indicate that, by connecting the two doping regions in a multilayer geometry, the proximity of the strong pairing interaction from the UD region may produce stronger superconductivity in the OD region.

Similar proposals were made in [136, 137]. Conversely, the phase stiffness of a superconductor with small superfluid density may be increased by coupling it to a good metal resulting in a higher T_c [138]. The effects of spatial inhomogeneity in (quasi-)two-dimensional systems have been investigated in the light of the nanoscale inhomogeneity often observed in UD cuprates [139–142]. In contrast to such in-plane inhomogeneity, the decoherence effect is small in SLs. Further, various conditions of the SL can be controlled by present experimental capabilities. In fact, enhanced superconductivity in such systems was reported recently [40–42].

In what follows, we investigate model SLs made up of UD and OD cuprate layers consisting of Hubbard and t-J type electronic interactions to provide insight into this problem. We use a variety of techniques, including the layer extensions of the cellular dynamical-mean-field theory (CDMFT) [107, 108] and the dynamic cluster approximation (DCA) [110, 111] as well as the slave-boson mean-field (SBMF) theory. We find that the superconducting (SC) order parameter is indeed increased in UD/OD SLs as compared to its value in the uniform system. For some combinations of dopings, the SC order parameter is found to become larger than its largest value in the uniform system. Furthermore, the SBMF calculation predicts a transition temperature in the SL that exceeds the maximum T_c in the uniform system (Fig. 3 of [75]). These results suggest that "higher-T_c" superconductivity may indeed be realized in SLs of cuprates.

First, we study the zero temperature SC order parameters of SLs and uniform systems using layer CDMFT. We treat the interlayer correlations on the mean-field level [71] and take care of short-ranged intralayer correlations by using a square 2×2 cluster. The CDMFT then maps the bulk lattice problem onto a number N_z of independent effective cluster problems, with N_z the number of layers in the unit-cell. The cluster problems are solved by using the Lanczos exact diagonalization (ED) technique [143]. For the uniform system, we compute the d-wave SC order parameter $\Psi = \langle c_{i\uparrow} c_{i+x\downarrow} \rangle = -\langle c_{i\uparrow} c_{i+y\downarrow} \rangle$ as a function of carrier density n. For the SLs, we modulate the potential along the z direction, so that the UD layers and the OD layers alternate, instead of solving the Poisson equation self-consistently. Then, we obtain layer z dependent carrier density $n(z)$ and d-wave order parameter $\Psi(z)$. The detail of the numerical calculation can be found in [75].

Figure 8.8 presents results for the d-wave SC order parameter. For the uniform system, we have a "dome" like shape which is consistent with previous calculations for the two-dimensional Hubbard model [112]. We find an optimal carrier density $n_{opt} = 0.91$ which maximizes Ψ. For SLs, we find that the d-wave order parameter in the OD layers is enhanced. The enhancement is stronger in AABB SLs (two UD layers and two OD layers alternate, filled squares) than in ABAB SLs (one UD and one OD layers alternate, open squares) possibly due to the dilution in the latter. We also considered thicker SLs consisting of several UD layers and OD layers sandwiching nearly optimal-doped layers, e.g. AA'BCC' SLs. Even in this case, we found enhancement in the OD layers. However, Ψ in the nearly optimal-doped layers is suppressed because the pairing interaction is reduced by the OD layers, and we never observed significantly larger order parameters than in the uniform system. An example is shown by open circles in Figure 8.8.

In order to see whether such a proximity effect can increase the critical temperature, we have performed the SBMF calculation for the quasi-two-dimensional t-J model. In this method, the spin degree of freedom is described by fermionic spinon f_σ (written as s_σ in Section 8.2) and the charge degree of freedom is described by bosonic holon b (written as h in Section 8.2).

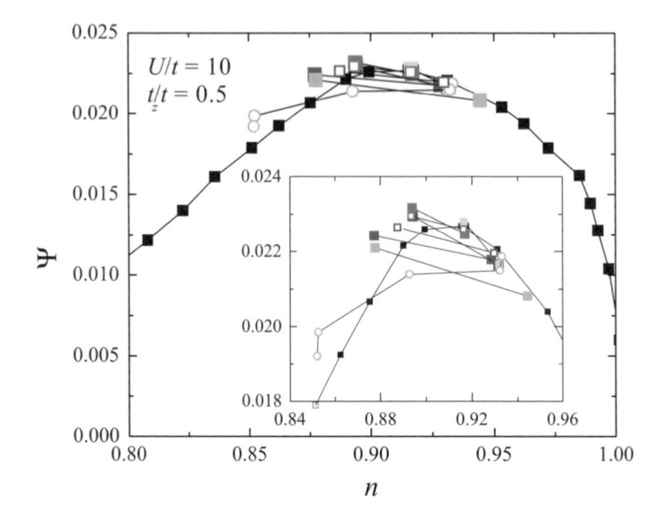

Fig. 8.8 The superconducting order parameter Ψ as a function of carrier density n for the quasi-two-dimensional Hubbard model computed by CDMFT with the ED impurity solver at $T = 0$. Black symbols are the results for the uniform systems. Shaded symbols connected by a thin line denote the results for the SL. The inset shows the magnification around the optimally doped region. Figure reproduced from [75]. This figure is reproduced in color in the color plate section.

We do not consider a doublon because the double occupancy is excluded in the model. There appear several order parameters in the SBMF, and the most important ones for our discussion are $\Delta = \langle f_{i\uparrow} f_{i+x\downarrow} \rangle = -\langle f_{i\uparrow} f_{i+y\downarrow} \rangle$ the singlet resonance valence bond (RVB) representing short-range singlet formation and $N_0 = \langle b_i \rangle^2$ the holon bose condensation representing the coherence of carriers.

For the uniform system, the transition temperature below which Δ (N_0) becomes finite decreases (increases) with decreasing electron density from $n = 1$ (Mott insulator). Since the d-wave SC order parameter Ψ becomes finite when both Δ and N_0 are finite in the mean-field approximation, T_c forms a dome-like shape as a function of n. An order parameter at low temperatures has a similar doping dependence as its transition temperature. For UD/OD SLs, we observe similar enhancement in the SC order parameter in OD layers at low temperatures over the maximum value of the uniform system [Fig. 2 (b) in [75]]. Such enhancement was also found to lead to a higher T_c than in the uniform system [Fig. 3 in [75]]. Detail of these calculations can be seen in [75]. These results originate from the proximity effect of the strong pairing scale from the UD layers. By further increasing the hole density in the OD layers, Δ starts to decrease while N_0 continues to increase. Thus, there appears to be an optimal combination between the carrier concentrations in UD and OD layers. As an example, we plot T_c of UD/OD SLs as a function of the carrier density in the OD layers n_{OD} with the fixed density in the UD layer $n_{UD} = 0.9$ in Figure 8.9. As expected, T_c increases with decreasing n_{OD} and becomes higher than the maximum T_c of the uniform system. The optimal OD density where T_c is maximized was found to be $n_{OD} \sim 0.7$ for $n_{UD} = 0.9$.

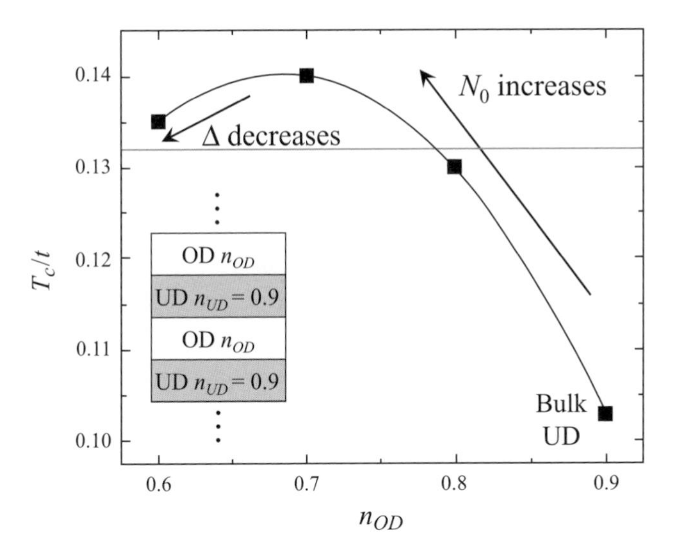

Fig. 8.9 T_c of over-doped/under-doped cuprates superlattices as a function of the carrier density in the over-doped layer n_{OD} computed by the slave-boson mean-field approximation for the quasi-two-dimensional t-J model. The carrier density in the under-doped layer is fixed as $n_{UD} = 0.9$. A thin horizontal line indicates the maximum T_c of the uniform system.

The present calculations suggest a "recipe" to increase T_c in SLs of cuprates: (1) Combine UD cuprates with large pairing scale and OD cuprates with strong coherence and (2) make the average carrier density n_{av} slightly smaller than the bulk optimal value n_{opt}. The latter condition is required to obtain a larger-order parameter, which gives rise to higher T_c even in the double-layer systems when there is enough phase stiffness. Requiring strong coherence, i.e., smaller n_{av} than n_{opt}, is related to the proposal in [138].

Finally, we check the finite temperature SBMF results using a layer extension of the DCA with the noncrossing approximation (NCA) impurity solver [111, 113]. Since the precise determination of the critical temperature of SLs is numerically involved, we present DCA results for the order parameter at a temperature below the T_c of the uniform and SL system, and argue that an enhancement of the order parameter in the SL close to T_c should translate to an increase in T_c over the uniform system. Figure 8.10 summarizes the results of the layer DCA calculations. In the uniform system we find an optimal doping $n_{opt} \approx 0.82$ with $T_c^{opt} \approx 0.043t$. The order parameter Ψ at a temperature $T = 0.041t$ slightly below T_c shows similar n dependence to T_c. For the AABB SL system, we again find a clear enhancement of the order parameter Ψ in the OD layers which can even exceed the optimal value in the uniform system. This behavior is consistent with the CDMFT at $T = 0$ and the SBMF.

A definite proof that the proximity effect of UD and OD layers in the SLs can give rise to higher T_c requires calculations with beyond-mean-field treatments within fully three-dimensional clusters, and such numerical calculations are in progress.

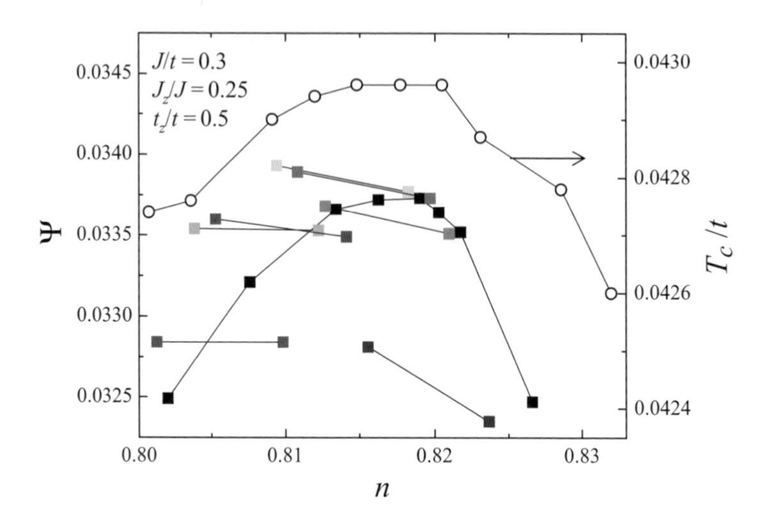

Fig. 8.10 The superconducting order parameter Ψ as a function of carrier density n for the quasi-two-dimensional t-J model computed by the DCA-NCA at a temperature $T = 0.041t$ close to T_C. The results for the uniform systems are indicated by the black symbols, while the shaded symbols connected by a thin line denote the result for the SL. Also shown is the doping dependence of T_C in the uniform system (open circles). Figure reproduced from [75]. This figure is reproduced in color in the color plate section.

8.5 Other directions

Strongly correlated electron systems have been providing novel phenomena which could revolutionize our current electronics. However, only a few "novel materials" have been commercialized. One of the few examples is a power cable made by high-T_c cuprates supplied by ULTERATM[2]. For realization of high-T_c cuprate power cables, controlling the grain boundaries has been the key [144]. Control of the interface properties is also the key to improving magnetoresistace ratio (MR) in tunneling magnetoresistance (TMR) junctions made by manganites [31, 32]. As discussed by Ogimoto and co-workers, not only the structural properties, but also the interfacial "spin" state and perhaps the "orbital" state would be of crucial importance. In addition, the electronic state could be modified significantly by applying an electric field or injecting carriers to operate an electronic device. The difficulty of dealing with strongly correlated systems is that the rigid-band picture does not normally apply. Such a situation may have already been realized in the Josephson junction made by high-T_c cuprates [38, 39].

In this section, we discuss the future directions of strongly correlated heterostructures along these lines by presenting two recent studies. First, we study the surface magnetic behavior of the double-exchange (DE) model for metallic manganites. We see that the surface magnetism decreases more rapidly than the bulk, even when the surface is "ideal." The thermal fluctuation

[2] Detail of the product, HTS Triax Superconducting Cable, can be found at http://www.supercables.com/index.html.

of spins plays a crucial role in reducing the surface magnetic moment. Second, we study the non-equilibrium steady state of two-terminal heterostructures involving strongly correlated systems. We see that a Mott-insulating system behaves slightly differently from a band-insulating system when it is used as a barrier in a metal/insulator/metal junction. Although results to be presented in this section are already published, these are parts of on-going projects. Therefore, we expect further developments.

8.5.1 Surface magnetism of double-exchange manganites

Because of the high spin polarization and relatively high Curie temperature T_C, perovskite manganites, $La_{1-x}Sr_xMnO_3$ with $0.2 < x < 0.5$, [30] could serve as ideal ferromagnetic leads for a TMR junction [28, 29] functioning at room temperature. Several attempts have been made to fabricate manganite TMR junctions [31, 32]. A very large TMR was indeed measured at low temperature, consistent with half-metallicity. However, the TMR decreases rapidly, and only small TMR is observed at room temperature [32]. On the basis of spin-resolved photoemission spectroscopy, it was suggested that the rapid decrease in TMR is due to the stronger temperature dependence of spin polarization at interfaces than in the bulk [145].

Surface magnetism has been theoretically studied within a classical Heisenberg model using the numerical Monte Carlo (MC) technique. Surface polarization was shown to decrease more rapidly than that in the bulk [78, 146]. However, the difference between bulk magnetism and interface magnetism in the DE-type model remains unresolved.

Here, we investigate the surface magnetic phase transition of the DE model by using the Schwinger-boson mean-field (SchBMF) method. We focus on metallic manganites possessing a relatively high T_C, such as $La_{1-x}Sr_xMnO_3$ with a doping concentration of $x \sim 0.3$. The SchBMF method has had success in describing the behavior of the quantum Heisenberg model in low dimensions [118] and has been applied to the bulk DE model [122, 147]. Since the SchBMF method correctly describes low-dimensional spin systems, it is also expected to provide a suitable description of surface and interface magnetic behaviors.

We consider the single-orbital DE model defined on an N layer lattice with the open-boundary condition in the z direction and the periodic-boundary condition in the xy plane. In the present SchBMF method, an electron operator c_σ is written as a product of a spin-less fermion f representing the charge degree of freedom and spinor boson b_σ representing the rotation of the local spin axis with the local constraint $\sum_\sigma b_\sigma^\dagger b_\sigma = 1$. After introducing various mean-field order parameters, we obtain single-particle mean-field Hamiltonians for spin-less fermions and spinor bosons. Then, self-consistent equations as well as the layer-dependent constraint are solved numerically. We consider a doping concentration $n = \sum_\sigma \langle c_\sigma^\dagger c_\sigma \rangle = 0.5$ and take a local spin as $S = 3/2$. Therefore, at $T = 0$, the magnetic moment saturates at $M = \langle \frac{1}{2} 0.5 + S \rangle = 1.75$. Details of the calculations can be found in [77].

Numerical results for the layer-dependent magnetization of an $N = 20$ layer system with uniform transfer are shown in the left panel of Figure 8.11 as a function of temperature. The lattice constant is taken to be unity: surface layers are located at $z = 1$ and 20, and the magnetization profile is symmetric with respect to the center of the system at $z = 10.5$. Clearly, the surface magnetization decreases faster than the bulk magnetization with increasing temperature, but all magnetizations disappear at the same temperature. We also find that layers at $4 \leq z \leq 17$ show roughly the same magnetization, and the thickness of the surface layers,

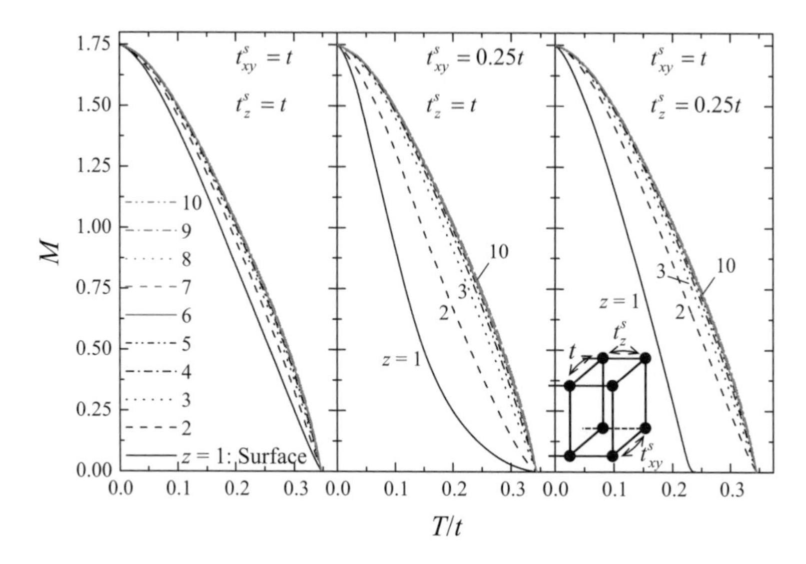

Fig. 8.11 Layer-dependent magnetization M for the 20-layer DE model as a function of temperature T using the Schwinger-boson mean-field approximation. Left panel: near-surface transfer is taken as $t^s_{xy} = t^s_z = t$, middle panel $t^s_{xy} = 0.25t$ and $t^s_z = t$, and right panel $t^s_{xy} = t$ and $t^s_z = 0.25t$. Inset shows the geometry of the near-surface transfer. Main figure reproduced from [77].

in which the magnetization drops, is about three unit-cells wide and it becomes thicker when approaching T_C.

Next we discuss the effect of surface condition on temperature-dependent magnetization. Since we are using the simple single-orbital DE model, we simulate various effects by changing the transfer intensity around the surface layers as indicated in the inset of Figure 8.11. The ferromagnetic Curie temperature remains unchanged, indicating that $N = 20$ is thick enough and the surface condition does not affect the bulk behavior.

First, we reduce the surface intraplane transfer t^s_{xy}. This may correspond to either the surface roughness or elongation of MnO_6 octahedron stabilizing the $d_{3z^2-r^2}$ orbital in the surface layers. Results are shown in the middle panel of Figure 8.11. The in-plane kinetic energy of electrons is reduced in the surface layers. However, coupling between the surface layers and the bulk region induces finite magnetization on the surface, resulting in the long tail of the magnetization curve. Next, the interlayer transfer t^s_z between the surface layer and its neighboring layer is reduced, the right panel of Figure 8.11. This roughly corresponds to contraction of the MnO_6 octahedron in the surface layers, resulting in an increase in $d_{x^2-y^2}$ orbital occupancy. In this case, strong ferromagnetic correlations remain in the surface layers preventing the rapid reduction of surface magnetization at low temperature. But, the interlayer magnetic correlation is rapidly reduced, and the surface layer becomes isolated from the rest of the system. Eventually, surface magnetism disappears below the bulk T_C. This transition is supposed to be an artifact of the mean-field treatment and the small surface magnetic moment is expected to survive up to the true bulk T_C within the theoretical model. If the antiferromagnetic interaction between the local t_{2g} spins (neglected in the present calculations) is included, the surface magnetic moment would change its relative orientation to the bulk moment as suggested in [32].

In the light of experimental results, surface magnetism reported by Park and co-workers [145] is similar to the theoretical results with smaller t^s_{xy} (middle panel on Figure 8.11). This may indicate that the surface MnO_6 octahedron is elongated along the perpendicular direction, but keeps the magnetic coupling between the surface layer and the bulk region. In the TMR junctions reported by Bowen *et al.* and Ogimoto *et al.*, manganite regions experience different strain from $SrTiO_3$ depending on the doping concentration. In the overdoped manganites, MnO_6 is expanded along the xy directions, and the $d_{x^2-y^2}$ orbital is stabilized; the situation may be closer to the theoretical result shown in the right panel of Figure 8.11. On the other hand in the underdoped manganites, MnO_6 is compressed along the xy directions. Therefore, the $d_{3z^2-r^2}$ orbital is stabilized at the interface, and the situation may be closer to the result shown in the middle panel in Figure 8.11.

In any event, we expect that the surface and interface magnetism of manganites degrades more rapidly than in the bulk. In order to see its effect on tunneling MR, we consider the "ideal" TMR junction, in which the dependence of the tunneling matrix on the Fermi velocity is weak and the two electrodes are identical, and use Jullière's formula:

$$\mathrm{MR} = \frac{2P^2}{1 - P^2}, \tag{8.17}$$

where P is the magnetization at the Fermi level. Note, however, that the assumptions to arrive at Jullière's formula are not in general satisfied, and one has to consider the realistic band structure in the presence of the interfaces and the dependence of the tunneling probability on the barrier height and thickness. But, it has been shown for a simple free electron model that the exact solution of MR approaches Jullière's model as the barrier thickness and height increase [148], which we call "ideal". Further, we approximate P by the bulk or surface magnetization with various conditions discussed before, normalized by its value at zero temperature, so $P(T) = M(T)/M(0)$. This is a rather good approximation for the DE model. The results of MR are summarized in Figure 8.12. Since TMR is proportional to P^2 at small P, the difference between the bulk magnetization and surface magnetization is pronounced near T_C. It should be noted that, even for the ideal surface (solid line), TMR near T_C is much smaller than what we expect from the bulk magnetization (dashed line). Besides various issues associated with the anisotropy of transfer, there is the thermal fluctuation of spins. The two-dimensional character of the surface geometry is another source of the rapid decrease in surface magnetism This can be seen in the cleanest surface, the left panel on Figure 8.11, resulting in a rapid suppression of MR, a solid line in Figure 8.12. Such low-energy fluctuation may be suppressed by introducing uniaxial spin anisotropy. For improving MR near T_C, using a magnetic insulator with higher transition temperature as a tunneling barrier may be another route. Recent synchrotron X-ray measurement for $LaMnO_3/SrTiO_3$ heterostructures revealed induced magnetic moments in $SrTiO_3$, which is a non-magnetic insulator in bulk [149]. Such induced moments could be another source of the rapid reduction of MR with the increase in temperature for $La_{1-x}Sr_xMnO_3$/ $SrTiO_3$/ $La_{1-x}Sr_xMnO_3$-type junctions. On the other hand, such induced moments may be utilized to improve the TMR characteristics.

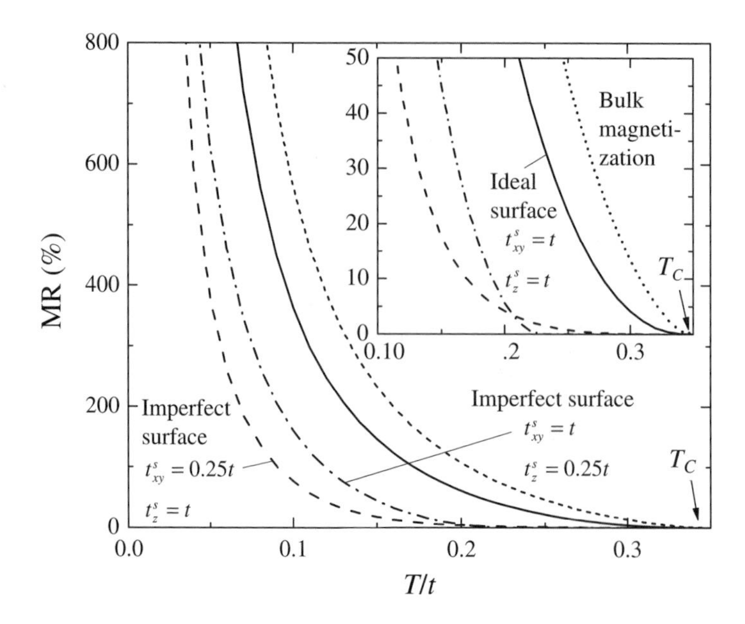

Fig. 8.12 Temperature dependence of tunneling MR computed by Jullière's formula Eq. (8.17) approximated by using the bulk or surface magnetization. Results of the ideal surface $t^s_{xy} = t^s_z = t$ are shown as a solid line, and imperfect surface with $t^s_{xy} = 0.25t$ and $t^s_z = t$ ($t^s_{xy} = 0.25t$ and $t^s_z = t$) broken (dash-dot) line. For comparison, MR expected from the bulk magnetization is indicated by dashed line. Inset shows the magnification where MR becomes small.

Including this effect is, therefore, important for more realistic calculations even for the simplest model systems. So far, we have neglected important issues focusing on high T_C manganites, such as the orbital degeneracy, electron–electron interactions, and the electron–lattice couplings. Including such effects, for example, within the DMFT treatment, is possible and is an important future direction. More realistic treatment for TMR including these effects at "high temperature" is also important to realize manganite-based TMR devices functioning at room temperature.

8.5.2 Transport through two-terminal strongly correlated heterostructures

As discussed in Section 8.1, any electronic device has an interface through which electrons are exchanged between the "sample" and a lead. Normally, one measures the electric current across the correlated region under various perturbations. Therefore, to realize "oxide electronics" devices utilizing such bulk properties, theoretical understanding of the transport properties of correlated heterostructures is of crucial importance. Transport properties of conventional metals and semiconductors have been thoroughly studied. However, theoretical techniques remain to be developed for the bulk-like effects of correlation on transport through heterostructures, including the correlation-induced Mott transition and symmetry breaking.

Here, we consider possibly the simplest problem concering the transport properties of strongly correlated heterostructures, i.e. a two-terminal junction consisting of several layers

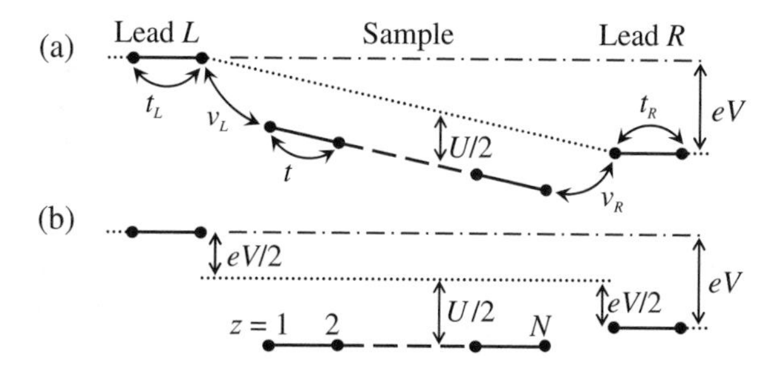

Fig. 8.13 Schematic view of the model heterostructure. (a) Linear potential profile for a sample with weak screening and (b) flat potential for strong screening. We consider the nearest-neighbor transfer t (t_α) of electrons in the sample (lead α), the hybridization v_α between the sample and lead $\alpha(= L$ or $R)$, and the layer-dependent potential $\varepsilon(z)$. Figure reproduced from Ref. [76].

of a strongly correlated system connected to two non-interacting reservoirs. We focus on the steady-state nonequilibrium properties of such structures under finite bias voltage. Our theoretical models are shown schematically in Figure 8.13. The basic idea is to combine the layer DMFT presented in Section 8.2 and Keldysh Green's function technique [150]. For details, see [76, 151]. Quantum impurity models appearing in the nonequilibrium layer DMFT are more complicated because of finite bias voltage. To solve such quantum impurity problems, we apply the non-crossing approximation [76, 152, 153].

Figure 8.14 plots the current versus bias voltage for the $N = 6$ heterostructure with several choices of on-site interaction: $U = 12t(< U_c)$, $15t(\lesssim U_c)$, $18t(> U_c)$, and 0 as a reference. Here, $U_c \sim 16t$ is the critical U for the metal–insulator transition of the bulk model estimated by the DMFT. Although there is quantitative difference, two choices of potential profile give similar curves for all finite Us. In particular, the two curves overlap at small bias voltage where the transport is governed by a (induced) quasiparticle band. In this region, crossover between the metallic and insulating regions can be seen in the linear conductance, G (inset of Fig. 8.14). Compared with the result for $U = 0$, the current and the conductance are substantially reduced corresponding to the reduction in the quasiparticle weight. By increasing a bias voltage, carriers are injected and the quasiparticle weight increases causing an upturn in the I-V curve for $U = 18t$. At the intermediate bias, there appear "plateaus" for small U. In this region, chemical potentials of two leads touch the Hubbard bands. Thus, electrons in the less-developed quasiparticle bands suffer from strong scattering.

These behaviors can be seen in Figure 8.15, where the layer-resolved spectral functions $A(z;\omega) = -\frac{1}{\pi}\mathrm{Im}G(z;\omega + i\delta)$ are presented for several choices of bias voltage with the linear potential. Shaded area shows the region occupied by electrons. In addition to the "rigid" shift of the layer-dependent spectral functions, quasiparticle-like features are visible in the spectral functions at $\omega \sim \pm eV/2$. These are the remnants of the nonequilibrium Kondo effect observed in the quantum dots [154].

Layer-resolved spectral functions deformed by an applied bias voltage are essentially measurable by photoemission spectroscopy and scanning tunneling microscopy (STM). Yet, in light of the available spatial resolution, the latter seems plausible. Since a current is already injected by an applied voltage, the insulating nature of the sample would not be a problem for

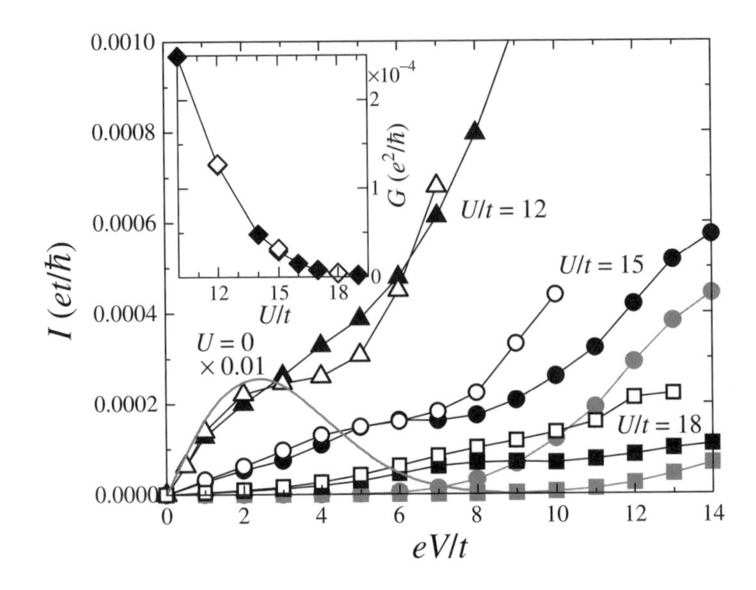

Fig. 8.14 Current-voltage characteristics of $N = 6$ heterostructure with $U = 12t$ (triangle), $15t$ (circle), $18t$ (square). Filled (open) symbols are the results for a linear (flat) potential, and light line corresponds to the linear potential with $U = 0$ multiplied by 0.01. For comparison, results using the equation-of-motion decoupling impurity solver are also shown for the linear potential with $U = 15t$ and $18t$ as light symbols. Inset: Linear conductance at $V = 0$ as a function of U. Filled (open) symbols are for the linear (flat) potential. Figure reproduced from Ref. [76].

STM. Optical conductivity measurements might also be useful to investigate deformed spectral functions, although it is an indirect measurement.

In order to realize oxide-based electronic devices utilizing novel behaviors that bulk strongly correlated systems exhibit, there are a lot of issues to overcome. For example, we have discussed surface magnetism of the double-exchange model for manganites and non-equilibrium transport though two-terminal strongly correlated heterostructures. For suppression of the rapid reduction in surface magnetism, suppressing the low-energy spin fluctuation seems to be key, in addition to making the ideal surface. Thus, to realize a practical TMR device made using manganites, one should suppress the spin fluctuation at an interface between manganite leads and an insulating barrier. Transport across a Mott insulator was found to be rather nonlinear as a function of bias voltage. Yet, if the Mott gap is so large or the correlated region is so thick that the induced quasiparticle band does not contribute much linear conductance, the Mott insulator can be used as an insulating barrier in a tunneling-type device. It would be extremely interesting if one could introduce collective behaviors, such as spin, orbital, charge symmetry breaking, and coupling with lattice, in small-Mott-gap systems and get highly nonlinear behavior, negative differential conductance for example. If this were to happen, such systems could be used as an interesting electronic device such as a logic circuit for computation.

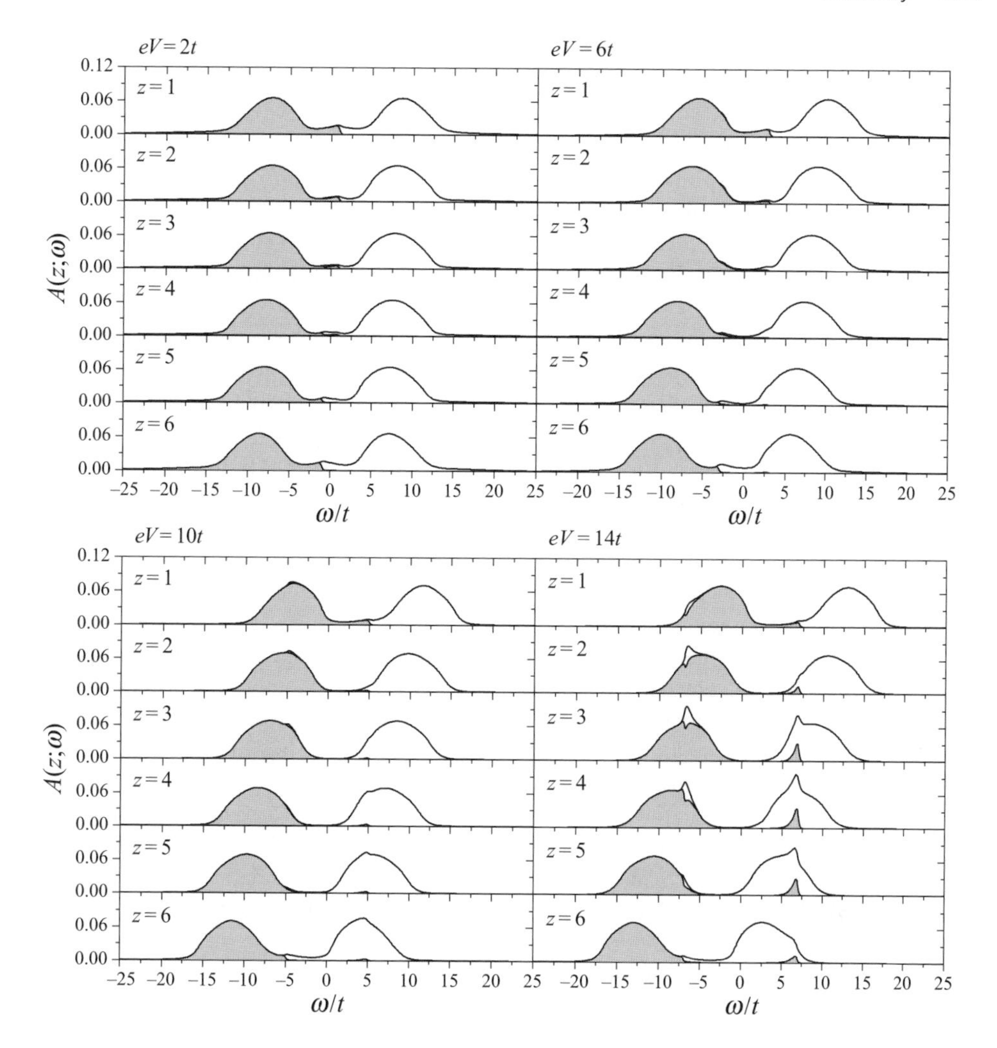

Fig. 8.15 Layer-resolved spectral function of electrons for an $N = 6$ Hubbard heterostructure with several choices of bias voltage indicated. On-site interaction is taken as $U = 15t$, and the potential is varied linearly. Shaded area shows the region occupied by electrons. Quasiparticle-like structures can be seen in $A(z; \omega)$ at $\omega = \pm eV/2$ corresponding to the chemical potentials of leads L and R.

8.6 Summary

In this chapter, we have reviewed recent topics in the field of oxide heterostructures. Recent developments by various groups, including both experiment and theory, are discussed briefly in the introduction. In the main part of the chapter, we have focused on the effect of strong correlations on the properties of such heterostructures.

In Section 8.2, we have described how to construct a model strongly correlated heterostructure. We also described how to analyze such a theoretical model by introducing the layer-extension of the dynamical-mean-field theory (DMFT) and auxiliary particle methods.

Section 8.3 focused on $[LaTiO_3]_m[SrTiO_3]_n$-type heterostructures. These systems are one of the cornerstones which have triggered an explosive research effort on oxide heterostructures. $[LaTiO_3]_m[SrTiO_3]_n$ heterostructures have two crucial issues: the strong screening effect coming from nearly-ferroelectric $SrTiO_3$ and the strong correlation effects in $LaTiO_3$, which is a Mott insulator in bulk. In order to address the issue of screening, we have performed a density-functional-theory calculation. We found that the strong ferroelectric-type lattice distortion indeed screens the electrostatic field from positively charged La ions and spread as the d electron density into the $SrTiO_3$ regions. By using the layer DMFT, we showed that an interface between a Mott insulator and a band insulator supports metallic behavior by charge redistribution, currently called *electronic reconstruction*.

In Section 8.4, we considered another strongly correlated electron system, high-T_c cuprates. Various experimental and theoretical studies have shown that the under-doped (UD) cuprates and the over-doped (OD) cuprates show distinct properties, especially, strong pairing tendency in the former and strong carrier coherency in the latter. Such reports made us speculate that superlattices (SLs), in which UD layers and OD layers alternate, can leverage both advantages, resulting in stronger superconductivity and higher T_c. We performed the layer-extension of the cellular DMFT and the dynamic-cluster approximation (DCA) for such SL's, and found that the proximity effect of the strong pairing scale can enhance the superconducting order parameter in the OD layers over the maximum value of the uniform bulk systems. Slave-boson mean-field and layer DCA calculations showed some hint that such SLs have higher critical temperature than the highest T_c of the bulk systems.

Section 8.5 was devoted to the future directions of strongly correlated heterostructures and was partly oriented to possible device applications. We discussed that there remain a lot of issues to overcome to realize oxide correlated electronic devices. Fundamental issues are that (interfacial) electronic properties are modified by a variety of perturbations including temperature and applied bias voltage. For double-exchange systems, the surface magnetic moment is rapidly suppressed with increasing temperature by the thermal spin fluctuation, and the surface moment is further reduced by degrading the surface. From our theoretical results, just making an ideal interface between manganite and an insulating barrier does not seem to improve the functionality of a manganite-based tunneling magnetoresistance junction. Instead, additional efforts may be necessary to suppress spin fluctuation. Using a Mott insulator as an insulating barrier for a tunneling junction may be possible as far as the induced quasiparticle band has negligible contribution to the linear conductance, this depends on the value of the Mott gap, the barrier thickness, and temperature. Another possibility would be utilizing collective behavior of strongly correlated systems, involving spin, orbital, charge orderings, and coupling with the lattice. Although such possibilities have not been fully theoretically explored, if they can generate highly nonlinear current–voltage characteristics, correlated-electron systems can be utilized as a variety of devices.

We hope that this article motivates both experimentalists and theorists, even outside this field, to start something new in the area of oxide heterostructures, and such efforts will lead to the realization of *oxide electronics*.

Acknowledgments

This article is based on several collaborations about strongly-correlated heterostructures. The author is grateful to his collaborators A. J. Millis, N. A. Spaldin, S. Yunoki, E. Dagotto, A. Fujimori, and T. A. Maier in the course of the research presented in this article. He also acknowledges very helpful conversations with H. Y. Hwang, A. Ohtomo, J. Chakhalian, H.-N. Lee, H. Kumigashira, M. Kawasaki, Y. Tokura, and D. J. Scalapino, and thanks A. J. Millis and E. Dagotto for critical reading of the manuscript. Work performed at Columbia University was supported by the Japan Society of Promotion of Science and NSF (Section 8.3). Work at Oak Ridge National Laboratory was supported by the Materials Sciences and Engineering Division, Office of Basic Energy Sciences, U.S. Department of Energy (Sections 8.4, 8.5). A portion of the research at ORNL was done at the Center for Nanophase Materials Sciences, sponsored by the Scientific User Facilities Division, Office of Basic Energy Sciences, U.S. Department of Energy. The author would like to thank the Kavli Institute for Theoretical Physics, the University of California, Santa Barbara, which is supported in part by the National Science Foundation under Grant No. PHY05-51164, for hospitality.

References

[1] Imada, M., Fujimori, A., and Tokura, Y. (1998). Metal-insulator transitions, Mod. Phys. **70**, 1039–1263.

[2] Tokura, Y. and Nagaosa, N. (2000). Orbital Physics in Transition-Metal Oxides, Science **288**, 462–468.

[3] Hesper, R., Tjeng, L. H., Heeres, A., and Sawatzky, G. A. (2000). Photoemission evidence of electronic stabilization of polar surfaces in K_3C_{60}, Phys. Rev. B **62**, 16046–16055.

[4] Moore, R. G., Zhang, J., Nascimento, V. B., *et al.* (2007). A Surface-Tailored, Purely Electronic, Mott Metal-to-Insulator Transition, Science **318**, 615–619.

[5] Potthoff, M. and Nolting, W. (1999). Surface metal-insulator transition in the Hubbard model, Phys. Rev. B **59**, 2549–2555.

[6] Potthoff, M. and Nolting, W. (1999). Metallic surface of a Mott insulator-Mott insulating surface of a metal, Phys. Rev. B **60**, 7834–7849.

[7] Schwieger, S., Potthoff, M., and Nolting, W. (2003). Correlation and surface effects in vanadium oxides, Phys. Rev. B **67**, 165408.

[8] Liebsch, A. (2003). Surface versus Bulk Coulomb Correlations in Photoemission Spectra of $SrVO_3$ and $CaVO_3$, Phys. Rev. Lett. **90**, 096401.

[9] Ohtomo, A., Muller, D. A., Grazul, J. L., and Hwang, H. Y. (2002). Artificial charge-modulation in atomic-scale perovskite titanate superlattices, Nature (London) **419**, 378–380.

[10] Takizawa, M., Wadati, H., Tanaka, K., *et al.* (2006). Photoemission from Buried Interfaces in $SrTiO_3/LaTiO_3$ Superlattices, Phys. Rev. Lett. **97**, 057601.

[11] Seo, S. S. A., Choi, W. S., Lee, H. N., *et al.* (2007). Photoemission from Buried Interfaces in $SrTiO_3/LaTiO_3$ Superlattices, Phys. Rev. Lett. **99**, 266801.

[12] Chahara, K., Ohno, T., Kasai, M., and Kozono, Y. (1993). Magnetoresistance in magnetic manganese oxide with intrinsic antiferromagnetic spin structure, Appl. Phys. Lett. **63**, 1990–1992.

[13] Tokura, Y., Urushibara, A., Moritomo, Y., *et al.* (1994). Giant Magnetotransport Phenomena in Filling-Controlled Kondo Lattice System: $La_{1-x}Sr_xMnO_3$, J. Phys. Soc. Jpn. **63**, 3931–3935.
[14] Izumi, M., Konishi, Y., Nishihara, T., *et al.* (1998). Atomically defined epitaxy and physical properties of strained $La_{0.6}Sr_{0.4}MnO_3$ films, Appl. Phys. Lett. **73**, 2497–2499.
[15] Konishi, Y., Fang, Z., Izumi, M., *et al.* (1999). Orbital-State-Mediated Phase-Control of Manganites, J. Phys. Soc. Jpn. **68**, 3790–3793.
[16] Koida, T., Lippmaa, M., Fukumura, T., *et al.* (2002). Effect of A-site cation ordering on the magnetoelectric properties in $[(LaMnO_3)_m/(SrMnO_3)_m]_n$ artificial superlattices, Phys. Rev. B **66**, 144418.
[17] Yamada, H., Kawasaki, M., Lottermoser, T., Arima, T., and Tokura, Y. (2006). $LaMnO_3/SrMnO_3$ interfaces with coupled charge-spin-orbital modulation, Appl. Phys. Lett. **89**, 052506.
[18] Smadici, S., Abbamonte, P., Bhattacharya, A., *et al.* (2007). Electronic Reconstruction at $SrMnO_3$-$LaMnO_3$ Superlattice Interfaces, Phys. Rev. Lett. **99**, 196404.
[19] Bhattacharya, A., May, S. J., and te Velthuis, S. G. E. (2008). Metal-Insulator Transition and Its Relation to Magnetic Structure in $(LaMnO_3)_{2n}/(SrMnO_3)_n$ Superlattices, Phys. Rev. Lett. **100**, 257203.
[20] May, S. J., Shah, A. B., and te Velthuis, S. G. E. (2008). Magnetically asymmetric interfaces in a $LaMnO_3/SrMnO_3$ superlattice due to structural asymmetries, Phys. Rev. B **77**, 174409.
[21] Adamo, C., Ke, X., Schiffer, P., *et al.* (2008). Electrical and magnetic properties of $(SrMnO_3)_n/(LaMnO_3)_{2n}$ superlattices, Appl. Phys. Lett. **92**, 112508.
[22] Aruta, C., Adamo, C., Galdi, A., *et al.* (2009). Evolution of magnetic phases and orbital occupation in $(SrMnO_3)_n/(LaMnO_3)_{2n}$ superlattices, Phys. Rev. B **80**, 140405(R).
[23] Yamada, H., Kawasaki, M., Ogawa, Y., and Tokura, Y. (2002). Perovskite oxide tricolor superlattices with artificially broken inversion symmetry by interface effects, Appl. Phys. Lett. **81**, 4793–4795.
[24] Yamada, H., Ogawa, Y., Ishii, Y., *et al.* (2004). Engineered Interface of Magnetic Oxides, Science **305**, 646–648.
[25] Izumi, M., Murakami, Y., Konishi, Y., *et al.* (1999). Structure characterization and magnetic properties of oxide superlattices $La_{0.6}sr_{0.4}MnO_3/La_{0.6}Sr_{0.4}FeO_3$, Phys. Rev. B **60**, 1211–1215.
[26] Takahashi, K. S., Kawasaki, M., and Tokura, Y. (2001). Interface ferromagnetism in oxide superlattices of $CaMnO_3/CaRuO_3$, Appl. Phys. Lett. **79**, 1324–1326.
[27] Yamada, H., Sato, H., Akoh, H., *et al.* (2008). Optical magnetoelectric effect at $CaRuO_3$-$CaMnO_3$ interfaces as a polar ferromagnet, Appl. Phys. Lett. **92**, 062508.
[28] Jullière, M. (1975). Tunneling between ferromagnetic films, Phys. Lett. **54A**, 225–226.
[29] Maekawa, S. and Gäfvert, U. (1982). Electron Tunneling Between Ferromagnetic Films, IEEE Trans. Magn. **18**, 707–708.
[30] Urushibara, A., Moritomo, Y., Arima, T., *et al.* (1995). Insulator-metal transition and giant magnetoresistance in $La_{1-x}Sr_xMnO_3$, Phys. Rev. B **51**, 14103.
[31] Bowen, M., Bibes, M., Barthélémy, A., *et al.* (2003). Nearly total spin polarization in $La_{2/3}Sr_{1/3}MnO_3$ from tunneling experiments, Appl. Phys. Lett. **82**, 233–235.

[32] Ogimoto, Y., Izumi, M., Sawa, A., *et al.* (2003). Tunneling Magnetoresistance above Room Temperature in $La_{0.7}Sr_{0.3}MnO_3/SrTiO_3/La_{0.7}Sr_{0.3}MnO_3$ Junctions, Jpn. J. Appl. Phys. **42**, L369–L372.
[33] Bednorz, J. G. and Müller, K. A. (1986). Possible High-T_c Superconductivity in the Ba-La-Cu-O System, Z. Physik, B **64**, 189–193.
[34] Bozovic, I., Eckstein, J. N., Virshup, G. F., *et al.* (1994). Atomic-layer engineering of cuprate superconductors, J. Supercond. **7**, 187–195.
[35] Sato, H. and Naito, M. (1997). Increase in the superconducting transition temperature by anisotropic strain effect in (001) $La_{1.85}Sr_{0.15}CuO_4$ thin films on $LaSrAlO_4$ substrates, Physica C **274**, 221–226.
[36] Locquet, J.-P., Perret, J., Fompeyrine, J., and Machler, E. (1998). Doubling the critical temperature of $La_{1.9}sr_{0.1}CuO_4$ using epitaxial strain, Nature (London) **394**, 453–456.
[37] Bozovic, I., Logvenov, G., Belca, I., Narimbetov, B., and Sveklo, I. (2002). Epitaxial Strain and Superconductivity in $La_{2-x}Sr_xCuO_4$ Thin Films, Phys. Rev. Lett. **89**, 107001.
[38] Bozovic, I., Logvenov, G., Verhoeven, M. A. J., *et al.* (2003). No mixing of superconductivity and antiferromagnetism in a high-temperature superconductor, Nature (London) **422**, 873–875.
[39] Bozovic, I., Logvenov, G., Verhoeven, *et al.* (2004). Giant Proximity Effect in Cuprate Superconductors, Phys. Rev. Lett. **93**, 157002.
[40] Asulin, O. I., Millo, O., Orgad, D., Iomin, L., and Koren, G. (2008). Enhancement of the Superconducting Transition Temperature of $La_{2-x}Sr_xCuO_4$ Bilayers: Role of Pairing and Phase Stiffness, Phys. Rev. Lett. **101**, 057005.
[41] Gozar, A., Logvenov, G., Fitting Kourkoutis, L., *et al.* (2008). High-temperature interface superconductivity between metallic and insulating copper oxides, Nature (London) **455**, 782–785.
[42] Smadici, S., Lee, J. C. T., Wang, S., *et al.* (2009). Superconducting Transition at 38 K in Insulating-Overdoped La_2CuO_4-$La_{1.64}Sr_{0.36}CuO_4$ Superlattices: Evidence for Interface Electronic Redistribution from Resonant Soft X-Ray Scattering, Phys. Rev. Lett. **102**, 107004.
[43] Logvenov, G., Gozar, A., and Bozovic, I. (2009). High-Temperature Superconductivity in a Single Copper-Oxygen Plane, Science **326**, 699–702.
[44] Sefrioui, Z., Arias, D., Pena, V., *et al.* (2003). Ferromagnetic/superconducting proximity effect in $La_{0.7}Ca_{0.3}MnO_3/YBa_2Cu_3O_{7-\delta}$ superlattices, Phys. Rev. B **67**, 214511.
[45] Varela, M., Lupini, A. R., Pennycook, S. J., Sefrioui, Z., and Santamaria, J. (2003). Nanoscale analysis of $YBa_2Cu_3O_{7-x}/La_{0.67}Ca_{0.33}MnO_3$ interfaces, Solid-State Electronics **47**, 2245–2248.
[46] Peña, V., Sefrioui, Z., Arias, D., *et al.* (2005). Giant Magnetoresistance in Ferromagnet/Superconductor Superlattices, Phys. Rev. Lett. **94**, 057002.
[47] Hoffmann, A., te Velthuis, S. G., Sefrioui, Z., *et al.* (2005). Suppressed magnetization in $La_{0.7}Ca_{0.3}MnO_3/YBa_2Cu_3O_{7-\delta}$ superlattices, Phys. Rev. B **72**, 140407.
[48] Chakhalian, J., Freeland, J. W., Srajer, G., *et al.* (2006). Magnetism at the interface between ferromagnetic and superconducting oxides, Nature Physics **2**, 244–248.
[49] Chakhalian, J., Freeland, J. W., Habermeier, H. U., *et al.* (2007). Orbital Reconstruction and Covalent Bonding at an Oxide Interface, Science **318**, 1114–1117.

[50] Nemes, N. M., García-Hernández, M., te Velthuis, S. G. E., *et al.* (2008). Origin of the inverse spin-switch behavior in manganite/cuprate/manganite trilayers, Phys. Rev. B **78**, 094515.
[51] Salafranca, J., and Okamoto, S. (2010). Unconventional proximity effect and inverse spin-switch behavior in a model manganite-cuprate-manganite trilayer system, Phys. Rev. Lett. **105**, 256804.
[52] Ueda, K., Tabata, H., and Kawai, T. (1998). Ferromagnetism in $LaFeO_3$-$LaCrO_3$ Superlattices, Science **280**, 1064–1066.
[53] Hotta, Y., Susaki, T., and Hwang, H. Y. (2007). Polar Discontinuity Doping of the $LaVO_3/SrTiO_3$ Interface, Phys. Rev. Lett. **99**, 236805.
[54] Chu, Y.-H., Martin, L. W., Holcomb, M. B., *et al.* (2009). Conduction at domain walls in oxide multiferroics, Nature Materials **8**, 229–234.
[55] Yu, P., Lee, J.-S., Okamoto, S., *et al.* (2010). Interface Ferromagnetism and Orbital Reconstruction in $BiFeO_3$-$La_{0.7}Sr_{0.3}MnO_3$ Heterostructure, Phys. Rev. Lett. **105**, 027201.
[56] Matzdorf, R., Fang, Z., Ismail, *et al.* (2000). All-Inorganic Field Effect Transistors Fabricated by Printing, Science **289**, 746–749.
[57] Wakabayashi, Y., Upton, M. H., Grenier, S., *et al.* (2007). Surface effects on the orbital order in the single-layered manganite $La_{0.5}Sr_{1.5}MnO_4$, Nature Materials **6**, 972–975.
[58] Ohtomo, A., and Hwang, H. Y. (2004). A high-mobility electron gas at the $LaAlO_3/SrTiO_3$ heterointerface, Nature (London) **427**, 423–426.
[59] Thiel, S., Hammerl, G., Schmehl, A., Schneider, C. W., and Mannhart, J. (2006). Tunable Quasi-Two-Dimensional Electron Gases in Oxide Heterostructures, Science **313**, 1942–1945.
[60] Reyren, N., Thiel, S., Caviglia, A. D., *et al.* (2007). Superconducting Interfaces Between Insulating Oxides, Science **317**, 1196–1199.
[61] Brinkman, A., Huijben, M., van Zalk, M., *et al.* (2007). Magnetic effects at the interface between non-magnetic oxides, Nature Materials **6**, 493–496.
[62] Caviglia, A. D., Gariglio, A., Reyren, N., *et al.* (2008). Electric field control of the $LaAlO_3/SrTiO_3$ interface ground state, Nature (London) **456**, 624–627.
[63] Cen, C., Thiel, S., Mannhart, J., and Levy, J. (2009). Oxide Nanoelectronics on Demand, Science **323**, 1026–1030.
[64] Potthoff, M. and Nolting, W. (1995). Surface magnetism studied within the mean-field approximation of the Hubbard model, Phys. Rev. B **52**, 15341–15354.
[65] Georges, A., Kotliar, B. G., Krauth, W., and Rozenberg, M. J. (1996). Dynamical mean-field theory of strongly correlated fermion systems and the limit of infinite dimensions, Rev. Mod. Phys. **68**, 13–125.
[66] Freericks, J. K. (2004). Dynamical mean-field theory for strongly correlated inhomogeneous multilayered nanostructures, Phys. Rev. B **70**, 195342.
[67] Helmes, R. W., Costi, T. A., and Rosch, A. (2008). Kondo Proximity Effect: How Does a Metal Penetrate into a Mott Insulator? Phys. Rev. Lett. **101**, 066802.
[68] Okamoto, S. and Millis, A. J. (2004). Electronic reconstruction at an interface between a Mott insulator and a band insulator, Nature (London) **428**, 630–633.
[69] Okamoto, S. and Millis, A. J. (2004). Theory of Mott insulator-band insulator heterostructures, Phys. Rev. B **70**, 075101.

[70] Okamoto, S., Millis, A. J., and Spaldin, N. A. (2006). Lattice Relaxation in Oxide Heterostructures: $LaTiO_3/SrTiO_3$ Superlattices, Phys. Rev. Lett. **97**, 056802.

[71] Okamoto, S. and Millis, A. J. (2004). Spatial inhomogeneity and strong correlation physics: A dynamical mean-field study of a model Mott-insulator-band-insulator heterostructure, Phys. Rev. B **70**, 241104(R).

[72] Hamann, D. R., Muller, D. A., and Hwang, H. Y. (2006). Lattice-polarization effects on electron-gas charge densities in ionic superlattices, Phys. Rev. B **73**, 195403.

[73] Lee, W.-C. and MacDonald, A. H. (2006). Modulation doping near Mott-insulator heterojunctions, Phys. Rev. B **74**, 075106.

[74] Lee, W.-C. and MacDonald, A. H. (2007). Electronic interface reconstruction at polar-nonpolar Mott-insulator heterojunctions, **76**, 075339.

[75] Okamoto, S. and Maier, T. A. (2008). Enhanced Superconductivity in Superlattices of High-T_c Cuprates, Phys. Rev. Lett. **101**, 156401.

[76] Okamoto, S. (2008). Nonlinear Transport through Strongly Correlated Two-Terminal Heterostructures: A Dynamical Mean-Field Approach, Phys. Rev. Lett. **101**, 116807.

[77] Okamoto, S. (2009). Surface magnetic phase transition of the double-exchange ferromagnet: Schwinger-boson mean-field study, J. Phys.: Condens. Matter **21**, 355601.

[78] Calderón, M. J., Brey, L., and Guinea, F. (1999). Surface electronic structure and magnetic properties of doped manganites, Phys. Rev. B **60**, 6698–6704.

[79] Brey, L. (2007). Electronic phase separation in manganite-insulator interfaces, Phys. Rev. B **75**, 104423.

[80] Lin, C. and Millis, A. J. (2008). Theory of manganite superlattices, Phys. Rev. B **78**, 184405.

[81] Dong, S., Yu, R., Yunoki, S., *et al.* (2008). Magnetism, conductivity, and orbital order in $(LaMnO_3)_{2n}/(SrMnO_3)_n$ superlattices, Phys. Rev. B **78**, 201102.

[82] Nanda, B. R. K. and Satpathy, S. (2008). Effects of strain on orbital ordering and magnetism at perovskite oxide interfaces: $LaMnO_3/SrMnO_3$, Phys. Rev. B **78**, 054427.

[83] Nanda, B. R. K. and Satpathy, S. (2009). Electronic and magnetic structure of the $(LaMnO_3)_{2n}/(SrMnO_3)_n$ superlattices, **79**, 054428.

[84] Fang, Z., Solovyev, I. V., and Terakura, K. (2000). Phase Diagram of Tetragonal Manganites, Phys. Rev. Lett. **84**, 3169–3172.

[85] Popovic, Z. S. and Satpathy, S. (2005). Wedge-Shaped Potential and Airy-Function Electron Localization in Oxide Superlattices, Phys. Rev. Lett. **94**, 176805.

[86] Pentcheva, R. and Pickett, W. E. (2007). Correlation-Driven Charge Order at the Interface between a Mott and a Band Insulator, Phys. Rev. Lett. **99**, 016802.

[87] Nanda, B. R. K., Satpathy, S., and Springborg, M. S. (2007). Electron Leakage and Double-Exchange Ferromagnetism at the Interface between a Metal and an Antiferromagnetic Insulator: $CaRuO_3/CaMnO_3$, Phys. Rev. Lett. **98**, 216804.

[88] Pentcheva, R. and Pickett, W. E. (2006). Charge localization or itineracy at $LaAlO_3/SrTiO_3$ interfaces: Hole polarons, oxygen vacancies, and mobile electrons, Phys. Rev. B **74**, 035112.

[89] Pardo, V. and Pickett, W. E. (2009). Charge localization or itineracy at $LaAlO_3/SrTiO_3$ interfaces: Hole polarons, oxygen vacancies, and mobile electrons, Phys. Rev. Lett. **102**, 166803.

[90] Banerjee, S., Singh R. R., Pardo, V., and Pickett, W. E. (2009). Tight-Binding Modeling and Low-Energy Behavior of the Semi-Dirac Point, Phys. Rev. Lett. **103**, 016402.

[91] Sugano, S., Tanabe, Y., and Kamimura, H. (1970). Multiplets of Transition-Metal Ions in Crystals, Academic, New York.
[92] Kubo, K. and Ohata, N. (1972). A Quantum Theory of Double Exchange. I, J. Phys. Soc. Jpn. **33**, 21–32.
[93] Furukawa, N. (1994). Transport Properties of the Kondo Lattice Model in the Limit $S = \infty$ and $D = \infty$, J. Phys. Soc. Jpn. **63**, 3214–3217.
[94] Furukawa, N. (1995). Magnetoresistance of the Double-Exchange Model in Infinite Dimension, **64**, 2734–2737.
[95] Furukawa, N. (1995). Magnetic Transition Temperature of (La,Sr)MnO_3, **64**, 2754–2757.
[96] Furukawa, N. (1995). Temperature Dependence of Conductivity in (La, Sr)MnO_3, **64**, 3164.
[97] Kasuya, T. (1956). A Theory of Metallic Ferro- and Anti-ferromagnetism on Zener's Model, Prog. Theor. Phys. **16**. 45–47.
[98] Aryasetiawan, F., Imada, M., Georges, A., *et al.* (2004). Frequency-dependent local interactions and low-energy effective models from electronic structure calculations, Phys. Rev. B **70**, 195104.
[99] Millis, A. J., Shraiman, B. I., and Mueller, R. (1996). Dynamic Jahn-Teller Effect and Colossal Magnetoresistance in $La_{1-x}Sr_xMnO_3$, Phys. Rev. Lett. **77**, 175–178.
[100] Kim, B. J., Jin, H., Moon, S. J., *et al.* (2008). Novel $J_{eff} = 1/2$ Mott State Induced by Relativistic Spin-Orbit Coupling in Sr_2IrO_4, Phys. Rev. Lett. **101**, 076402.
[101] Jackeli, G. and Khaliullin, G. (2009). Mott Insulators in the Strong Spin-Orbit Coupling Limit: From Heisenberg to a Quantum Compass and Kitaev Models, Phys. Rev. Lett. **102**, 017205.
[102] Yunoki, S., Moreo, A., Dagotto, E., *et al.* (2007). Electron doping of cuprates via interfaces with manganites, Phys. Rev. B **76**, 064532.
[103] Nakagawa, N., Hwang, H. Y., and Muller, D. A. (2006). Why some interfaces cannot be sharp, Nature Materials **5**, 204–209.
[104] Freericks, J. K., Nikolić B. K., and Miller, P. (2001). Tuning a Josephson junction through a quantum critical point, Phys. Rev. B **64**, 054511.
[105] Fleck, M., Lichtenstein, A. I., Pavarini, E., and Oleś, A. M. (2000). One-Dimensional Metallic Behavior of the Stripe Phase in $La_{2-x}Sr_xCuO_4$, Phys. Rev. Lett. **84**, 4962–4965.
[106] Lin, C., Okamoto, S., and Millis, A. J. (2006). Dynamical mean-field study of model double-exchange superlattices, Phys. Rev. B **73**, 041104(R).
[107] Kotliar, G., Savrasov, S. Y., Pálsson, G., and Biroli, G. (2001). Cellular Dynamical Mean Field Approach to Strongly Correlated Systems, Phys. Rev. Lett. **87**, 186401.
[108] Kotliar, G., Savrasov, S. Y., Haule, K., *et al.* (2006). Electronic structure calculations with dynamical mean-field theory, Rev. Mod. Phys. **78**, 865–951.
[109] Lichtenstein, A. I. and Katsnelson, M. I. (2000). Antiferromagnetism and d-wave superconductivity in cuprates: A cluster dynamical mean-field theory, Phys. Rev. B **62**, R9283–R9286.
[110] Hettler, M. H., Tahvildar-Zadeh, A. N., Jarrell, M., Pruschke, T., and Krishnamurthy, H. R. (1998). Nonlocal dynamical correlations of strongly interacting electron systems, Phys. Rev. B **58**, R7475–R7479.

[111] Maier, T. A., Jarrell, M., Pruschke, T., and Hettler, M. H. (2005). Quantum cluster theories, Rev. Mod. Phys. **77**, 1027–1080.
[112] Kancharla, S. S., Kyung, B., Sénéchal, D., *et al.* (2008). Anomalous superconductivity and its competition with antiferromagnetism in doped Mott insulators, Phys. Rev. B **77**, 184516.
[113] Maier, Th., Jarrell, M., Pruschke, Th., and Keller J. (2000). d-Wave Superconductivity in the Hubbard Model, Phys. Rev. Lett. **85**, 1524–1527.
[114] Barnes, S. E. (1976). New method for the Anderson model, J. Phys. F **6**, 1375–1383.
[115] Barnes, S. E. (1977). New method for the Anderson model: II. The $U = 0$ limit, J. Phys. F **7**, 2637–2647.
[116] Baskaran, G., Zou, Z., and Anderson, P. W. (1987). The resonating valence bond state and high-T_c superconductivity–A mean field theory, Solid State Commun. **63**, 973–976.
[117] Kotliar, G. and Liu, J. (1988). Superexchange mechanism and d-wave superconductivity, Phys. Rev. B **38**, 5142–5145.
[118] Arovas, D. P. and Auerbach, A. (1988). Functional integral theories of low-dimensional quantum Heisenberg models, Phys. Rev. B **38**, 316–332.
[119] Yoshioka, D. (1989). Slave-Fermion Mean Field Theory of the Hubbard Model, J. Phys. Soc. Jpn. **58**, 1516–1519.
[120] Kotliar, G. and Ruckenstein, A. E. (1986). New Functional Integral Approach to Strongly Correlated Fermi Systems: The Gutzwiller Approximation as a Saddle Point, Phys. Rev. Lett. **57**, 1362–1365.
[121] Rüegg, A., Pilgram, S., and Sigrist, M. (2007). Strongly renormalized quasi-two-dimensional electron gas in a heterostructure with correlation effects, Phys. Rev. B **75**, 195117.
[122] Sarker, S. K. (1996). Phase transition in the double-exchange model: a Schwinger boson approach, J. Phys.: Condens. Matter **8**, L515–L521.
[123] Yunoki, S., Hu, J., Malvezzi, A. L., *et al.* (1998). Phase Separation in Electronic Models for Manganites, Phys. Rev. Lett. **80**, 845–848.
[124] Sakudo, T. and Unoki, H. (1971). Dielectric Properties of $SrTiO_3$ at Low Temperatures, Phys. Rev. Lett. **26**, 851–853.
[125] Müller, K. A. and Burkard, H. (1973). $SrTiO_3$: An intrinsic quantum paraelectric below 4 K, Phys. Rev. B **19**, 3593–3602.
[126] Blochl, P. E. (1994). Projector augmented-wave method, Phys. Rev. B **50**, 17953–17979.
[127] Kresse, G. and Furthmuller, J. (1996). Efficient iterative schemes for *ab initio* total-energy calculations using a plane-wave basis set, Phys. Rev. B **54**, 11169–11186.
[128] Kresse, G. and Joubert, D. (1999). From ultrasoft pseudopotentials to the projector augmented-wave method, Phys. Rev. B **59**, 1758–1775.
[129] Liechtenstein, A. I., Anisimov, V. I., and Zaanen, J. (1995). Density-functional theory and strong interactions: Orbital ordering in Mott-Hubbard insulators, Phys. Rev. B **52**, R5467–R5470.
[130] Sai, N., Kolpak, A. M., and Rappe, A. M. (2005). Slow polarization relaxation in water observed by hyper-Rayleigh scattering, Phys. Rev. B **72**, 020101(R).
[131] Potthoff, M. (2001). Two-site dynamical mean-field theory, Phys. Rev. B **64**, 165114.
[132] Okamoto, S. and Millis, A. J. (2005). Electronic reconstruction in correlated electron heterostructures, Proc. SPIE Int. Soc. Opt. Eng. **5932**, 593218.

[133] Ishida, H. and Liebsch, A. (2008). Origin of metallicity of $LaTiO_3/SrTiO_3$ heterostructures, Phys. Rev. B **77**, 115350.

[134] Gomes, K. K., Pasupathy, A. N., Pushp, A., *et al.* (2007). Visualizing pair formation on the atomic scale in the high-Tc superconductor $Bi_2Sr_2CaCu_2O_{8+\delta}$, Nature (London) **447**, 569–572.

[135] Maier, T. A., Jarrell, M., and Scalapino, D. J. (2006). Pairing interaction in the two-dimensional Hubbard model studied with a dynamic cluster quantum Monte Carlo approximation, Phys. Rev. B **74**, 094513.

[136] Hayashi, M. and Ebisawa, H. (2003). Novel boundary effect in high temperature superconductors: superconductivity induced by pseudogap proximity effect, Physica C **392-396**, 48–52.

[137] Ginzburg, V. L. (2005). A few comments on superconductivity research, Physics-Uspekhi **48**, 173–176.

[138] Berg, E., Orgad, D., and Kivelson, S. (2008). Route to high-temperature superconductivity in composite systems, Phys. Rev. B **78**, 094509.

[139] Kivelson, S. A. and Fradkin, E. (2007). How optimal inhomogeneity produces high temperature superconductivity. In: Schrieffer, J. R. and Brooks, J. ed. *Treatise of High Temperature Sureconductivity*, chapter 15, Springer.

[140] Tsai, W.-F. and Kivelson, S. A. (2006). Superconductivity in inhomogeneous Hubbard models, Phys. Rev. B **73**, 214510; (2007). **76**, 139902(E).

[141] Tsai, W.-F., Yao, H., Läuchli, A., and Kivelson, S. A. (2008). Optimal inhomogeneity for superconductivity: Finite-size studies, Phys. Rev. B **77**, 214502.

[142] Doluweera, D. G. S. P., Jarrell, M., Maier, T. A., Macridin, A., and Pruschke, T. (2008). Suppression of *d*-wave superconductivity in the checkerboard Hubbard model, Phys. Rev. B **78**, 020504.

[143] Caffarel, M. and Krauth, W. (1994). Exact diagonalization approach to correlated fermions in infinite dimensions: Mott transition and superconductivity, Phys. Rev. Lett. **72**, 1545–1548.

[144] Hilgenkamp, H. and Mannhart, J. (2002). Grain boundaries in high-T_c superconductors, Rev. Mod. Phys. **74**, 485–549.

[145] Park, J, H., Vescovo, E., Kim, H. J., *et al.* (1998). Magnetic Properties at Surface Boundary of a Half-Metallic Ferromagnet $La_{0.7}Sr_{0.3}MnO_3$, Phys. Rev. Lett. **81**, 1953–1953.

[146] Binder, K. and Hohenberg, P. C. (1974). Surface effects on magnetic phase transitions, Phys. Rev. B **9**, 2194–2214.

[147] Arovas, D. P. and Guinea, F. (1998). Some aspects of the phase diagram of double-exchange systems, Phys. Rev. B **58**, 9150–9155.

[148] MacLaren, J. M., Zhang, X. G., and Butler, W. H. (1997). Validity of the Julliere model of spin-dependent tunneling, Phys. Rev. B **56**, 11827–11832.

[149] Garcia-Barriocanal, J., Cezar, J. C., Bruno, F. Y., *et al.* (2010). Spin and orbital Ti magnetism at $LaMnO_3/SrTiO_3$ interfaces, Nat. Commun. 1:82 doi: 10.1038/ncomms1080.

[150] Keldysh, L. V. (1965). Diagram Technique for Nonequilibrium Processes, Sov. Phys. JETP **20**, 1018–1026.

[151] Okamoto, S. (2007). Nonequilibrium transport and optical properties of model metal-Mott-insulator-metal heterostructures, Phys. Rev. B **76**, 035105.

[152] Pruschke, T. and Grewe, N. (1989). The Anderson model with finite Coulomb repulsion, Z. Phys. B **74**, 439–449.

[153] Pruschke, T., Cox, D. L., and Jarrell, M. (1993). Hubbard model at infinite dimensions: Thermodynamic and transport properties, Phys. Rev. B **47**, 3553–3565.

[154] Meir, Y., Wingreen, N. S., and Lee, P. A. (1993). Low-temperature transport through a quantum dot: The Anderson model out of equilibrium, Phys. Rev. Lett. **70**, 2601–2604.

9 Manganite multilayers

Anand Bhattacharya, Shuai Dong, and Rong Yu

9.1 Motivation

Historically, the successful synthesis of semiconductor heterostructures and superlattices triggered rapid developments in semiconductor physics, leading to a wide range of devices and fundamental discoveries in the field of low-dimensional quantum systems. In recent years, developments in the science of complex oxide heterostructures have aroused considerable interest. As discussed in the foregoing chapters, these heterostructures not only provide the possibility of fabricating new multifunctional devices, but also challenge our current understanding of how properties arising from strong correlations may manifest themselves at interfaces and be tuned with external fields [1, 2, 3]. In semiconductor heterostructures, electron correlations are relatively weak and hence the central problem can usually be reduced to modeling a single quantum mechanical particle moving through a potential landscape. The broken translational symmetry at the interface results in charge redistribution across it and the bending of the bands in its vicinity. In complex oxide heterostructures, these considerations are more challenging to understand, but in return may lead to a richer and more diverse set of phenomena. In fact, since strong correlations between electrons are very important in these systems, the band structure is often not rigid—it changes significantly with charge density and must be treated self-consistently. This alters how we view the ideas of charge transfer and band lineup within the non-interacting picture used to model semiconductors. In addition, complex oxides have much richer phase diagrams, including competing ground states that are close in their free energies but with strikingly different physical properties, such as superconducting versus Mott insulating phases in cuprates and ferromagnetic (FM) metallic versus charge-ordered (CO) insulating phases in manganites. In heterostructures, these competing phases are sensitive not only to external (magnetic/electric) fields, but also interfacial charge transfer, proximity effects, and substrate strain/stress. Thus, a better understanding of these systems may enable us to realize novel collective phases in the vicinity of an interface and to tune them by applying external fields, which is the definitive challenge of this area of research.

Among various complex oxide heterostructures, those made of manganites are especially interesting. These materials have been extensively studied for the wide variety of phases that arise as a result of the charge, spin, orbital, and lattice degrees of freedom being simultaneously at play. As is well known from analogous studies in thin films, lattice mismatch in epitaxially strained manganite heterostructures will bring about subtle changes in bond lengths and bond angles, which may result in dramatic changes in magnetism and transport behaviors, tipping the balance between competing phases. On the other hand, we would also like to understand how the transfer of charge across such interfaces may affect magnetism and orbital order, and how this in turn may affect electronic states and transport properties in the vicinity of such interfaces. We may then be able create interfaces where the different degrees of freedom can "reconstruct" in unique ways to give rise to novel collective states that may not occur in single-phase materials, such as an interfacial spin-polarized two-dimensional electron gas or a heterostructure with multiferroic properties. It may even be possible to tune these interfacial states through phase transitions (e.g. a metal–insulator transition) with gate electric fields, opening up a number of novel possibilities.

The organization of this chapter is as follows. In the introductory sections, we follow a brief description of the manganites with a discussion of theoretical models of manganite multilayers and methods for their synthesis and characterization. We then describe recent progress of manganites multilayers, including the effects of strain, cation ordering on the A-site, phase tuning via charge transfer, metal–insulator transitions, cation-ordered analogs, the consequences of band lineup, magnetic interactions at interfaces etc. In the end, a brief conclusion and outlook are presented.

9.2 Introduction to manganites

In this section, we give a brief introduction to manganites, especially $LaMnO_3$ (LMO), $SrMnO_3$ (SMO), and $La_{1-x}Sr_xMnO_3$ (LSMO). The $R_{1-x}A_xMO_3$ series form a perovskite structure, where rare earth R^{3+} and alkaline earth A^{2+} cations occupy the A-site, while Mn cations occupy the B-site, which is octahedrally coordinated with oxygen. After their discovery in the 1950s [4, 5], the study of manganites saw a resurgence due to the discovery of colossal magnetoresistance (CMR) in the early 1990s [6, 7], where it was found that the resistivity of thin films of these materials could be reduced by orders of magnitude upon application of a magnetic field, an effect that could be useful for magnetic sensors or storage media. However, these applications have not been realized, partially due to the fact that CMR usually occurs below ambient temperature, and needs relatively high fields of a few Tesla. Much of the later interest in this field was driven by fundamental interest in CMR phenomena, including ideas of electronic phase separation that could play an important role in many manganites [8, 9, 10].

LSMO is a wide-bandwidth manganite whose phase diagram has been studied in some detail, by a number of different groups, in bulk samples through the entire doping range [11, 12, 13]. The end-members are antiferromagnetic (AF) insulators (AFI). LMO with nominally Mn^{3+} $t_{2g}^3e_g^1$ occupancy (strongly hybridized with the O $2p$ states) is an AFI with strong Mott–Hubbard/charge-transfer Coulomb correlations in a half-filled e_g band [14, 15]. The Mn^{3+} sites are also Jahn–Teller active, which leads to an orthorhombic structure for bulk LMO at high temperatures and an A-type orbital-ordered (OO) layered antiferromagnet (ferromagnetic in plane, antiferromagnetic between planes) with a Néel temperature $T_N \sim$ 140 K [5]. At the other end, SMO is a more band-like insulator with a high-spin $Mn^{4+}t_{2g}^3e_g^0$ configuration. Bulk $SrMnO_{3-\delta}$ can be stabilized in hexagonal, orthorhombic, or cubic phases, depending upon oxygen content and temperature [16]. In the cubic perovskite phase, SMO has

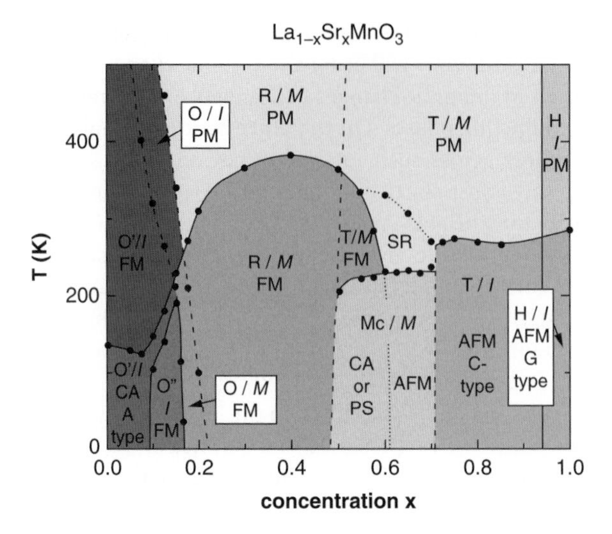

Fig. 9.1 Phase diagram for $La_{1-x}Sr_xMnO_3$. The crystal structures are Jahn–Teller distorted orthorhombic: O′; orthorhombic: O; orbital-ordered orthorhombic: O″; rhombohedral: R; tetragonal: T; monoclinic: MC; and Hexagonal: H. The magnetic structures are paramagnetic: (PM); short-range order (SR); canted (CA); *A*-type layered antiferromagnetic structure: (AFM); ferromagnetic: (FM); phase separated (PS); antiferromagnetic chain-type: AFM C-type. The electronic states are insulating: I; and metallic: M. Reproduced from Ref. [13]. Copyright 2002, American Physical Society.

a G-type antiferromagnetic (G-AF) order with $T_N \sim 240$ K. In LSMO, the A-site is populated randomly by La^{3+} and Sr^{2+} cations, allowing the Mn cation on the B-site to have a "mixed-valence" of Mn^{3+}/Mn^{4+}. As a result, several phases with a variety of magnetic and orbital ordering and crystal structures emerge as a function of doping, including a double-exchange (DE) mediated FM metal [17], and a number of antiferromagnets with different symmetries and orbital ordering, as summarized in Figure 9.1. For a more detailed description of these phases and their properties, we refer the reader to a number of comprehensive reviews [18, 19].

9.3 Theoretical description of manganite multilayers

In past years, several theoretical investigations, including both model simulations and density functional theory (DFT) calculations, have been performed on manganite heterostructures. Here, we briefly introduce the microscopic model for manganite heterostructures, which is the common kernel of most theoretical studies. Even for the DFT calculations which do not rely on model parameters, a proper model remains helpful to understand the physics behind the DFT results.

In perovskites, each Mn cation is surrounded by an oxygen octahedron, which splits Mn's 3*d* energy levels into two groups: a triplet t_{2g} (d_{xy}, d_{yz}, and d_{xz}) and a doublet e_g orbital ($d_{x^2-y^2}$ and $d_{3z^2-r^2}$). The energies of e_g orbitals are higher than t_{2g}'s because the spatial distribution of e_g electrons is closer to the oxygen anions which results in a stronger Coulomb repulsion. A strong

Hund coupling aligns all of the 3*d* spins, giving rise to a high-spin state. For $La_{1-x}Sr_xMnO_3$, the e_g charge density equals 1-*x* and the Fermi level is located around the e_g's spin-up band. The t_{2g}'s spin-up bands and oxygen's 2*p* bands are fully occupied, while La's 5*d*, 4*f*, Sr's 4*d* bands and Mn's 3*d* spin-down bands are all empty according to low temperature experimental data and DFT results. The t_{2g} electrons are not affected by doping and can be approximately treated as localized spins. Hence whether the material is metallic or insulating is determined by the mobility of e_g electrons. Strong Hund coupling ($J_H \sim 2$ eV) between the t_{2g} and e_g electrons aligns the spin of e_g electron to be parallel to the ones of the on-site t_{2g} electrons.

The double-exchange (DE) process of e_g electrons is essential to understanding the physics of manganites. It was first proposed by Zener in 1951 to explain qualitatively the coupling between ferromagnetism and conductivity [20, 21]. For perovskite manganites, the double-exchange process is sketched simply in Figure 9.2(a). When the t_{2g} spins of Mn cations are parallel, the e_g electron can hop from one Mn^{3+} to its nearest-neighbor (NN) Mn^{4+} via the intermediate oxygen anion. When the nearest-neighbor Mn spins are antiparallel, the large Hund coupling prohibits hopping of the e_g electron, as shown in Figure 9.2(b). For a finite angle φ between the core spins, the transfer integral $\sim \cos(\varphi/2)$, a consequence of the manner in which the amplitudes for a spin 1/2 particle change upon rotation [22]. When there are two orbitals involved, the hopping integrals are both orbital- and direction-dependent, as shown in Figure 9.2(c) [23]. The intensity of double-exchange hopping *t* is roughly estimated as 0.2–0.5 eV. Thus, the second-quantized Hamiltonian for the double-exchange process can be written as:

$$H_{DE} = -\sum_{<ij>,\gamma\gamma',\sigma} t^{\vec{r}}_{\gamma\gamma'}(c^{+}_{i\gamma\sigma}c_{j\gamma'\sigma} + H.c.) - J_H \sum_{i,\gamma} c^{+}_{i\gamma}\sigma c_{i\gamma} \cdot \vec{S}_i, \tag{9.1}$$

where *i* and *j* denote NN sites; γ and γ' run over orbitals *a* and *b*; σ are the Pauli matrices; and c^+ (c) is the creation (annihilation) operator of e_g electrons. $\vec{S}$ is the t_{2g} total spin which is large enough ($|\vec{S}| = 3/2$) to be considered as a classical vector. In the following, the normalized

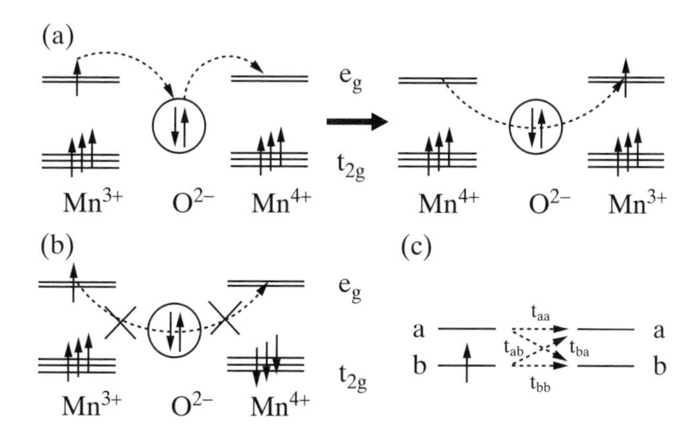

Fig. 9.2 Schematic representation of the double-exchange process. (a) The e_g electron can hop from a Mn^{3+} to its nearest-neighbor Mn^{4+} if their t_{2g} spins are parallel. (b) The hopping of the e_g electron is prohibited if the nearest-neighbor spins are antiparallel. (c) The hopping amplitude is orbital dependent.

$|\vec{S}| = 1$ will be used if not noted explicitly. Usually, this double-exchange Hamiltonian may further be simplified by assuming the Hund coupling (roughly estimated to be 2 eV [9, 10], much larger than t_0) to be infinite. In the infinite J_H limit, each e_g electron spin σ must be perfectly parallel to the on-site t_{2g} classical $\vec{S}$, while the antiparallel sector is discarded. Besides the double-exchange process that contributes to the kinetic energy of e_g electrons, there is an antiferromagnetic superexchange (SE) coupling between the neighboring localized t_{2g} spins which can be described using the Heisenberg model, $H_{SE} = J_{AF} \sum_{<ij>} \vec{S}_i \cdot \vec{S}_j$ with J_{AF} as the superexchange coefficient whose value is about $\sim 0.1t_0$ [9, 10].

In addition to the exchange interactions, the electron–lattice coupling also plays a key role in manganites, especially for charge-ordered phases [24]. By defining three distortion modes (Q_1, Q_2, Q_3), the electron–lattice contribution is written as [25]:

$$H_P = \lambda \sum_i (-Q_{1,i}n_i + Q_{2,i}\tau_i^x + Q_{3,i}\tau_i^z) + \frac{1}{2}\sum_i (BQ_{1,i}^2 + Q_{2,i}^2 + Q_{3,i}^2), \tag{9.2}$$

where λ is the dimensionless electron–phonon coupling coefficient; n_i is the e_g electron density operator at site i; τ^x and τ^z are the orbital (pseudo-spin) operators: $\tau^x = \sum_\sigma (c_{a,\sigma}^+ c_{b,\sigma} + c_{b,\sigma}^+ c_{a,\sigma})$ and $\tau^z = \sum_\sigma (c_{a,\sigma}^+ c_{a,\sigma} - c_{b,\sigma}^+ c_{b,\sigma})$. The first term is related to a Jahn–Teller-like distortion that can split the degenerate e_g levels. The second term of (9.2) is the elastic energy of lattice distortions, where only the harmonic terms have been taken into account. B is the stiffness ratio between the breathing mode (Q_1) and Jahn–Teller modes (Q_2 and Q_3), and usually $1 < B < 2$.

The above model Hamiltonians ($H = H_{DE} + H_{SE} + H_P$) have been widely used to study the phase diagram of bulk manganites and repeatedly confirmed them to be accurate. In this Hamiltonian, the inter- and intra-orbital on-site Coulomb repulsions have been neglected, these effects being replaced by the Hund coupling and Jahn–Teller distortion [26]. In some cases, further simplifications were made by either discarding certain terms in the above Hamiltonian, such as the superexchange interaction or electron–phonon coupling, or adopting the one-orbital approximation instead of the more realistic two-orbital one. Also, in some cases, other effects should be taken into account, such as the A-site cation disorder (to be discussed in detail in the Section 9.5).

In this chapter, the above double-exchange model Hamiltonian will be extended to study manganite superlattices/heterostructures, e.g. $RMnO_3/AMnO_3$. A key difference from the bulk case is that the electron density may be non-uniform in heterostructures, e.g. higher at the LMO side and lower at the SMO side. This leads to a charge redistribution across the interface, and the resulting long-range Coulomb interaction must be taken into account. This contains contributions from both the A-site cations (either R^{3+} or A^{2+} cations) and e_g electrons:

$$H_C = \frac{Ke^2}{\varepsilon}\left(-\sum_{i,k}\frac{n_i n_k^A}{|r_i - r_k|} + \frac{1}{2}\sum_{i\neq j}\frac{n_i n_j}{|r_i - r_j|}\right), \tag{9.3}$$

where n_k^A denotes the positive charge (in units of the elementary charge e) at A-site k, while n_i is the e_g electron number operator at Mn site i; K is Coulomb's constant and ε is the dielectric constant. Here, only the Coulomb interactions concerning the e_g electrons have been considered. In practice this interaction is usually handled within the mean-field approximation, i.e. by replacing $n_i n_j$ with $n_i < n_j > + < n_i > n_j - < n_i >< n_j >$ where $<>$ denotes the expectation value. Thus, (9.3) can be simplified as:

$$
\begin{aligned}
H_C &\approx \frac{Ke^2}{\varepsilon}\Big[\sum_i \Big(-\sum_k \frac{n_k^A}{|r_i - r_k|} + \sum_{i \neq j} \frac{< n_j >}{|r_i - r_j|}\Big) n_i - \frac{1}{2} \sum_{i \neq j} \frac{< n_i >< n_j >}{|r_i - r_j|}\Big], \\
&= \sum_i \phi_i n_i - \frac{Ke^2}{2\varepsilon} \sum_{i \neq j} \frac{< n_i >< n_j >}{|r_i - r_j|} \qquad (9.4)
\end{aligned}
$$

where ϕ_i is an effective Coulomb potential, that can be solved self-consistently with $< n_i >$.

To solve the above model Hamiltonian the fermionic sector including double-exchange, electron–lattice coupling, and Coulomb interactions, should be diagonalized. To handle the remaining sectors in the Hamiltonian, various numerical techniques can be used. Monte Carlo (MC) simulations or dynamic mean field theory (DMFT) can be adopted to study the properties at finite temperatures. For details of the MC technique used here, readers are referred to Chapter 7 of [10]. For the DMFT method, readers are referred to [27].

Besides these model calculations, DFT calculations which are independent of model details are also very useful to explore the ground states of manganite superlattices. Complementary to the MC simulation and DMFT, a zero-temperature (zero-T) relaxation method can be implemented in both model studies and DFT calculations to optimize crystal structures, distortion modes, and spin configurations in order to reach the real ground state. In addition, by directly comparing the energies of several candidate states, variational methods can be implemented in both model and DFT studies, which while being relatively simple remain very helpful.

9.4 Synthesis and structure of manganite multilayers

Superlattices and heterostructures incorporating manganites have been synthesized by a number of groups using both Pulsed Laser Deposition (PLD) and Molecular Beam Epitaxy (MBE) techniques. In PLD, films are deposited onto a substrate by ablating stoichiometric ceramic targets of the constituent materials using focused laser pulses in an oxidizing environment. Parameters such as the pressure of the oxidizing gas (typically oxygen, but also activated forms such as O_3 and NO_2), substrate temperature (anywhere between 500 °C and 800 °C), energy density of a laser pulse (in the range 0.5–3 J/cm^2), wait times between laser pulses and post-growth anneals in oxidizing atmospheres are varied to optimize the kinetics of the growth process, oxidation states, and stoichiometry. The pressure of oxidizing gases used during PLD is typically in the 10^{-3}–10^{-1} Torr regime. In oxide MBE, the films are obtained by thermally evaporating the metal atoms from their respective Knudsen cell (K-cell) furnaces onto a heated substrate in an oxidizing environment. Substrate temperatures are typically maintained in the range 650–750 °. Growth temperatures are chosen to ensure high mobility of the atomic species during growth and good crystallinity, while at the same time not being so high as to lead to intermixing between the various layers and re-evaporation/decomposition. The pressure of the oxidizing gas in the deposition chamber is typically in the 10^{-7}–10^{-5} Torr range, to allow for the longer mean-free paths needed for the metal atoms to reach the substrate from the furnace without collisions, and also to protect the hot filaments, crucibles etc. from oxidation. The oxidizing gas can be delivered through a water-cooled tube pointed at the substrate, which can lead to a flux of gas molecules on the substrate that is usually well over an order of magnitude higher than the cation flux at the substrate [28]. Thermodynamically, the stability of various oxidation states of Mn with different valences ranging from Mn^{2+} to Mn^{4+} depends upon the oxidation pressure and the temperature. In order to obtain high levels of oxidation in the film in MBE conditions, pure ozone which has much higher activity than molecular oxygen [29, 30], or a mix of O_3 and O_2 is often used [31]. Oxygen plasma, which has a mix of O and O_2, is also an option [32] though it has not been used in the work mentioned here. A detailed study of how these different oxidizing methods compare in the context of manganite thin films has yet to be carried out.

Reflection High Energy Diffraction (RHEED) allows *in-situ* diagnostics of the surface crystal structure and growth modes. While this technique is used routinely in oxide-MBE systems, differentially pumped electron guns and shorter path lengths from the gun to the RHEED screen allow the use of RHEED in PLD systems as well, despite the higher operating pressures. Typical energies for the electrons in RHEED are 10 keV during MBE growth, and can be significantly higher (20–30 keV) for PLD. At 10 keV, the electron wavelength is ~1.2 Å and the beam only probes the very first atomic layer. Periodic RHEED oscillations are routinely observed during the growth of superlattices with sharp interfaces, though the details of the shape of the oscillation pattern may depend upon the deposition rates, anneal time between layers, etc. A fully dynamical treatment of the scattering of electrons needs to take into account the angle of incidence of the electron beam, the coefficients of transmission and reflection at each interface [33], and this can make precise interpretation of the RHEED pattern challenging. Oscillations of the RHEED pattern intensity in the manganites are often interpreted in terms of a Frank van der Merwe growth mode [33], where the pattern periodically sharpens upon the completion of a full unit-cell layer with an atomically smooth surface; and becomes more diffuse when the layer is partially deposited and the surface has many islands and step edges. Layer-by-layer growth modes have been observed during synthesis of manganite superlattices by a number of groups [34, 35], and growth processes can be optimized to obtain nearly atomically abrupt interfaces between materials.

In order to create superlattices with atomically sharp interfaces and a precise number of integer unit-cells in each layer, we need to know the *absolute* number of atoms of each cation species deposited within a layer. The atomic fluxes are measured using a Quartz Crystal Monitor (QCM), which is calibrated against a Rutherford Backscattering (RBS) measurement of the stoichiometry of a manganite film deposited on an MgO substrate in realistic conditions [36]. Using this technique, it is possible to measure the *ratio* between the various cations in the film to within ~1% accuracy, given sufficient counts in the RBS measurement. The underlying assumption is that the sticking coefficient of the cations on the MgO substrate will be similar to that for the films grown on the substrate of choice, such as $SrTiO_3$ (STO) or $La_{0.3}Sr_{0.7}Al_{0.65}Ta_{0.35}O_3$ (LSAT) under growth conditions. Once the relative fluxes are known, the absolute flux of atoms is determined to better than 0.5% accuracy by measuring the thickness of a film grown on the substrate of choice, using X-ray reflectivity and diffraction measurements. In principle, this process allows determination of the absolute atomic fluxes to within ~1% accuracy. A detailed description of the calibration process and growth conditions for LSMO superlattices can be found in [28]. By far the greatest limiting factor in obtaining layers of precise cation stoichiometry and thickness is drift of the atomic fluxes of the various metals. It is not advisable to use the QCM to correct for this by measuring rates during growth, as shutters have to be opened and closed in succession, and thermal transients will limit its usefulness. However, it is possible to do this using an optical technique called Atomic Absorption Spectroscopy (AAS) [37, 38] for a number of different elements, e.g. Ca, Ba, Sr atomic fluxes can be measured with better than 0.1% accuracy, though for other elements such as Cu this may only be at the 1% level, and the accuracy can be much lower for elements such as Bi, La and Ti. For LSMO, the La, and Mn sources cannot be measured with AAS with as high an accuracy as Sr, but they tend to have drifts that are typically $\sim$1% /hr level or lower. Thus, with AAS for Sr, the cation stoichiometry can be controlled at the $< 1\%$ level if the La and Mn sources remain stable.

In determining the structural properties of LMO/SMO superlattices, we are interested in the thickness of the constituent layers, crystallinity, and abruptness of the interfaces. These have been chararacterized using both X-ray and transmission electron microscopy (TEM) based techniqiues. X-ray diffraction is used to ascertain the average lattice parameters and strain state of the films, and SL fringes of the primary film peaks may be used to determine its period. X-ray reflectivity (XRR) at near glancing angle provides quantitative information about the SL period, film thickness, and roughness at the surface and at interfaces. At these angles, the scattering is dynamical, i.e. the effects of transmission and reflection at every interface and interference between these waves have to be taken into account in a self-consistent manner. Superlattice peak positions are shifted towards higher scattering vectors from the expected values for kinematic scattering, the effect being especially important near the critical edge. A number of different software packages [39] enable simulations and refinement by fitting to the data using the Parratt formalism [40, 41], which takes dynamical scattering into account. In this manner, layer thicknesses and roughness at interfaces and at the surfaces can be extracted from the fitting parameters in a quantitative manner (Figure 9.3).

Both high-resolution transmission electron microscopy (HRTEM) [34] and aberration corrected scanning transmission electron microscopy (STEM) have been used to obtain real-space structural information in manganite heterostructures with atomic resolution [42, 43]. The electron probes can also be used for local spectroscopy at the relevant elemental absorption edges to obtain information about electronic structure at the atomic scale using electron energy loss spectroscopy (EELS). These local probes combined with scattering and transport can lead to numerous insights into local structure and doping, and its relation to magnetism.

Fully strained epitaxial films of LMO and SMO have been grown on STO (001) substrates with oxide-MBE and PLD based techniques. LMO films are found to have a *c*-axis parameter of 3.95Å while SMO has 3.78 Å. Compared to the bulk, LMO is under about 2.2% compressive strain while SMO is under 2.6% tensile strain. The films are found to be single phase with well-defined film peaks in X-ray diffraction. SMO films of 80 unit-cell (u.c.) thickness were found to develop cracks that grew over time at room temperature, while films 40 u.c. thick were found to be stable. Both materials were found to be insulating with activated transport, i.e. $\rho \sim \exp(\Delta/k_B T)$, where ρ is the resistivity. For SMO, $\Delta = 160$ meV while for LMO $\Delta = 128$ meV (energy gap $E_g = 2\Delta \sim 0.3$ eV $\sim$ Jahn–Teller splitting of two e_g bands in bulk LMO). At low temperatures (< 100 K), this translates to $\rho > 2 \times 10^3 \Omega$-cm for both materials. Magnetically, LMO films on STO are found to be ferromagnetic, with a $T_C \sim$145 K. Earlier

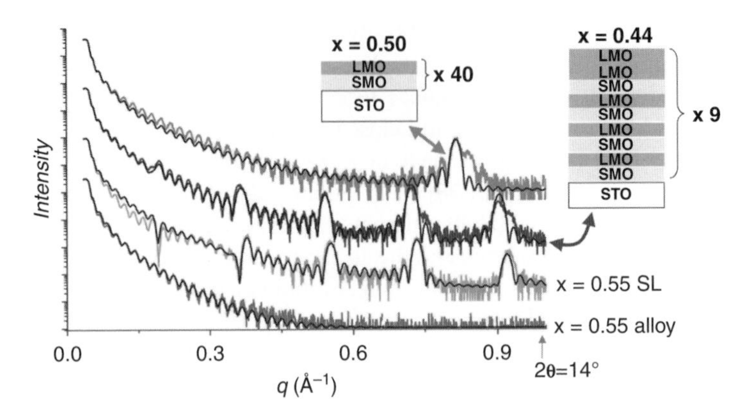

Fig. 9.3 X-ray reflectivity of digitally synthesized SMO/LMO superlattices, with $x = 0.50$ alternating SMO and LMO layers; $x = 0.55$ by inserting an additional SMO and $x = 0.44$ by inserting an additional LMO. The black lines for each spectrum are the fit calculated using Parratt's dynamical formalism. The spectra are offset for clarity. Reproduced from [28]. Copyright 2009, American Physical Society.

studies of PLD grown films of LMO found that deficiencies in the La/Mn ratio can cause it to behave like ferromagnetic and metallic LSMO, with a Curie temperature as high as 250 K [44]. The magnetic structure of strained SMO thin films have not yet been measured, though they show no magnetization in SQUID measurements. Oxygen excess/deficiency will also cause hole/electron doping, and affect the resistivity and magnetism of both LMO [45] and SMO thin films. The c-axis lattice parameter depends upon the ionic radius of $Mn^{(3+x)+}$ as a function of the hole doping level x, and can be used to infer the La/Mn ratios or oxygen off-stoichiometry.

9.5 Recent progress on manganite multilayers

9.5.1 Phase transitions and orbital order driven by strain

Epitaxial strain is known to play an important role in the properties of manganites, especially for compositions where there are competing phases with comparable free energies. Nominally, when the film is strained to the substrate, the out-of-plane lattice parameter will decrease (increase) for tensile (compressive) biaxial strain in the plane of the substrate. To address the strain effects in manganite heterostructures, the $(LMO)_1/(SMO)_1$ superlattice is an interesting model system. In this superlattice, the e_g electron density is uniformly 0.5, since every Mn cation is sandwiched between a LaO and a SrO sheet. In other words, the spatial modulation of physical properties, which occurs in most other heterostructures, does not exist here. Besides, in the bulk phase diagram, $La_{0.5}Sr_{0.5}MnO_3$ is located near the boundary between FM and A-type antiferromagnetic (A-AF) phases, providing an ideal playground to tune its phases using substrate strain. Yamada *et al.* synthesized the $(LMO)_1/(SMO)_1$ superlattice on various substrates [46] and found that the ground state of this superlattice could be tuned in this manner. As shown in Figure 9.4, the spin order in the $(LMO)_1/(SMO)_1$ superlattice is strongly coupled to the orbital order, which is determined by the strain. In the lattice matched case, namely with (nearly) equal lattice constants between a and c, e.g. $c/a = 1.01$ on LSAT substrate,

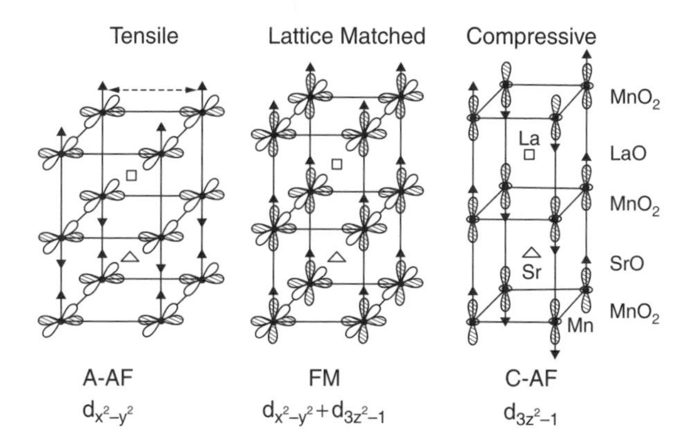

Fig. 9.4 The possible spin order and orbital order of a $(LMO)_1/(SMO)_1$ superlattice in the conditions of in-plane tensile, lattice matched, and in-plane compressive case conditions. The oxygen atoms at the midpoint between two NN Mn cations are not shown. Reproduced from Ref. [47]. Copyright 2008, the American Physical Society.

the A-site ordered $(LMO)_1/(SMO)_1$ is FM and orbital disordered with the same electron occupation in $d_{x^2-y^2}$ and $d_{3z^2-r^2}$. With in-plane tensile strain ($a > c$), the $d_{x^2-y^2}$ orbital is lower in energy and has a higher e_g electron population. This enhances the in-plane double-exchange, while the out-of-plane double-exchange is suppressed. Thus, the total effects are the enhancement of in-plane ferromagnetic exchange and the out-of-plane antiferromagnetic exchange, which result in the *A*-AF order when the strain is large enough, e.g. $c/a = 0.98$ on STO substrate [46]. Conversely, the in-plane compressive strain prefers the $d_{3z^2-r^2}$ orbital, which enhances the out-of-plane double-exchange and reduces the in-plane one. Therefore, the *C*-type antiferromagnetic (*C*-AF) order appears in this superlattice grown on $LaAlO_3$ (LAO) ($c/a = 1.05$) [46]. Note that the *C*-AF state does not appear in the bulk phase diagram of LSMO with cubic symmetry around electron density 0.5. Hence this state as well as the $d_{3z^2-r^2}$ orbital occupancy in the $(LMO)_1/(SMO)_1$ superlattice is exclusively driven by substrate strain. These ideas are captured by theoretical calculations by Nanda and Satpathy, who performed a DFT study on the $(LMO)_1/(SMO)_1$ superlattice [47]. Similar phase tuning has also been predicted in the $(LMO)_1/(SMO)_3$ superlattices where, although the SMO region has a robust *G*-AF spin order, the interfacial bilayer may be tuned between *A*-AF, FM, and *C*-AF types by changing the in-plane lattice constant a [47]. Thus, superlattices where the interfacial e_g electron density is at 0.5 or a little less are found to be highly susceptible to strain.

Besides the e_g density near 0.5, other cases have also been studied. For example, a recent experimental study of LSMO films near $x = 0.3$ measured ε_{xx} and ε_{zz} (the in-plane and out-of-plane strain values) and the relation between them, for both signs of ε_{xx} using a broad range of substrates (Figure 9.5). They found a ratio $\varepsilon_{xx}/\varepsilon_{zz} = -0.85$ [48], while an earlier work for films near $x = 0.4$ found $\varepsilon_{xx}/\varepsilon_{zz} = -0.7$ [35]. For an incompressible (volume conserving) isotropic material $\varepsilon_{a-b}/\varepsilon_c = -0.5$. Thus, the change in the out-of-plane lattice parameter is smaller than expected from the volume conserving approximation. The effects of biaxial strain on the FM transition temperature T_C have been predicted in terms of a model [49] where T_C

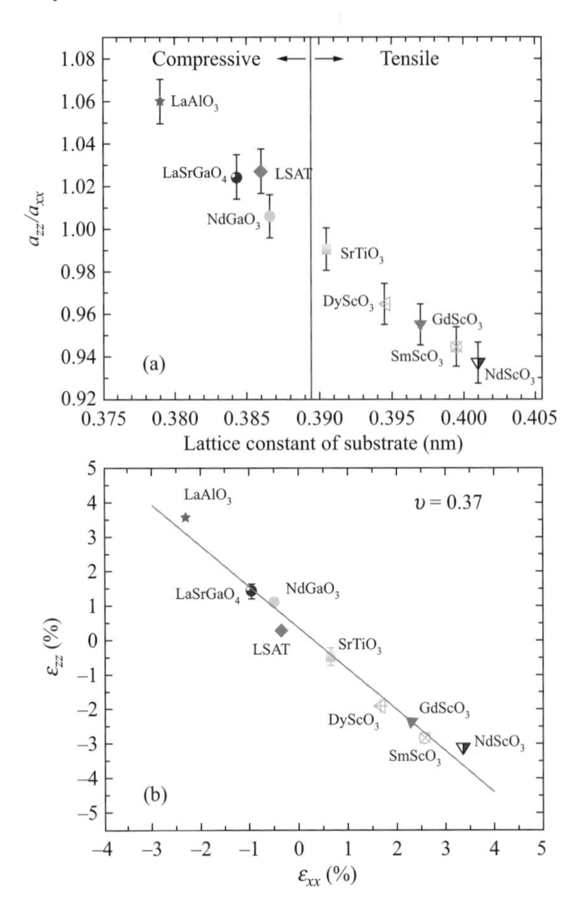

Fig. 9.5 (a) Ratio between the out-of-plane ε_{zz} and in-plane ε_{xx} lattice spacing of commensurately $La_{0.7}Sr_{0.3}MnO_3$ films as a function of the pseudocubic in-plane substrate lattice parameter. (b) The out-of-plane lattice strain as a function of the in-plane lattice strain. Reproduced from Ref. [48]. Copyright 2009, American Institute of Physics.

varies linearly with volume strain (which affects the bandwidth), but decreases quadratically with volume-preserving Jahn–Teller distortions which tend to localize carriers. Random alloy films in the FM metallic regime have been tuned through a metal–insulator transition with biaxial strain, both compressive and tensile. Metallic behavior with onset of ferromagnetism at $T_C \sim 355$ K is found for $\varepsilon_{xx} = +0.6\%$ (STO substrate), while insulating films with suppressed T_C and weak ferromagnetism have been observed for a compressive strain of $\varepsilon_{xx} = -2.3\%$ (LAO substrate), and also for tensile strains of $\varepsilon_{xx} > +2.3\%$ [50]. In fully strained short-period superlattices of (LMO)/(SMO) on STO, the "average" *c*-axis lattice parameter is very near that of randomly alloyed films of the same composition. Thus, it is not too surprising that when the effects of biaxial strain have been studied in $(LMO)_3/(SMO)_2$ [46], the response was found to be quite similar to equivalent random alloy thin films near $x = 0.4$ [46] and also at $x = 0.3$ [50].

Theoretical works have predicted many possible phase transitions in different strain conditions in this regime of doping [51, 52, 53].

For higher period superlattices such as $(LMO)_6/(SMO)_4$, the equivalence to random alloy films of the same nominal composition is no longer true, and these are electronically and magnetically quite different from the short period superlattices. When we consider superlattices with a period $>\sim 8$ u.c., the bulk-like properties of LMO and SMO begin to manifest themselves separately from the interfaces, and strong modulations of orbital and magnetic order are anticipated within the superlattice. Using resonant X-ray scattering, Kiyama *et al.* [54] have observed an orbital superlattice in $[(La_{0.45}Sr_{0.55}MnO_3)_{10}/(La_{0.60}Sr_{0.4}MnO_3)_3]_{20}$ films on STO, where $La_{0.45}Sr_{0.55}MnO_3$ is an A-type antiferromagnet with $d_{x^2-y^2}$ orbital occupancy while $La_{0.60}Sr_{0.4}MnO_3$ is a ferromagnetic metal with both $d_{x^2-y^2}$ and $d_{3z^2-r^2}$ orbitals equally occupied. It should be noted that when the charge modulation is present in manganites heterostructures, strain could tune the local phase by affecting not only the orbital occupation but also the local e_g charge density [47]. Tailoring heterostructures in this manner opens up new possibilities in "orbital band engineering" where entirely new states could arise from changes in orbital order/occupancy at interfaces as a result of charge transfer and strain.

9.5.2 Charge transfer and spin-polarized two-dimensional electron gas

In the semiconductor $p-n$ junction, electron leakage from the n-type side to the p-type side determines the asymmetric nonlinear voltage–current characteristics underlying semiconductor devices. Similarly, charge transfer may also lead to electronic reconstruction in complex oxide heterostructures. If band alignment allows, the e_g electrons in LMO near the interface will hop to the nearest empty e_g level in the neighboring SMO layer, with an amplitude t. However, this comes at the cost of a Coulomb energy of $e^2/\varepsilon a$ for leaving behind a hole, where ε is the dielectric function and a in the lattice spacing. This gives rise to a natural length scale that is analogous to a Thomas-Fermi length $\lambda_{TF} \sim a(\frac{t_o}{e^2/\varepsilon a}) = a/\alpha$. When the LMO/SMO interfaces are closer than this length scale, the "charge leakage" envelopes will overlap causing a screening of the attractive Coulomb potential that confines the electrons and a Mott-like transition is anticipated, where the carriers become delocalized along the growth direction (c-axis). Though we assume perfect band lineup between LMO and SMO in this simple picture, the argument can be modified along the lines of mini-bands formation along the growth direction in semiconductor superlattices [55, 56].

In the early stages, theorists adopted the one-orbital double-exchange model to study charge transfer in manganite heterostructures [57, 58, 59, 60]. In this model, only one isotropic orbital is considered in the Hamiltonian, implying that the hopping amplitude has a single value in all three directions. The Jahn–Teller phonons, which split the degenerate e_g orbitals, are also not included and only the spin and charge degrees of freedom are involved. Later, Lin *et al.* performed DMFT studies of both one-orbital and two-orbital double-exchange models for $(RMnO_3)_m/(AMnO_3)_n$ superlattices [61]. The J_H in (9.1) was set to be finite and the superexchange term was not included. Thus, the only model parameter is α [57], which measures the relative strength of the Coulomb screening relative to the electron hopping. It is straightforward to expect that the spatial extent of "leakage" of e_g electrons from LMO (λ_{TF}) will decrease with α, which has been confirmed numerically in the DMFT study [57, 61] and MC simulation (with the added superexchange term) [59]. With a proper value for α, e.g. $0.2 < \alpha < 2$, the charge transfer in manganite heterostructures is mostly restricted within 2 – 4 MnO_2 layers around the interface [61].

Most interestingly, due to the half-metallic properties of manganites, this charge transfer can form a spin-polarized two-dimensional electron gas (2DEG) in long-period $(LMO)_1(SMO)_7$ manganite heterostructures, predicted by both the DFT and two-orbital model studies [62]. Here (Figure 9.6), the ratio between LMO and SMO being relatively large (1:7), with the SMO being much thicker than λ_{TF}. The in-plane lattice constant is 3.802 Å, which is the optimized (GGA without U) lattice constant for the bulk cubic SMO. The atom position along the z-axis of $(LMO)_1/(SMO)_7$ is optimized and the electronic structure for the relaxed lattice was calculated using the same LMTO method, with the LSDA+U approximation as in the above section.

The DFT calculated spin configurations are marked in Figure 9.6. There are two FM layers with the nominal chemical formula $La_{0.5}Sr_{0.5}MnO_3$. Due to the charge transfer to neighbor SMO layers, the local e_g electron densities of these two layers are about 0.35 per Mn. In bulk LSMO, this e_g density corresponds to the A-AF phase. However, a bilayer FM order occurs in this heterostructure in accordance with the DFT calculation. The spin order of the next two (left and right) MnO_2 layers from the interfaces is worth discussing, since there could be some spin canting from an ideal G-AF order given the fact that the e_g densities take finite, albeit small,

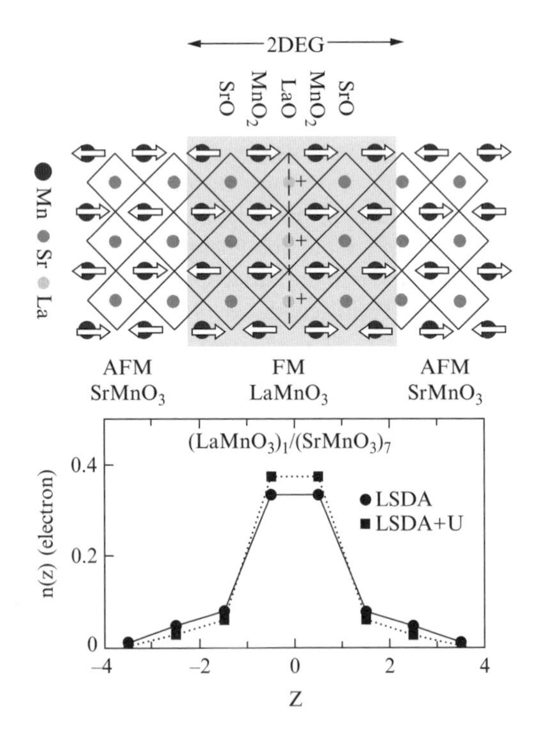

Fig. 9.6 Upper: By inserting a monolayer of LaO into the SMO bulk, a spin-polarized 2DEG is formed and confined to the interface region, as indicated by the shadowed region. Lower: The DFT calculated layer occupancy $n(z)$ of the e_g electrons surrounding the LaO monolayer ($z = 0$) inserted in the SMO bulk. Reproduced from Ref. [62]. Copyright 2008, the American Physical Society.

values (about 0.07 per Mn). Using the one-orbital model ($J_{AF}/t_0 \approx 0.047$) with an effective Coulomb potential approximation, the spin canting effect is ruled out, suggesting an ideal *G*-AF as the ground state for the next MnO_2 layers. Therefore, the one extra electron per LaO unit in this heterostructure is mainly restricted near the LaO sheet. Provided there is strong Hund coupling and FM order within the interfacial layers, which allows the double-exchange process, a spin-polarized 2DEG state is formed in the MnO_2 layers on either side of the inserted LaO sheet. For other geometries, e.g. heterostructures with thick LMO and SMO, a 2DEG also emerges at the interface, as predicted by Nanda and Satpathy in a two-orbital double-exchange model calculation [62]. There, a tri-layer spin-polarized 2DEG is predicted near the interface, sandwiched by *A*-AF LMO and *G*-AF SMO.

It should be noted that interfaces between hole-rich and electron-rich correlated materials are much more complex than semiconductor *p*-*n* junctions. The reason, as mentioned earlier, is that the energy bands in these materials are dynamic i.e. they respond to changes in doping and the accompanying changes in long range magnetic and orbital ordering. Characterizing the charge distribution across an LMO/SMO interface, and how it influences or responds to long-range orbital/magnetic order in its vicinity is thus central to our understanding of these heterostructures. This has been probed in a $(LMO)_8/(SMO)_4$ superlattice using resonant X-ray scattering (RXS) [63]. In the absence of any interfacial charge transfer, the Mn^{3+}/Mn^{4+} states in LMO/SMO will follow the chemical profile, with three different kinds of Mn sites: those in the "bulk" of the LMO and SMO regions, and those at the interface, situated between an LaO and an SrO layer. In superlattices such as these where the LMO/SMO ratio is 2:1, the structure factor for the 3rd superlattice (SL3) peak in reflectivity is strongly suppressed. In the absence of any interfacial charge transfer between LMO and SMO, this remains true. However, if the charge in the e_g levels gets smeared due to charge transfer, the spectral weights at this energy will violate the condition for the forbidden peak and the corresponding will rise in intensity. Furthermore, if the Mn spins align ferromagnetically, this may alter the hopping amplitude for charge and consequently the distribution of holes. This may further increase SL3. Upon doping LMO with holes, the spectral weights in both the Mn-3*d* and O-2*p* states are altered due to hybridization, and SL3 at both of these energies was measured. The intensity at SL3 was found to increase below the Curie temperature at both the Mn-*L* edge and the O-*K* pre-edge feature, consistent with charge (holes) spreading further from the SMO into the LMO in response to ferromagnetism near the interfaces (Figure 9.7). This is particularly intriguing in the light of theoretical calculations that assert that the charge distribution near an interface is determined by 'alpha' alone [61]. This decides the phase of magnetic/orbital order that is obtained at the interface, and the establishment of magnetic/orbital order will not alter the charge distribution in a significant manner. Further studies of these interfaces that are more quantitative about the extent to which charge is redistributed, and how this relates to magnetic and orbital order in the vicinity of the interface, are needed to clarify this issue.

9.5.3 A-site ordering in short-period superlattices

Electronic doping is a frequently-used method to obtain novel phases and exotic phenomena in strongly correlated electronic materials. The best examples are superconductivity in doped cuprates and colossal magnetoresistance in doped manganites. In bulk materials, electronic doping is usually achieved through doping A-site cations, in which disorder is unavoidable

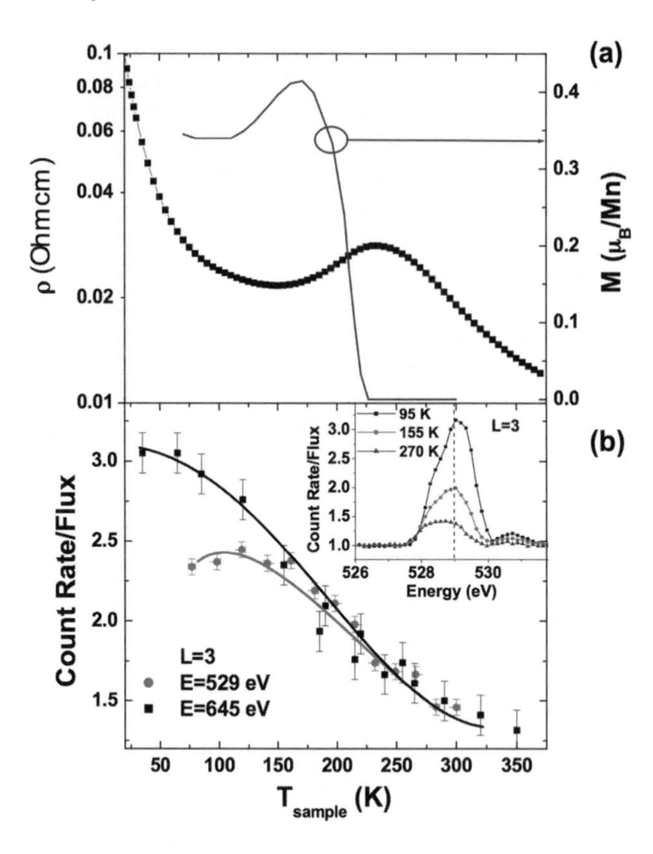

Fig. 9.7 (a) T dependence of the conductivity and magnetization of a $(SrMnO_3)_4/(LaMnO_3)_8$ superlattice when cooled in 50 Oe. (b) T dependence of the $L = 3$ reflection at the O K and Mn L_3 edge and $L = 3$ (corresponding to the SL3 reflection) for the same sample. Inset shows T dependent scans at the O K edge. Reproduced from [63]. Copyright 2007, American Physical Society.

due to the random distribution of the doped cations. Even in stoichiometric and structurally "perfect" single crystals of materials such as LSMO, there is intrinsic disorder that arises as a result of random distribution of the rare earth (3+) and alkaline earth (2+) cations on the A-site.

Broadly speaking, the A-site disorder in manganites has two consequences. The first is a random Coulomb potential arising from the distribution of 2+ and 3+ charges on the A-sites [64]. This introduces a random disorder potential on the Mn sites that can serve to localize carriers whose bandwidth is comparable to the amplitude of the potential. The second is a variance in the A-site cation radii, which results in a random distribution of Mn-O-Mn bond angles and bond lengths [65]. This disorder in nearest-neighbor Mn-O-Mn bonds causes variation in the overlap integral, $t \sim \cos^2(\omega/2)/d^{3.5}$, where ω is the deviation of the Mn-O-Mn bond angle away from 180°, and d is the Mn-O bond length [66]. This implies that both the ferromagnetic double exchange, $J_{DE} \sim t$, and the antiferromagnetic superexchange, $J_{SE} \sim t^2/U$ (U is the on-site Coulomb energy), would vary locally as ω and d are varied.

The consequences of this disorder can be quite profound for both magnetism and transport. Localization effects due to disorder in the relatively narrow bands of the 3-d transition-metal oxides are well known, and are reviewed in [67]. As is also well known from seminal studies of two-dimensional Ising systems by Griffiths [68], disorder (site vacancies in this case) can suppress the transition temperature T_C for the onset of long-range FM order to below that for the "pure" phase T_G (Griffiths temperature), and gives rise to short-range order and non-analytical behavior of the free energy in the regime $T_C < \mathrm{T} < T_G$. This is often referred to as the "Griffiths phase" and has been studied in some detail in manganites in bilayer [69] and three-dimensional [70, 71] materials.

While disorder due to cation doping occurs naturally in the manganites, this can be mitigated in ordered analogs that have also been synthesized and studied over the past decade. The earliest reported results on cation-ordered manganites were in bulk $La_{0.5}Ba_{0.5}MnO_3$ [72] where under the appropriate synthesis conditions single crystals of the material formed a $(BaMnO_3)_1/(LMO)_1$ superlattice, driven by the large size difference between Ba and La cations on the A-site. These were reported to have their Curie temperatures enhanced by as much as 80° K over their cation-disordered (random alloy) analogs. It was subsequently shown that the cation-disordered compounds favored the formation of polarons at higher temperatures, while in the random alloy, this is avoided [73]. These polarons were said to localize carriers and truncate double-exchange mediated ferromagnetism, leading to a lower Curie temperature for the random alloy compounds.

In the manganites, the competition between different ground states, often with free energies in close proximity to one another makes disorder a very powerful tuning parameter that can locally favor one phase over another, and the effects of cation-ordering can go far beyond changes in transition temperature. This is regarded as a possible origin for "mixed phase" behavior, where fluctuations in the disorder potential can cause phases with very different electronic and magnetic properties to coexist over different length scales. The delicate balance between these phases may be perturbed easily with magnetic, electric, and strain fields, giving rise to "colossal" effects, as one phase is swept out at the expense of the other. In some instances, cation-ordered and cation-disordered analogs of the same nominal composition can have *entirely different* ground states, with distinct symmetries. This was demonstrated in an extensive study carried out on a series of compounds of composition $Ln_{0.5}Ba_{0.5}MnO_3$ (Ln = Lanthanide). Particularly, it was shown that (a) the CO phase is strongly favored in cation-ordered materials, an effect that was attributed to better long-range phase coherence of the CO order parameter in these materials compared with the random alloy, and (b) magnetic ordering temperatures were significantly enhanced for the ordered analogs [74]. Thus, cation-ordered analogs are in effect "new" materials, with properties that may be qualitatively different from their cation-disordered counterparts.

Theoretically, the A-site disorder effects have already been extensively simulated by introducing some random fields into the model Hamiltonian, such as various random potential fields [75, 76, 77, 78, 79, 80, 81, 82], or random exchange fields [83, 84, 85], or both [86]. In these studies, it is generally found that the A-site disorder may destroy long-range order, especially in those AF phases with complex structures. Even for states with robust long-range orders, e.g. the FM phase, the ordering temperatures may be strongly suppressed by disorder [75, 81].

In contrast with bulk materials, the electronic doping in manganite heterostructures is achieved via charge transfer across the interface. When the superlattice period is short enough compared to the characteristic length of charge transfer, e.g. in $(LMO)_1/(SMO)_1$, $(LMO)_2/(SMO)_1$, or $(LMO)_1/(SMO)_2$, all Mn cations are adjacent to the interfaces, and are homogeneously doped. Therefore, the e_g charge density is (almost) uniform in these superlattices. The A-site cations are regularly distributed, and the system is electronically doped but nominally free of disorder. Some phases, which are inaccessible or fragile in alloy-mixed bulks, may survive in these short-period superlattices till higher temperatures.

Cation-ordered $(LMO)_{2n}/(SMO)_n$ superlattices ($x = 0.33$) are very similar in their properties to randomly alloyed samples of the same composition [87]. This is also found to be true near $x = 0.5$ [28], where both the cation-ordered and random alloy samples have the onset of long-range magnetic order occuring at approximately the same temperature, although the random alloy samples are found to have a higher resistivity at low temperatures. This is consistent with charge spreading uniformly through the material, making the ordered sample have the same nominal overall composition but with lower levels of localization by disorder. However, cation order has a striking effect on long-range magnetic ordering and orbital occupancy near $x = 0.67$ [30]. In bulk form, $La_{1/3}Sr_{2/3}MnO_3$ ($x = 0.67$) is close to a phase boundary between an *A*- and a *C*-type antiferromagnet with Néel temperatures (T_N) $\sim$ 230 K [12, 13, 88], and the material also shows some evidence for short-range order above T_N. The resistivity of these materials is significantly higher than those near $x = 0.5$ or $x = 0.33$, and are insulating at low temperatures. Thin films of this material grown on $SrTiO_3$ (STO) are under tensile strain and are tetragonal ($a = b > c$), which splits the degeneracy of the e_g states leading to an enhanced occupancy of the in-plane $d_{x^2-y^2}$ orbitals [89]. This promotes *A*-type antiferromagnetism [51, 90], shown schematically in Figure 9.8, in which the spins couple ferromagnetically within the *ab* plane and antiferromagnetically along the *c*-axis (the growth direction of the films). The in-plane ferromagnetic coupling is mediated by the double-exchange mechanism, in which the spins on near-neighbor $Mn^{3+/4+}$ ions are aligned in order to promote

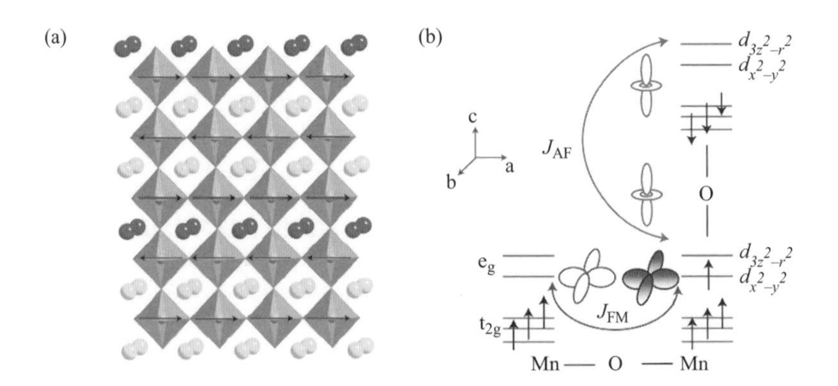

Fig. 9.8 (a) Chemical and magnetic structure of a $(SrMnO_3)_2/$ $(LaMnO_3)_1$ superlattice, with the La atoms (dark) and the Sr atoms (light). The *A*-AF spin structure is shown with alternate layers of Mn spins aligned FM within the plane and AF out of plane. (b) The in-plane FM is due to double-exchange mediated by the e_g electrons while the out-of-plane AF is due to superexchange coupling between the core t_{2g} Mn spins. Reproduced from [92]. Copyright 2009, Macmillan Publishers Limited.

delocalization of carriers within the $d_{x^2-y^2}$ orbitals. The out-of-plane antiferromagnetic coupling is mediated by the superexchange interaction between the filled $t_{2g} - t_{2g}$ orbitals along the c-axis [91].

Transport measurements show that the A-site ordering strongly affects the in-plane resistivity in $(SMO)_{2m}/(LMO)_m$ superlattices (Figure 9.9(a)) [30]. At 5 K, the resistivity of the $m = 1$ superlattices is less than that of the random alloy by more than an order of magnitude. Additionally, local resistivity maxima, which often coincide with the onset of ferromagnetic order in manganites, are present at 320 and 300 K in the $m = 1$ and 2 samples respectively, although no signature of ferromagnetic order was observed (Figure 9.9(b)). The observed decrease in resistivity was thus consistent with the onset of A-AF order, where in-plane ferromagnetism increases carrier mobility, and out-of-plane antiferromagnetism makes the net moment vanish.

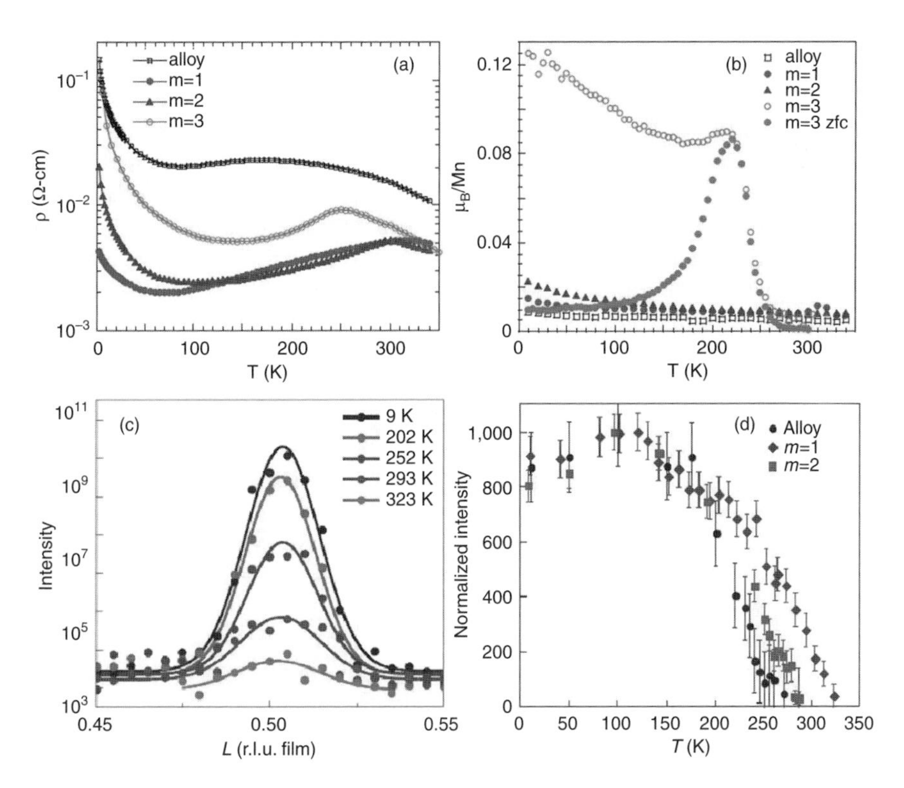

Fig. 9.9 (a) Resistivity versus T for $(LaMnO_3)_m/(SrMnO_3)_{2m}$ superlattices and a random alloy of the same composition ($x = 0.67$). (b) Magnetization versus T for the same samples. Field cooling was done in 300 Oe, cooling down from 350 K. All measurements were made while warming up using a measurement field of 50 Oe. (c) Neutron diffraction peak near (0,0,1/2) for the $m = 1$ superlatice at different temperatures. (d) Integrated intensity of the (0,0,1/2) peak as a function of temperature for the $m = 1$, $m = 2$, and alloy film. (a) and (b) are reproduced from Ref. [30]. Copyright 2007, American Institute of Physics. (c) and (d) are reproduced from Ref. [92]. Copyright 2009, Macmillan Publishers Limited.

Neutron diffraction measurements were used to investigate the role of cation-site order on the magnetic order of alloyed $La_{1/3}Sr_{2/3}MnO_3$ films and $(SMO)_{2m}/(LMO)_m$ superlattices (Figure 9.9(c) and 9.9(d)). Both the alloy film and superlattices exhibit A-type antiferromagnetism with the alloy having a $T_N \sim 240$ K, similar to that of bulk $La_{1/3}Sr_{2/3}MnO_3$, while the $m = 1$ and 2 superlattices maintain A-AF magnetic order up to ~ 320 K and 285 K, respectively [92]. Structurally, the c-axis parameter (measured with X-ray diffraction) was found to have an enhanced rate of contraction at T_N, which was attributed to increased occupancy of the in-plane $d_{x^2-y^2}$ orbitals at the expense of the out-of-plane $d_{3z^2-r^2}$ orbitals. The observed decrease is not as sharp as in bulk samples, maybe due to the fact that the film is fully strained to the substrate.

Under these conditions, the $a - b$ plane parameters of the film are considered to be nominally clamped to the substrate. However, an in-plane transverse modulation is present in all samples, the periodicity of which evolves systematically from short- to long-range coincident with the increased T_N value (Figure 9.10). This was measured using synchrotron X-ray diffraction in grazing incidence geometry around the $(2\ 2\ \delta)$ reflection ($\delta \sim 0.05$). The main film peak is found to be coincident with the substrate, indicating that the $a - b$ parameters of the film and substrate are equal. However, in addition to the $(2\ 2\ \delta)$ reflection, two first-order satellite peaks are also observed for the films. These arise from atomic displacements along the scattering vector, with a modulation wavevector perpendicular to it (i.e. transverse). For example, for satellite peaks around the $(2\ 2\ \delta)$ reflection, the atomic displacements are along the [110] direction with a modulation wavevector along the $[1\bar{1}0]$ direction. Crosscuts along the $H = -K$ of the reciprocal space maps are shown in Figure 9.10 and illustrate the changes in the satellite feature, with increase in T_N. In the random alloy, the period of displacement (1.63 nm) is ~ 3 u.c. diagonals along $[1\bar{1}0]$, with a coherence length of 14.4 nm. This period is increased to ~ 4.5 u.c. (2.5 nm) in the $m = 2$ superlattice, and is strongly suppressed in the $m = 1$ superlattice. Also, a longer-period ordering (~ 58 u.c.) emerges along the same directions in the $m = 1$ superlattice. This can be seen in a higher resolution reciprocal space map around the $(2\ 2\ \delta)$ reflection of the $m = 1$ superlattice (Figure 9.10(b) and (d)). The $m = 2$ superlattice, which has an intermediate T_N, has coexisting short-period (~ 4.5 u.c.) and long-period modulations. All periodicities measured in the alloy and superlattices, are purely structural in nature, are not affected by magnetic ordering, and are temperature independent.

The fact that the observed transverse modulations may be incommensurate with the underlying lattice is very intriguing [93, 94]. Incommensurations may arise in a number of different ways. In conducting materials such as cuprates and nickelates [95], this may be a result of the lattice coupling to a density wave instability of spin or charge, which often occurs due to nesting of the Fermi surface. This may also occur for purely structural reasons, such as a mismatch between a material and its underlying substrate [96, 97] or a "displacive incommensurability" where distortions of a periodic lattice may themselves have a period that is incommensurate with the underlying lattice due to next-nearest-neighbor interactions or anharmonic forces [98]. While the origins of the observed modulations in $(LMO)_m/(SMO)_{2m}$ superlattices and equivalent random alloy films are not understood at this time, incommensurations consistent with long-wavelength octahedral tilts have been observed in cuprates [99], and may provide an energy lowering instability in these systems as well. The ability to tailor the strength of

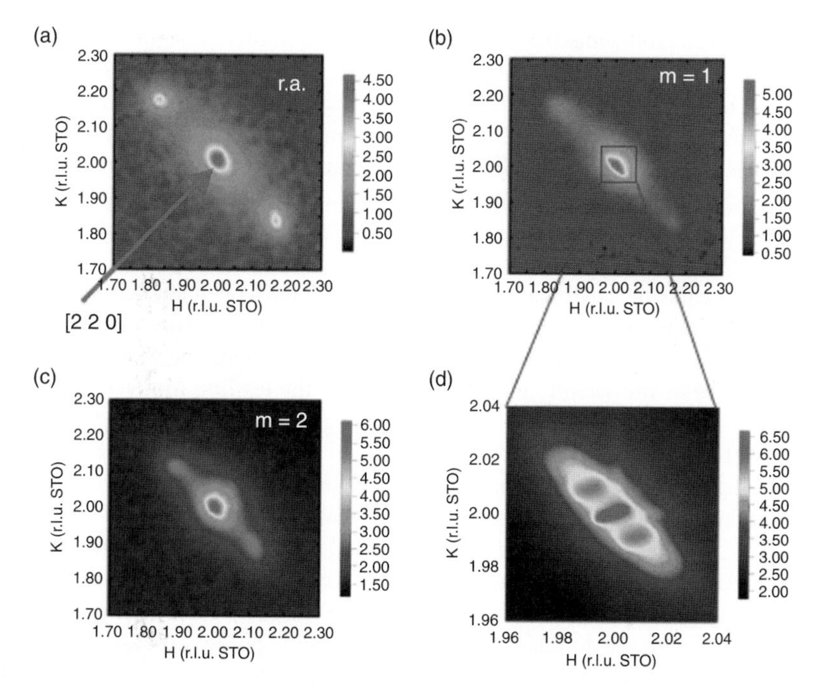

Fig. 9.10 Reciprocal space maps around the (2 2 δ) reflection in grazing incidence geometry for the (a) random alloy (r.a.) (b) $m = 1$ and (c) $m = 2$ superlattices. The red arrow in (a) shows the scattering vector q along (2 2 0). Satellite peaks are seen with Δq orthogonal to q, indicating a transverse modulation that is strongest for *r.a.* and weakest for $m = 1$. The region near the 220 peak in the $m = 1$ sample is magnified in (d) to reveal a transverse modulation peak at much longer wavelength. The transverse modulation wavelength for *r.a.* is ~ 1.6 nm, for $m = 2$, is ~ 2.5 nm, and for $m = 1$, ~ 32 nm. Reproduced from [92]. Copyright 2009, Macmillan Publishers Limited. This figure is reproduced in color in the color plate section.

such instabilities using a layered superlattice approach may provide pathways to new materials properties, as has been observed in this work.

9.5.4 Tuning between ferromagnetism and antiferromagnetism

Manganites near $x = 0.5$ doping have been of considerable interest because they allow sharp transitions between diverse phases such as metal/insulator or antiferromagnet/ferromagnet for small changes in carrier doping. Such materials may have a high or even "colossal" susceptibility to external electric, magnetic, and strain fields. For example, we have already discussed the phase tuning of $(LMO)_1/(SMO)_1$ by substrate strain in Section 9.5.1. In high bandwidth manganites such as LSMO, there is an *A*-AF phase found in this regime, as described in the previous section, where the spins are ferromagnetic in plane and antiferromagnetic out of plane. In such materials, the in-plane conductivity is double-exchange mediated, and can be nearly metallic at low temperatures, while the out-of-plane conductivity is insulating due to a "spin-valve" effect [100], making it a two-dimensional antiferromagnetic metal (this is in contrast with a range of other lower bandwidth manganites, where a robust charge-ordered phase is found near $x = 0.5$). Transition to the antiferromagnetic state is also accompanied by a transition to $d_{x^2-y^2}$ orbital occupancy. There are a number of intriguing possibilities with this class of materials, which is half metallic in plane, but antiferromagnetic overall. These include superconductivity due to proximity effects [101], and a volume spin torque effect (as opposed to an interface effect) that may cause the spins through the entire material to cant upon driving a current through it [102].

Bulk single crystals of A-AF manganites near $x = 0.5$ have been studied in both 3D [107] and bilayer [108] materials, and the magnetic structure, along with the anisotropic nature of both transport and magnetic properties, [109] have been verified. The ability to tune between the different states in the vicinity of $x = 0.5$ is one of the leading motivations for studying materials in this composition range [110]. In a striking demonstration, $Pr_{0.5}Sr_{0.5}MnO_3$ films grown on [011] oriented LSAT substrates were found to undergo a transition between a 2D insulating antiferromagnet and a 3D metallic ferromagnetic state by applying an in-plane magnetic field of a few Tesla [111]. This dimensional crossover is a consequence of a transition from preferential $d_{x^2-y^2}$ orbital-occupancy to a two-orbital liquid, and is distinct from transitions where a "mixed phase" near a phase boundary gives rise to a colossal response.

In de Gennes' seminal model for the transition between the insulating superexchange antiferromagnetic and metallic double-exchange ferromagnetic phases [103], mobile carriers are doped into the conduction/valence band of an insulating antiferromagnetic parent compound. These carriers mediate a ferromagnetic coupling between spins via the double-exchange interaction, and give rise to a canted antiferromagnet. The magnitude of the kinetic energy gain (and thus the canting spins) is proportional to the number of doped carriers, and with increased doping, a fully ferromagnetic metallic material is realized. This provides an intriguing path to controlling ferromagnetism in a continuous manner with doping, which in principle can also be accomplished reversibly via capacitive charging with gate electric fields. de Gennes' descriptions does not hold for the transition between antiferromagnetic insulator LMO and ferromagnetic $La_{1-x}Sr_xMnO_3$ near $x \sim 0.16$ because Jahn–Teller effects and Mott–Hubbard correlations, not included in his model, lead to mixed phase behavior in this regime. However, near $x = 0.5$ these effects are not as important, as the material undergoes a transition from the A-AF to a three-dimensional (3D) ferromagnetic via canting, and the orbital occupancy goes from being purely $d_{x^2-y^2}$-like to a more isotropic state where both $d_{x^2-y^2}$ and $d_{3z^2-r^2}$ orbitals are occupied. Thus, this transition/crossover is more along the lines of what was proposed in the classical paper by de Gennes [103], but where the added electrons go into the bottom of the band formed by $d_{3z^2-r^2}$ orbitals, and the additional requirement that we consider both orbitals explicitly in a model description [104, 105]. In recent work, a "delta-doped" quasi-2D FM region was realized experimentally within a parent A-type AF (LaMnO3)1/(SrMnO3)1 superlattice, where an extra "dopant" layer (1 unit-cell) of LaMnO3 was added to the AF parent material. Using neutron scattering, it was established that the extra electrons added by this layer cause the A-type AF spin structure to cant towards ferromagnetic alignment in a region of approximately $\pm$ 3 unit-cells in the vicinity of the dopant layer [106]. This establishes a fundamental length scale analogous to a "screening length" for charge in a two-orbital material, involving the charge, spin, and orbital degrees of freedom.

Since the transport in these A-AF materials is quasi-two dimensional, we would expect that cation disorder could suppress in-plane double-exchange ferromagnetism and enhance fluctuations. Cation-ordered bulk single crystals are known to have enhanced ordering temperatures for magnetic and CO phases. In cation-ordered $Nd_{0.5}Ba_{0.5}MnO_3$ single crystals, the magnetic ordering temperatures are enhanced and a marked preference for the A-AF phase over the FM phase is found at low temperatures [74], and these properties are very similar to those of the higher-bandwidth $La_{0.5}Sr_{0.5}MnO_3$ single crystals [13]. Thus, removing cation disorder can lead to new properties for materials near $x = 0.5$, and nucleate phases like A-AF that seem to be fragile to disorder. However, bulk synthesis routes for ordered analogs are limited to the $Ln_{0.5}Ba_{0.5}MnO_3$ systems, and thin film based techniques provide a means to explore a broader range of ordered manganites near this composition. One might ask, what happens to the high bandwidth $La_{0.5}Sr_{0.5}MnO_3$ when we remove cation disorder? Cation-ordered superlattices of $(LMO)_p/(SMO)_p$ have been grown by both PLD [112] and MBE [28] based techniques. Using MBE based techniques, it is possible to grow samples with compositions on either side of $x = 0.5$ by inserting an extra SMO or LMO layer within a $(LMO)_p/(SMO)_p$ superlattice depending on the desired composition, without introducing cation disorder. It should be noted that this requires control of both cation and oxygen stoichiometry to prevent inadvertent doping of the material away from the $x = 0.5$ composition. These are found to *not* have a higher magnetic ordering temperature than their disordered counterparts, with onset of weak ferromagnetism near 300 K and metallic behavior at lower temperatures [28]. However, the low temperature resistivity of the cation-ordered films in the vicinity of $x = 0.5$ was found to be lower by a factor of 3–10, the effect being weaker for $x < 0.5$ (Fig. 9.11). The weak ferromagnetism in these films was attributed to canting of the A-AF spins due to the competition between double-exchange and superexchange interactions, and this has been confirmed with neutron diffraction [28]. Furthermore, it was shown that in both cation-ordered and cation-disordered films, the A-AF

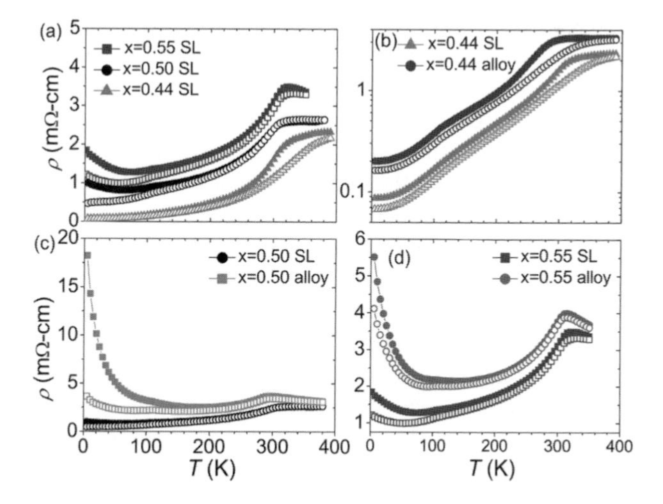

Fig. 9.11 (a) Resistivity versus T of superlattices in the vicinity of $x = 0.5$ in zero (open symbols) and 7 Tesla (closed symbols) magnetic fields. (b)–(d) resistivity compared between superlattices and their equivalent random alloy films. Reproduced from [28]. Copyright 2009, American Physical Society.

phase was stabilized all the way up to the highest magnetic ordering temperature [28]. Since these films were grown on STO substrates, they are under biaxial tensile strain in the *a*-*b* plane, and the *c*-axis parameter is consequently lower than in bulk samples. This causes an increase in the antiferromagnetic superexchange coupling between planes, as has been discussed earlier, and thus an enhancement of the Néel temperature at the expense of the FM order seen in bulk samples. This further demonstrates the sensitivity of the FM/AF order parameters near this composition to external fields, and points to future directions for exploration.

9.5.5 Interfacial magnetism

As mentioned earlier, charge transfer at an LMO/SMO interface can lead to long range ferromagnetic order near the interface. In a study that combined STEM and PNR to probe both the chemical and magnetic structure of an LMO/SMO superlattice, it was found that the magnetization near an LMO/SMO interface is sensitive to chemical roughness. In one of the large-period LMO/SMO superlattices, the layers were not integer unit-cells thick, and STEM and X-ray reflectivity measurements showed that the film was consistent with a $[(LMO)_{11.8}/(SMO)_{4.4}]_6$ structure. Furthermore, the STEM images showed that while the SMO/LMO interface in the growth direction was atomically sharp (i.e. the intermixing is over a length scale of well under one unit-cell) and smooth over tens of nanometers laterally, the LMO/SMO interface, while also atomically abrupt, had undulating plateaus and valleys with a peak-to-valley amplitude of about two unit-cells, with single unit-cell steps every 5–15 nm. The exact reasons for this asymmetry are unknown, although the kinetics of film growth, surface, and interface energies and wetting-related phenomena may all play a role. For example, the fact that SMO is under a large tensile strain may be a reason why a free surface of SMO would prefer to lower its surface area by becoming relatively flat, while a free surface of LMO being under compressive strain may behave conversely. Upon refinement of the magnetization profile to the PNR data (taken in applied parallel fields of ~ 6 kG), the local magnetization within the superlattice was seen to be very strongly dependent on the interfacial morphology (Figure 9.12). The magnetization peaked at the smooth interfaces with a value of 3.8 μ_B/Mn, while it was strongly suppressed at the "rough" interface with a value of $\sim 0.1\mu_B$/Mn, close to the value in the middle of the SMO layer. The middle of the LMO layer was found to have a magnetization of approximately 2.6μ_B/Mn. A detailed theoretical study of these interfaces that takes this kind of roughness into account is yet to be undertaken. However, there are a number of observations that can be made on the strength of the data. First, it underlines the need for atomically abrupt interfaces, if we are to realize states that depend on interfacial charge transfer that occurs on the scale of 1–2 unit-cells. Second, the Curie temperature of the superlattice is about 80 K higher than just LMO, and this may be due to the magnetization at the interface. Thus, there exists the possibility that at temperatures higher than the T_C of LMO, and below that of the superlattice, only the moment at the interface survives. Measurements of second harmonic generation in LMO/SMO heterostructures have observed evidence for an interfacial electronic state [113], and also interfacial magnetism as characterized by the nonlinear magneto-optic Kerr effect [50]. These measurements imply that, depending upon the extent and magnitude of charge transfer that is responsible for the observed electronic states and interfacial magnetism, we may be able to tune the magnetization at the interface, if we can control the charge transfer with a capacitively coupled electric field.

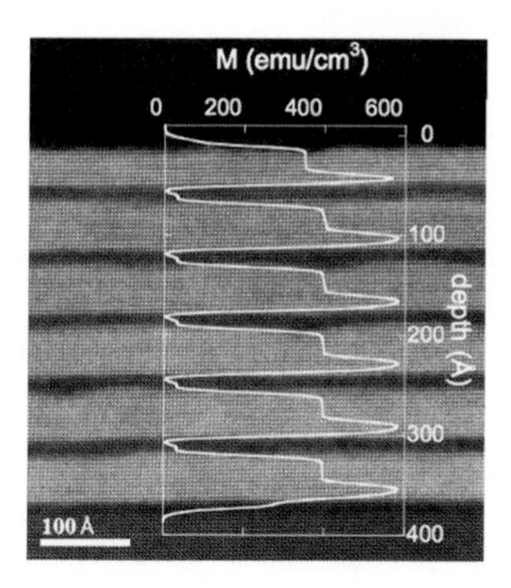

Fig. 9.12 Magnetization profile for an LMO/SMO superlattice superimposed on a STEM image. The rough interface has a strongly suppressed magnetization. (Figure courtesy Steve May and Amish Shah.)

Until now, most theoretical studies have focused on the LMO/SMO superlattices, although not much is known about the interfacial state of the narrow or intermediate bandwidth manganites, such as (La,Ca)MnO_3 (LCMO) or (Pr,Ca)MnO_3 (PCMO). Their phase diagram is distinct from LSMO, in that a charge/orbital ordered CE state is stabilized at electron density ~ 0.5. Such a charge order originates from the much stronger coupling between electrons and Jahn–Teller lattice distortions in these materials.

A zero-T relaxation method was adopted by Yu *et al.* to explore the possible ground state of the PMO/CMO heterostructure [114]. The model system consists of four PMO layers (labeled by index $Z = 1-4$) plus four CMO layers ($Z = 5-8$). Here, only the optimized spin patterns at $\alpha = 1.0$ are presented in Figure 9.13. It is clear that the layers close to the interface (layers 4 and 5) exhibit the CE-type spin order. This coincides with the phase diagram of bulk narrow bandwidth manganites, in which the CE state is stabilized at electronic density $n \sim 0.5$ and has a staggered charge order. However, states significantly different from the bulk-like ones also exist. For instance, at layer 3, the e_g density is close to 1, but it still presents the CE spin order. This suggests that the robust CE state at layer 4 is strongly pinned by its neighbor. Moreover, the spins of adjacent zigzag chains in layer 3 are not perfectly antiferromagnetically coupled, as in a conventional CE state, but are canted. Even for layers 4 and 5, which exhibit almost a perfect CE order individually, the spins in the two layers are not perfectly AFM aligned with respect to one another, compared with what they should be in a bulk CE phase. In the layers $Z = 4$ and 5, the relative spin orientation is at approximately 90 degrees, different from the 180 degrees stacking in the bulk CE phase.

Besides the $RMnO_3/AMnO_3$ series, other narrow bandwidth manganite heterostructures have also been studied. Here, both charge and orbital order are found to propagate across interfaces in very interesting ways. Calderón *et al.* studied the $(La_{0.5}Ca_{0.5}MnO_3)_4/(CMO)_4$ superlattices using the two-orbital double-exchange model [115]. They found that there may be one or more FM metallic layers between these two robust antiferromagnetic insulators.

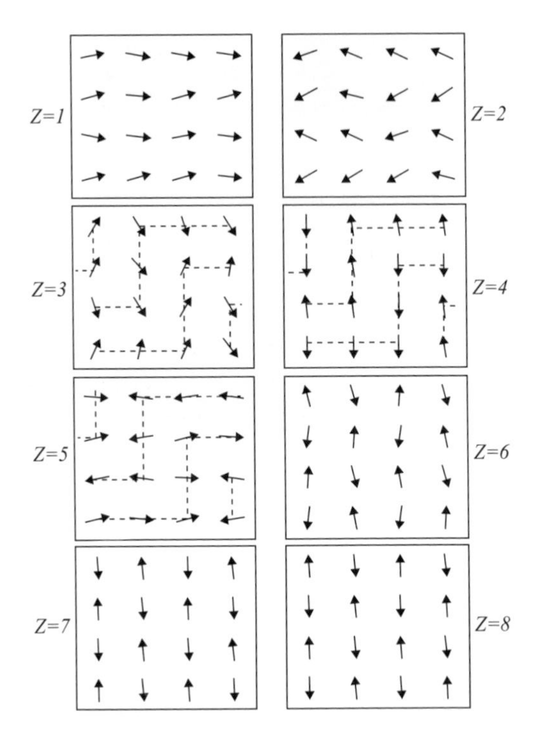

Fig. 9.13 Spin configurations for each layer with $a = 1$ obtained via model Hamiltonians and optimization techniques. The FM zigzag chains of the CE state are highlighted by dashed lines. Reproduced from [114]. Copyright 2009, the American Physical Society.

Furthermore, influenced by the neighbor CE phase, these FM layers have similar CE-type zigzag charge order and a trace of orbital order, as shown in Figure 9.14. Conversely, in the CE type LCMO layer next to the FM layers, there is more charge on the zigzag chains parallel to the FM plane. Similarly, for the G-AF CMO layer next to the FM layer, there is some charge at the sites with spin parallel to the neighboring FM plane (only on the $d_{3z^2-r^2}$ orbitals, in agreement with the idea of "hand-in-hand" orbital connection). Thus, the manner in which charge/charge order propagates across an interface depends on the way in which the orbitals connect across the interface.

9.5.6 Metal–insulator transitions

In recent years, a number of groups have observed metal–insulator transitions (MIT) in LMO/SMO superlattices as the period of the superlattice was increased [61]. Since magnetic and orbital order are intimately linked to the itinerancy of charge, it is of great interest to understand how these degrees of freedom evolve within the superlattices with increasing period, particularly as this might enable us to control the MIT with external fields. The earliest observation of this MIT was in ferromagnetic $(LMO)_p/(SMO)_q$ superlattices [116] where p and q are the number of unit-cells in each layer, and the nominal doping level $x = q/(p+q) = 0.26$

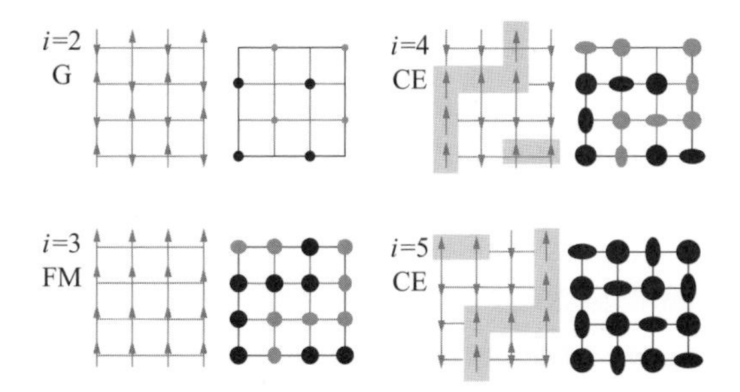

Fig. 9.14 Sketch of the magnetism and charge distribution in different MnO_2 layers of LCMO/CMO heterostructures. Here only one interfacial layer ($i = 3$) is ferromagnetic by tuning the superexchange intensity. The size and shape of dots represents the local e_g charge density and orbital shape. Plane 3 is the interfacial MnO_2 layer. Plane 2 is in the CMO side, while planes 4, 5 are in the LCMO side. Reproduced from Ref. [114]. Copyright 2008, the American Physical Society.

(here, p and q were non integers). They found that when the period of the superlattice was low i.e. when $p + q < 6$ unit-cells, the films were metallic down to the lowest temperatures measured (< 10 K), with a Curie temperature (T_C) well in excess of 300 K. Upon increasing the period to where $p + q > 10$ u.c., the samples become insulating at the lowest temperatures. They also found that T_C decreased monotonically with increasing superlattice period, and the saturation magnetization M_S was suppressed.

Subsequently, a number of groups have synthesized "digital" superlattices of $(LMO)_{2n}/(SMO)_n$, where n is an integer, which would in principle lead to atomically sharp interfaces between cation-ordered LMO and SMO regions (Figure 9.15) [48, 87]. X-ray reflectivity and STEM measurements confirm that the intermixing of the La/Sr cations at the SMO/LMO interface in these materials is at the 0.5 unit-cell level or lower [87]. The resistivity of a $(LMO)_2/(SMO)_1$ sample ($n = 1$) was found to be $\sim 35 \times 10^{-6}\Omega$-cm, slightly lower than a random alloy film of the same composition (Figure 9.16). Measurements of the magnetization, Hall effect, and spectroscopy at the Mn-L and the O-K edges all indicate that the $n = 1$ samples are nearly identical electronically and magnetically to their random alloy counterparts. The lower resistivity values for the $n = 1$ superlattice are thus presumed to be due to lower levels of disorder scattering. The metallic state persists for $n \leq 2$, but for $n \geq 3$ an insulating state is obtained at low temperatures, with the resistivity at 2 K changing by $> 10^8$ between $n = 1$ and $n = 5$. T_C and M_S are both significantly reduced upon increasing n, and by $n = 8$, the magnetic properties begin to resemble that of strained LMO thin films [31], i.e. the bulk of the LMO layers begin to dominate over the interfacial regions. Measurements of X-ray linear dichroism (XLD) have also observed a change in the magnetic anisotropy from having an in-plane easy axis to an out-of-plane easy axis, for antiferromagnetically aligned spins in larger n insulating superlattices [117]. While the $n = 1$ and the random alloy both share very similar magnetic properties, it was shown using polarized neutron reflectometry (PNR) techniques that the evolution of the metal into the insulating state is accompanied by an increasingly modulated magnetization within the films, where the LMO regions have a higher magnetization, while it is strongly suppressed within the SMO regions (similar to the magnetic profile shown in

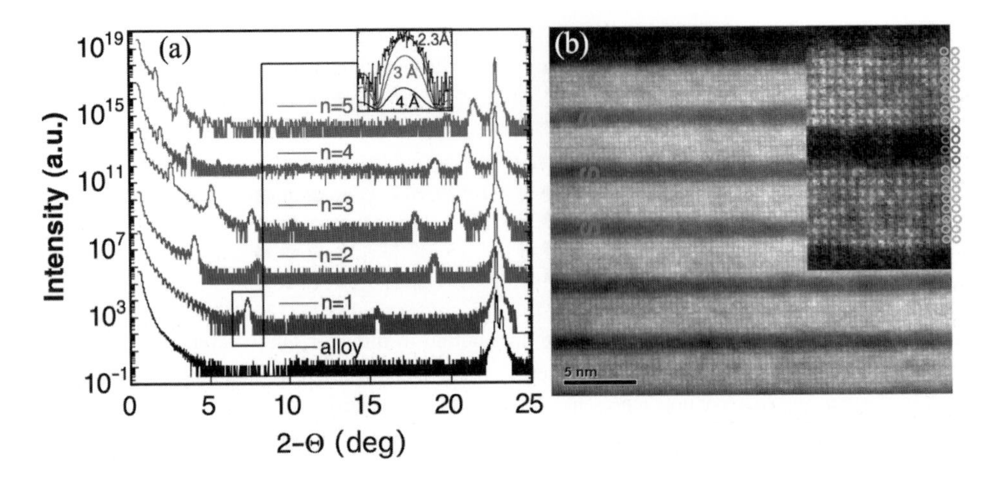

Fig. 9.15 (a) X-ray reflectivity and diffraction for $n = 1-5$. Simulations of the first superlattice peak for $n = 1$ (inset) are shown for roughness of 2.3, 3, and 4 Å. (b) STEM z-contrast images for $n = 4$ superlattice showing LMO and SMO regions, labeled L and S. Rows of atoms are shown on the right as a guide to the eye. Reproduced from [87]. Copyright 2008, American Physical Society.

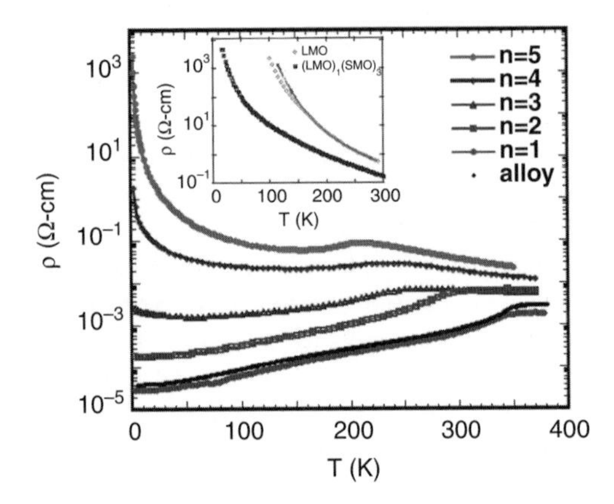

Fig. 9.16 Resistivity of $La_{0.67}Sr_{0.33}MnO_3$ random alloy film and corresponding $(SrMnO_3)_n/(LaMnO_3)_{2n}$ superlattices, $1 \leq n \leq 5$. The inset shows the resistivity of a pure $LaMnO_3$ thin film with Δ = 125 meV, and also the resistivity of a $(SrMnO_3)_3/(LaMnO_3)_1$ superlattice. Reproduced from [87]. Copyright 2008, American Physical Society.

Figure 9.12) [87]. The MIT with increasing superlattice period is observed in compositions with $0.2 < x < 0.7$ of the LSMO phase diagram, wherever metallic or metal-like behavior is seen at low temperatures in random alloy samples [46, 112].

Theoretically, Lin and Millis calculated the optical conductivity for LMO/SMO superlattices using DMFT based techniques and compared them to their results for 'bulk' samples. Measurements of optical conductivity for different LMO/SMO layering and superlattice periods would yield a definitive test of these calculations. They also calculated the dc conductivities of $(LMO)_n/(SMO)_1$ by extrapolating the optical conductivities at the limit of zero-frequency [61]. The in-plane conductance of a superlattice is the parallel sum of the conductance of the different layers, each of which has a conductivity equal to that of bulk LSMO, with the same charge density. In contrast, the out-of-plane conductivity is more complex and there is no simple analytical rule. Furthermore, they find that for all $n = 1$ superlattices, the metal–insulator transitions (MITs) occur near the Curie T_Cs, similar to the alloy-mixed bulk $La_{2/3}Sr_{1/3}MnO_3$. However, there are several disagreements between these calculations and the experimental data. First, the calculated T_C (0.1–$0.12t_0 \sim 750$–800 K) is higher than the experimental ones (about 350 K for $La_{2/3}Sr_{1/3}MnO_3$ thin film) [87]. This over-estimated T_C may originate from over-estimated t_0, neglecting the antiferromagnetic superexchange interactions, or due to approximations in DMFT. Second, the calculated resistivity ρ of $(LMO)_2/(SMO)_1$ is higher than that of alloy-mixed $La_{2/3}Sr_{1/3}MnO_3$, while in fact the latter is higher due to the disorder effects [87]. Third, all $(LMO)_{2n}/(SMO)_n$ superlattices that were also discussed in [61] were found to be metallic ($d\rho/dT > 0$) below T_C. However, it has been shown that at low temperatures the ground state turns insulating when $n \geq 3$ as shown above.

Dong *et al.* consider the effects of superexchange interactions since it is essential to stabilize an insulating antiferromagnetic phase inside the SMO layers when $n > 1$ [118]. They studied the two-orbital double-exchange model using the MC technique. At the same time, Nanda and Satpathy performed the DFT calculation on the same series [52]. In contrast with previous model studies, which usually determined the Coulomb potential using the Hatree–Fork approximation, Dong *et al.* apply a fixed "pseudo-potential" profile to the superlattices. Each onsite potential is determined by its eight nearest-neighbor A-site cations, an approximation used in the past to investigate the correlated disorder effects in manganites [82]. More specifically, the potential for those Mn layers between two LaO sheets is 0, while it is V between two SrO sheets, and $V/2$ at the interface. This positive constant V is the only parameter used to regulate the charge distribution, and plays a role similar to α in the previous model. The Coulomb screening by the e_g electron redistribution is also taken into account in part by regulating the value of V. In a DFT study, the calculated potential profiles qualitatively agree with the fixed potential used in the model simulation, as shown in Figure 9.17.

The MC results of e_g electron density distributions and spin orders at low-T ($0.01t_0$) are shown in Figure 9.18(a) and 9.18(b). The electron density in the $n = 1$ superlattice is almost uniform for a large range of values of V (0.3–$0.9t_0$). For higher period superlattices, the e_g electron density in the SMO regions are always lower than 0.5, except for $V < 0.3t_0$. The local charge densities in other regions are intermediate between 0.5 and 1, corresponding to the FM phase in the bulk. Taking $V = 0.9t_0$, for example, the magnetic structures for all these superlattices ($1 \leq n \leq 4$) are $n-1$ G-AF layers in the SMO region plus $2n + 1$ FM layers.

For comparison, the DFT calculated magnetic structure is shown in Figure 9.17. For the $n = 1$ case, the DFT study confirms the full FM phase to be the ground state, in agreement with the above model calculations. All effective exchanges between NN Mn-O-Mn bonds are strongly FM (-11 to -39 meV). As shown in Figure 9.19, the total and partial spin-resolved density of states (DOSs) of this superlattice with the FM spin order show all layers are partially occupied, suggesting a uniform metallic superlattice. For $n = 2$, the calculated magnetic interactions are no longer pure FM. The in-plane and out-of-plane effective exchanges of the SMO layer are strongly antiferromagnetic (10–17 meV) and the out-of-plane exchange deep-inside the LMO region is very weak (-4 meV), even though it is still ferromagnetic. Thus, the magnetic structure of the $n = 2$ superlattice consists of five ferromagnetic layers and one

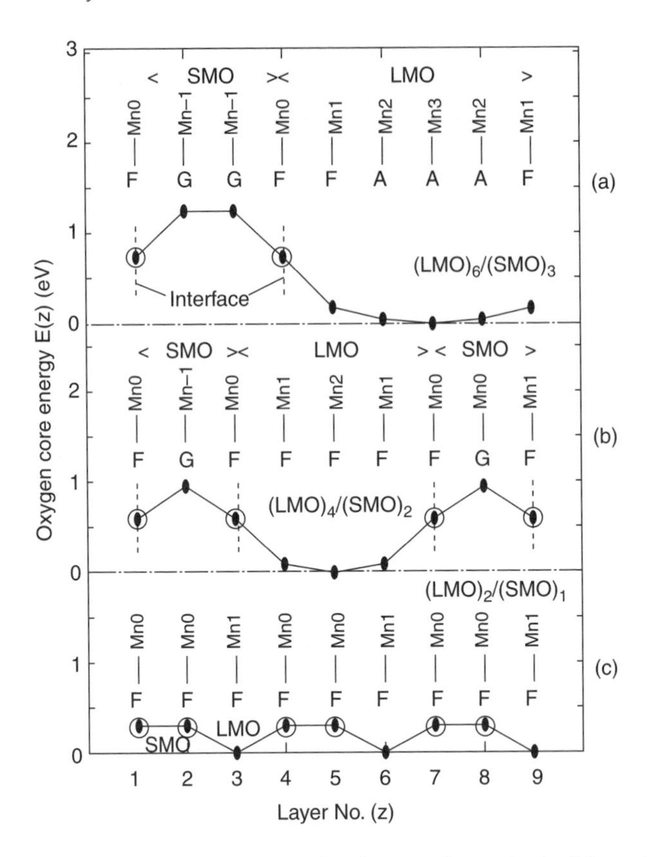

Fig. 9.17 Variations in the oxygen 1*s* core energy of each MnO_2 layer, obtained from the layer-projected wave-function characters. The interfacial Mn layers are shown by open circles with vertical dashed lines. The DFT calculated ground state magnetic orders are also marked. Reproduced from Ref. [52]. Copyright 2009, the American Physical Society.

G-AF layer, which still coincides with the model simulation result. Further increasing n leads to a disagreement between the model simulation and DFT calculation. According to the DFT, three *A*-AF layers exist in the central LMO when $n = 3$. The DOS shows an energy gap at the Fermi level for the Mn cations in these *A*-AF layers, as shown in Figure 9.19. In other words, these inner LMO layers restore the bulk *A*-AF insulating state. However, the MC simulation of the two-orbital model gives a FM + *G*-AF result, without any *A*-AF layer inside the LMO region. Moreover, the experimental data show a strong magnetization in the whole LMO region for larger n [43], which seems to support the model results. This question will be further addressed later in the Conclusions and Outlook section.

In model simulations, the conductivity was calculated using the Kubo formula. As shown in Figure 9.18(c), the in-plane conductivities all show metallic signatures, in agreement with the idea that the in-plane conductance is the sum of the conductance of each individual layer in parallel [61]. For interfacial layers, the local charge densities fall in the range of a FM metallic state in the bulk phase diagram, resulting in a metallic interface. Previous theoretical studies on the $LaTiO_3/SrTiO_3$ heterostructure also found the formation of a metallic interface via charge

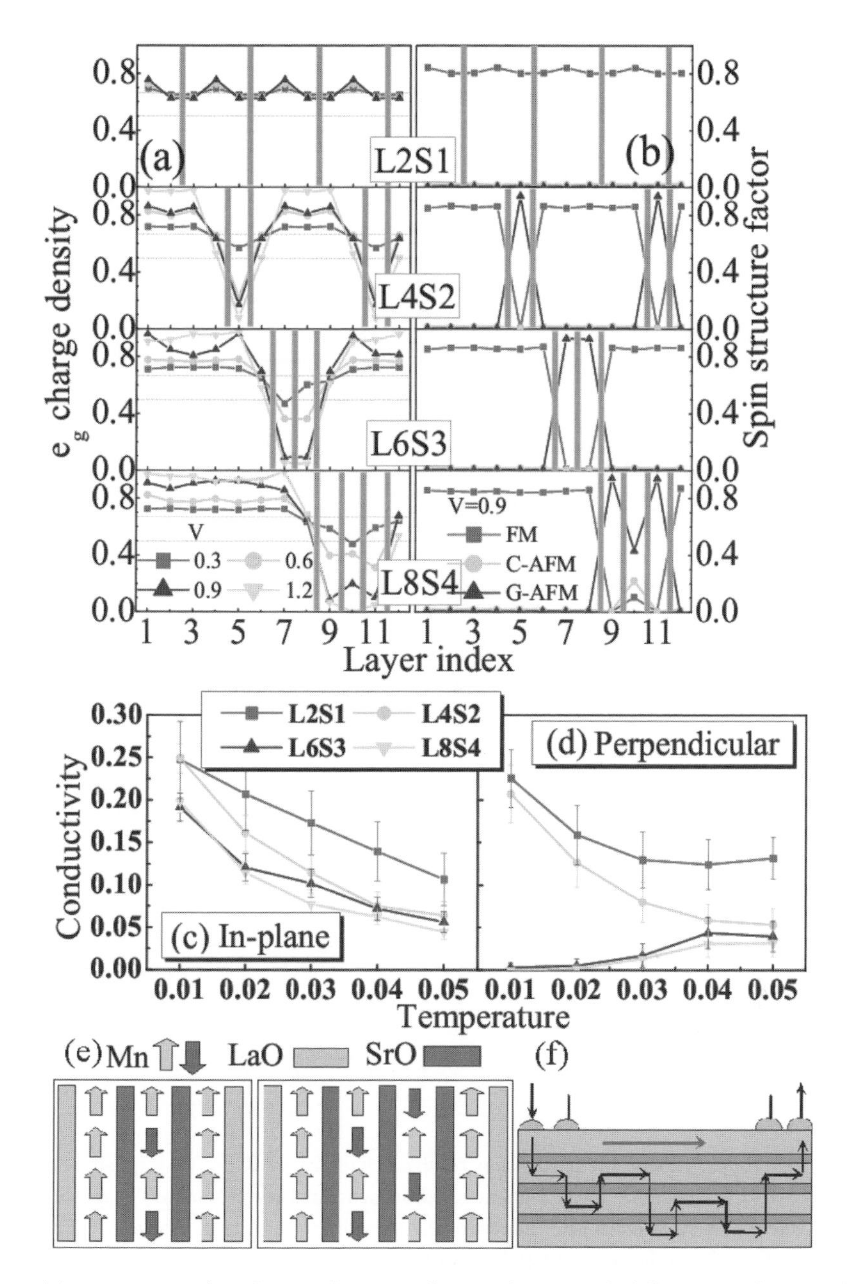

Fig. 9.18 (a) Layer-averaged e_g electron density in the superlattices with different Vs. The case of $n = 1$ with $V = 1.2$ cannot be obtained due to phase separation. Two reference values 0.5 and 2/3 are marked as the horizontal lines. (b) In-plane spin structure factor at $V = 0.9$. In (a) and (b), pink bars denote SrO sheets in the superlattice, while LaO sheets are not shown. $L_{2n}S_n$ denotes $(LMO)_{2n}/(SMO)_n$. Conductivities as a function of T: (c) in-plane one (d) out-of-plane (perpendicular) one. (e) Sketch of the spin order at interfaces for the $n = 2$ (left) and $n = 3$ (right) cases. (f) Sketch of experimental setup for resistance measurements. Pink bars are SMO layers. Typical conducting paths via the double-exchange process (black curves) connect NN interfaces, which are broken when $n \geq 3$. Reproduced from [118]. Copyright 2008, the American Physical Society. This figure is reproduced in color in the color section.

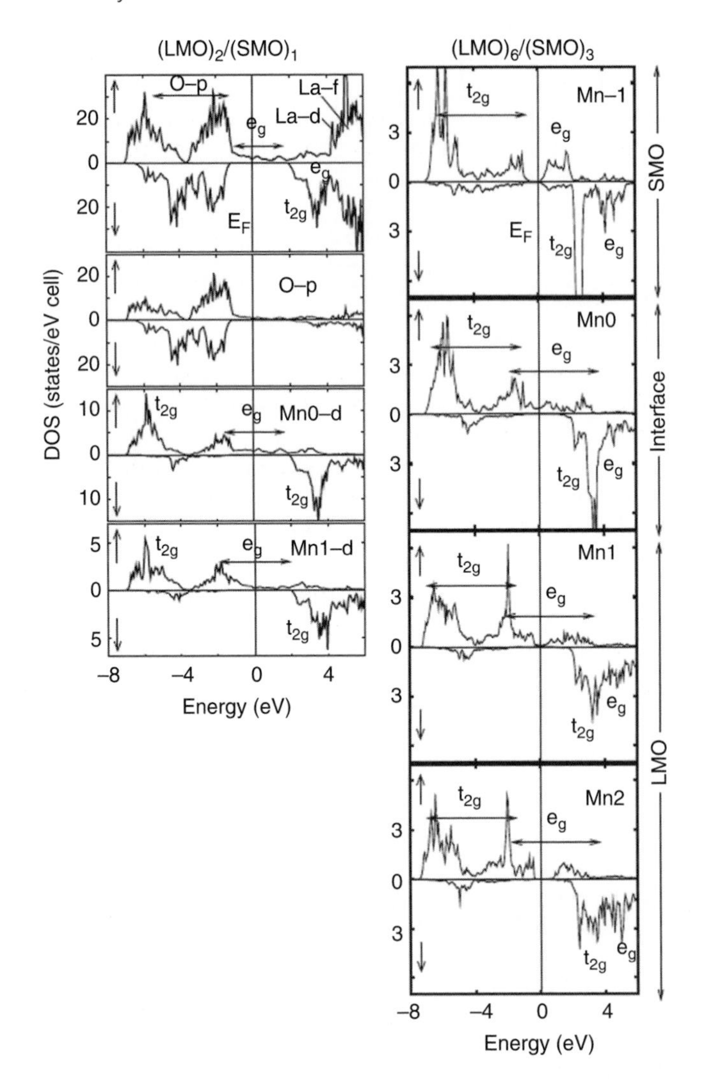

Fig. 9.19 Total (upper panels) and partial spin-resolved DOSs for $(LMO)_{2n}/(SMO)_n$ superlattices. Left: $n = 1$; Right: $n = 3$. Upper and lower segments within each panel correspond to the spin-up and spin-down bands respectively. The x of Mnx stands for the layer distance from the interface, e.g $x = 0$ means the interfacial layer. Reproduced from Ref. [52]. Copyright 2009, the American Physical Society.

transfer between a Mott insulator and a band insulator [119, 120]. In contrast with the in-plane transport properties, the conductivities perpendicular to the interfaces are more complex. For the low-T states, in Dong *et al.*'s model, an MIT occurs at $n = 3$, as shown in Fig. 9.18(d). The key to understanding this MIT is the spin configuration in the SMO region. As shown in Figure 9.18(e), for the $n = 2$ case, the single G-AF layer breaks half of the double-exchange "green channels" across the interface, while the rest of the channels are still conducting. But for the $n = 3$, the G-AF bi-layer breaks all double-exchange channels across the SMO region, giving rise to an insulating behavior.

Thus far, transport measurements have only been on the in-plane direction, as sketched in Figure 9.18(f). However, the perpendicular conductance also plays an important role in these measurements. When the double-exchange channels are all disabled in the SMO region, the in-plane conductance becomes quasi-two-dimensional, which restricts the current near the interfacial layers. In this case, the disorder that generally exists in real heterostructures becomes important, because insulating behavior is likely induced due to the effect of Anderson localization in quasi 2D interfaces. In short, although the in-plane conductivity shows metallic behavior in the simulation, a MIT can still be observed in real superlattices by taking into account disorder effects that may occur due to defects, or intrinsic reasons such as frustrated or pinned spins at AF/FM interfaces [87]. Interestingly, this MIT is associated with the MIT of the perpendicular conductance in the simulation.

The FM + G-AF spin configuration also affects the orbital order at the interfaces. To optimize the double-exchange process between the FM and G-AF layers, more electron population on the $d_{3z^2-r^2}$ orbital is preferred at the interface, to form "hand-in-hand" orbital connections, especially for the SMO layer close to the interface. In addition to the double-exchange process, the substrate strain may also intensively affect orbital occupation, as discussed in Section 9.5.1. If the superlattices are grown on STO substrates, the shrinking c-axis of SMO prefers the $d_{x^2-y^2}$ orbital. Therefore, the tensile strain suppresses the "hand-in-hand" double-exchange connections, and thus reduces the e_g density in the SMO layer, as found in the DFT study. Therefore, orbital orders in real superlattices are determined by the competition between the strain and double-exchange effects.

9.5.7 Half-manganite heterostructures: band lineup and magnetic interactions at interfaces

In the sections presented before, heterostructures involving only manganites have been described in detail. However, there are also several so-called half-manganite heterostructures that consists of a combination of manganites and some other transition-metal oxides (such as high-T_C superconducting cuprates, metallic $CaRuO_3$, multiferroic $BiFeO_3$, or insulating STO). In general, different materials have different work functions (or in other words, different chemical potentials), which drive the itinerant electrons to transfer from the high potential side to the low potential side. Fortunately, in the aforementioned full-manganite heterostructures, the potential difference mainly come from the difference in electron densities on the two sides of the heterostructure, and are fully covered by the theoretical model. However, in half-manganite heterostructures, the potential modulation is more complex. Usually, the $3d$ bands of different cations are intrinsically different in energy, in addition to the Coulomb contribution from A-site cations. When charge transfer is involved, the experimental work functions of the different materials have to be known in order to accurately model the interfaces. In Figure 9.20, the experimental data, obtained by contact potential and photoemission spectroscopy measurements, is summarized for several oxides. With these work functions data, researchers can predict, approximately, the direction of the transfer of charge. In general, the charge transfer across interfaces in half-manganite heterostructures has similar effects as in full-manganite interfaces. For example, by electronic doping via the transfer of charge from

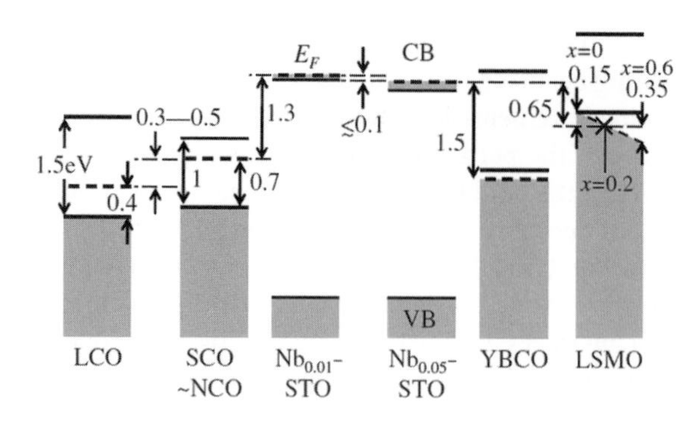

Fig. 9.20 Schematic band diagrams of La_2CuO_4 (LCO), Sm_2CuO_4 (SCO), Nd_2CuO_4 (NCO), 1% Nb doped STO, 5% doped STO, $YBa_2Cu_3O_y$ (YBCO), and LSMO, based on diffusion voltage measurements and photoemission spectroscopy. VB: valence band. CB: conduction band. Chemical potentials are indicated by dashed lines. Reproduced from Ref. [58]. Copyright 2007, the American Physical Society.

manganites, the superconducting state could possibly be turned on/off on the cuprate side in a manganite/cuprate bilayer [58]. While this has yet to be demonstrated, it is interesting to note that a magnetic interaction has been observed between the Cu and Mn atoms at such an interface, which has been attributed to a Cu-O-Mn bond that would be absent in an ideal structure, but seems to be favored under the growth conditions used. Such effects can only come to light with measurements on real systems. [121] Similar predictions for phase tuning via transfer of charge were also reported in other half-manganite heterostructures involving $CaRuO_3$ [62] and STO [122]. Readers are referred to the original literature for more details.

The ideas of band bending and charge transfer at interfaces, as applied to conventional semiconductors and insulators, have to be entirely re-examined in the light of the strong electronic correlations inherent to the 3*d* transition metals. However, these ideas still provide a starting point and are surprisingly resilient. In the perovskites, the valence bands are typically strongly O-2p like in character, while the conduction bands for electrons are formed from the strongly correlated 3*d*-states in the transition metal. Metal/Mott-insulator interfaces were studied by Oka *et al.* using a DMFT approach [123]. They found that the ideas of band bending as used in conventional insulators and semiconductors can be carried forward into the Mott-insulator system, where the lower and upper Hubbard bands play the role of valence and conduction bands. Depending on the details of how the bands line up, one might be able to create either (i) the equivalent of a two-dimensional correlated electron gas in an inversion layer, (ii) Ohmic contacts or (iii) a Schottky barrier. While ideas related to band bending in Mott insulators have also been explored theoretically by a number of other groups for idealized interfaces, in real samples the location of the Fermi level in a material that is nominally an insulator may be entirely determined by defect states that arise due to to lack of control of cation stoichiometry or oxygen vacancies at the 0.1% level, chemical interdiffusion at interfaces during film growth, etc. Thus, it is of primary importance that the relevant electronic properties related to band lineup be determined experimentally in the lab in materials both at surfaces and at interfaces.

In order to carry out controlled studies of band lineup, *n*-STO is a good candidate for a reference material, since its surface can be treated such that we have single atomic layer control on the interface termination, and there are well-established recipes for doping STO in a controlled manner. Thus, one might study junctions between n-STO and a range of correlated oxides that are epitaxially compatible with it. In early work, Tanaka *et al.* fabricated junctions between n-type Nb-STO (1.1×10^{18} /cm^3) and hole-doped $La_{0.9}Ba_{0.1}MnO_3$($T_C > 300$ K on STO) and observed rectifying behavior [124]. The temperature dependence of the junction resistance R_J was found to be a strong function of applied bias, and furthermore dR_J/dT was positive (metallic) in forward bias beyond a threshold voltage of $\sim +1.0$V, while it was negative (insulating) in reverse bias. They also found a striking increase in the temperature at which the resistance maximum in R_J versus T occurs, going from 290 K at $V_{\text{bias}} = 1.0$ V to 340 K at $V_{\text{bias}} = 1.8$ V. Further, a large change in the *in-plane* resistance was observed for a 10-nm-thick film of $La_{0.9}Ba_{0.1}MnO_3$ on n-STO, upon changing the bias voltage of the $p - n$ junction. This method offers an alternative to the usual metal-oxide-semiconductor (MOS) or analogous device paradigms for carrying out electric field-effect gating experiments. While these measurements clearly indicate a strong dependence of the interfacial electronic states on the applied bias, measurements of the low bias interface potential (V_{bi}) may be carried out using internal-photoemission based techniques (Figure 9.21) [125], and have shown that V_{bi} is strongly dependent on the exact termination at the interface [126]. The manner in which the states line up can also affect optical properties, as has been revealed by the observation of new optical transitions attributed to interfacial states in LMO/STO superlattices [127].

In manganite heterostructures, interfacial magnetic coupling and charge transfer are intimately related. For example, a number of groups have explored exchange bias in epitaxial heterostructures and superlattices between antiferromagnetic (typically near $x \sim 0.7$) and ferromagnetic ($x \sim 0.3$) manganite layers [128, 129], where presumably the band lineup is quite good. While the chemical interface between these systems can be atomically sharp [130], the length scale for charge transfer is anticipated to be of order 1–2 unit-cells, and the magnetic exchange interactions that act between nearest neighbors are sensitive to the local doping level. Thus, even though these interfaces may be atomically sharp, the boundaries between different

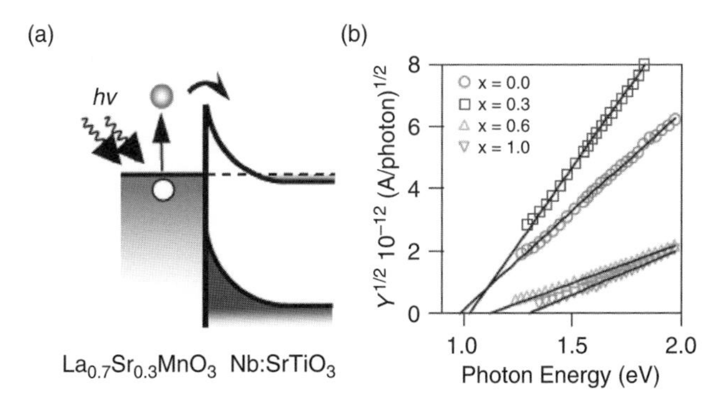

Fig. 9.21 (a) Schematic of band lineup and internal photoemission (IPE). The electrons surmounting the barrier are detected as a photocurrent (b) IPE spectra of $La_{0.7}Sr_{0.3}MnO_3/(SMO)_x$/Nb: $SrTiO_3$ junction at room temperature. The square root of the photo yield is plotted versus photon energy. Reproduced from Ref. [125]. Copyright 2009, American Physical Society.

magnetic structures are presumably more gradual. Furthermore, depending upon the doping level in the AF, the symmetry of the AF and concomitant orbital order can vary widely, and the configuration of spins at a given interface may affect relevant details, such as the degree of compensation or frustration between spins. With first principles approaches, it may be possible to tailor these interface structures by design. In studies where the doping level of the AF and FM layer were systematically varied, it was found that the exchange bias field and coercivity at low temperatures depend systematically on the "average" chemical composition of the interfacial region [131]. While it is not clear if these structures will have technological relevance, they may serve as model systems to study AF/FM interfaces.

In a striking demonstration of the effects of magnetic structure on electronic properties in all-manganite FM/insulator/FM heterostructures, Salafranca *et al.* studied a trilayer [132], composed of two FM $La_{2/3}Sr_{1/3}MnO_3$ electrodes with a thin (2 or 3 atom units) antiferromagnetic barrier $Pr_{2/3}Ca_{1/3}MnO_3$. This spin valve like structure shows high values of tunneling magnetoresistance (TMR). If the spins in the two LSMO electrodes are parallel, the electronic state of the PCMO slab is *metallic*, while it becomes *insulating* if the spins in the two electrodes are antiparallel. Therefore, it is possible to reversibly switch the system through a metal insulator transition by altering the magnetic alignment of the FM electrodes. In addition, the resistance of this heterostructure decreases with the applied magnetic field, presenting a large negative magnetoresistance. In related theoretical work, *A*-AF LMO was studied as the barrier in a manganite spin valve. Yunoki *et al.* studied the LSMO-LMO-LSMO trilayer using a one-orbital model [60]. In the ground state, the spins in the two LSMO leads are parallel (antiparallel) when odd (even) number *A*-AF LMO presents. The antiparallel-lead configuration has a large resistance which can be suppressed drastically by a small magnetic field, giving rise to a large negative TMR, as shown in Figure 9.22.

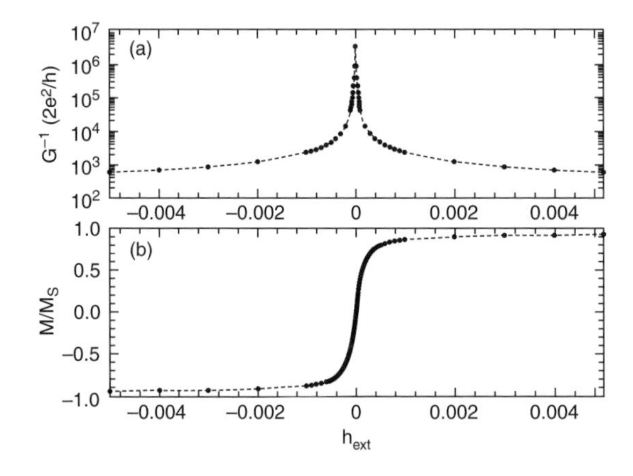

Fig. 9.22 (a) Resistance versus magnetic field h_{ext} for an even number of LMO sites in the central barrier at zero T. The couplings used are $J_{\text{H}} = 8.0$ and $a = 1.0$. Notice that the large changes in resistance occur at very small magnetic fields. (b) Total magnetization M of the classical t_{2g} spins versus h_{ext}. M_S is the saturation magnetization. Reproduced from Ref. [60]. Copyright 2008, the American Physical Society.

Manganites have also been incorporated in superlattices with $LaNiO_3$, which is a paramagnetic metal, to yield structures that have very diverse magnetic and electronic properties. There is significant evidence to indicate that the Ni-O-Mn bonds formed at the interface allow charge transfer and magnetic exchange interactions between the Mn and Ni sites. Oscillatory magnetic couplings between FM metallic manganite layers, which change sign as a function of the thickness of the intervening metallic nickelate layer, have been observed in manganite-nickelate superlattices [133], reminiscent of coupling in conventional metallic heterostructures. On the other hand, in LMO/LNO [111] superlattices, which when fully ordered constitute a double perovskite structure, it is believed that an electron is transferred from Mn^{3+} to Ni^{3+} leading to a Mn^{4+}-O-Ni^{2+} bond, and in accordance with the Goodenough–Kanamori rules, a ferromagnetic insulator is obtained with a $T_C \sim 300$ K [134, 135].

In addition to charge-transfer driven phase tuning, there are phenomena that involve interfacial exchange interactions in these half-manganite heterostructures. For example, the exchange bias effect has been observed in the $SrRuO_3$/SMO [136, 137, 138] LSMO/$BiFeO_3$ [139, 140] and more recently in LMO/LNO [111] heterostructures [146]. Here, the ground states of SMO and $BiFeO_3$ (BFO) are both *G*-AF, while LSMO ($x \sim 0.3$) and $SrRuO_3$ (SRO) are ferromagnetic at low temperature. In contrast, the exchange bias effect has not been observed in the above studied $(LMO)_{2n}/(SMO)_n$ superlattices [87] although at least the $n = 2$ case also involves FM + *G*-AF structures, as predicted by both the model and DFT calculations [47, 118]. Usually, in the case of a compensated antiferromagnetic interface, such as the [001] plane of *G*-AF materials, a tiny (frozen) uncompensated magnetic moment originating from extrinsic factors, such as interface roughness or spin canting, is considered to be responsible for exchange bias [141]. However, Dong *et al.* recently proposed two (related) alternative possible mechanisms, which are driven by the intrinsic Dzyaloshinskii–Moriya interaction and ferroelectric polarization, to rationalize the existence of exchange bias effects in perovskites with G-type antiferromagnetism [142].

As shown in Figure 9.23(a), although the staggered AF spins $\vec{S}_i$ cancel the total superexchange interaction ($\sim \vec{S}_i \cdot \vec{S}_j$, $\vec{S}_j$ denote the ferromagnetic spins) at the *G*-AF interface, the Dzyaloshinskii–Moriya interaction ($\sim \vec{D}_{ij} \cdot (\vec{S}_i \times \vec{S}_j)$, $\vec{D}_{ij}$ is also staggered) gives rise to a uniform effective field which breaks the inversion symmetry. This effective field is perpendicular to the antiferromagnetic spin's direction as well as to the $\vec{D}_{ij}$ vectors. Therefore, if the spins in both the FM and antiferromagnetic sides are in-plane as in the $(LMO)_{2n}/(SMO)_n$ superlattices, this effective field does not contribute to the exchange bias. However, it works in the SRO/SMO superlattices, in which the spins of Ru are out-of-plane, perpendicular to the in-plane spins of Mn.

If a multiferroic material is involved, there is a finite ferroelectric polarization, implying a uniform shift for all oxygen anions relative to the cations. In the ferroelectric state, the bond angles across the interfaces are no longer symmetric considering the original staggered distortion, as shown in Figure 9.23(b). The normal superexchanges of these staggered bent bonds give rise to one more effective field to induce exchange bias, because the strength of the superexchange J is modulated accompanying the antiferromagnetic spins.

Besides these half-manganite heterostructures, the electronic reconstruction at the surface of manganites (or in other words the interface between the manganites and the vacuum) has also been studied using the double-exchange model [143, 144] and DFT [145]. Both the existence of a ferromagnetic surface in antiferromagnetic manganites, or an antiferromagnetic surface in ferromagnetic manganites have been predicted, depending on the details. For more

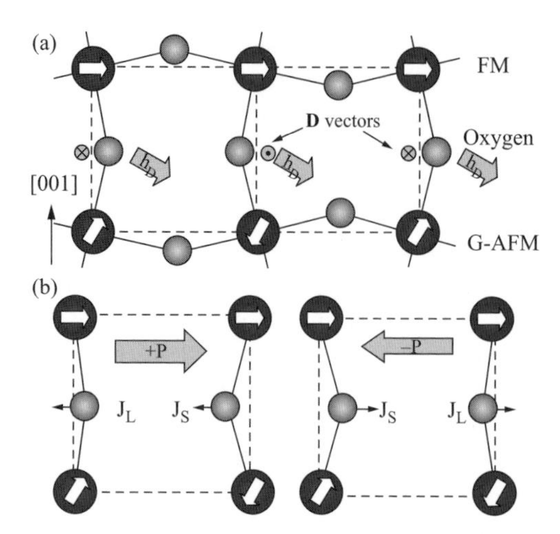

Fig. 9.23 (a) Sketch of the interface between FM and *G*-AF perovskites, including oxygen octahedral tilting. The staggered directions of the $\vec{D}_{ij}$ vectors at the interface are marked as in- and out-arrows, which are determined by the oxygen movements from bond midpoints. The uniform effective field is also plotted as arrows (h_D) near the oxygens. (b) FE-polarization-driven asymmetric bond angles and modulated normal superexchange strengths at the interface. A switch of the FE polarization (from left to right or vice versa) also switches the modulation of superexchange strengths and thus the effective field. Reproduced from Ref. [141]. Copyright 2009, the American Physical Society.

information and references regarding surface reconstructions, readers are referred to the original literature.

9.6 Conclusions and outlook

In the sections given above, we have provided a brief overview of recent developments in manganite multilayer structures. Theoretically, the appeal of incorporating manganites in heterostructures lies in the fact that they can be addressed using models that have been widely applied and tested in bulk materials. Experimentally, manganites have been incorporated into a broad range of heterostructures, involving many different kinds of materials including titanates, cuprates, and nickelates. The work reviewed here has covered only a limited selection of current research in this very active area, and is by no means complete. Nonetheless, some general conclusions do emerge about how we understand these heterostructures. First, the central phenomenon that drives interfacial physics is the charge redistribution driven by the difference of work functions and orbital and spin configurations at the interface. This leads to charge transfer across the interface until the work function difference is eventually compensated by the electrostatic potential of the redistributed charges. Hence the long-range Coulomb interactions among electrons and between electrons and A-site cations play a very central role, and must be properly treated in heterostructures. Second, in general, an electronically homogeneous state does not exist in a heterostructure as a consequence of the intrinsic length scales for screening and charge redistribution. In situations where the period of a superlattice is shorter than this intrinsic length scale, a relatively homogenous state may be obtained, whose properties may be quite different from that of the constituent materials. In general, the properties of the *RMO*/*AMO* heterostructure are closely associated with the phase diagram of the bulk compound (*R*,*A*)MO. While some local states in the heterostructure are directly related to bulk phases with the same e_g density, others may not have a counterpart in the bulk phase diagram, but can be regarded as exotic interpolations between bulk-like phases. Third, besides charge redistribution, substrate strain also provides a way towards tuning collective states in manganite heterostructures, in a manner that is very different from conventional semiconductor superlattices. Lattice deformation caused by substrate strain may affect orbital order via changing the Jahn–Teller splitting of e_g orbitals, which effectively changes the band structure and charge itinerancy at the interface. Furthermore, it may also alter the t_{2g} superexchange pathway and consequently the magnetic structure at the interface, which also influences the electronic properties of the electrons in the e_g states through Hund's coupling.

Theoretical investigation on manganite heterostructures is currently at its early stages. Several important issues are still awaiting further study. The reader should realize that even for the studies reviewed in the previous sections, the theoretical results and interpretations do not provide the final answer to the many issues found experimentally. Given the broad range of possibilities, more theoretical work should be performed in the future, adapting to the rapidly developing experimental results in this exciting area of research. In previous studies, some details of the crystal lattice of these heterostructures, especially the Mn-O-Mn bonds lengths and angles, have not been fully taken into account, neither in model studies, nor in DFT calculations, although they may play very important roles in some heterostructures. Modeling these realistic distortions introduces extra challenges to current theory, as well as to the current computational resources.

Until now, most theoretical and experimental efforts have been focused on wide bandwidth LMO/SMO superlattices. As we have already emphasized, it is only the beginning of the story. Given their richer phase diagrams, more diverse phenomena would be expected in narrow bandwidth manganite superlattices. For example, in narrow bandwidth compounds such as $TbMnO_3$, $DyMnO_3$, and $HoMnO_3$, their multiferroicity is attracting much attention. It is expected that exciting physics will be revealed in potential multifunctional devices based on heterostructures and thin films synthesized with these types of materials.

Acknowledgments

A.B. was supported by the US Department of Energy, Office of Science, Office of Basic Energy Sciences, under contact No DE-AC02-06CH11357. R.Y. was supported by NSF Grant No. DMR-1006985 and the Robert A. Welch Foundation Grant No. C-1411. S.D. was supported by the 973 Project of China (2011CB922101), NSFC (11004027), and NCET (10-0325).

References

[1] E. Dagotto, Science **318**, 1076 (2007).
[2] H. Takagi and H. Y. Hwang, Science **327**, 1601 (2010).
[3] J. Mannhart and D. G. Schlom, Science **327**, 1607 (2010).
[4] G. H. Jonker and J. H. Vansanten, Physica (Utrecht) **16**, 337 (1950).
[5] E. O. Wollan and W. C. Koehler, Phys. Rev. **100**, 545 (1955).

[6] R. Vonhelmolt, J. Wecker, B. Holzapfel, L. Schultz, and K. Samwer, Phys. Rev. Lett. **71**, 2331 (1993).
[7] S. Jin, T. H. Tiefel, M. Mccormack, *et al.*, Science **264**, 413 (1994).
[8] A. Moreo, S. Yunoki, and E. Dagotto, Science **283**, 2034 (1999).
[9] E. Dagotto, T. Hotta, and A. Moreo, Phys. Rep. **344**, 1 (2001).
[10] E. Dagotto, *Nanoscale Phase Separation and Colossal Magnetoresistance.* Springer (Berlin), (2002).
[11] A. Urushibara, Y. Moritomo, T. Arima, *et al.* Phys. Rev. B **51**, 14103 (1995).
[12] O. Chmaissem, B. Dabrowski, S. Kolesnik, *et al.* Phys. Rev. B **67**, 094431 (2003).
[13] J. Hemberger, A. Krimmel, T. Kurz, *et al.* Phys. Rev. B **66**, 094410 (2002).
[14] I. Loa, P. Adler, A. Grzechnik, *et al.* Phys. Rev. Lett. **87,** 125501 (2001).
[15] A. Yamasaki, M. Feldbacher, Y.-F. Yang, O. K. Andersen, and K. Held, Phys. Rev. Lett. **96**, 166401 (2006).
[16] T. Negas and R. S. Roth, J. Sol. State. Chem. **1**, 409 (1970).
[17] Y. Tokura and Y. Tomioka, J. Magn. Magn. Mater. **200**, 1 (1999).
[18] Y. Tokura, Rep. Prog. Phys. **69**, 797 (2006).
[19] M. B. Salamon and M. Jaime, Rev. Mod. Phys. **73**, 583 (2001).
[20] C. Zener, Phys. Rev. **81**, 440 (1951).
[21] C. Zener, Phys. Rev. **82**, 403 (1951).
[22] P. W. Anderson and H. Hasegawa, Phys. Rev. **100**, 675 (1955).
[23] J. C. Slater and G. F. Koster, Phys. Rev. **94**, 1498 (1954).
[24] A. J. Millis, Nature **392**, 147 (1998).
[25] J. Kanamori, J. Appl. Phys. **31**, 14S (1960).
[26] T. Hotta, A. L. Malvezzi, and E. Dagotto, Phys. Rev. B **62**, 9432 (2000).
[27] A. Georges, G. Kotliar, W. Krauth, and M. J. Rozenberg, Rev. Mod. Phys. **68**, 13 (1996).
[28] T. S. Santos, S. J. May, J. L. Robertson, and A. Bhattacharya, Phys. Rev. B **80**, 155114 (2009).
[29] D. D. Berkley, A.M. Goldman, B. R. Johnson, J. Morton, and T. Wang, Rev. Sci. Instrum. **60**, 3769 (1989).
[30] A. Bhattacharya, X. Zhai, M. Warusawithana, J. N. Eckstein, and S.D Bader, Appl. Phys. Lett. **90**, 222503 (2007).
[31] C. Adamo, X. Ke, P. Schiffer, *et al.* Appl. Phys. Lett. **92**, 112508 (2008).
[32] J.-P. Locquet and E. Machler, J. Vac. Sci. Tecnol. A **10**, 3100 (1992).
[33] A. Ichimiya and P. I. Cohen, *Reflection High Energy Electron Diffraction*, Cambridge University Press, (2004).
[34] B. Mercey, P. A. Salvador, Ph. Lecoeur, *et al.* J. Appl. Phys. **94**, 2716 (2003).
[35] M. Kawasaki, M. Izumi, Y. Konishi, T. Manako, and Y. Tokura, Mater. Sci. Eng. B **63**, 49 (1999).
[36] W.-K. Chu, J. W. Mayer, and M.-A. Nicolet, *Backscattering Spectrometry*, Academic Press Inc., (1978).
[37] M. E. Klausmeier-Brown, J. N. Eckstein, I. Bozovic, and G. F. Virshup, Appl. Phys. Lett. **60**, 657 (1992).
[38] Y. Kasai and S. Sakai, Rev. Sci. Instrum. **7**, 2850 (1997).
[39] Panalytical has an X-ray reflectivity analysis package with its XRD suite of software.
[40] L. G. Parratt, Phys. Rev. **95**, 359 (1954).

[41] E. Fullerton, Ivan K. Schuller, H. Vanderstraeten, and Y. Bruynseraede, Phys. Rev. B **45**, 9292 (1992).
[42] J. Verbeeck, O.I. Lebedev, G. Van Tendeloo, and B. Mercey, Phys. Rev. B **66**, 184426 (2002).
[43] S. J. May, A. B. Shah, S. G. E. te Velthuis, *et al.* Phys. Rev. B **77**, 174409 (2008).
[44] A. Gupta, T. R. McGuire, P. R. Duncombe, *et al.* Phys. Lett. **67**, 3494 (1995).
[45] P. Orgiani, C. Aruta, R. Ciancio, A. Galdi, and L. Maritato, Appl. Phys. Lett. **95**, 013510 (2009).
[46] H. Yamada, M. Kawasaki, T. Lottermoser, T. Arima, and Y. Tokura, Appl. Phys. Lett. **89**, 052506 (2006).
[47] B. R. K. Nanda and S. Satpathy, Phys. Rev. B **78**, 054427 (2008).
[48] C. Adamo, X. Ke, H. Q. Wang, *et al.* Appl. Phys. Lett. **95**, 112504 (2009).
[49] A. J. Millis, T. Darling, and A. Migliori, J. Appl. Phys. **83**, 1588 (1998).
[50] H. Yamada, Y. Ogawa, Y. Ishii, *et al.* Science **305**, 646 (2004).
[51] Z. Fang, I. V. Solovyev, and K. Terakura, Phys. Rev. Lett. **84**, 3169 (2000).
[52] B. R. K. Nanda and S. Satpathy, Phys. Rev. B **79**, 054428 (2009).
[53] J. H. Lee and K. M. Rabe, preprint arXiv:0910.5438.
[54] T. Kiyama, Y. Wakabayashi, H. Nakao, *et al.* J. Phys. Soc. Jpn. **72**, 785 (2003).
[55] L. L. Chang, L. Esaki, and R. Tsu, Appl. Phys. Lett. **24**, 593 (1974).
[56] D. Mukherji and B. R. Nag, Phys. Rev. B **12**, 4338 (1974).
[57] C. W. Lin, S. Okamoto, and A. J. Millis, Phys. Rev. B **73**, 041104(R) (2006).
[58] S. Yunoki, A. Moreo, E. Dagotto, *et al.* Phys. Rev. B **76**, 064532 (2007).
[59] I. González, S. Okamoto, S. Yunoki, A. Moreo, and E. Dagotto, J. Phys.: Condens. Matter **20**, 264002 (2008).
[60] S. Yunoki, E. Dagotto, S. Costamagna, and J. A. Riera, Phys. Rev. B **78**, 024405 (2008).
[61] C. W. Lin and A. J. Millis, Phys. Rev. B **78**, 184405 (2008).
[62] B. R. K. Nanda, S. Satpathy, and M. S. Springborg, Phys. Rev. Lett. **98**, 216804 (2007).
[63] Smadici, P. Abbamonte, A. Bhattacharya, *et al.* Phys. Rev. Lett. **99**, 196404 (2007).
[64] W. E. Pickett and D. J. Singh, Phys. Rev. B **55**, R8642 (1997).
[65] J. P. Attfield, Chem. Mater. **10**, 3239 (1998).
[66] J.-S. Zhou and J. B. Goodenough, Phys. Rev. B **77**, 132104 (2008).
[67] M. Imada, A. Fujimori, and Y. Tokura, Rev. Mod. Phys. **70**, 1039 (1998).
[68] R. B. Griffiths, Phys. Rev. Lett. **23**, 17 (1969).
[69] J. Y. Gu, S. D. Bader, H. Zheng, J. F. Mitchell, and J. E. Gordon, Phys. Rev. B **70**, 054418 (2004).
[70] M. B. Salamon, P. Lin, and S. H. Chun, Phys. Rev. Lett. **88**, 197203 (2002).
[71] J. Deisenhofer, D. Braak, H.-A. Krug von Nidda, *et al.* Phys. Rev. Lett. **95**, 257202 (2005).
[72] F. Millange, V. Caignaert, B. Domenges, and B. Raveau, Chem. Mater. **10**, 1974 (1998).
[73] T. J. Sato, J. W. Lynn, and B. Dabrowski, Phys. Rev. Lett. **93**, 267204 (2004).
[74] D. Akshoshi, M. Uchida, Y. Tomioka, *et al.* Phys. Rev. Lett. **90**, 177203 (2003).
[75] Y. Motome, N. Furukawa, and N. Nagaosa, Phys. Rev. Lett. **91**, 167204 (2003).
[76] C. en, G. Alvarez and E. Dagotto, Phys. Rev. B **70**, 064428 (2004).
[77] C. en, G. Alvarez, H. Aliaga, and E. Dagotto, Phys. Rev. B **73**, 224441 (2006).
[78] S. Kumar, A. P. Kampf, and P. Majumdar, Phys. Rev. B. **75**, 014209 (2007).
[79] S. Kumar and P. Majumdar, Phys. Rev. Lett. **91**, 246602 (2003).

[80] S. Kumar and P. Majumdar, Phys. Rev. Lett. **96**, 176403 (2006).
[81] J. Salafranca and L. Brey, Phys. Rev. B **73**, 214404 (2006).
[82] G. Bouzerar and O. Cepas, Phys. Rev. B **76**, 020401(R) (2007).
[83] H. Aliaga, D. Magnoux, A. Moreo, *et al.* Phys. Rev. B **68**, 104405 (2003).
[84] J. Burgy, A. Moreo, and E. Dagotto, Phys. Rev. Lett. **92**, 097202 (2004).
[85] G. Alvarez, H. Aliaga, C. en, and E. Dagotto, Phys. Rev. B **73**, 224426 (2006).
[86] S. Kumar and A. P. Kampf, Phys. Rev. Lett. **100**, 076406 (2008).
[87] A. Bhattacharya, S. J. May, S. G. E. te Velthuis, *et al.* Phys. Rev. Lett. **100**, 257203 (2008).
[88] H. Fujishiro, T. Fukase, and M. Ikebe, J. Phys. Soc. Jpn. **67**, 2582 (1998).
[89] Y. Tokura and N. Nagaosa, Science **288**, 462 (2000).
[90] Y. Konishi, F. Zhong, M. Izumi, *et al.* J. Phys. Soc. Jpn. **68**, 3790 (1999).
[91] J. B. Goodenough, *Magnetism and the Chemical Bond*, Interscience Publishers, (1963), Chap. 3.
[92] S. J. May, P. J. Ryan, R. L. Robertson, *et al.* Nature Mater. **8**, 892 (2009).
[93] R. Pynn, Nature **281**, 433 (1979).
[94] P. Bak, Rep. Prog. Phys. **45**, 587 (1982).
[95] E. D. Isaacs, G. Aeppli, P. Zschack, *et al.* Phys. Rev. Lett. **72**, 3421 (1994).
[96] J. K. Kjems, L. Passell, H. Taub, J. G. Dash, and A. D. Novaco, Phys. Rev. B **13**, 1446 (1976).
[97] M. D. Chinn and S. C. Fain. Jr., Phys Rev. Lett. **39**, 146 (1977).
[98] M. Izumi, J. D. Axe, and G. Shirane, Phys. Rev. B **15**, 4392 (1977).
[99] S. Wakimoto, H. Ki,ura, M. Fujita, *et al.* J. Phys. Soc. Jpn. **75**, 074714 (2006).
[100] H. Kuwahara, T. Okuda, Y. Tomioka, A. Asamitsu, and Y. Tokura, Phys. Rev. Lett. **82**, 4316 (1999).
[101] L. P. Gorkov and V. Z. Kresin, Phys. Rep. **400**, 149 (2004).
[102] A. S. Nunez, R. A. Duine, P. Haney, and A. H. MacDonald, Phys. Rev. B **73**, 214426 (2006).
[103] P.-G. de Gennes, Phys. Rev. **118**, 141 (1960).
[104] J. van den Brink and D. Khomskii, Phys. Rev. Lett. **82**, 1016 (1999).
[105] I. V. Solovyev and K. Terakura, Phys. Rev. Lett. **83**, 0031 (1999).
[106] T.S. Santos, B.J. Kirby, S. Kumar, *et al. Phys. Rev. Lett.* **107**, 167202 (2011).
[107] H. Kawano, R. Kajimoto, H. Yoshizawa, *et al.* Phys. Rev. Lett. **78**, 4253 (1997).
[108] M. Konoto, T. Kohashi, K. Koike, *et al.* Phys. Rev. Lett. **93**, 107201 (2004).
[109] V. V. Krishnamurthy, J. L. Robertson, R. S. Fishman, M. D. Lumsden, and J. F. Mitchell, Phys. Rev. B **73**, 060404 (2006).
[110] P. K. Muduli, S. K. Bose, and R. C. Budhani, J. Phys.: Condens. Matter. **19**, 226204 (2007).
[111] Y. Uozu, Y. Wakabayashi, Y. Ogimoto, *et al.* Phys. Rev. Lett. **97**, 037202 (2006).
[112] T. Koida, M. Lippmaa, T. Fukumura, *et al.* Phys. Rev. B **66**, 144418 (2002).
[113] T. Satoh, K. Miyano, Y.Ogimoto, H. Tamaru, and S. Ishihara, Phys. Rev. B **72**, 224403 (2005).
[114] R. Yu, S. Yunoki, S. Dong, and E. Dagotto, Phys. Rev. B **80**, 125115 (2009).
[115] M. J. Calderon, J. Salafranca, and L. Brey, Phys. Rev. B **78**, 024415 (2008).
[116] P. A. Salvador, A. M. Haghiri-Gosnet, B. Mercey, M. Hervieu, and B. Raveau, Appl. Phys. Lett. **75**, 2638 (1999).
[117] C. Aruta, C. Adamo, A. Galdi, *et al.* Phys. Rev. B **80**, 140405(R) (2009).

[118] S. Dong, R. Yu, S. Yunoki, *et al.* Phys. Rev. B **78**, 201102 (R) (2008).
[119] S. Okamoto and A. J. Millis, Nature **428**, 630 (2004).
[120] S. S. Kancharla and E. Dagotto, Phys. Rev. B **74** , 195427 (2006).
[121] J. Chakhalian, J. W. Freeland, G. Srajer, *et al.* Nature Phys. **2**, 244 (2006).
[122] L. Brey, Phys. Rev. B **75**, 104423 (2007).
[123] T. Oka and N. Nagaosa, Phys. Rev. Lett. **95**, 266403 (2005).
[124] H. Tanaka, J. Zhang, and T. Kawai, Phys. Rev. Lett. **88**, 027204 (2002).
[125] Y. Hikita, Y. Kozuka, T. Susaki, H. Takagi, and H. Y. Hwang, Appl. Phys. Lett. **90**, 143507 (2007).
[126] Y. Hikita, M. Nishikawa, T. Yajima, and H. Y. Hwang, Phys. Rev. B **79**, 073101 (2009).
[127] X. Zhai, C. S. Mohapatra, A. B. Shah, J.-M. Zuo, and J. N. Eckstein, Adv. Mater. **22**, 1136 (2010).
[128] I. Panagiotopoulos, C. Christides, M. Pissas, and D. Niarchos, Phys. Rev. B **60**, 485 (1999).
[129] K. R. Nikolaev, I. N. Krivorotov, W. K. Cooley, *et al.* Appl. Phys. Lett. **76**, 6951 (2000).
[130] A. L. Kobrinskii, A. M. Goldman, M. Varela, and S. J. Pennycook, Phys. Rev. B **79**, 094405 (2010).
[131] N. Moutis, C. Christides, I Panagiotopoulos, and D. Niarchos, Phys. Rev. B **64**, 094429 (2001).
[132] J. Salafranca, M. J. Calderon, and L. Brey, Phys. Rev. B **77**, 014441 (2008).
[133] K. R. Nikolaev, A. Yu. Dobin, I. N. Krivorotov, *et al.* Phys. Rev. Lett. **85**, 3728 (2000).
[134] H. Guo, J. Burgess, S. Street, *et al.* Appl. Phys. Lett. **89**, 022509 (2006).
[135] M. P. Singh, K. D. Truong, S. Jandl, and P. Fournier, Phys. Rev. B **79**, 224421 (2009).
[136] P. Padhan and W. Prellier, Phys. Rev. B **72**, 104416 (2005).
[137] P. Padhan and W. Prellier, Appl. Phys. Lett. **88**, 263114 (2006).
[138] Y. Choi, Y. Z. Yoo, O. Chmaissem, *et al.* Appl. Phys. Lett. **91**, 022503 (2007).
[139] P. Yu, J.-S. Lee, S. Okamoto, *et al.* Phys. Rev. Lett. **105**, 027201 (2010).
[140] S. M.Wu, S. A. Cybart, P. Yu, *et al.* Nature Mater. **9**, 756 (2010).
[141] M. Kiwi, J. Magn. Magn. Mater. **234**, 584 (2001).
[142] S. Dong, K. Yamauchi, S. Yunoki, *et al.* Dagotto, Phys. Rev. Lett. **103**, 127201 (2009).
[143] M. J. Calderon, L. Brey, and F. Guinea, Phys. Rev. B **60**, 6698 (1999).
[144] S. Dong, R. Yu, S. Yunoki, J.-M. Liu, and E. Dagotto, Phys. Rev. B **78**, 064414 (2008).
[145] Z. Fang and K. Terakura, J. Phys. Soc. Jpn. **70**, 3356 (2001).
[146] M. Gilbert, P. Zubko, R. Scherwitzl *et al.*, Nature Materials **11**, 195 (2011).

10
Thermoelectric oxides: films and heterostructures

Hiromichi Ohta and Kunihito Koumoto

10.1 Introduction

Today, many energy resources are discharged as waste heat into the environment without being used. Such exhaust heat comprises approximately 60% of primary energy. Thermoelectric energy conversion technology attracts great attention to convert waste heat into electricity. The principle of thermoelectric energy conversion was first discovered by T. J. Seebeck in 1821 [1]. He found that a voltage is generated between two ends of a metal bar by introducing a temperature difference in the bar. Thus, when electric loads are connected at both ends of the metal bar, an electric current can be obtained.

In order to realize efficient thermoelectric energy conversion, three physical properties are required for thermoelectric materials, as follows:

(1) Low thermal conductivity (κ), which is required to introduce a large temperature difference between both ends of the material.
(2) High electrical conductivity (σ), which is required to reduce the internal resistance of the material.
(3) Large thermopower (Seebeck coefficient, S), which is required to obtain a high voltage.

Generally, the performance of thermoelectric materials is evaluated in terms of a dimensionless figure of merit, $ZT = S^2 \cdot \sigma \cdot T \cdot \kappa^{-1}$, where Z and T are the figure of merit and the absolute temperature, respectively. For practical thermoelectric application, thermoelectric materials having $ZT > 1$ are fundamentally required [2].

Figure 10.1 summarizes ZT versus T curves for conventional "heavy metal based" thermoelectric materials. The ZT values of several materials such as Bi_2Te_3 and PbTe are $\sim$1, enough for practical applications [2–8]. However, these materials are not attractive, particularly operating at high temperatures ($T \sim 1000$ K), because decomposition, vaporization, or melting of the constituents can easily occur at high temperatures. Furthermore, the use of these heavy metals should be limited to specific environments, such as space, since they are mostly toxic and low in abundance as natural resources, and thus not environmentally benign.

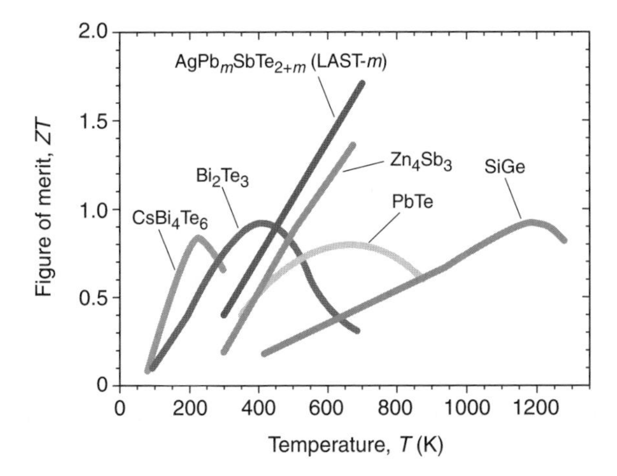

Fig. 10.1 Thermoelectric figure of merit, ZT versus temperature for conventional heavy metal based materials. The ZT values of several materials such as Bi_2Te_3 and PbTe are $\sim$1, enough for practical applications.

Based on this background, recently, metal oxides have attracted much attention, especially in Japan, as thermoelectric power generation materials at high temperature, on the basis of their potential advantages over heavy metallic alloys in terms of chemical and thermal robustness [9]. Although there are number of reports on the thermoelectric properties of metal oxides, intrinsic properties are very hard to obtain because most of the reported materials comprise very small single crystals and/or polycrystalline ceramics, which are composed of non-oriented grains with many grain boundaries, pores, and impurities.

In order to clarify the intrinsic thermoelectric properties of oxides, high-quality epitaxial films of several thermoelectric oxides, including Na_xCoO_2 [10, 11, 12], Li_xCoO_2 [13], Sr_xCoO_2 [14], Ca_xCoO_2 [15], $Ca_3Co_4O_9$ [15, 16], $SrTiO_3$;Nb [17], TiO_2:Nb [18], and $SrO(SrTiO_3)$:Nb [19] have been fabricated.

In this chapter, the thermoelectric properties of two representative oxide epitaxial films are reviewed: p-type $Ca_3Co_4O_9$ and n-type $SrTiO_3$:Nb, which exhibit the best thermoelectric ZT among oxide thermoelectric materials so far reported. Two recent topics are also reviewed: the giant S of two-dimensional electrons confined within a unit-cell layer thickness ($\sim$0.4 nm) of $SrTiO_3$ [20–22] and the field modulation of S for $SrTiO_3$.

10.2 p-type layered cobalt oxide: $Ca_3Co_4O_9$ films

Figure 10.2 shows schematic crystal structures of (left) A_xCoO_2 (A = Li, Na, Ca, Sr, x: x-value of Na_xCoO_2 varies from 0.3 to 0.9) [23, 24] and (right) $Ca_3Co_4O_9$. The A_xCoO_2 crystal is composed of an alternating stack of CdI_2-type CoO_2 layer and an A^+ or A^{2+} layer along the c-axis, while the $Ca_3Co_4O_9$ crystal is composed of a CoO_2 layer and a rock-salt-type $Ca_2CoO_3^+$ layer [25]. Since positive hole carrier conduction occurs predominantly in the CoO_2 layer [26], c-axis-oriented epitaxial films are preferable to utilize their intrinsic thermoelectric properties.

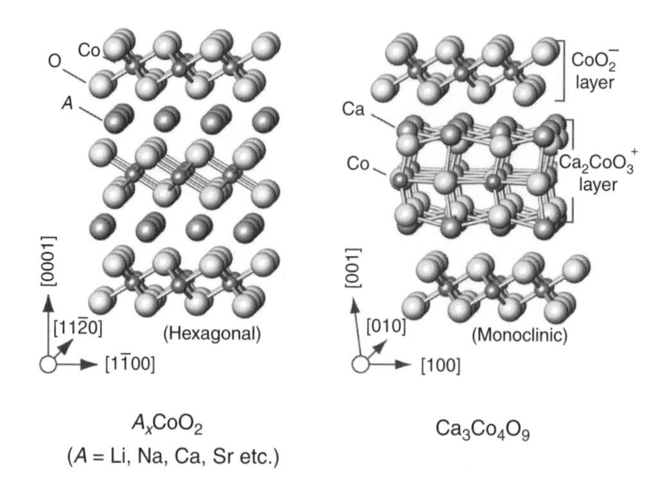

Fig. 10.2 Schematic crystal structures of layered cobalt oxides [left: A_xCoO_2 (A = Li, Na, Ca, Sr), right: $Ca_3Co_4O_9$]. Both crystals have a CoO_2 layer composed of edge-shared CoO_6 octahedra. Note that positive hole carrier conduction occurs predominantly in the CoO_2 layer.

In order to fabricate c-axis-oriented epitaxial films of layered cobalt oxides, Ohta *et al.* developed a specialized film growth method for the reactive solid-phase epitaxy (R-SPE) [10], which is a powerful method for fabricating a single crystalline film of layered oxides [27, 28], followed by topotactic ion exchange. First, a high-quality epitaxial film of CoO was deposited on a (0001)-face of α-Al_2O_3 substrate at 700 °C by the pulsed laser deposition (PLD) method, using a Co_3O_4 sintered disk as the target. Then, the PLD-deposited CoO film was heated, together with $NaHCO_3$ powder, at 700 °C in air. After that, we obtained the c-axis-oriented Na_xCoO_2 (x ~0.8) epitaxial film. The Na_xCoO_2 film can then be converted into Li_xCoO_2 [13], Sr_xCoO_2 [14], Ca_xCoO_2 [15], and $Ca_3Co_4O_9$ [15, 16] epitaxial films by appropriate ion-exchange treatment. Details of the film growth method are described elsewhere.

Figure 10.3 (left panel) shows out-of-plane X-ray Bragg diffraction patterns of (a) Na_xCoO_2 and (b) $Ca_3Co_4O_9$ films. Intense Bragg diffraction peaks of 000l Na_xCoO_2 are seen together with 0006 α-Al_2O_3 in Figure 10.3(a), indicating high c-axis orientation of the film. The chemical composition of the film was evaluated to be $Na_{0.83}CoO_2$ by X-ray fluorescence (XRF, ZSX 100e, Rigaku Co.) analysis. After the ion-exchange treatment of Na^+ with Ca^{2+} [29], the XRD pattern changed dramatically, as shown in Figure 10.3(b), which indicates that the $Na_{0.83}CoO_2$ epitaxial film was converted into highly c-axis-oriented $Ca_3Co_4O_9$ film by the ion-exchange treatment.

Figure 10.3(right panel) shows in-plane X-ray Bragg diffraction patterns of (c) Na_xCoO_2 and (d) $Ca_3Co_4O_9$ films. Although only one intense Bragg peak of 11-20 $Na_{0.83}CoO_2$ is seen, together with 3-300 α-Al_2O_3 in Figure 10.3(c), two independent Bragg peaks at $2\theta\chi/\varphi$ ~40 and ~ 66° are seen in Figure 10.3(d). These peaks correspond to the 020 diffractions of the rock-salt-type Ca_2CoO_3 layer (b_1 = 0.455 nm) and that of the CdI_2-type CO_2 layer (b_2 = 0.282 nm), respectively, due to the fact that $Ca_3Co_4O_9$ is composed of these two layers, with different lattice constants along the b-axis [25]. A sixfold symmetrical in-plane rocking curve is clearly

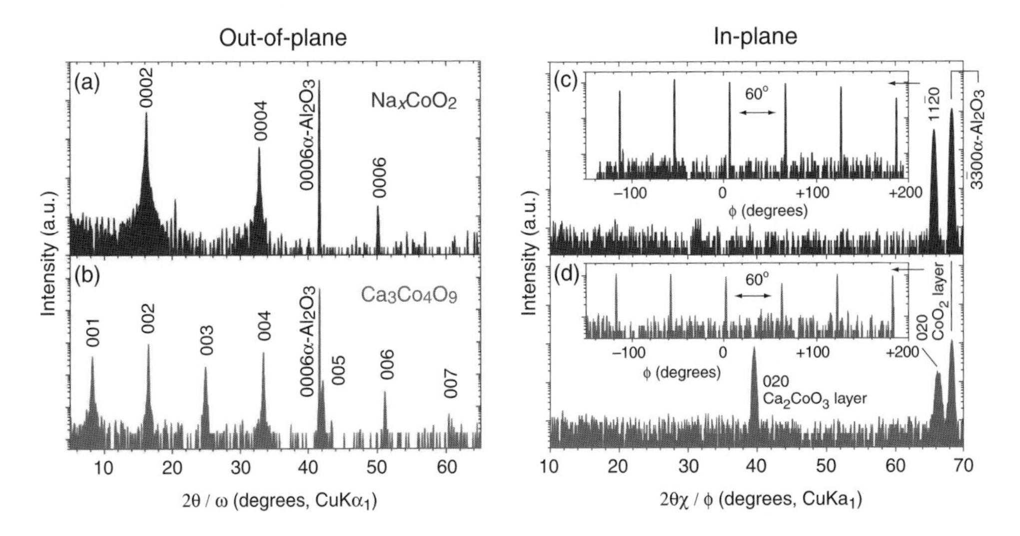

Fig. 10.3 (left) Out-of-plane and (right) in-plane X-ray diffraction patterns of (a), (c)$Na_{0.8}CoO_2$ and (b), (d)$Ca_3Co_4O_9$ epitaxial films. Highly *c*-axis-oriented epitaxial film of $Na_{0.8}CoO_2$ can be obtained by the R-SPE method. The $Na_{0.8}CoO_2$ epitaxial film can be converted into highly *c*-axis-oriented $Ca_3Co_4O_9$ film by ion-exchange treatment.

seen in both Figure 10.3(c) and (d), indicating that topotactic ion exchange from Na^+ to Ca^{2+} has occurred.

Figure 10.4 shows a topographic AFM image of (a) $Na_{0.8}CoO_2$ and (b) $Ca_3Co_4O_9$ films. A step-like structure composed of several flake-like domains is seen in both (a) and (b). Although only hexagonal-shaped domains are seen in Figure 10.4(a), several square-shaped domains ($\sim 1\,\mu m^2$) are clearly seen in Figure 10.4(b) due to the fact that the crystal symmetry of $Ca_3Co_4O_9$ is monoclinic. These features in the $Ca_3Co_4O_9$ film are very similar to those in the $Na_{0.83}CoO_2$ film, indicating that the framework composed of the epitaxial CoO_2 layer

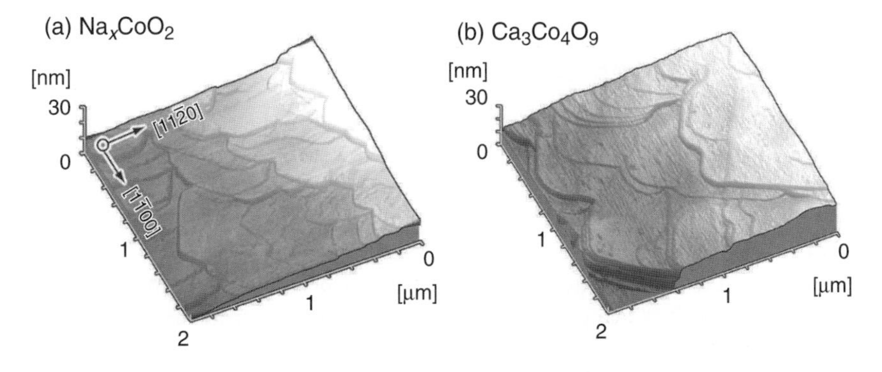

Fig. 10.4 Topographic AFM images of (a) $Na_{0.8}CoO_2$ and (b) $Ca_3Co_4O_9$ epitaxial films.

is maintained during this process, and that the $Na_{0.83}CoO_2$ epitaxial film was successfully converted to $Ca_3Co_4O_9$ epitaxial film by the topotactic ion-exchange method.

Figure 10.5 summarizes the thermoelectric properties of p-type $Ca_3Co_4O_9$ epitaxial film [(a)$|S|-T$ curves, (b)$\sigma-T$ curves, (c)$S^2\sigma-T$ curves, (d)$\kappa-T$ curves, and (e)$ZT-T$ curves]. Data for $Ca_3Co_4O_9$ from several publications [1: single crystal (Masset, 2000 [30] and Limelette, 2005 [31]), 2: single crystal (Shikano, 2003 [32]), 3: ceramic (Miyazaki, 2000 [33]), 4: ceramic (Xu, 2002 [34]), 5: ceramic (Itahara, 2004 [35]), 6: film (Hu, 2005 [36]), 7: single crystal (Satake, 2004 [37]), 8: ceramic (Li, 2000 [38])] are also plotted in the figure for comparison.

The S values are positive, indicating p-type conductivity [Figure 10.5(a)]. The S values of the epitaxial film are almost similar to those of bulk samples and other films, suggesting that the hole concentration is almost the same among them. Although a small difference is seen between the epitaxial film and polycrystalline ceramics, this is probably due to the orientation of the crystal, because the $Ca_3Co_4O_9$ has some anisotropy in S [39]. On the other hand, the σ values have a fairly large difference among the samples [Figure 10.5(b)]. Those of the epitaxial film are several times higher than those of the ceramics, and almost comparable to those of the single crystal, suggesting that carrier scattering by grain boundaries would be mostly eliminated in the film. Furthermore, the σ values of the fabricated film are higher than any other reported values in films grown by conventional methods [36, 39–41], indicating that the fabricated film by R-SPE is a high quality epitaxial film. The high conductivity of the epitaxial film leads to a high power factor ($S^2\sigma$), which is $\sim 10^{-3}$ $Wm^{-1}K^{-2}$, over the temperature range 100–1000 K, as shown in Figure 10.5(c).

The κ values of $Ca_3Co_4O_9$ are almost independent of temperature from the results for bulk materials ($\kappa \sim 3$ $Wm^{-1}K^{-1}$ in single crystal, $\kappa < 2$ $W m^{-1}K^{-1}$ in poly crystal) [Figure 10.5(d)]. Thus, we calculated the ZT-values of the epitaxial film assuming the κ value to be 3 $Wm^{-1}K^{-1}$ in the temperature range 100–1000 K. The ZT-value of the epitaxial film increases gradually with temperature, and reaches ~ 0.3 at 1000 K [Figure 10.5(e)]. This value is lower than that of conventional alloyed-metals such as p-type Si-Ge ($ZT \sim 0.8$) [42] even though this is higher than any p-type oxides' ZT-value. Further improvement of the thermoelectric performance in p-type $Ca_3Co_4O_9$ would be essential through, for example, nano-structural design such as a multi-quantum-well structure.

10.3 Heavily electron doped $SrTiO_3$ films

Among several n-type oxides so far proposed, including Al-doped ZnO [43], $In_2O_3(ZnO)_m$ (m = integer) [44], $CaMnO_3$ [45], and $SrTiO_3$ [46–48], heavily electron-doped $SrTiO_3$ is the most promising candidate, because it exhibits a rather large $|S|$ due to the large density of states (DOS) effective mass (m_d^* = carrier effective mass, $m^* \times$ band degeneracy $\times$ spin degeneracy = $6 \sim 10\ m_0$) [46]. Also, bulk single crystals of heavily La-doped $SrTiO_3$ ($n_e \sim 10^{21}$ cm^{-3}) have recently been found to have a large power factor ($PF = S^2\sigma$) of 3.6×10^{-3} $Wm^{-1}K^{-2}$ at room temperature [47], which is comparable to that of practical Peltier material Bi_2Te_3 with low carrier concentration ($n_e \sim 10^{19}$ cm^{-3}) [2]. Furthermore, since the melting point of $SrTiO_3$ is very high (2080°C), electron-doped $SrTiO_3$ crystal may be applicable at high temperatures ($\sim$1000 K).

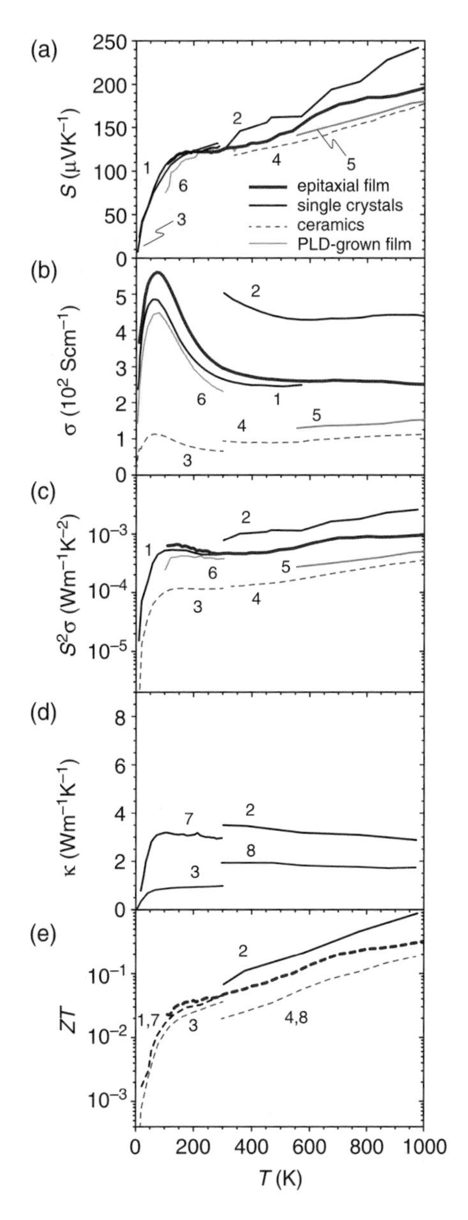

Fig. 10.5 Thermoelectric properties of [(a)$|S|-T$ curves, (b)$\sigma-T$ curves, (c)$S^2\sigma-T$ curves, (c)$\kappa-T$ curves, and (d)$ZT-T$ curves] of *p*-type $Ca_3Co_4O_9$. [1: single crystal (Masset, 2000 [30] and Limelette, 2005 [31]), 2: single crystal (Shikano, 2003 [32]), 3: ceramic (Miyazaki, 2000 [33]), 4: ceramic (Xu, 2002 [34]), 5: ceramic (Itahara, 2004 [35]), 6: film (Hu, 2005 [36]), 7: single crystal (Satake, 2004 [37]), 8: ceramic (Li, 2000 [38])]. The reliable ZT-value of $Ca_3Co_4O_9$ (red dotted line) increases gradually with temperature, and reaches ~0.3 at 1000 K.

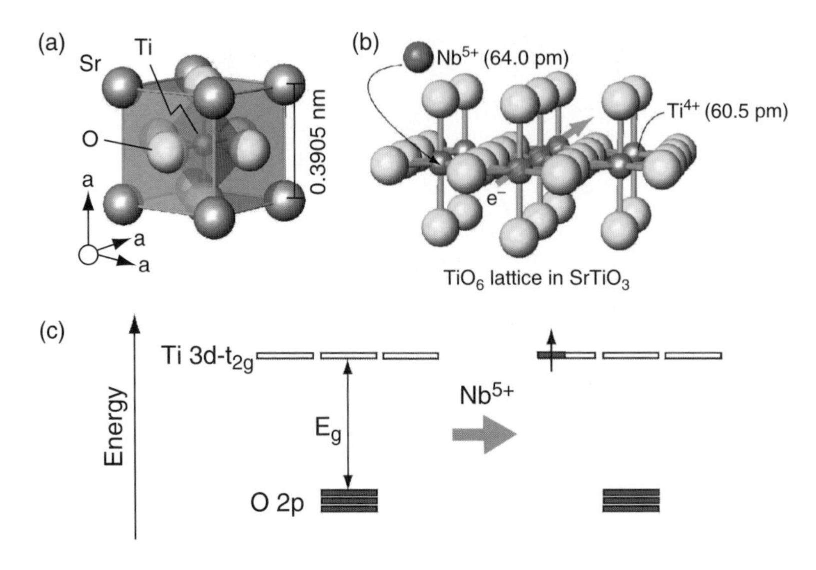

Fig. 10.6 (a),(b) Schematic crystal structure and (c) Nb doping mechanism of $SrTiO_3$. This figure is reproduced in color in the color plate section.

$SrTiO_3$ is a popular metal oxide with a cubic perovskite structure (lattice parameter, $a = 0.3905$ nm), as shown in Figure 10.6(a). All the constituents of $SrTiO_3$ are rich in natural resources. Also, the electrical conductivity of $SrTiO_3$ can easily be controlled from insulator to metal by the substitutional doping of La^{3+} or Nb^{5+} [Figure 10.6(b)(c)] (detailed photo emission study of electron doped $SrTiO_3$: see [49]). We have examined the carrier transport properties of Nb- and La-doped $SrTiO_3$ single crystals ($n_e \sim 10^{20}$ cm^{-3}) at high temperatures ($\sim$1000 K) to clarify their intrinsic thermoelectric properties [48]. Although the experimental data suggest that a fairly high ZT is expected in heavily Nb-doped $SrTiO_3$, the maximum ZT could not be clarified because the solubility of the Nb^{5+} ions in the $SrTiO_3$ lattice is substantially smaller than an optimal concentration. In order to overcome this problem, we fabricated $SrTiO_3$ epitaxial films with $\sim 10^{22}$ cm^{-3} Nb as dopant and clarified that 20% ($n_e \sim 2 \times 10^{21}$ cm^{-3}) Nb-doped $SrTiO_3$ epitaxial film exhibits $ZT \sim 0.37$ at 1000 K, which is the largest value among n-type oxide semiconductors so far reported [17].

Epitaxial films of Nb-doped $SrTiO_3$ were grown on the (001)-face of $LaAlO_3$ single-crystalline substrates at 700°C by PLD, using $SrTiO_3$ targets containing up to 40% Nb as dopant. Only an intense X-ray Bragg diffraction peak of 002 $SrTiO_3$ is observed, together with 002 $LaAlO_3$ (Figure 10.7). Pendellousung fringes are clearly observed around the 002 $SrTiO_3$ peak, indicating high-crystalline qualities of the films. The lattice parameter of Nb-doped $SrTiO_3$ increases proportionally to the Nb concentration, indicating that Nb^{5+} (64.0 pm) is substituted at the Ti^{4+} (60.5 pm) site. This site selectivity of Nb is also supported by the fact that Nb^{5+} ions act as donor. Intense streak patterns are observed in the reflection high energy electron diffraction (RHEED) pattern of the films (Figure 10.7 upper), indicating that the $SrTiO_3$:Nb films are heteroepitaxially grown on the (001)-face of $LaAlO_3$. Atomically flat terraces and steps, which correspond to the unit-cell height ($\sim$ 0.4 nm) of

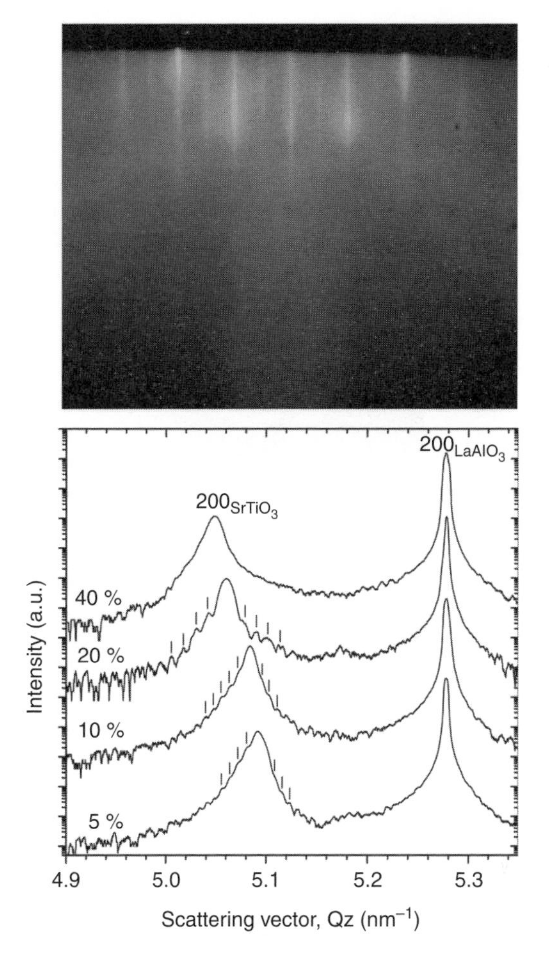

Fig. 10.7 Out-of-plane XRD patterns of Nb-doped $SrTiO_3$ epitaxial films on the (001)-face of $LaAlO_3$ substrate. RHEED pattern (azimuth <100>) is also shown. Reprinted with permission from [17]. Copyright 2005, American Institute of Physics.

$SrTiO_3$:Nb, are clearly seen in the topographic AFM image of a $SrTiO_3$:Nb film (Figure 10.8). From these results, we conclude that the film quality is sufficient to evaluate thermoelectric properties.

Figure 10.9 summarizes the thermoelectric properties of *n*-type $SrTi_{0.8}Nb_{0.2}O_3$ epitaxial film [(a)$|S|-T$ curves, (b)$\sigma-T$ curves, (c)$S^2\sigma-T$ curves, (d)$\kappa-T$ curves, and (e)$ZT-T$ curves]. For comparison, the properties of several electron-doped $SrTiO_3$ [1: $Sr_{0.9}La_{0.1}TiO_3$ single crystal (Okuda, 2001 [47]), 2: $Sr_{0.9}Y_{0.1}TiO_3$ (Obara, 2004 [50]), 3: $Ba_{0.3}Sr_{0.6}La_{0.1}TiO_3$ ceramic (Muta, 2004 [51]), 4: $Sr_{0.95}La_{0.05}TiO_3$ single crystal (Muta, 2005 [52]), 5: $Ce_{0.2}Sr_{0.8}TiO_3$ epitaxial film (Ohtomo, 2007 [53]), 6: $SrTi_{0.8}Nb_{0.2}O_3$ ceramic, Kato, 2007 [54]] are also plotted in the figure $|S|$-value of the $SrTi_{0.8}Nb_{0.2}O_3$ epitaxial film gradually increases

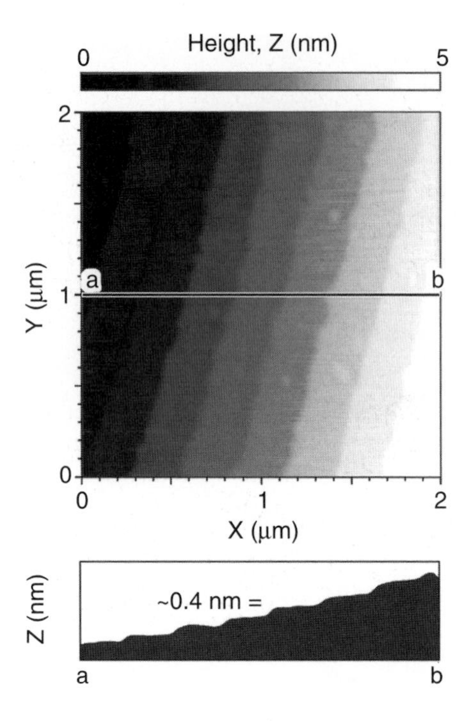

Fig. 10.8 Topographic AFM image of 20%-Nb-doped $SrTiO_3$ epitaxial film. Atomically flat terraces and steps, which correspond to the unit-cell height (~0.4 nm) of $SrTiO_3$:Nb, are clearly seen.

with an increase in temperature [Figure 10.9(a)]. The $|S|$-value of $SrTiO_3$ is well expressed by the following equations:

$$S = -\frac{k_B}{e}\left(\frac{(r+2)F_{r+1}(\xi)}{(r+1)F_r(\xi)} - \xi\right), \tag{10.1}$$

where $k_B.\xi.r$ [roughly $r = 0.5$ ($T < 750$ K) or $r = 0$ ($T > 750$ K)], and F_r are the Boltzmann constant, chemical potential, scattering parameter of relaxation time, and Fermi integral, respectively. The value of F_r is given by

$$F_r(\xi) = \int_0^\infty \frac{x^r}{1+e^{x-\xi}}dx. \tag{10.2}$$

The value of n is given by

$$n = 4\pi\left(\frac{2m_d^* k_B T}{h^2}\right)^{3/2} F_{1/2}(\xi), \tag{10.3}$$

where h, T, and m_d^* (= [orbital degeneracy] × [spin degeneracy] × [effective mass] $\approx 7m_0$) are the Planck constant, absolute temperature, and DOS effective mass, respectively.

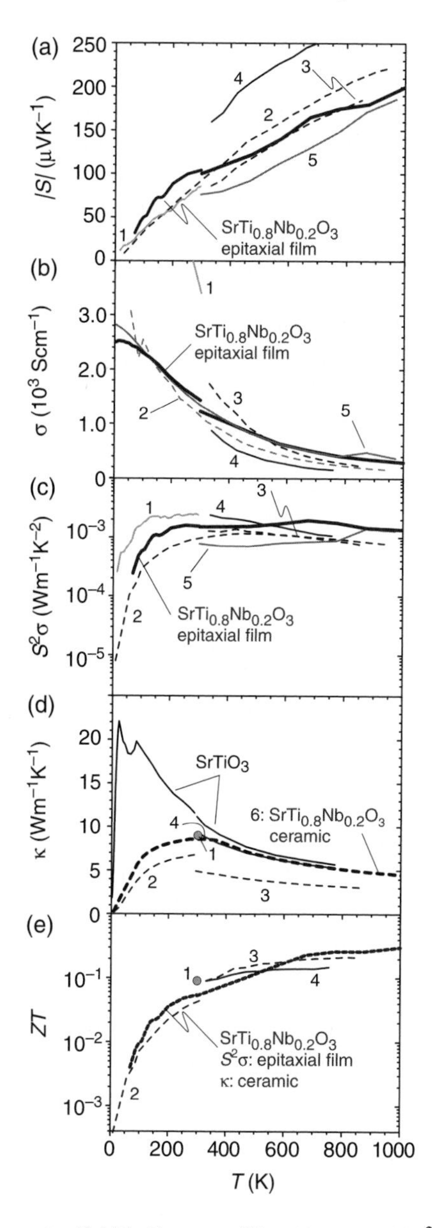

Fig. 10.9 Thermoelectric properties [(a)$|S|$–T curves, (b)σ–T curves, (c)$S^2\sigma$–T curves, (d)κ–T curves, and (e)ZT–T curves] of n-type electron doped $SrTiO_3$ [1: $Sr_{0.9}La_{0.1}TiO_3$ single crystal (Okuda, 2001 [47]), 2: $Sr_{0.9}Y_{0.1}TiO_3$ (Obara, 2004 [50]), 3: $Ba_{0.3}Sr_{0.6}La_{0.1}TiO_3$ ceramic (Muta, 2004 [51]), 4: $Sr_{0.95}La_{0.05}TiO_3$ single crystal (Muta, 2005 [52]), 5: $Ce_{0.2}Sr_{0.8}TiO_3$ epitaxial film (Ohtomo, 2007 [53]), 6: $SrTi_{0.8}Nb_{0.2}O_3$ ceramic (Kato, 2007 [54])]. At 1000 K, the ZT value of $SrTi_{0.8}Nb_{0.2}O_3$ is ~0.3 (red dotted line).

Contrary to this, the σ-value decreases with temperature [Figure 10.9(b)]. The carrier electron concentration (n_e) of the $SrTi_{0.8}Nb_{0.2}O_3$ epitaxial film is $\sim 2 \times 10^{21}$ cm^{-3}, almost independent of temperature. Thus, carrier mobility (μ) decreases proportionally to $T^{-2.0}$ below $\sim$750 K, and $T^{-1.5}$ above $\sim$750 K, most likely because the dominant mechanism of carrier scattering changes with increasing temperature. The power factor ($S^2\sigma$) of the $SrTi_{0.8}Nb_{0.2}O_3$ epitaxial film above 200 K is almost independent of temperature ($\sim$ 1.5)[Figure 10.9(c)].

Since κ-measurement of the $SrTi_{0.8}Nb_{0.2}O_3$ epitaxial film is quite difficult, we measured κ-values of the $SrTi_{0.8}Nb_{0.2}O_3$ dense ceramic with relative density 92%, which was prepared by conventional solid-state sintering at 1500°C in an Ar atmosphere. Although a Schottky-type barrier at the grain boundary of dense ceramics significantly scatters conduction electrons, the barrier does not affect thermal conduction, because most of the heat transport of $SrTi_{0.8}Nb_{0.2}O_3$ would be dominated by the phonon. Therefore we used the dense ceramics of $SrTi_{0.8}Nb_{0.2}O_3$ for κ measurements. The κ values of the $SrTi_{0.8}Nb_{0.2}O_3$ ceramic below 300 K were measured by a conventional steady-state method using a *Physical Properties Measurement System* (PPMS, Quantum Design), while, above 300 K, they were measured using a laser flash method (thermal diffusivity) together with calorimetry (heat capacity).

Figure 10.9(d) shows κ versus T curves for the $SrTi_{0.8}Nb_{0.2}O_3$ dense ceramic. The κ-values for an undoped $SrTiO_3$ single crystal [52, 55] are also plotted for comparison. The κ-value of the $SrTi_{0.8}Nb_{0.2}O_3$ ceramic at room temperature is 8.8 $Wm^{-1}K^{-1}$, which is 20% lower than that of the undoped $SrTiO_3$ single crystal, indicating that the dopant Nb^{5+} ion effectively reduces κ. The κ-value decreases proportional to $T^{0.5}$, indicating that the main mechanism of thermal transport in these samples is the lattice vibration (phonon scattering)

ZT-values of the $SrTi_{0.8}Nb_{0.2}O_3$, which were calculated using $S^2\sigma$-values of the epitaxial film [Figure 10.9(c)] and κ-values of the ceramic [Figure 10.9(d)], increase gradually with temperature [Figure 9(e)]. At 1000 K, the ZT value of $SrTi_{0.8}Nb_{0.2}O_3$ is $\sim$0.3, smaller than that of heavy-metal-based materials. Thus, significant improvement in ZT is required for the practical thermoelectric application of $SrTiO_3$. In order to improve the ZT of $SrTiO_3$, essentially, reduction of the κ-value without reducing the $S^2\sigma$ is required. Muta *et al* [51, 56] have reported that Sr^{2+}-site substitution of $SrTiO_3$ with Ca^{2+} or Ba^{2+} may be a good way to reduce the κ-value of $SrTiO_3$, most likely because introduction of defects such as site substitution and/or layered structure are effective in reducing the κ-value. However, Yamamoto *et al.* [57] clarified that the $S^2\sigma$ drastically decreases when Ca^{2+} and/or Ba^{2+} are substituted for Sr^{2+} in $SrTi_{0.8}Nb_{0.2}O_3$, indicating that Sr^{2+}-site substitution affects the thermoelectric performance of $SrTi_{0.8}Nb_{0.2}O_3$ negatively. Thus, further improvement of the ZT value of $SrTi_{0.8}Nb_{0.2}O_3$ is almost impossible in the conventional three-dimensional (3D) bulk state.

10.4 Two-dimensional electron gas

Two-dimensionally confined electrons in extremely narrow quantum wells (narrower than the thermal de Broglie wavelength, $\lambda_D = h / \sqrt{3 \cdot m^* \cdot k_B \cdot T}$, where h, m^*, and k_B are the Planck constant, effective mass of conductive electron or hole, and Boltzmann constant, respectively) exhibit exotic electron transport properties in comparison with the bulk materials, due to the fact that the density of states (DOS) near the bottom of the conduction band and/or top of the valence band increases with a decrease in the thickness of the quantum well [58]. This phenomenon is called the "Quantum size effect", and is widely applied for optoelectronic devices such as light emitting diodes and laser diodes.

In 1993, Hicks and Dresselhaus predicted theoretically that two-dimensional thermoelectric figure of merit, $Z_{2D}T$ of a quantum well for thermoelectric semiconductors can be dramatically enhanced by using superlattices, because only the S value increases with the DOS of the quantum well, while σgnd κgemain the same [59]. This model is based on the assumption that the enhancement of S^2 arises mainly from an increase in the DOS near the conduction band edge, when the carrier electrons are confined in such a narrow space. Their prediction has been partly confirmed experimentally with the use of a PbTe (1.5 nm)/$Pb_{0.927}Eu_{0.073}Te$ (45 nm) multiple quantum well (MQW) [60], which exhibited an $|S|$ value ~2.5 times larger than that of the corresponding 3D-bulk. It is considered that the conduction carrier electrons are localized more strongly in $SrTiO_3$ than in the heavy-metal semiconductors, because $SrTiO_3$ is basically an insulator: two-dimensional electron confinement may be effective in significantly enhancing $|S|$ value of $SrTiO_3$ [20–22].

In 2007, Ohta *et al* [20] fabricated several superlattices composed of insulating $SrTiO_3$ (undoped, conduction electron concentration: $n_e \ll 10^{15}\text{cm}^{-3}$) and highly conductive Nb-doped $SrTiO_3$ ($SrTi_{.8}Nb_{0.2}O_3$, $n_e = 2.4 \times 10^{21}\text{cm}^{-3}$) on the (001)-face of a $LaAlO_3$ single-crystal substrate by PLD at 900°C in an oxygen atmosphere (oxygen pressure $PO_2 = 3 \times 10^{-3}$Pa). During film growth of the superlattices, we monitored the intensity oscillation of the RHEED pattern to precisely control the layer thickness.

Figure 10.10 shows the Cs-corrected high-angle angular dark-field scanning transmission electron microscope (HAADF-STEM) image of a resultant superlattice. Stripe-shaped contrast is clearly seen in the HAADF-STEM image. Further to this, the intensity of the Ti

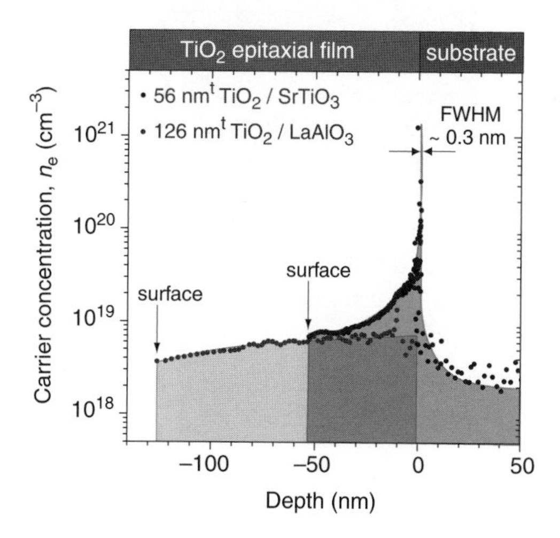

Fig. 10.10 (left) Depth profile around the interface between a 56-nm-thick TiO_2 epitaxial layer and the $SrTiO_3$ substrate. The profile for the interface between a 126-nm-thick epitaxial layer and insulating $LaAlO_3$ is also plotted for comparison (red). An intense carrier concentration peak ($n_e \sim 1.4 \times 10^{21}$ cm^{-3}) with a full-width at half maximum of ~0.3 nm is seen at the $TiO_2/SrTiO_3$ interface. [H. Ohta *et al., Nature Mater.* **6**, 129 (2007)]. This figure is reproduced in color in the color plate section.

column in one-unit-cell's thickness of $SrTi_{0.8}Nb_{0.2}O_3$ shows a higher intensity than that of the $SrTiO_3$ barrier layer, confirming that the doped Nb^{5+} ions are exclusively confined to the $SrTi_{0.8}Nb_{0.2}O_3$ layer.

A 2DEG is also formed at the heterointerface in $TiO_2/SrTiO_3$, because electrons in the conduction band are expected to be localized at the heterointerface between TiO_2 and $SrTiO_3$, as a result of possible bending of the $SrTiO_3$ conduction band. Epitaxial films of TiO_2 were deposited on the (001)-face of $SrTiO_3$ single-crystal plates by PLD with the use of a ceramic TiO_2 (rutile) target at 700 °C in an oxygen atmosphere (oxygen pressure $PO_2 = 3 \times 10^{-3}$ Pa). Oxygen-deficient $TiO_{2-\delta}$ is likely formed by the PLD process under the low-oxygen atmosphere. The deposited film may extract oxide ions (O^{2-}) from the $SrTiO_3$ substrate, resulting in the formation of carrier electrons confined at the interface of the $TiO_2/SrTiO_3$.

Figure 10.11 shows the depth profile for the $TiO_2/SrTiO_3$ heterointerface. A steep peak with n_e of $\sim 1.4 \times 10^{21}$ cm^{-3} is seen at the heterointerface of $TiO_2/SrTiO_3$. The full width at half maximum (FWHM) of the peak is $\sim$ 0.3 nm, which agrees well with the lattice parameter of $SrTiO_3$ ($a = 0.3905$ nm), indicating that high density carrier electrons ($n_e \sim 7.0 \times 10^{20}$ cm^{-3}) are localized within a unit-cell layer thickness of $SrTiO_3$ at the heterointerface, clearly demonstrating that a 2DEG is formed at the heterointerface in $TiO_2/SrTiO_3$.

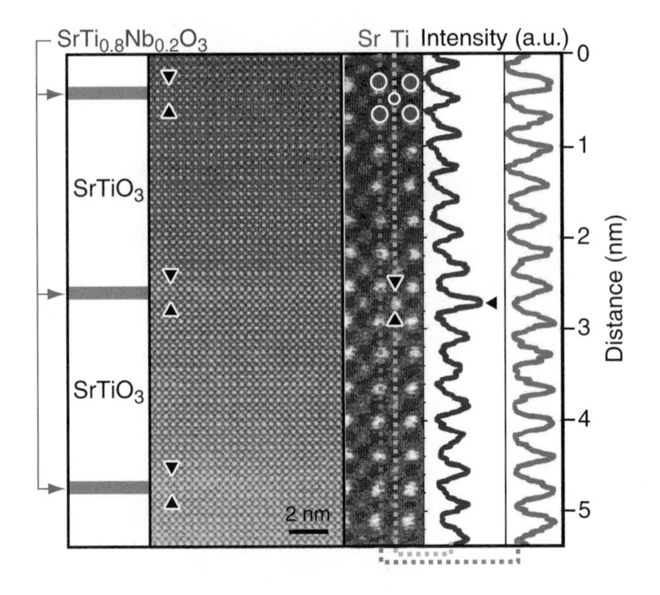

Fig. 10.11 HAADF-STEM image of $[(SrTiO_3)_{24}/(SrTi_{0.8}Nb_{0.2}O_3)_1]_{20}$ superlattice and intensity profiles across Ti and Sr atomic columns. Stripe-shaped contrast is clearly seen. Also, the intensity of the Ti column in one-unit-cell's thickness of $SrTi_{0.8}Nb_{0.2}O_3$ shows higher intensity than that of the $SrTiO_3$ barrier layer, confirming that the doped Nb^{5+} ions are exclusively confined in the $SrTi_{0.8}Nb_{0.2}O_3$ layer. [H. Ohta *et al.*, *Nature Mater.* **6**, 129 (2007)]. This figure is reproduced in color in the color plate section.

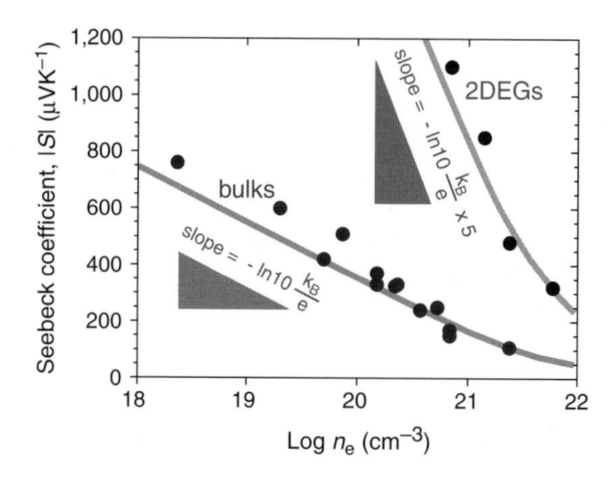

Fig. 10.12 $|S|_{300K}$ versus Log n_e plots for the 2DEGs and the $SrTiO_3$-bulk samples. The slope of $|S|_{300K}$ for 2DEGs is $\sim -1000 \mu VK^{-1}$, which is around five times larger than those for $SrTiO_3$-bulk (-198 μVK^{-1}). [H. Ohta *et al.*, *Nature Mater.* **6**, 129 (2007).]

The $|S|$ versus Log n_e plots are shown in Figure 10.12, which verify that the experimental data points for the 2DEGs and the $SrTiO_3$-bulk samples form two straight lines with different slopes. The relationship between $|S|$ and n_e is simply explained by the following equation [61]:

$$|S| = -k_B/e \cdot \ln 10 \cdot A \cdot (\log n_e + B), \qquad (10.4)$$

where k_B is the Boltzmann constant. A and B are parameters that depend on the types of materials and their energy band structures: for instance, A is equal to 1 for a 3D energy band with a parabolic DOS near the Fermi surface, which clearly leads to difficulty in enhancing the $S^2\sigma$ value for $SrTiO_3$-bulks, because the slope of the $|S|$ versus Log n_e line gives a constant value of -k_B/e·ln10 (-198 μVK^{-1}). On the other hand, the slope of the $|S|$ versus Log n_e line for the 2DEGs is -1,000 μVK^{-1}. Since the slope provides the A value, this clearly demonstrates that the $|S|$ value for the 2DEGs was enhanced by a factor of ~5 compared to that for the $SrTiO_3$-bulk.

10.5 Field effect thermopower modulation

A field-effect transistor (FET) structure on single-crystalline materials would be a powerful tool in exploring thermoelectric materials, because it provides the charge carrier dependence of both S- and σ-values, simultaneously [62]. In this section, $SrTiO_3$-based FET is selected to demonstrate the effectiveness of an FET structure for exploring thermoelectric materials.

In order to unequivocally measure the field-modulated S-values, $SrTiO_3$-FETs with excellent transistor characteristics are required. Although several attempts have been made towards the realization of $SrTiO_3$-FETs, and several gate insulators including MgO, [63]

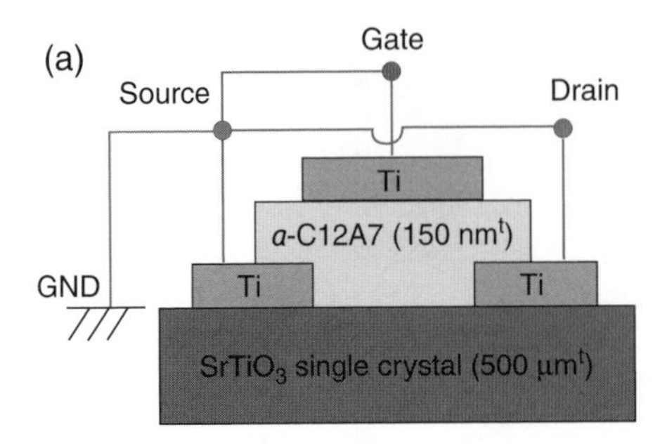

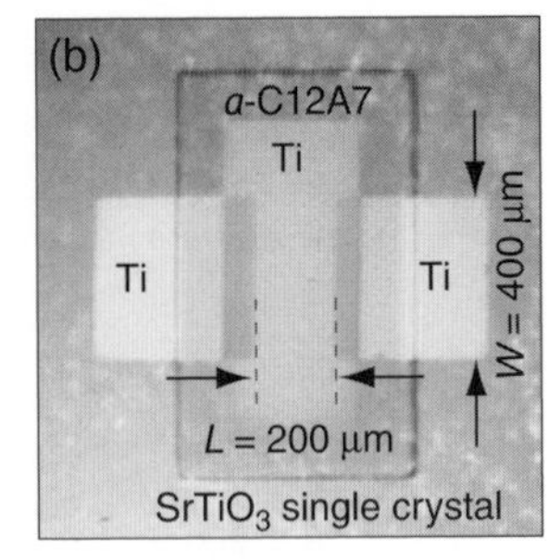

Fig. 10.13 (a) Schematic device structure and (b) photograph of the $SrTiO_3$FET. Ti films (20 nm thick) are used as the source, drain, and gate electrodes. A 150-nm-thick *a*-C12A7 film is used as the gate insulator. Channel length (L) and channel width (W) are 200 and 400 μm, respectively. Reprinted with permission from [71]. Copyright 2009, American Institute of Physics.

amorphous (*a*-) Al_2O_3, [64] $CaHfO_3$ [65, 66] and parylene [67] have been proposed to date, it is still difficult to fabricate $SrTiO_3$-FET with excellent transistor characteristics. Ohta *et al.*, tested various gate insulator materials to obtain $SrTiO_3$-FETs with superior transistor characteristics and found *a*-$12CaO{\cdot}7Al_2O_3$ [68–70] (*a*-C12A7, dielectric constant $\varepsilon_r = 12$) to be an excellent gate insulator for $SrTiO_3$ [71].

A schematic device structure and a photograph of the $SrTiO_3$-FET are shown in Figure 10.13. First, 20-nm-thick metallic Ti films, used as the source and drain electrodes, were deposited through a stencil mask by electron beam (EB) evaporation (base pressure $\sim 10^{-4}$ Pa, no substrate heating) onto a stepped $SrTiO_3$ substrate (10 mm × 10 mm × 0.5 mm), treated with NH_4-buffered HF (BHF) solution [72]. Second, 150-nm-thick *a*-C12A7 film was deposited through a stencil mask by PLD (KrF excimer laser, fluence ~3 Jcm^{-2} $pulse^{-1}$, oxygen pressure ~0.1 Pa) using dense polycrystalline C12A7 ceramic as target. Finally, a 20-nm-thick metallic Ti film, used as the gate electrode, was deposited through a stencil mask by EB evaporation. After deposition processes, the devices were annealed at 200 °C for 30 min in air to reduce the off current.

Figure 10.14 shows typical (a) transfer and (b) output characteristics of the resultant FET. The drain current (I_d) of the FET increased markedly as the gate voltage (V_g) increased, hence the channel was *n*-type, and electron carriers were accumulated by positive V_g [Figure 10.14(a)]. A small hysteresis (~0.5 V) in I_d, probably due to traps (mid-10^{11} cm^{-2}) at the *a*-C12A7/$SrTiO_3$ interface, was also seen [Figure 10.14(a)]. We observed a clear pinch-off and saturation in I_d [Figure 10.14(b)], indicating that the operation of this FET conformed to standard FET theory. The on-to-off current ratio, *S*-factor, and threshold gate voltage (V_{gth}), which were obtained from a linear fit of an $I_d^{0.5}$ versus V_g plot (data not shown), are $>10^6$, ~0.3 V·$decade^{-1}$, and +1.1 V, respectively.

The field-modulated *S* of the $SrTiO_3$-FET can be measured as follows. First, a temperature difference (ΔT = 0.2–1.5 K) was introduced between the source and drain electrodes by using two Peltier devices. Then, thermo-electromotive force (V_{TEMF}) was measured during

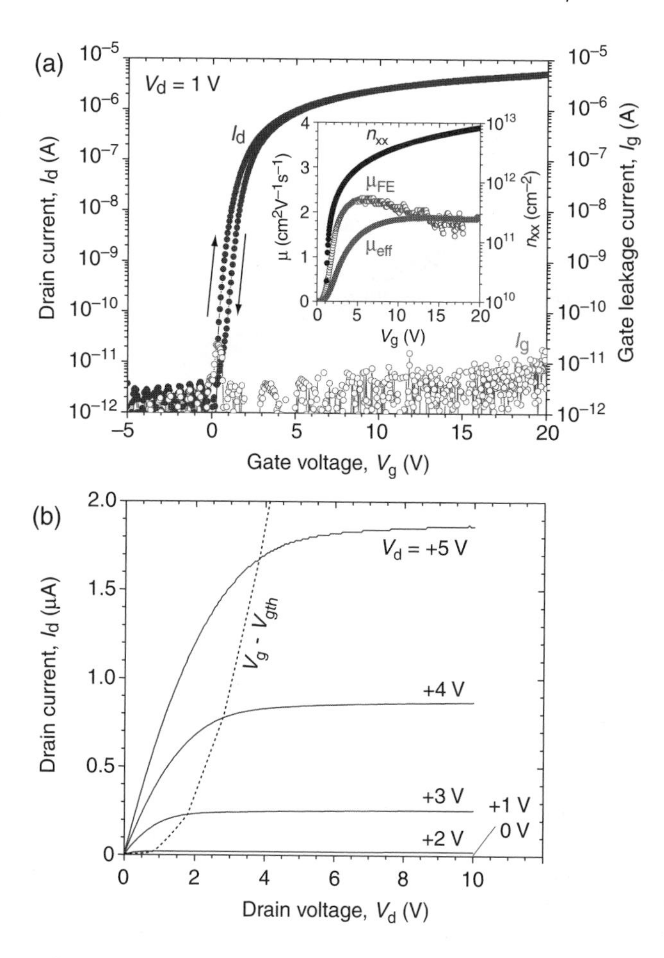

Fig. 10.14 (a) Typical transfer and (b) output characteristics of the $SrTiO_3$-FET with 150-nm-thick *a*-C12A7 ($\varepsilon_r = 12$) gate insulator at room temperature. Channel length (L) and channel width (W) are 200 and 400 μm, respectively. Effective mobility (μ_{eff}), field-effect mobility (μ_{FE}) and sheet charge density (n_{xx}) versus V_g plots for the $SrTiO_3$-FET are also shown in the inset of (a). The dotted line in (b) indicates $V_g - V_{gth}$ value. Reprinted with permission from [71]. Copyright 2009, American Institute of Physics.

the V_g-sweeping. The values of S were obtained from the slope of V_{TEMF}–ΔT plots (data not shown). Figure 10.15 shows S versus V_g plots for the $SrTiO_3$-FET. The S-values are negative, confirming that the channel is n-type. The $|S|$-value gradually decreases from 900 to 580 μVK^{-1}, which corresponds to an increase in the volume charge density (n_{3D}) from $\sim 2 \times 10^{17}$ to $\sim 8 \times 10^{18}$ cm^{-3} (see Figure 10.15 inset), due to the fact that electron carriers are accumulated by positive V_g (up to +30 V).

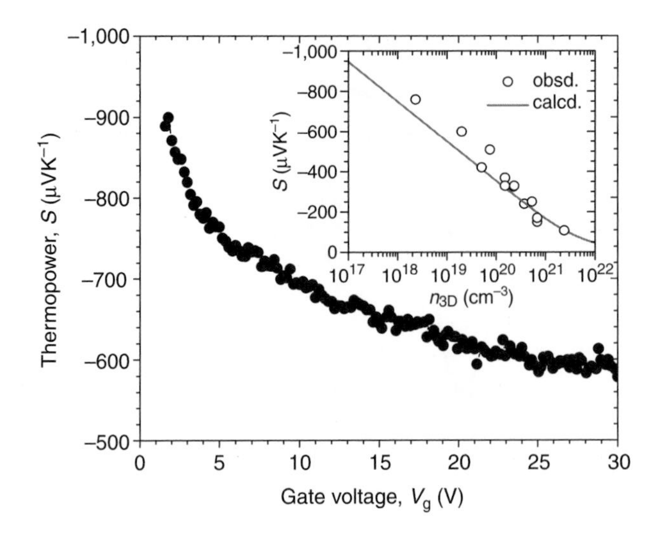

Fig. 10.15 Field-modulated thermopower (S) for the $SrTiO_3$-FET channel. Sn_{3D} (log scaled) plots for electron doped $SrTiO_3$ are also shown in the inset for comparison. Reprinted with permission from [71]. Copyright 2009, American Institute of Physics.

10.6 Summary

In this chapter, thermoelectric properties of two representative oxide epitaxial films were reviewed; p-type $Ca_3Co_4O_9$ and n-type $SrTiO_3$:Nb, which exhibit the best thermoelectric figures of merit, ZT ($=S^2{\cdot}\sigma{\cdot}T{\cdot}\kappa^{-1}$, S: Seebeck coefficient, σ: electrical conductivity, κ: thermal conductivity, and T: absolute temperature) among oxide thermoelectric materials so for reported. Although very high ZT values ~1 have been reported so far, reliable ZT values of both materials were clarified to be ~0.05 at 300 K and ~0.3 at 1000 K. Thus, nano-structural control, such as artificial superlattice or 2DEG, may be necessary to obtain high-ZT thermoelectric oxides. For example, the present approach utilizing a 2DEG in $SrTiO_3$ may provide a new route towords realizing practical thermoelectric materials without the use of toxic heavy elements. Furthermore, the effectiveness of field-effect transistor (FET) structure for the exploration of thermoelectric materials was demonstrated.

References

[1] Seebeck, T.J. (1823). Magnetische Polarisation der Matalle und Erze durch Temperatur-Differenz *Abh. K. Akad. Wiss.* 265–373.

[2] Tritt, T.M., Subramanian, M.A., Bottner, H., *et al.* (2006). Thermoelectric materials, phenomena, and applications: a bird's eye view. *MRS Bull.* **31**, 188–98.

[3] Mahan, G., Sales, B., and Sharp J. (1997). Thermoelectric Materials: New Approaches to an Old Problem. *Physics Today* **50**, 42–47.

[4] DiSalvo, F.J. (1999). Thermoelectric Cooling and Power Generation. *Science* **285**, 703–6.

[5] Rupprech, J. and Maier, R.G. (1965). Neuere Untersuchungen an halbleitenden Mischkristallen unter besonderer Berücksichtigung von Zustandsdiagrammen. *Phys. Status Solidi* **8**, 3–39.

[6] Wright, D.A. (1958). Thermoelectric Properties of Bismuth Telluride and its Alloys. *Nature* **181**, 834.

[7] Hsu, K.F., Loo, S., Guo, F., *et al.* Cubic $AgPb_mSbTe_{2+m}$: Bulk Thermoelectric Materials with High Figure of Merit. (2004). *Science* **303**, 818–21.

[8] Abeles, B. and Cohen, R.W. (1964). Ge-Si Thermoelectric Power Generator. *J. Appl. Phys.* **35**, 247–48.

[9] Koumoto, K., Terasaki, I., and Funahashi, R. (2006). Complex Oxide Materials for Potential Thermoelectric Applications. *MRS Bull.* **31**, 206–10.

[10] Ohta, H., Kim, S.W., Ohta, S., *et al.* (2005). Reactive Solid-Phase Epitaxial Growth of Na_xCoO_2 ($x \sim 0.83$) via Lateral Diffusion of Na into a Cobalt Oxide Epitaxial Layer. *Cryst. Growth Des.* **5**, 25–8.

[11] Ohta, H., Mizutani, A., Sugiura, K., *et al.* (2006). Surface Modification of Glass Substrates for Oxide Heteroepitaxy: Pasteable Three-Dimensionally Oriented Layered Oxide Thin Films. *Adv. Mater.* **18**, 1649–52.

[12] Ishida, Y., Ohta, H., Fujimori, A., and Hosono, H. (2007). Temperature Dependence of the Chemical Potential in Na_xCoO_2: Implications for the Large Thermoelectric Power. *J. Phys. Soc. Jpn.* **76**, 103709 (1–4).

[13] Mizutani, A., Sugiura, K., Ohta, H., and Koumoto, K. (2008). Epitaxial Film Growth of Li_xCoO_2 ($0.6 \leq x \leq 0.9$) via Topotactic Ion Exchange of $Na_{0.8}CoO_2$. *Cryst. Growth Des.* **8**, 755–8.

[14] Sugiura, K., Ohta, H., Nomura, K., *et al.* (2006). Fabrication and thermoelectric properties of layered cobaltite, γ-$Sr_{0.32}Na_{0.21}CoO_2$ epitaxial films. *Appl. Phys. Lett.* **88**, 082109 (1–3).

[15] Sugiura, K., Ohta, H., Nomura, K., *et al.* (2007). High electrical conductivity of layered cobalt oxide $Ca_3Co_4O_9$ epitaxial films grown by topotactic ion-exchange method. *Appl. Phys. Lett.* **89**, 032111 (1–3).

[16] Sugiura, K., Ohta, H., Nomura, K. *et al.* (2007). Thermoelectric properties of the layered cobaltite $Ca_3Co_4O_9$ epitaxial films fabricated by topotactic ion-exchange method. *Mater. Trans.* **48**, 2104–7.

[17] Ohta, S., Nomura, T., Ohta, H., *et al.* (2005). Large thermoelectric performance of heavily Nb-doped $SrTiO_3$ epitaxial film at high temperature. *Appl. Phys. Lett.* **87**, 092108 (1–3).

[18] Kurita, D., Ohta, S., Sugiura, K., Ohta, H., and Koumoto, K. (2006). Carrier generation and transport properties of heavily Nb-doped anatase TiO_2 epitaxial films at high temperatures. *J. Appl. Phys.* **100**, 096105 (1–3).

[19] Lee, K.H., Ishizaki, A., Kim, S.W., Ohta, H., and Koumoto, K. (2007). Preparation and thermoelectric properties of heavily Nb-doped $SrO(SrTiO_3)_1$ epitaxial films. *J. Appl. Phys.* **102**, 033702 (1–4).

[20] Ohta, H., Kim, S.W., Mune, Y. *et al.* (2007). Giant thermoelectric Seebeck coefficient of a two-dimensional electron gas in $SrTiO_3$. *Nature Mater.* **6**, 129–34.

[21] Mune, Y., Ohta, H., Koumoto, K., Mizoguchi, T., and Ikuhara, Y. (2007). Enhanced Seebeck coefficient of quantum-confined electrons in $SrTiO_3/SrTi_{0.8}Nb_{0.2}O_3$ superlattices. *Appl. Phys. Lett.* **91**, 192105 (1–3).

[22] Lee, K.H., Mune, Y., Ohta, H., and Koumoto, K. (2008). Thermal Stability of Giant Thermoelectric Seebeck Coefficient for $SrTiO_3$ / $SrTi_{0.8}Nb_{0.2}O_3$ Superlattices at 900 K. *Appl. Phys. Express* **1**, 015007 (1–3).
[23] Fouassier, C., Matejka, G., Reau, J.M., and Hagenmuller, P. (1973). Sur de nouveaux bronzes oxygénés de formule $Na_xCoO_2\square(x<1)$. Le système cobalt-oxygène-sodium. *J. Solid State Chem.* **6**, 532–7.
[24] Huang, Q., Foo, M.L., Pascal, R.A. *et al.* (2004). Coupling between electronic and structural degrees of freedom in the triangular lattice conductor Na_xCoO_2 *Phys. Rev. B* **70**, 184110 (1–4).
[25] Miyazaki, Y., Onoda, M., Oku, T. *et al.* (2002). Modulated Structure of the Thermoelectric Compound $[Ca_2CoO_3]_{0.62}CoO_2$. *J. Phys. Soc. Jpn.* **71**, 491–7.
[26] Terasaki, I., Sasago, Y., and Uchinokura, K. (1997). Large thermoelectric power in $NaCo_2O_4$ single crystals. *Phys. Rev. B* **56**, R12685–87.
[27] Ohta, H., Nomura, K., Orita, M. *et al.* (2003). Single-Crystalline Films of the Homologous Series $InGaO_3(ZnO)_m$ Grown by Reactive Solid-Phase Epitaxy. *Adv. Funct. Mater.* **13**, 139–44.
[28] Nomura, K., Ohta, H., Ueda, K., *et al.* (2003). Thin-Film Transistor Fabricated in Single-Crystalline Transparent Oxide Semiconductor. *Science* **300**, 1269–72.
[29] Cushing B.L. and Wiley, J.B. (1998). Topotactic Routes to Layered Calcium Cobalt Oxides. *J. Solid State Chem.* **141**, 385–91.
[30] Masset, A.C., Michel, C., Maignan, A. *et al.* (2000). Misfit-layered cobaltite with an anisotropic giant magnetoresistance: $Ca_3Co_4O_9$. *Phys. Rev. B* **62**, 166–75.
[31] Limelette, P., Hardy, V., Auban-Senzier, P., *et al.* (2005). Strongly correlated properties of the thermoelectric cobalt oxide $Ca_3Co_4O_9$ *Phys. Rev. B* **71**, 233108 (1–4).
[32] Shikano, M. and Funahashi, R. (2003). Electrical and thermal properties of single-crystalline $(Ca_2CoO_3)_{0.7}CoO_2$ with a $Ca_3Co_4O_9$ structure. *Appl. Phys. Lett.* **82**, 1851–3.
[33] Miyazaki, Y., Kudo, K., Akoshima, M., *et al.* (2000). Low-Temperature Thermoelectric Properties of the Composite Crystal $[Ca_2CoO_{3.34}]_{0.614}[CoO_2]$. *Jpn. J. Appl. Phys.* **39**, L531–33.
[34] Xu, G.J., Funahashi, R., Shikano, M., Matsubara, I., and Zhou, Y.Q. (2002). Thermoelectric properties of the Bi- and Na-substituted $Ca_3Co_4O_9$ system. *Appl. Phys. Lett.* **80**, 3760–62.
[35] Itahara, H. (2004). Processing design, synthesis and thermoelectric properties of textured cobaltite ceramics. PhD. thesis, Nagoya University.
[36] Hu, Y.F., Si, W.D., Sutter, E., and Li, Q. (2005). In situ growth of c-axis-oriented $Ca_3Co_4O_9$ thin films on Si (100). *Appl. Phys. Lett.* **86**, 082103 (1–3).
[37] Satake, A., Tanaka, H., Ohkawa, T., Fujii, T., and Terasaki, I. (2004). Thermal conductivity of the thermoelectric layered cobalt oxides measured by the Harman method. *J. Appl. Phys.* **96**, 931–3.
[38] Li, S.W., Funahashi, R., Matsubara, I., *et al.* (2000). Synthesis and Thermoelectric Properties of the New Oxide Materials $Ca_{3-x}Bi_xCo_4O_{9+\delta}$ $(0.0 < x < 0.75)$. *Chem. Mater.* **12**, 2424–7.
[39] Sakai, A., Kanno, T., Yotsuhashi, S., Odagawa, A., and Adachi, H. (2005). Control of Epitaxial Growth Orientation and Anisotropic Thermoelectric Properties of Misfit-Type $Ca_3Co_4O_9$ Thin Films. *Jpn. J. Appl. Phys.* **44**, L966–9.

[40] Minami, H., Itaka, K., Kawaji, H., *et al.* (2002). Rapid synthesis and characterization of $(Ca_{1-x}Ba_x)_3Co_4O_9$ thin films using combinatorial methods. *Appl. Surf. Sci.* **197**, 442–7.

[41] Yoshida, Y., Kawai, T., Takai, Y., and Yamaguchi, M. (2002). Thermoelectrical properties and the microstructure of $[Ca_2CoO_3]_xCoO_2$ thin films fabricated on the various substrates using the pulsed laser deposition. *J. Ceram. Soc. Jpn.* **110**, 1080–3.

[42] Dismukes, J.P., Ekstrom, L., Steigmeier, E.F., Kudman, I., and Beers, D.S. (1964). Thermal and Electrical Properties of Heavily Doped GeSi Alloys up to 1300 °K. *J. Appl. Phys.* **35**, 2899–907.

[43] Ohtaki, M., Tsubota, T., Eguchi, K., and Arai, H. (1996). High-temperature thermoelectric properties of $(Zn_{1-x}Al_x)O$. *J. Appl. Phys.* **79**, 1816–18.

[44] Ohta, H., Seo, W.-S., and Koumoto, K. (1996). Thermoelectric Properties of Homologous Compounds in the $ZnO–In_2O_3$ System. *J. Am. Ceram. Soc.* **79**, 2193–6.

[45] Matsubara, I., Funahashi, R., Takeuchi, T., *et al.* (2001). Fabrication of an all-oxide thermoelectric power generator. *Appl. Phys. Lett.* **78**, 3627–9.

[46] Frederikse, H.P.R., Thurber, W.R., and Hosler, W.R. (1964). Electronic Transport in Strontium Titanate. *Phys. Rev.* **134**, A442–5.

[47] Okuda, T., Nakanishi, K., Miyasaka, S., and Tokura, Y. (2001). Large thermoelectric response of metallic perovskites: $Sr_{1-x}La_xTiO_3$ $(0<\sim x<\sim 0.1)$. *Phys. Rev. B* **63**, 113104 (1–4).

[48] Ohta, S., Nomura, T., Ohta, H., and Koumoto, K. (2005). High-temperature carrier transport and thermoelectric properties of heavily La- or Nb-doped $SrTiO_3$ single crystals. *J. Appl. Phys.* **97**, 034106 (1–4).

[49] Ishida, Y., Eguchi, R., Matsunami, M., *et al.* (2008). Coherent and Incoherent Excitations of Electron-Doped $SrTiO_3$. *Phys. Rev. Lett.* **100**, 056401 (1–4).

[50] Obara, H., Yamamoto, A., Lee, C.H., *et al.* (2004). Thermoelectric Properties of Y-Doped Polycrystalline $SrTiO_3$. *Jpn. J. Appl. Phys.* **43**, L540–2.

[51] Muta, H., Kurosaki, K., and Yamanaka, S. (2004). Thermoelectric properties of doped $BaTiO_3$-$SrTiO_3$ solid solution. *J. Alloy. Compd.* **368**, 22–4.

[52] Muta, H., Kurosaki, K., and Yamanaka, S. (2005). Thermoelectric properties of reduced and La-doped single-crystalline $SrTiO_3$. *J. Alloy. Compd.* **392**, 306–09.

[53] Yamada, Y.F., Ohtomo, A., and Kawasaki, M. (2007). Parallel syntheses and thermoelectric properties of Ce-doped $SrTiO_3$ thin films. *Appl. Sur. Sci.* **254**, 768–71.

[54] Kato, K., Yamamoto, M., Ohta, S., *et al.* (2007). The effect of Eu substitution on thermoelectric properties of $SrTi_{0.8}Nb_{0.2}O_3$. *J. Appl. Phys.* **102**, 116107 (1–3).

[55] Suemune, Y. (1965). Thermal Conductivity of $BaTiO_3$ and $SrTiO_3$ from 4.5° to 300°K. *J. Phys. Soc. Jpn.* **20** 174–5.

[56] Muta, H., Ieda, A., Kurosaki, K., and Yamanaka, S. (2004). Substitution effect on the thermoelectric properties of alkaline earth titanate. *Mater. Lett.* **58**, 3868–71.

[57] Yamamoto, M., Ohta, H., and Koumoto, K. (2007). Thermoelectric phase diagram in a $CaTiO_3$-$SrTiO_3$-$BaTiO_3$ system. *Appl. Phys. Lett.* **90**, 072101 (1–3).

[58] Meyers, M.A. and Inal, O.T. (Eds.) (1985). *Frontiers in Materials Technologies*. Elsevier, Amsterdam-Oxford-New York-Tokyo.

[59] Hicks, L.D. and Dresselhaus, M.S. (1993). Effect of quantum-well structures on the thermoelectric figure of merit. *Phys. Rev. B* **47**, 12727–31.

[60] Hicks, L.D., Harman, T.C., Sun, X., and Dresselhaus, M.S. (1996). Experimental study of the effect of quantum-well structures on the thermoelectric figure of merit. *Phys. Rev. B* **53**, R10493–6.
[61] Jonker, G.H. (1968). Application of combined conductivity and Seebeck-effect plots for analysis of semiconductor properties *Philips Res. Rep.* **23**, 131–8.
[62] Liang, W., Hochbaum, A.I., Fardy, M., *et al.* (2009). Field-effect modulation of Seebeck coefficient in single PbSe nanowires. *Nano Lett.* **9**, 1689–93.
[63] Pallecchi, I., Grassano, G., Marrié, D., *et al.* (2001). $SrTiO_3$-based metal-insulator-semiconductor heterostructures. *Appl. Phys. Lett.* **78**, 2244–6.
[64] Ueno, K., Inoue, I.H., Akoh, H., *et al.* (2003). Field-effect transistor on $SrTiO_3$ with sputtered Al_2O_3 gate insulator. *Appl. Phys. Lett.* **83**, 1755–7.
[65] Shibuya, K., Ohnishi, T., Uozumi, T., *et al.* (2006). Field-effect modulation of the transport properties of nondoped $SrTiO_3$. *Appl. Phys. Lett.* **88**, 212116 (1–3).
[66] Shibuya, K., Ohnishi, T., Lippmaa, M., Kawasaki, M., and Koinuma, H. (2004). Single crystal $SrTiO_3$ field-effect transistor with an atomically flat amorphous $CaHfO_3$ gate insulator. *Appl. Phys. Lett.* **85**, 425–7.
[67] Nakamura, H., Takagi, H., Inoue, I.H., *et al.* (2006). Low temperature metallic state induced by electrostatic carrier doping of $SrTiO_3$. *Appl. Phys. Lett.* **89**, 133504 (1–3).
[68] Hayashi, K., Matsuishi, S., Kamiya, T., Hirano, M., and Hosono, H. (2002). Light-induced conversion of an insulating refractory oxide into a persistent electronic conductor. *Nature* **419**, 462–5.
[69] Watauchi, S., Tanaka, I., Hayashi, K., Hirano, M., and Hosono, H. (2002). Crystal growth of $Ca_{12}Al_{14}O_{33}$ by the floating zone method. *J. Cryst. Growth* **237-239**, 801–05.
[70] Kim, S.-W., Toda, Y., Hayashi, K., Hirano, M., and Hosono, H. (2006). Synthesis of a Room Temperature Stable $12CaO{\cdot}7Al_2O_3$ Electride from the Melt and Its Application as an Electron Field Emitter. *Chem. Mater.* **18**, 1938–44.
[71] Ohta, H., Masuoka, Y., Asahi, R. *et al.* (2009). Field-modulated thermopower in $SrTiO_3$-based field-effect transistors with amorphous $12CaO{\cdot}7Al_2O_3$ glass gate insulator. *Appl. Phys. Lett.* **95**, 113505 (1–3).
[72] Kawasaki, M., Takahashi, K., Maeda, T., *et al.* (1994). Atomic Control of the $SrTiO_3$ Crystal Surface. *Science* **266**, 1540–2.

Part IV

Applications

11 High-κ gate dielectrics for advanced CMOS

SUMAN DATTA AND DARRELL G. SCHLOM

11.1 Introduction

More than 40 years later, Gordon Moore's accurate observation [1] that the number of transistors in an integrated circuit doubles roughly every eighteen months continues to be the guiding principle of the semiconductor industry. We have almost taken for granted the apparent corollary; as transistor count increases, each transistor becomes smaller, faster, and cheaper. Today, the transistor gate length (L_G) in production is less than 30 nanometers and a modern multi-core microprocessor may consist of up to three billion field effect transistors (FETs) per chip; this transistor type that now dominates semiconductor technology was first proposed by J.E. Lilienfeld in 1928 [2], but all early attempts to realize it in various semiconductor materials, including extensive attempts at Bell Labs by W.B. Shockley and G.L. Pearson, failed. This was due to the presence of surface states (electron traps) at the interface between the semiconductor and the overlying insulator [3]. The breakthrough came in 1960 when D. Kahng made an FET using the interface between silicon and its native oxide, SiO_2 [4]. Inspite of the superior transport properties of many other semiconductors (e.g. Ge, GaAs, InGaAs), silicon-based FETs have dominated modern electronics because of the low density of traps that can be achieved at the silicon/SiO_2 interface. Over the last four decades, this low-trap interface has been key to the performance and reliability of metal oxide semiconductor field-effect transistors (MOSFETs).

It was recognized by the semiconductor industry a dozen years ago that further geometric scaling of conventional silicon MOSFETs with no materials changes to the SiO_2-based gate stack faces fundamental challenges. These include excessive gate leakage current, gate stack reliability and channel mobility degradation from increasing electric field, exponentially increasing source-to-drain sub-threshold leakage current, and rising dynamic power dissipation (CV^2f) from non-scaled supply voltages. In the 1970s, it was identified that, in order to continue the physical scaling of the silicon-based MOSFETs, the gate oxide needs to be continually scaled [74]. SiO_2 itself faces two fundamental challenges when scaled to thin layers of close to one nanometer. The first is the large and exponentially increasing leakage currents that arise from the direct quantum mechanical tunneling of electrons through such thin SiO_2 (or nitrided SiO_2 gate dielectrics). The second is the loss of the bulk electronic properties of SiO_2 when it becomes thinner than about 0.7 nm [5]. The only solution is to replace SiO_2 with a dielectric having significantly higher dielectric constant (κ) than amorphous SiO_2 ($\kappa =$ high-3.9). Provided that the silicon/high-κ interface has sufficiently low trap density, such a high-κ gate dielectric layer could be a factor of κ/κ_{SiO2} thicker than that of the SiO_2 layer it replaces and provide equivalent FET inversion carrier density modulation. The increased thickness of the gate dielectric, made possible by its higher-κ, solves the electron and hole tunneling problems, while allowing for reduced equivalent oxide thickness (EOT) and enabling physical scaling of FETs. The depletion layer at the degenerately doped polysilicon gate electrode close to the gate oxide also contributes about 0.4 nm to the EOT of the entire gate stack, resulting in less inversion channel charge modulation. Thus, in addition to replacing SiO_2 with higher-κ dielectrics, further scaling to the smallest possible EOT also requires the polysilicon gate electrode to be replaced with metal gate electrodes, which have much higher carrier concentrations than degenerately doped polysilicon and, hence, much thinner (<0.05 nm) depletion layers.

The replacement dielectric and metal materials, whatever their choice might be, needed to have the incredible electrical properties of the SiO_2 (or nitrided SiO_2) gate dielectric and polysilicon gate stack in order to be supplanted. Further, they needed to be the electrical equivalent of an SiO_2 layer having a physical thickness, Tox, of at most 1 nm [6–8]. Many dielectrics are known with κ >3.9, so at first glance, this materials challenge might seem trivial to solve. The initial common sense approach was to try the high-κ materials being investigated for use in dynamic random-access memories (DRAMs): Ta_2O_5, TiO_2, and (Ba,Sr)TiO_3 since they have already been introduced in a semiconductor processing environment [6, 7]. Unfortunately, all of these materials react with silicon [9–19]. Further, the reaction products all involve unwanted low-κ dielectrics, namely, SiO_2, $SrSiO_3$, and $BaSiO_3$ [20]. When a capacitance corresponding to a total SiO_2 physical thickness of < 1 nm is needed for the replacement of the SiO_2 (or nitrided SiO_2) gate dielectric in MOSFETs, such low-κ reaction layers in series with the desired high-κ dielectrics rapidly nullify the benefits of the high-κ dielectrics. The reactivity between these initial high-κ materials and silicon is readily apparent from thermodynamics, as demonstrated by the reaction in (11.1) between silicon and Ta_2O_5[20, 21]:

$$\frac{13}{2}Si + Ta_2O_5 \xrightarrow{\Delta G^{o}_{1000K} = 413.3\,kj/mol} 2TaSi_2 + \frac{5}{2}SiO_2 \tag{11.1}$$

where ΔG^{o}_{1000K} is the free energy change of the system when the reaction between the reactants and products, all taken to be in their standard state, proceeds in the direction indicated at a temperature of 1000 K. Similar reactions occur between silicon and TiO_2 and (Ba,Sr)TiO_3 [20].

The foremost motivation for seeking a high-κ gate oxide was one of simply reducing the leakage current without reducing the oxide layer capacitance. Much of the early work was therefore focused on the measurement of leakage currents compared to SiO_2 for different high-κ oxides. The results were quite successful, with up to four to five orders of magnitude reduction in leakage at the same equivalent electrical thicknesses, Tox. Indeed, working transistors fabricated with Al_2O_3 as the gate dielectric using a 200-mm process in a standard complementary metal oxide semiconductor (CMOS) line were reported as early as 2000 [21]. Two critical problems, however, immediately plagued the functioning of the fabricated MOS capacitors and high-κ MOSFETs. First, many of the dielectrics, including Al_2O_3, exhibited flat-band voltages in capacitance–voltage measurements that were shifted from their ideal positions, indicating the presence of uncompensated charge or dipoles in the oxides. Second, it was often found that the high-κ oxide (typically a transition metal oxide) catalytically promoted the growth of a thick interfacial silicon oxide layer between the high-κ layer and the silicon substrate, increasing Tox to unacceptable levels. Controlling these undesired effects occupied much of the gate stack research activity in subsequent years.

The first step in the process of down-selecting potential high-κ dielectrics was to identify binary oxides [22, 23, 24] and nitrides [24] that are thermodynamically stable in contact with silicon. The oxides thus identified are more numerous and have much higher dielectric constants than the nitrides [20, 25]. The results of a comprehensive study [23, 24] on the thermodynamic stability of binary oxides with silicon are graphically summarized by the periodic table in

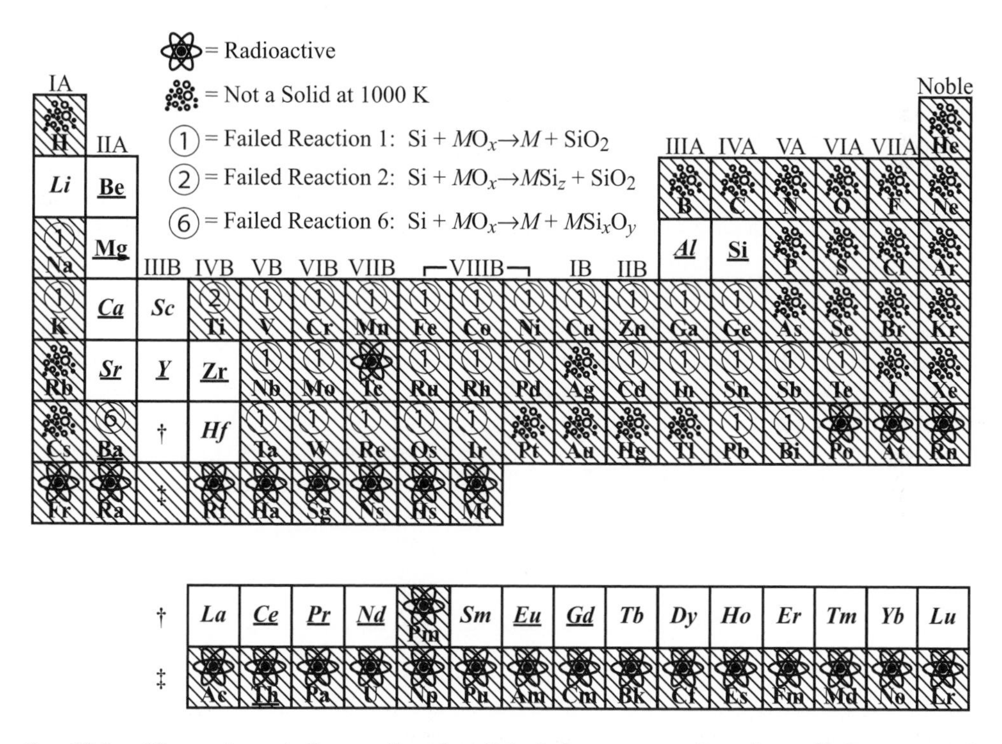

Fig. 11.1 Periodic table showing which elements have oxides that might be stable in contact with silicon, based on calculations over the full range of temperatures for which relevant thermodynamic data are available, as much as 300 K to 1600 K (27° C to 1327° C). Elements M having no thermodynamically stable or potentially thermodynamically stable oxide (MO_X) are shaded (hatched), and the reason for their elimination is given. Also shown are the elements M having an oxide (MO_X) that has been experimentally demonstrated to be stable in direct contact with silicon. (This is an updated version of the plot in [24]).

Figure 11.1. All oxides of the crossed-out elements on the periodic table are thermodynamically unstable in contact with silicon. Figure 11.1 also shows the elements having oxides that have been experimentally demonstrated to be stable in direct contact with silicon. The agreement with a simple bulk thermodynamic analysis is apparent; the only discrepancy is BaO, which can be deposited at low temperature with a reaction-free interface [26, 27], but when heated to above about 500° C, reactions between BaO and silicon are observed in agreement with thermodynamic expectations [28–30].

11.2 High-κ dielectric materials

In addition to being stable in contact with silicon, a high κ and a high bandgap (to yield >1 eV band offsets for both electrons and holes) [31–34] are needed for suitable alternative gate dielectrics. The results of which binary oxides are compatible with silicon were used to identify potential multicomponent oxides that should be compatible with silicon. A way of assessing promising candidates from among these materials is to plot their bandgaps as a function of κ, as has been done in Figure 11.2 for silicon-compatible dielectrics. Materials with high κ and high bandgap (the upper right corner of the figure) are desired. Of the binary oxides, HfO_2 is well-positioned. For many potential alternative dielectrics, especially multicomponent oxides,

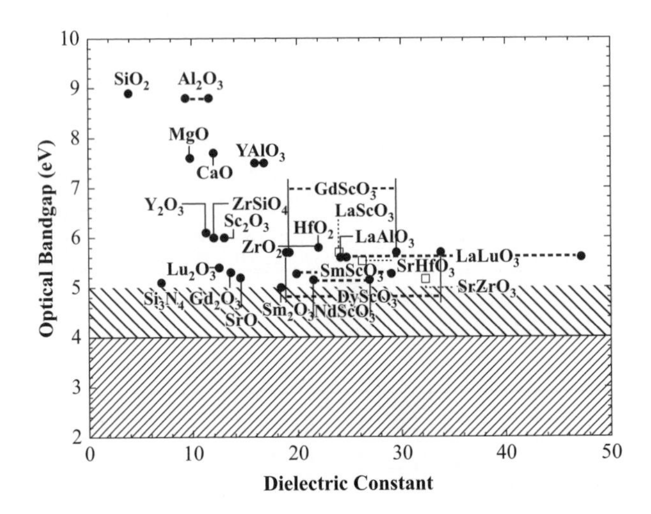

Fig. 11.2 Plot of dielectric constant (κ) versus optical bandgap (Eg, optical) for silicon-compatible (and also III-V compatible) gate dielectrics. Materials where the full dielectric tensor is known are denoted with solid circles (•) and include SiO_2, Si_3N_4, Al_2O_3, MgO, Y_2O_3, CaO, $ZrSiO_4$, Lu_2O_3, Sc_2O_3, Gd_2O_3, SrO, $YAlO_3$,$SmScO_3$, Sm_2O_3, $DyScO_3$, $GdScO_3$, ZrO_2, HfO_2, and $LaAlO_3$. The variation of κ with orientation is indicated by the dashed line between two solid circles, one denoting the orientation where κ is minimal and the second where κ is maximal. Open squares indicated materials where the complete dielectric constant tensor is not known and include $LaLuO_3$, $LaScO_3$, and $SrZrO_3$. The shading indicates the recommendations of [6] a bandgap of at least 4 eV and preferably exceeding 5 eV. (This is an updated version of the plot in [24]).

the dielectric constants and bandgaps were not known. Multicomponent oxides with figures of merit exceeding that available in HfO_2 (for example, $LaLuO_3$) have emerged [35–37]. For instance, Coh *et al.* [75] identifies over two dozen silicon-compatible multicomponent oxides that, in theory, should have κ greater than HfO_2 and could play an important role in the future.

11.3 Metal-gate electrodes

11.3.1 Poly-depletion elimination

Conventional CMOS transistors use heavily doped polycrystalline silicon as electrodes for both p-channel and n-channel FETs. With the transistor channel in inversion, a depletion layer develops in the polysilicon electrodes at its junction with the metal oxide. Inversion is the condition where the applied electric field is sufficient to induce a conductive channel at the silicon surface of opposite carrier type as the silicon is chemically doped. The capacitance due to this depletion layer at the polysilicon/dielectric interface is equivalent to that due to a ~0.4-nm-thick layer of SiO_2, and it reduces the overall capacitance of the gate stack. This is the fundamental reason for replacing polycrystalline silicon electrodes with metal gates.

11.3.2 Interfacial layer control

A further practical impetus forcing the examination of metal gates was felt because of the difficulties encountered in integrating HfO_2 with polysilicon—from the perspectives of poor mobility, threshold voltage, and thickness control. It was recognized early on that all of the high-κ metal oxides under consideration, including HfO_2, had high oxygen diffusivities that resulted in the diffusion of oxygen through the films and the formation of a SiO_x layer at the high-κ/silicon interface. This was observed in fine-grained Al_2O_3 polycrystalline films, [38] as well as in epitaxial films of $(La,Y)_2O_3$ that were lattice matched to silicon substrates and free of extended defects [39]. Interfacial SiO_x formation was directly related to oxygen exposure at high temperatures, and the SiO_x thicknesses were higher than expected for oxidation of a bare silicon surface, suggesting catalytic effects. For bulk HfO_2, the expected mechanism for oxygen diffusion would be through vacancy exchange on the oxygen sub-lattice, based on expectations from the closely related compound ZrO_2. Adequate data for oxygen diffusivity in HfO_2, however, do not exist, particularly for films with nanometer-scale thicknesses. The high diffusivity of oxygen in the dielectric layer imposes additional constraints on the gate electrode (for both polysilicon and metal gate) that can be used. Dissolved oxygen in the polysilicon or amorphous silicon electrode or metal electrode, over time and in the course of thermal treatments, diffuses from the metal to the silicon/HfO_2 interface, resulting in SiO_x formation. Considering that the dielectric layer thickness spans fewer than a few tens of atomic hops, the entire gate stack and the variation of the oxygen chemical potential across this stack need to be taken into consideration when evaluating interfacial SiO_x formation, since the oxygen will travel to the region of lowest chemical potential. Unfortunately, many of the gate electrodes containing platinum and tungsten are incapable of providing devices with very low electrical thicknesses. Electrodes such as TiN [40] and to some extent TaN [41], or TaC [42] on the other hand, have an oxygen-scavenging effect that allows the preservation of a very thin dielectric stack and very little interfacial SiO_x growth. This is the reason for the extensive use of these electrodes in transistor prototypes, even though they have non-ideal band lineups for either p-channel or n-channel Silicon MOSFETs, necessitating other adaptations to adjust these values. Nevertheless, an ultra-thin TiN, TaN, or TaC layer could still be incorporated as an oxygen scavenging layer and as a work function transparent layer in between the actual threshold voltage setting n-type and p-type work-function metal layers and the high-κ dielectric.

11.3.3 High-κ phonon screening

The choice between polysilicon or a metal as the gate electrode for the high-κ dielectric is crucial. The combination of a high-κ dielectric and a polysilicon gate is not suitable for high performance logic applications, since the resulting high-κ/polysilicon transistors have high threshold voltages and degraded channel mobility, as shown in Figure 11.3 and, hence, poor drive current performance [43]. It has been proposed that the high threshold voltage is caused by Fermi level pinning at the polysilicon/high-κ dielectric interface [44] and that Fermi level pinning is most likely caused by defect formation at that very interface [45], arising from the oxygen vacancies at the surface of the high-κ dielectric. Furthermore, it has been demonstrated both experimentally [46] and theoretically [47] that, surface phonon scattering in high-κ dielectrics is the primary cause of channel mobility degradation. Significantly, the collective oscillations of the free electrons in the metal gate electrodes (i.e. gate plasmons) are effective in screening remote phonon scattering centers in the high-κ dielectric from coupling to the channel electrons when under inversion conditions [46, 47], as shown in Figure 11.4. This results in improved channel mobility as shown in Figure 11.3. The high-κ dielectric film attributes its high dielectric constant to its polarizable metal–oxygen bonds, which also give rise to low energy optical phonons. These phonons can be modeled as oscillating dipoles as shown in Figure 11.4. These oscillating dipoles couple strongly with the channel electrons when the

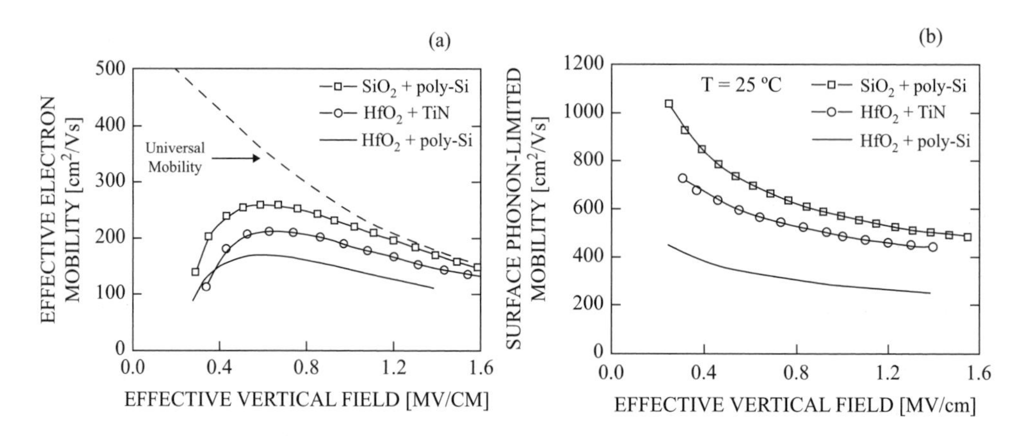

Fig. 11.3 (a) Effective electron mobility μ_{eff} as a function of effective vertical field E_{eff} comparing SiO_2/polysilicon, HfO_2/polysilicon, and HfO_2/TiN metal-gate field effect devices. (b) Surface phonon-limited mobility component at room temperature extracted from inverse modeling of experimental data as a function of effective vertical field E_{eff}. The use of midgap TiN metal-gate significantly improves channel mobility in high-κ gate dielectrics compared to conventional poly-Silicon gates. Reprinted from [43]. Copyright 2005, with permission from Elsevier.

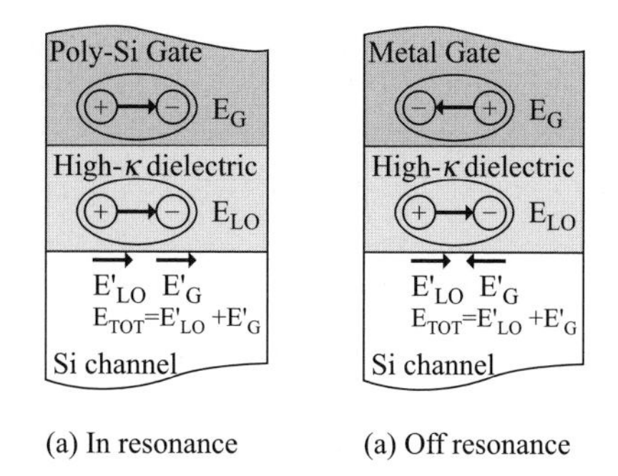

Fig. 11.4 (a) Schematic of high-κ/poly-Si gate stack depicting gate plasma oscillations and high-κ phonons as electric dipoles in resonance, which results in degradation of surface phonon-limited mobility. (b) Schematic of high-κ/metal-gate stack depicting gate plasma oscillations and high-κ phonons as opposing electric dipoles, which reduce the coupling of high-κ phonons with carriers in the Si channel. In this case, the coupling is off-resonance and the surface phonon-limited mobility is recovered. Reprinted from [43]. Copyright 2005, with permission from Elsevier.

gate plasma oscillations and the phonons in the high-κ dielectric are in resonance, as shown in Figure 11.4(a). This resonance occurs when the gate carrier density is $\sim 1 \times 10^{18}$ cm^{-3}, as is the case with a standard doped polysilicon gate in depletion, and the corresponding gate plasmon energy falls within the dominant LO (longitudinal optical) and TO (transverse optical) energy modes of the high-κ dielectric [47]. This resonance condition leads to significant degradation of surface phonon-limited mobility. In the case of the metal-gate electrode, where the free carrier density exceeds $1 \times 10^{20} cm^{-3}$, the resonance condition is not satisfied, as shown in Figure 11.4(b). This weakens the carrier phonon coupling and leads to a recovery of the surface phonon-limited mobility.

11.3.4 Metal gates with "correct" work function

Metal gates with "correct" work functions can be used to provide the right transistor threshold voltages, alleviate the mobility degradation problem, and enable high-performance high-κ/metal-gate transistors with low gate dielectric leakage current for logic applications. The initial screening identified metals with suitable work functions and thermal compatibility with the rest of the gate stack. It is desirable for metals for n-channel FETs to have a work function of ~4 eV, because they need to be aligned with the silicon conduction band, and for metals for p-channel FETs to have a work function of ~5 eV to be aligned with the silicon valence band. Most metal electrodes for n-channel FETs (such as aluminum, titanium) were found to be thermally unstable at the temperatures required for conventional CMOS processing, whereas many of the metals for p-channel FETs (such as ruthenium, rhenium) were stable to the 800–1000° C range [48]. Regardless of the high-work-function metal used, the band bending in the silicon was altered such that the threshold voltage would shift in a negative direction—as though the electrode were effectively aligned mid-gap to the silicon, particularly when p-channel high-κ/metal-gate transistors were fabricated using a conventional high temperature process. This effect was observed in various laboratories, but only for high-work-function electrodes,

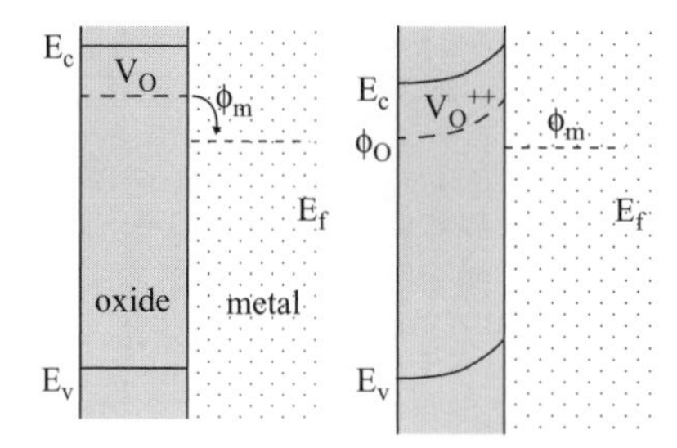

Fig. 11.5 Charge transfer from a neutral vacancy level in the dielectric oxide slab and into the metal-gate electrode results in an uncompensated positively charged oxygen vacancy and a voltage drop across the dielectric. E_C, E_V, and Ef refer to the conduction band edge, valence band edge, and metal Fermi levels, respectively. V_O is the oxygen vacancy, the dashed line is the positional dependence of its energy level with a value of φ_O far away from the metal interface, and φ_m is at the metal–oxide interface.

including thermally stable electrodes such as platinum, rhenium, and p-doped polysilicon. The shifts in the threshold voltage (and, equivalently in the flat-band voltage) implied the formation of unbalanced positive charge in the dielectric layer and a corresponding voltage drop across it (Figure 11.5) resulting from the transfer of electrons from the dielectric into the high-work-function metal-gate electrode. The shifts were consistent with the positively charged oxygen vacancy induced native defects often encountered in the high-κ dielectrics.

11.4 High-κ/metal-gate silicon FETs

11.4.1 Integration

Replacing the polysilicon gate with metal electrodes eliminates many of the issues related to the incompatibility between the high-κ dielectric and the polysilicon electrode. Hafnium-based high-κ dielectrics were successfully integrated with TiN metal gates, strained silicon, and silicon–germanium channels, yielding transistors with improved mobility [49], albeit with a midgap work function gate electrode. A significant research effort was focused on finding metal electrodes with the correct work function [48]. To correctly target the threshold voltages of the NMOS and PMOS devices, metals with work functions of 4.2 eV for NMOS and 5.2 eV for PMOS are needed. Ideally, such metals must withstand the high thermal budget of conventional CMOS integration flows where dopants in the source–drain regions are activated at elevated temperatures. At high temperatures, many metals are thermally unstable: they either spike through the thin high-κ dielectric, as in the case of metals with work functions appropriate for NMOS, or they interact with the oxygen vacancies in the high-κ dielectric, resulting in large flat-band voltage shifts, as in the case of metals with work functions appropriate for PMOS. Two alternate integration strategies dubbed the "gate-last" and "gate-first" approaches are being pursued today to integrate high-κ/metal gate stack with high-performance CMOS transistors.

Gate-last integration

The gate-last integration approach is considered as the low-temperature process because the work-function metals comprising the gate electrodes are not exposed to the high-temperature

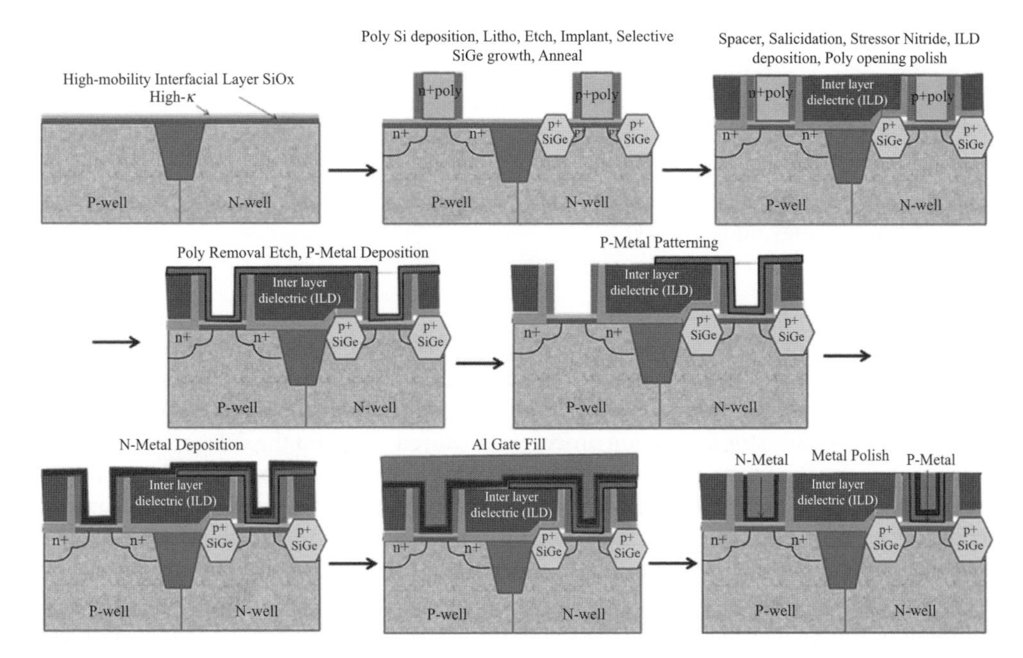

Fig. 11.6 Gate-last approach to the integration of high-κ dielectrics into silicon-based transistors. The gate-last integration approach uses a low-temperature metal-gate process. The high-κ dielectric is deposited prior to the polysilicon layer and undergoes the standard high-temperature process flow. After interlayer dielectric (ILD) deposition and a polysilicon opening polish, the polysilicon layers from both n- and p-channel metal oxide semiconductor (NMOS and PMOS, respectively) devices are removed simultaneously. After blanket deposition of the gate metal for the p-channel field-effect transistors (FETs), one lithography step is required to pattern this metal. Then, the metal for the n-channel FETs is deposited. Finally, the gate trench is filled by an aluminum reflow process, followed by a final metal polishing step.

source–drain dopant activation process. As illustrated in Figure 11.6, a high-κ film is deposited along with the SiO_X layer, followed by a sacrificial polysilicon layer. The high-κ dielectric and dummy polysilicon stack goes through the standard CMOS process flow (including selective epitaxial source–drain regions for PMOS devices) up to salicidation and the processing of stress inducing nitride layers. An interlayer dielectric is then deposited to fill the gaps between polysilicon lines and then polished to expose the dummy polysilicon. The polysilicon is removed simultaneously for both NMOS and PMOS using a standard wet etch process without affecting the high-κ layer. This is followed by deposition of the metal for the PMOS devices over the n-channel and p-channel transistor areas, followed by patterning and selective removal of the metal for the PMOS devices from the NMOS gate trench. Then, the metal for the NMOS devices is deposited as a blanket film. The last metal deposition step involves deposition of aluminum as the final gate fill metal to completely fill the gate trenches without any void and seam, and is followed by a polishing step. This reduces the gate line resistance. Interestingly, the stress-inducing nitride layers that uniaxially strain the NMOS channel for mobility enhancement are planarized in this gate-last process, which decreases the channel strain. On the other hand, the stress-inducing embedded SiGe source–drain regions that compressively strain the PMOS channel experience a strain enhancement effect during the dummy polysilicon removal process [50, 51]. Although the gate-last flow enjoys the fundamental advantage of maintaining the low thermal budget suitable for high-κ dielectric and metal gates with appropriate work functions, the challenge arises in being able to fill the scaled gate trench structures in future technology nodes. The gate-last approach is the one used by Intel for its high-κ metal-gate transistors at the 45-nm technology node [50, 51].

Gate-first integration

In the "gate-first" integration approach the high-κ/metal-gate stack is exposed to the high temperature source and drain dopant activation process. In the gate-first integration flow, as shown in Figure 11.7, with the exception of inserting the high-κ, cap layer and metal gate, a standard CMOS transistor fabrication process is employed. To avoid the deleterious influence of the high thermal budget on the work function of the metal, a midgap, thermally stable thin metal layer such as TiN or TaN is used for both NMOS and PMOS devices. Vt is adjusted by interface dipole engineering of the dielectric independent of the refractory midgap metal electrode. A hafnium (Hf) based dielectric is deposited and then followed by a dielectric cap layer that intermixes with the HfO_2 upon high temperature annealing and creates a dipole, shifting the threshold voltage, Vt. In general, for a Hf-based high-κ dielectric, a relatively more electropositive element such as La, Sc, Er, Sr and other rare earth elements will shift the Vt more towards NMOS, while a more electronegative element such as Al, Re, Ir shifts the work function towards PMOS [52–56]. Lithography steps then follow to remove the initial dielectric cap layer from the second (PMOS) side. The PMOS dielectric cap layer is then

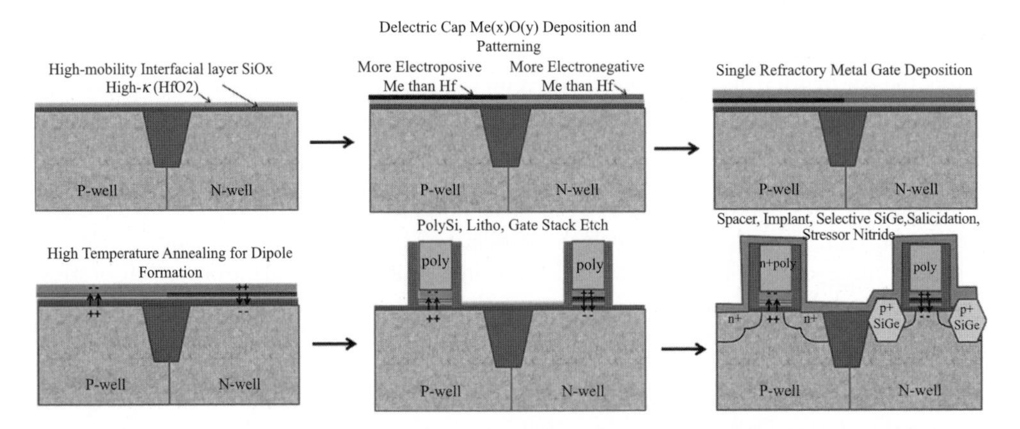

Fig. 11.7 Gate-first integration approach integrates the high-κ/metal-gate steps into the standard process flow. The dielectric cap layers of opposite electronegativity types relative to the Hf in HfO_2 high-κ are deposited on N and P areas through a patterning step after the blanket high-κ film deposition on the oxynitride. This is followed by a midgap thermally stable metal deposition. Vt is adjusted in an anneal step which leads to the intermixing of the cap layer with the HfO_2 and SiOxNy leading to the formation of an interface dipole of opposite polarity to adjust NMOS and PMOS threshold voltages.

deposited and subsequently removed from the NMOS side. The dipole is activated by a thermal activation anneal process when the cap layer intermixes with the high-κ layer and the SiO_x layer underneath. The net dipole orientation and the magnitude of its moment is decided by the electronegativity or electropositivity character and the ionic radius of the cap element, respectively. Since Vt is tuned by the respective dielectric cap layers, a single metal is used for both NMOS and PMOS, followed by polysilicon deposition to complete the gate stack. The rest of the CMOS processing is conventional. The advantages of the "gate-first" process are the elimination of complex polishing steps, retention of channel strain in both NMOS and PMOS transistors, and the potential of gate length scalability for future technology nodes. The disadvantage of the "gate-first" integration approach employing dipole formation techniques is the strong possibility of channel mobility degradation due to remote Coulomb-type scattering mechanism. The disadvantage for gate-first integration is that, because of the requirements for high-temperature compatibility, there are restrictions on the choice of electrode and gate dielectric materials that can be used. The gate-first process is the one being developed by IBM and their partners for its 32 nm node CMOS technology [54].

11.4.2 Devices

Starting with the 45-nm process technology node, the leading-edge semiconductor logic device manufacturers have started introducing high-κ/metal-gate stack in their CMOS transistor architecture. These companies will adopt either the gate-last low temperature approach or the gate-first high-temperature approach, described earlier. The introduction of a high-κ/metal-gate stack allows scaling of the equivalent electrical thickness, Tox, from 1.85 nm (SiO_2/polysilicon) used in the 65 nm technology node to less than 1.3 nm (HfO_2/metal gate), which includes 0.4–0.5 nm of interfacial SiO_x. This allows a significant increase (up to 25% or more) in the transistor channel conductance at the same operating voltage over the previous-generation MOSFETs, which translates to a proportional reduction in the loaded ring oscillator stage delay [51]. The high-κ/metal-gate stack also results in a reduction by over three orders of magnitude in the PMOS gate leakage current and over 50 times in the NMOS gate leakage current, compared to the polysilicon/SiO_2 gate stack. The asymmetry results from the differences in the conduction-band offset (1.5 eV) and the valence-band offset (4 eV) between the high-κ (HfO_2) dielectric and the silicon substrate. This makes future scaling of the high-κ/metal-gate stack for future CMOS generations an interesting materials and device integration challenge, involving band offset engineering, interfacial SiO_x layer engineering, and channel material engineering.

11.4.3 Reliability

Inspite of the reduced gate leakage with a high-κ/metal-gate stack, the 30% increase in the effective electric field poses challenges for gate stack reliability in terms of the fundamental dielectric integrity (measured by time-dependent dielectric breakdown (TDDB)) as well as a progressive shift in transistor parameters such as threshold voltage and transconductance (called bias-temperature instability (BTI)). The existence of a bilayer (high-κ and interfacial SiO_x) and a large mismatch in dielectric constant between the bilayer constituents results in bias-dependent variation in the tunneling leakage mechanism. Thus, dielectric breakdown could occur through the generation of electrical defects called traps, either in the high-κ layer or in the interfacial SiO_x depending on the material properties and bias conditions [57]. In comparison, it is generally accepted that NMOS BTI is driven by electron trapping within the high-κ dielectric that directly tunnels across the interfacial SiO_x layer. This also explains the absence of a reduction in the NMOS transconductance peak in experiments, because the trap generation and electron trapping occur farther from the interface [58]. PMOS degradation due to BTI is found to be very similar to that of conventional SiON and is driven by positive charge trapping near the silicon/dielectric interface due to Pb (Si–H dangling bonds at the interface) defect centers [59]. The reliability degradation in high-κ dielectrics has been attributed primarily to the presence of oxygen vacancies and strongly correlates with the thermal history of the gate stack, which, in turn, influences the process integration strategy, whether high-κ dielectric first or high-κ dielectric last.

Although hafnium-based dielectrics have replaced SiO_2 (and nitrided SiO_2) as the gate dielectric for high-performance MOSFETs and enabled functional transistors to have thicknesses, Tox, of less than 1 nm, the new challenge is to continue to scale the high-κ/metal-gate FETs to ever-thinner dimensions. Ways to accomplish this are to reduce (or eliminate) the interfacial SiO_x layer under the HfO_2 dielectric layer and/or to switch to a high-κ dielectric with an even higher κ than HfO_2. Both of these alternatives are possible. High-κ dielectrics can be deposited on silicon without a SiO_x interfacial layer, and as Figure 11.2 shows, there are several silicon-compatible dielectrics with significantly higher dielectric constants than HfO_2 [75]. Further, several of these candidates, namely, the rare-earth scandates and $LaLuO_3$, are known [60, 75] to have band offsets of >1 eV to both the valence and conduction bands of silicon. This is why such materials are being actively studied for the next generation of high-κ materials for silicon-based MOSFETs. Elimination or thickness scaling of the interfacial SiO_x layer between the high-κ dielectric and the silicon channel will almost invariably be accompanied by mobility and reliability degradation. The mobility reduction primarily results from the increased coupling between the remote soft optical phonon modes of the high-κ dielectric and the carriers in the channel. The reliability impact arises from the increased bias-temperature instabilities in the threshold voltages of the transistors, which are due to increased tunneling of the carriers through the SiO_x layer into the border traps residing at the high-κ/SiO_x interface. Recently, there has been significant research interest in using high-mobility materials, such as germanium [61], carbon nanotubes [62], and III–V quantum wells [63] as the channel materials for future high performance CMOS logic transistors, as discussed in the next section.

11.5 High-κ/metal-gate nonsilicon FETs

11.5.1 Integration

Ultra-high mobility compound semiconductor-based (e.g. indium antimonide, indium arsenide and $In_xGa_{1-x}As$) MOSFETs and quantum-well FETs, are under research today for the next generation of logic transistors operating at low supply voltages, since these materials exhibit excellent low-field and high-field electron transport properties [64, 65]. For these emerging nonsilicon nanoelectronic devices, a high-κ/metal-gate is required to achieve the low equivalent oxide thickness for high performance and low gate oxide leakage. Two parallel approaches, "*in situ*" and "*ex situ*," are being pursued today to demonstrate high-quality dielectrics on nonsilicon channel materials, particularly for III–V channels including $In_xGa_{1-x}As$, InAs, and InSb. In the *in situ* approach, after growth of the high-mobility channel material by molecular-beam epitaxy (MBE) and before exposure to air, the channel layer is capped with an amorphous arsenic or antimony layer under ultrahigh vacuum (UHV). The layer and conditions minimize both the interfacial defect density and native oxide formation upon exposure to air. Such capped high-mobility channel layers can then be transferred in air to another MBE chamber dedicated to the growth of oxides. There, the group V capping layer is desorbed in UHV just before the high-κ dielectric is deposited. In the *ex situ* approach, the native oxide and organic contaminants on the III–V surface are removed in the oxide deposition chamber using a reactive chemical etch before the oxide deposition (using MBE, molecular-beam deposition (MBD), metal–organic chemical vapor deposition (MOCVD), or atomic layer deposition (ALD)). Both *in situ* and *ex situ* approaches have resulted in the demonstration of insulated gate III–V transistors working in inversion and flat-band operation mode with high mobilities. Effective channel mobility as a function of the transverse effective electric field or inversion carrier density is an important metric for characterizing the performance of III–V based MOSFETs, since it not only affects the long channel MOSFET performance directly but also determines indirectly the short channel MOSFET performance in the quasiballistic regime by influencing the source side injection velocity. In the next section, we discuss the challenges in accurate measurement of the field effect mobility in nonsilicon FETs with high-κ/metal-gate, which significantly affects the future development of the III–V MOSFET technology.

11.5.2 Devices and characterization

The split C-V measurement of the MOSFET inversion capacitance is the standard technique for extracting the effective channel mobility of MOSFETs, which involves direct estimation of the mobile inversion charge density (N_{inv}) through the gate to channel capacitance (C_{gc}) as a function of the gate to source voltage (Vg) as given by

$$Q_{inv} = \int_{-\alpha}^{Vg} C_{gc}(V)dv \tag{11.2}$$

While this method is reliable and highly accurate for silicon FETs with high-κ/metal-gate stack, it is less straightforward in the case of III–V MOSFETs, for example, $In_{0.53}Ga_{0.47}As$-based MOSFETs. In $In_{0.53}Ga_{0.47}As$-based MOSFETs, the complex nature of the semiconductor–dielectric interface with relatively high density of interface states, Dit, can exhibit a capacitance, Cit, that contributes significantly to the measured Cgc, even in inversion leading to an overestimation of extracted Ninv. This can lead to incorrect evaluation of the effective channel mobility. $In_{0.53}Ga_{0.47}$ As and high-dielectric interfaces are known to possess interface defects. Although the exact origin of the defects is still under debate, there is evidence that compound semiconductors exhibit interface states that arise from the native defects, such as Ga or As dangling bonds, as well as Ga–Ga or As–As like-atom bonds created by unwanted oxidation during the process of gate dielectric formation. It has been proposed that the As–As anti bonding states due to local excess arsenic created during gate oxide deposition can lead to a distribution of states that extend into the conduction band [66, 67]. The presence of interface states near the conduction band leads to fast trap response as the Fermi level approaches and enters the conduction band in the inversion regime. Many recent publications of III–V MOSFETs have reported split C-V measurements and the resultant mobility calculated from those measurements [68–70]. However, frequency dispersion due to Cit as well as lumped and distributed resistance effects in the inversion regime have strongly influenced the Cgc versus Vg (or C-V) curves resulting in incorrect mobility calculations. While the conductance method is pursued by most researchers in evaluating the interface state density at the III–V/high-κ dielectric interface, a small signal equivalent circuit modeling method has been proposed by Ali *et al.* [71] which self-consistently solves the capacitance–voltage (C-V) and conductance–voltage (G-V) measurement data as a function of gate bias and small signal AC frequency to uniquely determine the interface state, D_{it}, response as well as the true inversion carrier response for a given gate voltage.

Full conductance versus equivalent circuit method

The conductance method, proposed by Nicollian and Goetzberger [73] in the sixties, is generally accepted as the most sensitive method to determine the interface state density, D_{it}. Recently, Martens *et al.* [72] proposed the full conductance technique which is suitable for extracting the interface state density across the band gap for any nonsilicon material system. However, Martens' approach requires detecting the exact location of the conductance peak due to interface states, which depends on the measurement AC frequency and the temperature. Further, the conductance method technique does not allow direct extraction of the inversion carrier density necessary for effective mobility extraction. Instead, the equivalent circuit technique directly extracts the interface state density, trap time constant, and the frequency independent inversion channel capacitance by directly solving an equivalent circuit model from the measured admittance values. Another key difference in the equivalent circuit method versus the full conductance technique [72, 73] is that, no information about the peak position in the measured conductance versus frequency is needed; rather the conductance and the capacitance contributions of Dit are solved in a self-consistent manner over the entire frequency and voltage range. This allows for extraction of the Dit distribution over a wider range of energy than given by the peak conductance method for a given frequency range of the impedance measuring instrument at a given temperature (Figure 11.8). This also allows for extraction of the true inversion capacitance, Cinv, free from any frequency dispersion. Figure 11.9 shows the equivalent circuit used to analyze the C-V and G-V data for InGaAs MOSFETs. The equivalent circuit enables one to extract the true inversion capacitance (Cinv) as a function of temperature and gate bias in the inversion regime. The impact of parameters such as oxide capacitance, tunnel conductance, fixed series resistance, distributed channel resistance, and interface state capacitance and conductance on the extraction of true inversion carrier density are systematically accounted for and de-embedded from the measurements. Figure 11.10 shows the measured C-V characteristics exhibiting the effects described above and the extracted true inversion capacitance characteristics for three different temperatures for an $In_{0.53}Ga_{0.47}As$ MOSFET. The equivalent circuit modeling technique allows one to extract the true inversion capacitance (Cinv) as a function of temperature and gate bias in the inversion regime for III–V and other nonsilicon channel-based MOSFETs where the interface state response markedly affects the inversion charge response. The true mobile inversion charge density as a function of gate voltage, Vg, is plotted in Figure 11.11(a). The slope of each of the curves is verified

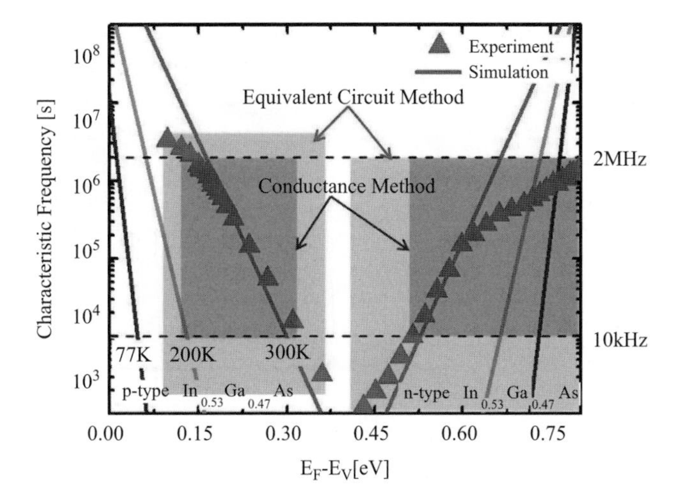

Fig. 11.8 Full conductance versus equivalent circuit methods for extracting the interface state density, Dit, and the trap response time at InGaAs/high-κ dielectric interface. This figure is reproduced in color in the color plate section.

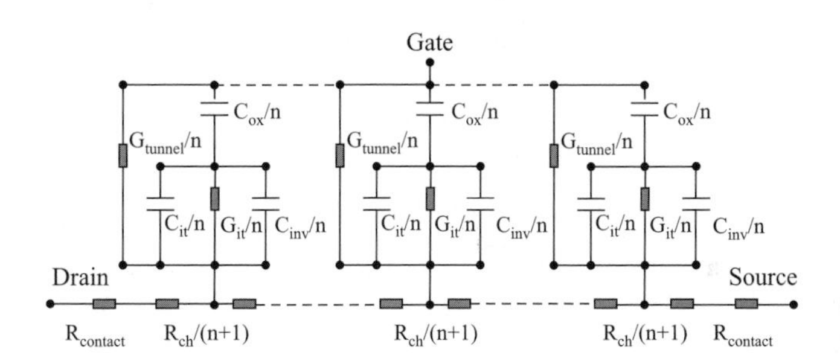

Fig. 11.9 Equivalent circuit model of MOSFET in weak and strong inversion: C_{ox} = oxide capacitance, G_{tunnel} = tunnel conductance, C_{it} = interface trap capacitance, G_{it} = interface trap conductance, C_{inv} = semiconductor inversion capacitance, R_{ch} = the gate bias dependent inversion channel resistance, $R_{contact}$ = series resistance associated with implanted source/drain regions, contacts, and metal pads. © 2010 IEEE. Reprinted, with permission, from [71].

to be equivalent to the series combination of the In0.53Ga0.47As inversion channel capacitance (limited by the density of states) and the insulator capacitance which is 0.65μF/cm^2 (EOT of 5.45 nm). Finally, the effective inversion channel mobility of In0.53Ga0.47As MOSFETs is extracted (Figure 11.11(b)) using the low-field drain conductance and the inversion charge extracted using the model. The extracted field effect mobility of high-κ/metal-gate In0.53Ga0.47As MOSFET far exceeds that of high-κ/metal-gate silicon MOSFETs, and, hence, provides a promising device architecture for future technology nodes at 15 nm and beyond.

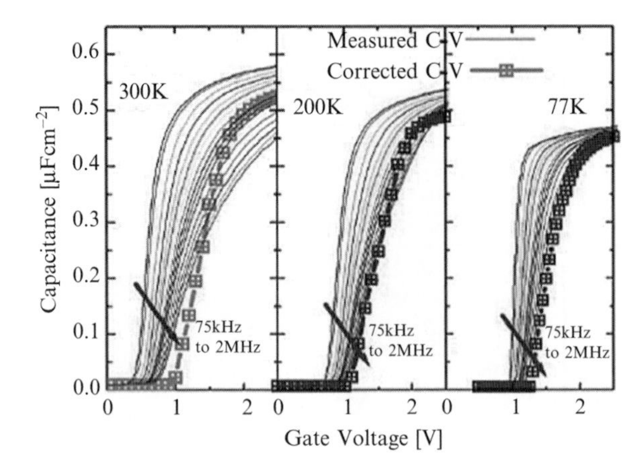

Fig. 11.10 Measured C-V and extracted true C-V characteristics of $In_{0.53}Ga_{0.47}As$ MOSFETs with high-κ/metal gate at 300° K, 200° K, and 77° K. © 2010 IEEE. Reprinted, with permission, from [71].

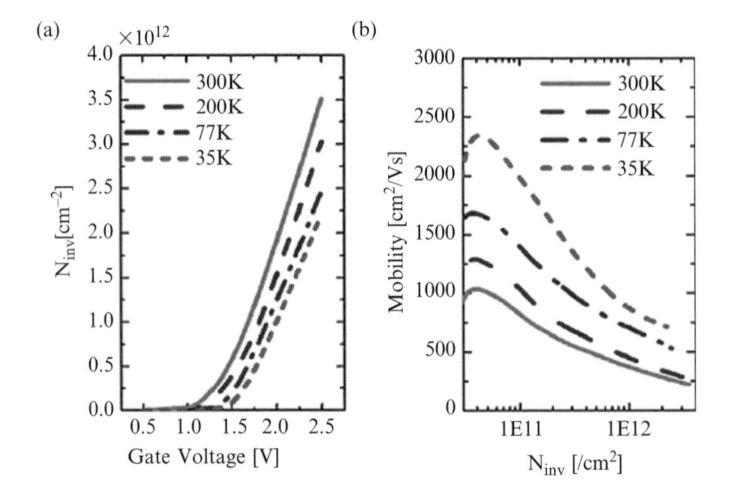

Fig. 11.11 (a) Extracted inversion charge density, N_{inv}, from the C-V and G-V measurements and (b) measured field effect effective mobility as a function of inversion charge density, N_{inv}, for high-κ/metal-gate $In_{0.53}Ga_{0.47}As$ MOSFETs. © 2010 IEEE. Reprinted, with permission, from [71].

Acknowledgments

We gratefully acknowledge our colleagues and collaborators for sharing their insights and making the high-κ materials revolution a reality. We especially thank Robert Chau, Jack Kavalieros, Gilbert Dewey, Wilman Tsai, Marc Heyns, Ashlar Ali, Matthias Passlack, Supratik Guha, Paul McIntyre, Jeff Haeni, Lisa Edge, Tassilo Heeg, Jürgen Schubert, Valery Afanas'ev, André Stesmans, Jon-Paul Maria, Hans Christen, John Freeouf, Bob Wallace, Moon Kim, David Gilmer,

Tom Jackson, Gerry Lucovsky, Frank Lichtenberg, John Robertson, Roy Gordon, Susanne Stemmer, and Jochen Mannhart. D.G.S. and S.D. gratefully acknowledge financial support from Intel.

References

[1] Moore, G. (1965). Cramming more components onto integrated circuits. *Electronics* **38**, 114–117.

[2] Lilienfeld, J.E. U.S. Patent 1,900,018 (March 7, 1933).

[3] Bardeen, J. (1947). Surface states and rectification at a metal semi-conductor contact. *Phys. Rev.* **71**, 717–27.

[4] Kahng, D. U.S. Patent 3,102,230 (August 27, 1963).

[5] Muller, D.A., Sorsch, T., Moccio, S., *et al.* (1999). The electronic structure at the atomic scale of ultrathin gate oxides. *Nature* **399**, 758–61.

[6] The National Technology Roadmap for Semiconductors. (1997). Semiconductor Industry Association, San Jose, CA, pp. 59–60, 70–72.

[7] Kingon, A.I., Maria, J.-P., and Streiffer, S.K. (2000). Alternative dielectrics to silicon dioxide for memory and logic devices. *Nature* **406**, 1032–8.

[8] Wilk, G.D., Wallace, R.M., and Anthony, J.M. High-κ gate dielectrics: Current status and materials properties considerations. *J. Appl. Phys.* **89**, 5243–75 (2001).

[9] Zaima, S., Furuta, T., Yasuda, Y., and Iida, M. (1990). Preparation and properties of Ta_2O_5 films by LPCVD for ULSI application. *J. Electrochem. Soc.* **137**, 1297–1304.

[10] Alers, G.B., Werder, D.J., Chabal, Y., *et al.* (1998). Intermixing at the tantalum oxide/silicon interface in gate dielectric structures. *Appl. Phys. Lett.* **73**, 1517–19.

[11] Mao, A.Y., Son, K.A., White, J.M., *et al.* (1999). Effects of vacuum and inert gas annealing of ultrathin tantalum pentoxide films on Si(100). *J. Vac. Sci. Technol. A* **17**, 954–60.

[12] Gilmer, D.C., Colombo, D.G., Taylor, C.J., *et al.* (1998). Low temperature CVD of crystalline titanium dioxide films using tetranitratotitanium(IV). *Chem. Vap. Deposition* **4**, 9–11.

[13] Pennebaker, W.B. (1969). RF sputtered Strontium Titanate films. *IBM J. Res. Dev.* **13**, 686–95.

[14] Panitz, J.K.G. and Hu, C.C. (1979). Radio-frequency-sputtered tetragonal barium titanate films on silicon. *J. Vac. Sci. Technol.* **16**, 315–19.

[15] Dharmadhikari, V.S. and Grannemann, W.W. (1983). AES study on the chemical composition of ferroelectric $BaTiO_3$ thin films rf sputter-deposited on silicon. *J. Vac. Sci. Technol.* A **1**, 483–6.

[16] Matsubara, S., Sakuma, T., Yamamichi, S., Yamaguchi, H., and Miyasaka, Y. (1990). In Mater. Res. Soc. Symp. Proc. 200, Eds. E.R. Myers, A.I. Kingon, (Materials Research Society, Warrendale, PA), pp. 243–53.

[17] Sakuma, T., Yamamichi, S., Matsubara, S., Yamaguchi, H., and Miyasaka, Y. (1990). Barrier layers for realization of high capacitance density in $SrTiO_3$ thin-film capacitor on silicon. *Appl. Phys. Lett.* **57**, 2431–3.

[18] Yamaguchi, H., Matsubara, S., and Miyasaka, Y. (1991). Reactive Coevaporation Synthesis and Characterization of $SrTiO_3$ Thin Films. *Jpn. J. Appl. Phys.* **30**, 2197–9.

[19] Nagata, H., Tsukahara, T., Gonda, S., Yoshimoto, M., and Koinuma, H. (1991). Heteroepitaxial Growth of $CeO_2(001)$ Films on Si(001) Substrates by Pulsed Laser Deposition in Ultrahigh Vacuum. *Jpn. J. Appl. Phys.* Part 2 **30**, L1136–8.

[20] Schlom, D.G., Billman, C.A., Haeni, J.H., *et al.* (2005). In *Thin Films and Heterostructures for Oxide Electronics*, Ed. S.B. Ogale (Springer, New York) pp. 31–78.

[21] Barin, I. (1995). *Thermochemical Data of Pure Substances.* ed. 3, vol. 1–2 (VCH, Weinheim, Germany).

[22] Gusev, E.P., Buchanan, D.A., Cartier, E., *et al.* (2002). In IEDM Technical Digest 2001 (IEEE, Piscataway, NJ) pp. 451–4.

[23] Hubbard, K.J. and Schlom, D.G. (1996). Thermodynamic stability of binary oxides in contact with silicon. *J. Mater. Res.* **11**, 2757–76.

[24] Schlom, D.G. and Haeni, J.H. (2002). A Thermodynamic Approach to Selecting Alternative Gate Dielectrics. *MRS Bull.* **27**, 198–204.

[25] Billman, C.A., Tan, P.H., Hubbard, K.J., and Schlom, D.G. (1999). Alternate Gate Oxides for Silicon MOSFETS Using High-κ Dielectrics. In "Ultrathin SiO_2 and High-κ Materials for ULSI Gate Dielectrics" *Mater. Res. Soc. Symp. Proc.* **567**, Eds. H.R. Huff, C.A. Richter, M.L. Green, G. Lucovsky, T. Hattori, (Materials Research Society, Warrendale) pp. 409–14.

[26] McKee, R.A., Walker, F.J., and Chisholm, M.F. (1998). Crystalline Oxides on Silicon: The First Five Monolayers. *Phys. Rev. Lett.* **81**, 3014–17.

[27] McKee, R.A., Walker, F.J., and Chisholm, M.F. (2001). Physical Structure and Inversion Charge at a Semiconductor Interface with a Crystalline Oxide. *Science* **293**, 468–71.

[28] Lettieri, J., Haeni, J.H., and Schlom, D.G. (2002). Critical issues in the heteroepitaxial growth of alkaline-earth oxides on silicon. *J. Vac. Sci. Technol. A* **20**, 1332–40.

[29] Il'chenko, V.V., Kuznetsov, G.V., Strikha, V.I., and Tsyganova, A.I. (1998). The formation of a barium silicate layer on silicon. *Mikroelektronika* **27**, 340–5; (1998). *Russ. Microelectron.* **27**, 291.

[30] Il'chenko, V.V., and Kuznetsov, G.V. (2001). Effect of Oxygen on the Chemical Reactions and Electron Work Function in Ba–Si and BaO–Si Structures. *Sov. Tech. Phys. Lett.* **27**, 333–5; (2001). *Pis'ma Zh. Tekh. Fiz.* **27**, 58.

[31] Robertson, J. (2000). Band offsets of wide-band-gap oxides and implications for future electronic devices. *J. Vac. Sci. Technol. B* **18**, 1785–94.

[32] Peacock, P.W. and Robertson, J. (2002). Band offsets and Schottky barrier heights of high dielectric constant oxides. *J. Appl. Phys.* **92**, 4712–21.

[33] Robertson, J. (2006). High dielectric constant gate oxides for metal oxide Si transistors. *Rep. Prog. Phys.* **69**, 327–95.

[34] Afanas'ev, V.V. and Stesmans, A. (2007). Internal photoemission at interfaces of high-κ insulators with semiconductors and metals. *J. Appl. Phys.* **102**, 081301 [1–28].

[35] Ovanesyan, K.L., Petrosyan, A.G., Shirinyan, G.O., Pedrini, C., and Zhang, L. (1998). Single crystal growth and characterization of $LaLuO_3$. *Opt. Mater.* **10**, 291–5.

[36] Lopes, J.M.J., Roeckerath, M., Heeg, T., *et al.* (2006). Amorphous lanthanum lutetium oxide thin films as an alternative high-κ gate dielectric. *Appl. Phys. Lett.* **89**, 222902 [1–3].

[37] Schubert, J., Trithaveesak, O., Zander, W., *et al.* (2008). Characterization of epitaxial lanthanum lutetium oxide thin films prepared by pulsed-laser deposition. *Appl. Phys. A* **90**, 577–9.

[38] Copel, M., Cartier, E., *et al.* (2001). Robustness of ultrathin aluminum oxide dielectrics on Si(001). *Appl. Phys. Lett.* **78**, 2670–72.

[39] Narayanan, V., Guha, S., Copel, M., *et al.* (2002). Interfacial oxide formation and oxygen diffusion in rare earth oxide-silicon epitaxial heterostructures. *Appl. Phys. Lett.* **81**, 4183–5.

[40] Chau, R., Datta, S., Doczy, M., *et al.* (2004). High-κ/metal-gate stack and its MOSFET characteristics. *IEEE Electron Device Lett.* **25**, 408–10.

[41] Doris, B., Park, D.G., Settlemyer, K., *et al.* (2005). Ultra-thin SOI replacement gate CMOS with ALD TaN/high-κ gate stack. *International Symposium on VLSI Technology* (VLSI-TSATECH) (IEEE, Piscataway, NJ) pp. 101–102.

[42] Taylor, W.J., Capasso, C., Min, B., *et al.* (2006). Single Metal Gate on High-κ Gate Stacks for 45nm Low Power CMOS. *IEDM Technical Digest* (IEEE, Piscataway, NJ, 2007) pp. 1–4.

[43] Chau, R., Brask, J., Datta, S., *et al.* (2005). Application of high-κ gate dielectrics and metal gate electrodes to enable silicon and non-silicon logic nanotechnology, *Journal of Microelectronic Engineering* **80** (17), 1–6.

[44] Hobbs, C., Fonseca, L., Dhandapani, V., *et al.* (2003). Fermi-level pinning at the polySi/metal oxide interface. *Proc. Symp. VLSI Tech. Dig.* 9–10.

[45] Chau, R. (2004). Advanced Metal Gate/High-κ Dielectric Stacks for High-Performance CMOS Transistors. *Proceedings of the American Vacuum Society 5th International Conference on Microelectronics and Interfaces* 1–3.

[46] Chau, R., Datta, S., Doczy, M., *et al.* (2004). High-κ/Metal-Gate Stack and Its MOSFET Characteristics. *IEEE Electron Dev. Lett.* **25**, 408–410.

[47] Kotlyar, R., Giles, M. D., Matagne, P., *et al.* (2004). Inversion mobility and gate leakage in high-κ/metal gate MOSFETs. *IEDM Tech. Dig.*, 391–4.

[48] Gusev, E.P., Narayanan, V., and Frank, M.M. (2006). Advanced high-κ dielectric stacks with poly-Si and metal gates: Recent progress and current challenges. *IBM J. Res. Dev.* **50**, 387–410.

[49] Datta, S., Dewey, G., Doczy, M., *et al.* (2003). High mobility Si/SiGe strained channel MOS transistors with HfO2/TiN gate stack. *IEDM Technical Digest 2003* (IEEE, Piscataway, NJ, 2004), pp. 653–6.

[50] Mistry, K., Allen, C., Auth, C., *et al.* (2007). A 45nm Logic Technology with High-κ+Metal Gate Transistors, Strained Silicon, 9 Cu Interconnect Layers, 193nm Dry Patterning, and 100% Pb-free Packaging. in *IEDM Technical Digest 2007* (IEEE, Piscataway, NJ, 2008) pp. 247–50.

[51] Auth, C., Buehler, M., Cappellani, A., *et al.* (2008). 45nm High-κ+Metal Gate Strain-Enhanced Transistors. *Intel Technol. J.* **12**, 77–85.

[52] Narayanan, V., Paruchuri, V.K., Bojarczuk, N.A., *et al.* (2006). Band-Edge High-Performance High-κ/Metal Gate n-MOSFETs Using Cap Layers Containing Group IIA and IIIB Elements with Gate-First Processing for 45 nm and Beyond. *2006 Symposium on VLSI Technology* (IEEE, Piscataway, NJ), pp. 178–9.

[53] Guha, S., Paruchuri, V.K., Copel, M., *et al.* (2007). Examination of flatband and threshold voltage tuning of HfO_2/TiN field effect transistors by dielectric cap layers. *Appl. Phys. Lett.* **90**, 092902 [1–3].

[54] Chudzik, M., Doris, B., Mo, R., *et al.* (2007). High-performance high-κ/metal gates for 45nm CMOS and beyond with gate-first processing. *2007 Symposium on VLSI Technology* (IEEE, Piscataway, NJ, 2007), pp. 194–5.

[55] Sivasubramani, P., Böscke, T.S., Huang, J., *et al.* (2007). Dipole Moment Model Explaining nFET Vt Tuning Utilizing La, Sc, Er, and Sr Doped HfSiON Dielectrics. *2007 Symposium on VLSI Technology* (IEEE, Piscataway, NJ, 2007), pp. 68–9.

[56] Kirsch, P.D., Sivasubramani, P., Huang, J., *et al.* (2008). Dipole model explaining high-κ/metal gate field effect transistor threshold voltage tuning. *Appl. Phys. Lett.* **92**, 092901[1–3].

[57] McPherson, J.W., Kim, J., Shanware, A., Mogul, H., and Rodriguez, J. (2003). Trends in the Ultimate Breakdown Strength of High Dielectric-Constant Materials. *IEEE Trans. Electron Devices* **50**, 1771–8.

[58] Ribes, G., Mitard, J., Denais, M., *et al.* (2005). Review on High-κ Dielectrics Reliability Issues. *IEEE Trans. Device Mater. Reliability* **5**, 5–19.

[59] Kang, A.Y., Lenahan, P.M., and Conley, Jr., J.F. (2003). Electron spin resonance observation of trapped electron centers in atomic-layer-deposited hafnium oxide on Si. *Appl. Phys. Lett.* **83**, 3407–09.

[60] Afanas'ev, V.V. and Stesmans, A. (2007). Internal photoemission at interfaces of high-κ insulators with semiconductors and metals. *J. Appl. Phys.* **102**, 081301[1–28].

[61] Shang, H., Frank, M.M., Gusev, E.P., *et al.* (2006). Germanium channel MOSFETs: Opportunities and challenges. *IBM J. Res. Dev.* **50**, 377–86.

[62] Zhang, G., Wang, X., Li, X., *et al.* (2006). Carbon nanotubes: From growth, placement and assembly control to 60mv/decade and sub-60 mv/decade tunnel transistors. *IEDM Technical Digest 2006* (IEEE, Piscataway, NJ, 2007), pp. 1–4.

[63] Datta, S., Ashley, T., Brask, J., *et al.* (2005). 85 nm gate length enhancement and depletion mode InSb quantum well transistors for ultra high speed and very low power digital logic applications. *IEDM Technical Digest 2005* (IEEE, Piscataway, NJ, 2006), pp. 763–6.

[64] Chau, R., Doyle, B., Datta, S., Kavalieros, J., and Zhang, K. (2007). Integrated nanoelectronics for the future? *Nature Materials* **6(11)**, 810–12.

[65] Datta, S., Dewey, G., Fastenau, J., *et al.* (2007). Ultrahigh-Speed 0.5 V Supply Voltage $In_{0.7}Ga_{0.3}As$ Quantum-Well Transistors on Silicon Substrate. *IEEE Electron Device Lett.* **28(8)**, 685–7.

[66] Robertson, J. (2009). Model of interface states at III-V oxide interfaces. *Appl. Phys. Lett.* **94**, 152104 [1–3].

[67] O'Reilly, E. and Robertson, J. (1986). Electronic structure of amorphous III-V and II-VI compound semiconductors and their defects. *Phys. Rev. B* **34(12)**, 8684–95.

[68] Xuan, Y., Wu, Y., and Ye, P. (2008). High-performance inversion-type enhancement-mode InGaAs MOSFET with maximum drain current exceeding 1 A/mm. *IEEE Electron Device Lett.* **29(4)**, 294–6.

[69] Koveshnikov, S., Goel, N., Majhi, P., *et al.* (2008). High Electron Mobility (2270 cm^2/Vsec) $In_{0.53}Ga_{0.47}As$ Inversion Channel N-MOSFETs with ALD ZrO_2 Gate Oxide Providing 1 nm EOT. *66th Device Research Conf. Digest*, 43–4.

[70] Zhao, H., Chen, Y., Yum, J.H., *et al.* (2009). High performance $In_{0.7}Ga_{0.3}As$ metal-oxide-semiconductor transistors with mobility >4400 cm^2/V?s using InP barrier layer. *Appl. Phys. Lett.* **94(19)**, 193502 (1–3).

[71] Ali, A., Madan, H., Koveshnikov, S., *et al.* (2010). Small-signal response of inversion layers in high-mobility $In_{0.53}Ga_{0.47}As$ MOSFETs made with thin high-κ dielectrics. *IEEE Trans. Electron Devices* **57(4)**, 742–8.

[72] Martens, K., Chui, C., Brammertz, G., *et al.* (2008). On the correct extraction of interface trap density of MOS devices with high-mobility semiconductor substrates. *IEEE Trans. Electron Devices* **55(2)**, 547–56.

[73] Nicollian, E. H. and Brews, J. R. (1982). *MOS (Metal Oxide Semiconductor) Physics and Technology*. Wiley, New York.

[74] Dennard, R.H., Gaensslen, F.H., Yu, H-N., *et al.* (1974). Design of Ion-Implanted MOS-FETs with Very Small Physical Dimensions. *IEEE Journal of Solid-State Circuits* **sc-9**, 256–68.

[75] Coh, S., Heeg, T., Haeni, J.H., *et al.* (2010). Si-compatible candidates for high-κ dielectrics with the *Pbnm* perovskite structure. *Phys. Rev. B* **82**, 064101 [1–16].

12
FeFET and ferroelectric random access memories

Hiroshi Ishiwara

12.1 Overview of ferroelectric random access memories (FeRAMs)

Ferroelectric random access memories (FeRAMs) are being mass-produced at present and widely used in IC (integrated circuits) tags and smart cards. Their features are (1) nonvolatile data storage (the stored data do not disappear even if the electricity is turned off), (2) the lowest power consumption among various semiconductor memories, and (3) operation speed as fast as that of DRAMs (dynamic RAMs). The idea of ferroelectric memories was first presented by researchers in Bell Laboratory, in 1955. In their patents various structures composed of ferroelectric films and semiconductors were proposed and a prototype of the current ferroelectric-gate field-effect transistor (FeFET) was also included. The device structure illustrated in the patent by Ross [1] is shown in Figure 12.1. It is evident that the device operates as an n-channel enhancement-type FET, if the electrical properties at the ferroelectric/semiconductor interface are good.

Si-based FeFETs were first fabricated by Wu in 1974 [2]. He deposited a $Bi_4Ti_3O_{12}$ film on a Si(100) substrate as the gate insulator of an FET and observed hysteresis loops in I_D–V_{GS} (drain current versus gate voltage) characteristics. However, the rotation direction of the loops was opposite to the direction expected from the polarization of the ferroelectric film, which means that the charge injection phenomenon at the ferroelectric/semiconductor interface was more pronounced than the polarization effect. The charge injection phenomenon was found to be sufficiently suppressed by inserting a thin SiO_2 layer between the $Bi_4Ti_3O_{12}$ film and Si substrate, that is, by forming a MFIS (M: metal, F: ferroelectric, I: insulator, S: semiconductor) structure [3]. This improvement stimulated many studies on FeFETs. However, since it was difficult to form ferroelectric/semiconductor interfaces with good electrical properties, and since the semiconductor industry was conservative in introducing novel materials containing uncommon elements such as Pb and Bi, these studies almost stopped in the 1980s.

In the meantime, a new type of FeRAM, in which data are stored by the polarization direction in ferroelectric capacitors (MFM capacitors) and read out using the polarization reversal

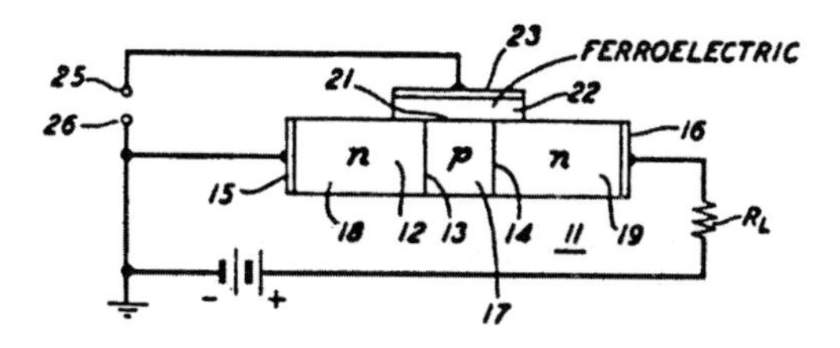

Fig. 12.1 Semiconductor translating device drawn in the patent by I.M. Ross [1].

current, was proposed and successfully operated in the late 1980s [4, 5]. Since the operation of this capacitor-type FeRAM was more stable than that of an FeFET, the studies on this type of FeRAM became very popular in the USA, Japan, and Korea in the 1990s. In the mid 1990s, the reliability of ferroelectric capacitors was much improved by optimization of the deposition conditions of the ferroelectric films, development of passivation films for preventing hydrogen penetration, development of conductive oxide films, such as IrO_2 and $SrRuO_3$, for preventing polarization fatigue of the ferroelectric films, and so on.

By using optimized processes and materials, it became possible to rewrite data more than 10^{12} times and mass-production of FeRAMs was started. At present, the maximum memory capacity of the commercially available chip is 4 Mbit and the operation voltage is 1.5 V in the chips using $PbZr_XTi_{1-X}O_3$ (PZT) capacitors, and 0.9 V in the chips using $SrBi_2Ta_2O_9$ (SBT) capacitors. The main applications of FeRAMs are in IC tags and smart cards. After the success of the capacitor-type FeRAM, studies on FeFETs have again become popular, in which not only the oxide ferroelectric films but also organic ferroelectric films such as P(VDF-TrFE) (polyvinyliden fluoride-trifluoroethylene) are used as the gate insulator [6].

As described above, FeRAMs are classified into two categories: capacitor-type FeRAMs and FET-type FeRAMs [7] A typical cell structure in the capacitor-type FeRAM is a 1T1C-type cell, as shown in Figure 12.2(a), while a typical cell structure in the FET-type FeRAM is a 1T-type cell, as shown in Figure 12.2(b). The cell structure of the 1T1C-type is similar to that of DRAM, except that the cell is connected to the third line (the plate line) in addition to the bit line and

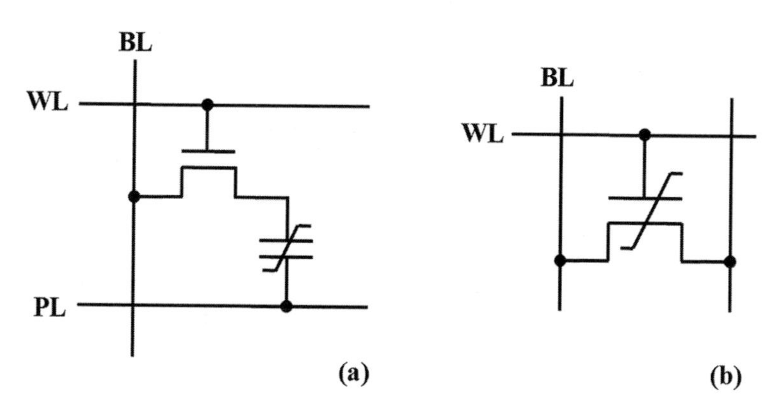

Fig. 12.2 Classification of FeRAMs. (a) 1T1C-type and (b) 1T-type.

the word line. In this cell, since the polarization reversal current of the ferroelectric capacitor is detected, the readout method is destructive and the "rewrite" operation is necessary. In the 1T-type FeRAM, on the other hand, the memory cell is composed of a single FeFET and therefore the cell size can be shrunk using the proportionality rule. It is also advantageous that the stored data can be non-destructively read out using the drain current of FET.

12.2 Ferroelectric films used for FeRAMs

12.2.1 Properties necessary for FeRAMs

A ferroelectric material exhibits a polarization (an electric dipole moment per unit volume) even in the absence of an external electric field, and the direction of the spontaneous polarization can be reversed by an external electric field. In the ferroelectric state, the center of the positive charge in a unit-cell in the crystal does not coincide with the center of negative charge. A typical plot of polarization versus electric field (P-E) in a ferroelectric film is shown in Figure 12.3, in which the coercive field E_C is the reverse field necessary to bring the polarization to zero, and the remanent polarization P_r is the value of P at $E = 0$.

In a capacitor-type FeRAM cell, data are stored by the polarization direction in a ferroelectric film and the stored data are read out using the polarization reversal current. Thus, the following characteristics are desired for a ferroelectric film. The remanent polarization should be large, so that a large polarization reversal current can be derived from a small-area capacitor. The dielectric constant should be low, because a high dielectric constant material produces a large displacement current (linear response) and hinders detection of the polarization reversal current. The coercive field should be low for low-voltage operation of the FeRAM. Degradation of the ferroelectric film such as fatigue, imprint, and retention loss, which is caused during the operation of FeRAMs as well as in the fabrication process, should be as low as possible. On the other hand, in the case of the FET-type FeRAM, since the ferroelectric film is used as the gate insulator of an FET, the large remanent polarization is not necessarily important, but the

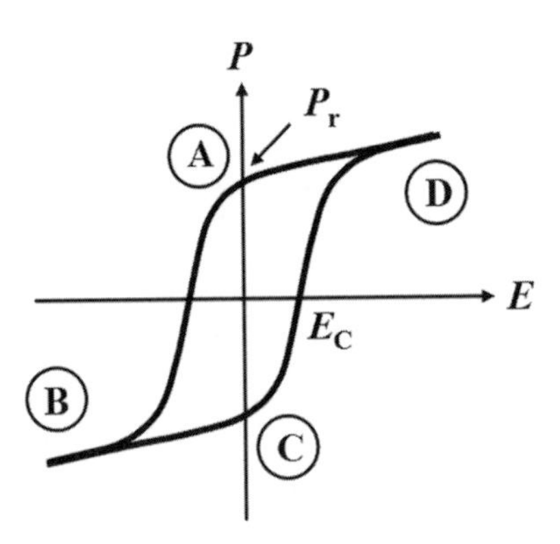

Fig. 12.3 A schematic drawing of a P-E hysteresis loop in a ferroelectric film. P_r: remanent polarization, E_C: coercive field.

low reactivity of the ferroelectric film with the semiconductor substrate or with the insulating buffer layer is more important.

In the following, typical degradation mechanisms in the ferroelectric films are discussed. Polarization fatigue describes that the remanent polarization P_r becomes smaller when a ferroelectric film experiences a large number of polarization reversals. Variation of the hysteresis loop due to fatigue is shown schematically in Figure 12.4(a). Because of the decrease of P_r, the charge difference between logic "0" and "1" becomes smaller and this phenomenon can lead to a failure in the "read" operation. The physical origin of fatigue is not very clear, but the following factors will be related to the phenomenon: domain wall pinning by charged defects, inhibition of domain nucleation by injected charges, and voltage drop at the interfacial layer between the ferroelectric film and the electrode. The fatigue endurance in FeRAMs is known to be typically 10^{12} switching cycles. Thus, it is difficult at present to use FeRAMs in such applications as a cache memory, in which data are continuously rewritten during "write/read" operations.

Imprint describes such a phenomenon. When a ferroelectric film experiences a high DC voltage or repeated unipolar pulses for a long time, particularly at high temperature, its polarization is not fully reversed by application of a single voltage pulse with the opposite polarity. Imprint leads to a shift of the *P-E* hysteresis loop on the electric field axis, as well as to a loss of P_r, which is shown in Figure 12.4(b). Hence, imprint can lead to either "read" or "write" failure of the memory cell. Retention loss describes a decrease of P_r during an absence period of the external voltage, as shown in Figure 12.4(c). Similarly to fatigue, the difference between switching and non-switching charges becomes smaller. The fatigue, imprint, and retention loss characteristics have been greatly improved by optimizing the materials in the ferroelectric capacitors, as well as fabrication processes.

So far, many ferroelectric materials have been investigated, and at present the following three materials are known to be most important for fabricating FeRAMs: PZT, SBT, and $(Bi.La)_4Ti_3O_{12}$ (BLT). Their typical characteristics as polycrystalline films are summarized in Table 12.1. Fabrication methods for ferroelectric films are CSD (chemical solution deposition), RF(radio frequency)-sputtering, MOCVD (metal-organic chemical vapor deposition), and so on. Concerning the bottom electrodes for ferroelectric capacitors, noble metals such as Pt and Ir, or conductive oxides such as IrO_2 and $SrRuO_3$ are usually used, since the ferroelectric films

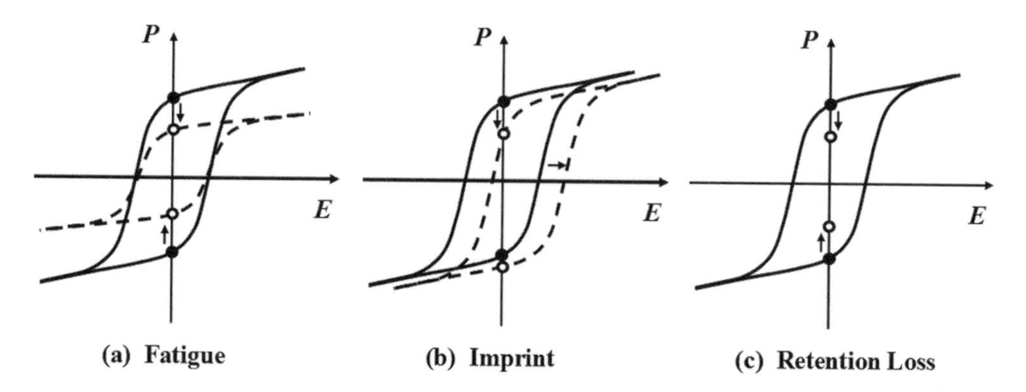

Fig. 12.4 Degradation of hysteresis loops by (a) fatigue, (b) imprint, and (c) retention loss.

Table 12.1 Properties of typical ferroelectric thin films used for FeRAMs.

Materials	P_r (μC/cm^2)	E_C (kV/cm)	Crystallization temperature (°C)
$Pb(Zr,Ti)O_3$ (PZT)	30	60	600
$SrBi_2Ta_2O_9$ (SBT)	10	40	750
$(Bi,La)_4Ti_3O_{12}$ (BLT)	15	80	700

are crystallized in oxidizing gas at an elevated temperature. Some of the bottom electrode films show the strong preferred orientation of polycrystalline grains, even on amorphous SiO_2. A typical example is a Pt film. The preferred <111> orientation of the Pt film is effectively used to control the orientation of grains in the overgrown ferroelectric film.

12.2.2 $Pb(Zr,Ti)O_3$ and Bi-layer structured ferroelectrics

$PbZr_XTi_{1-X}O_3$ (PZT) is a typical ferroelectric material with a perovskite crystal structure, and its large P_r value is advantageous for fabricating high-density FeRAMs. PZT has the morphotropic phase boundary (MPB) between tetragonal ($PbTiO_3$-rich) and rhombohedral ($PbZrO_3$-rich) crystal structures at a Zr composition (X) of 0.52 and it has high dielectric and piezoelectric constants in the vicinity of the MPB composition. In FeRAM applications, since the dielectric constant of the ferroelectric film is not needed to be high, a composition X of 0.3 to 0.4 is often used. The crystallization temperature of PZT films is lower than 650 °C, which is suitable for implementing PZT capacitors on CMOS (complementary MOS) circuits. Some of the largest problems with PZT were fatigue and imprint, which were conspicuous with Pt electrodes. However, these problems have almost been solved through the use of conductive oxide electrodes such as IrO_2 and $SrRuO_3$.

It is well known that the properties of PZT such as resistivity, ferroelectricity, piezoelectricity, electro-optical effect, and photovoltaic effect can be improved by partially substituting impurity atoms for the host atoms A typical example is $(Pb,La)(Zr,Ti)O_3$ (PLZT), which is a well-known transparent ceramic used for high-speed optical switches and image memories Other impurity atoms substituting for the A-site Pb atoms are Mg, Ca, Sr, and Ba, and those for the B-site Zr or Ti atoms are Nb, Ta, and W.

Kijima *et al.* have reported that the co-doping of Si and Nb atoms is effective for decreasing the leakage current and for improving the fatigue and imprint characteristics of PZT films [8]. It is known that in a PZT crystal a large amount of Pb vacancies are generated due to evaporation of Pb atoms, and oxygen vacancies are also generated to maintain charge neutrality in the crystal. As a result, these vacancy pairs drastically affect lattice stability and increase the leakage current. To avoid generation of oxygen vacancies, charge compensation has been attempted by substituting Nb atoms for Zr or Ti atoms, because Nb and Zr (or Ti) act as 5+ and 4+ ions, respectively. However, it has been reported that different crystal phases, such as the dielectric pyrochlore phase, appear when the Nb concentration exceeds 5 at% [9], and that the charge compensation effect is insufficient even at a concentration of 5 at%.

In the experiment by Kijima *et al.*, 150 nm-thick $PbZr_{0.2}Ti_{0.8}O_3$ and $PbZr_{0.2}Ti_{0.6}Nb_{0.2}O_3$ films were prepared by sol-gel spin-coating and 1 to 3 mol% Si was co-doped in the latter film as $PbSiO_3$. The Nb fraction was determined so that the charge neutrality condition was satisfied for the 10 at% deficiency of Pb atoms. The co-doped film crystallized at 650 °C, showed a single perovskite structure in spite of the high Nb concentration, and its leakage current decreased by more than three orders of magnitude (less than 1×10^{-8} A/cm^2 at 3V) compared to that in the pure PZT film. The P_r value (30 μC/cm^2) was comparable to that of the pure PZT film and fatigue, imprint, and data retention characteristics were much improved.

It has also been reported that the ferroelectric properties of PZT are improved by forming solid solutions with other ferroelectrics with the same perovskite structure. An example is the solution with $BiFeO_3$, which is discussed in the next section as a novel material with a large remanent polarization. A P_r value as large as 32 μC/cm^2 has been reported in a 100 nm-thick $[PZT]_{0.95}$-$[BiFeO_3]_{0.05}$ film at an applied voltage of 2 V [10]. Another example is a solid solution with $BiZn_{0.5}Ti_{0.5}O_3$. In the experiment by Tang *et al.*, approximately 200 nm-thick $PbZr_{0.4}Ti_{0.6}O_3$ and $[PbZr_{0.4}Ti_{0.6}O_3]_{0.95}$-$[BiZn_{0.5}Ti_{0.5}O_3]_{0.05}$ films were deposited by spin-coating and crystallized at 600 °C for 30 min in an O_2 atmosphere [11]. Figure 12.5 shows a comparison of the *P-E* hysteresis loops of MFM capacitors composed of the pure PZT and the solid-solution films. It can be seen from the figure that the P_r value increases from 35 μC/cm^2 to 45 μC/cm^2 by forming the solid solution. It has also been found that the fatigue endurance cycles, at which the switching charge becomes half of its initial value is prolonged from 1×10^5 cycles to 6×10^7 cycles.

SBT and BLT are typical Bi-layer structured ferroelectrics (BLSF). The crystal structure of SBT is shown in Figure 12.6, in which the spontaneous polarization is directed along the *a*-axis of the crystal. One of the largest advantages of an SBT film is that it does not show a fatigue phenomenon anywhere up to 10^{13} switching cycles, even if Pt electrodes are used. It is also known that the imprint and retention characteristics at high temperatures are superior to those

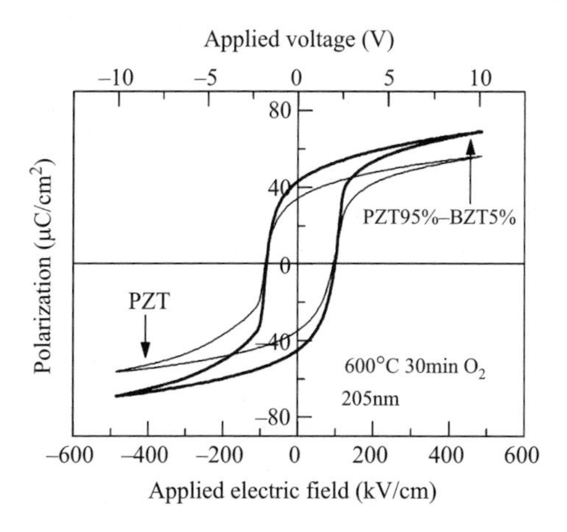

Fig. 12.5 Comparison of *P-E* hysteresis loops of $PbZr_{0.4}Ti_{0.6}O_3$ and $[PbZr_{0.4}Ti_{0.6}O_3]_{0.95}$-$[BiZn_{0.5}Ti_{0.5}O_3]_{0.05}$ films. The capacitor diameter is 200 μm and the measurement frequency is 10 kHz.

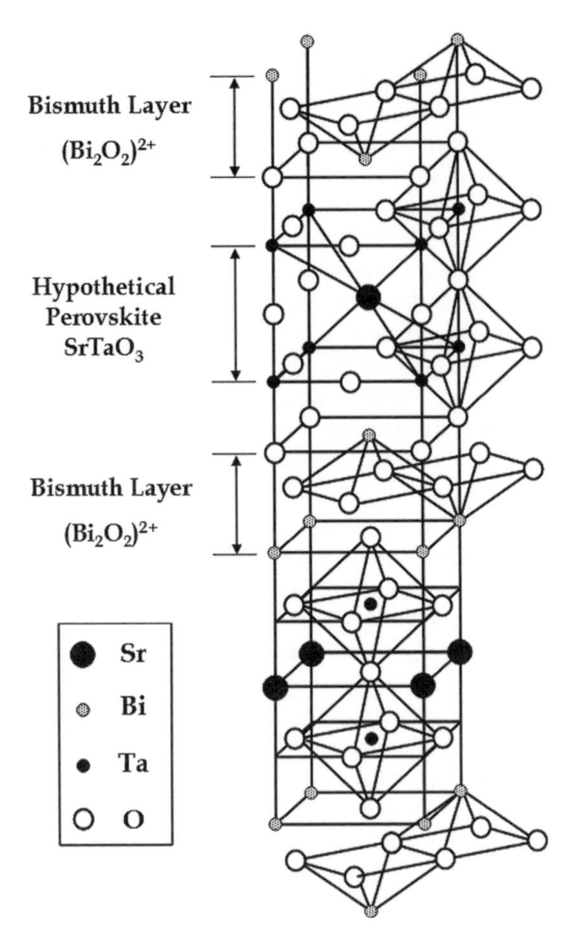

Fig. 12.6 A crystal structure of $SrBi_2Ta_2O_9$.

of PZT. Contrary to this, it is disadvantageous that the crystallization temperature of BLSF is generally higher than 700°C. In some cases, Nb atoms are added to SBT at up to 20 to 30%. The Nb addition increases the switched charge density $2P_r$ typically from 18 μC/cm^2 to 24 μC/cm^2, but the coercive field E_C also increases, typically from 40 to 63 kV/cm. For similar reasons, Sr-deficient (20–30% less than the nominal stoichiometry) and Bi-rich (10–15%) compositions are often used to increase the remanent polarization and the switched charge [12].

12.2.3 Novel ferroelectric films with large remanent polarization

In order to fabricate future capacitor-type FeRAMs with high packing density and low operation voltage, a ferroelectric film with a large P_r and a low E_C is needed. $BiFeO_3$(BFO) is one of the most promising candidates for this purpose. BFO is a multiferroic material exhibiting ferroelectricity and antiferromagnetism at room temperature (RT), and its crystal structure is a rhombohedrally distorted perovskite structure. Recently a remanent polarization as large as

90 μC/cm^2 was found in a single crystalline BFO film grown on a $SrRuO_3$-coated $SrTiO_3$(111) substrate [13], which was more than 10 times larger than the values (3 to 6 μC/cm^2) reported for bulk ceramics. BFO has another advantage in the fabrication of FeRAMs in that the crystallization temperature is as low as 550 °C. However, the coercive field is still higher than 200 kV/cm and the leakage current density in polycrystalline BFO films at a high electric field is very high at RT. In this section, it is shown that the leakage current characteristic of BFO films deposited on polycrystalline Pt electrodes is significantly improved by substituting Mn atoms for Fe atoms.

$BiFe_{1-x}Mn_xO_3$ (x = 0 to 0.5) thin films were formed on Pt/Ti/SiO_2/Si(100) structures by chemical solution deposition. Chemical solutions with exact metal composition ratios at 0.2 mol/l concentrations were spin-coated on to the substrate, dried at 240 °C for 3 min, and prefired at 350 °C for 10 min in air. This process was repeated to several tens of times until a film of the desired thickness was obtained. Then, the film was annealed at 550 °C for 10 min in a N_2 atmosphere using a rapid thermal annealing (RTA) furnace. Pt top electrodes were deposited by electron beam evaporation through a shadow mask and the samples were finally annealed at 550 °C for 10 min in a N_2 atmosphere.

Figure 12.7 shows *J-E* (current density versus electric field) characteristics of the Mn-substituted BFO (BFMO) films measured using Pt/BFMO/Pt capacitors at RT [14]. The current density in a pure BFO film is very low (on the order of 10^{-8} A/cm^2) at a lower electric field than 0.3 MV/cm, but it increases sharply when the electric field exceeds 0.3 MV/cm and reaches the range of 10^{-2} A/cm^2 at 1 MV/cm. The figure also shows that the current density in the low electric field region steadily increases with increase of Mn concentration, and that the critical electric field at which current increases sharply shifts to a field higher than 1 MV/cm by Mn substitution. As a result, the leakage current densities at 1 MV/cm are lower in the 3 and 5 at% Mn-substituted films than in a pure BFO film.

The current conduction mechanism in BFMO films can be explained as follows, using the theory of space-charge-limited current in an insulator with traps. In an insulator with traps,

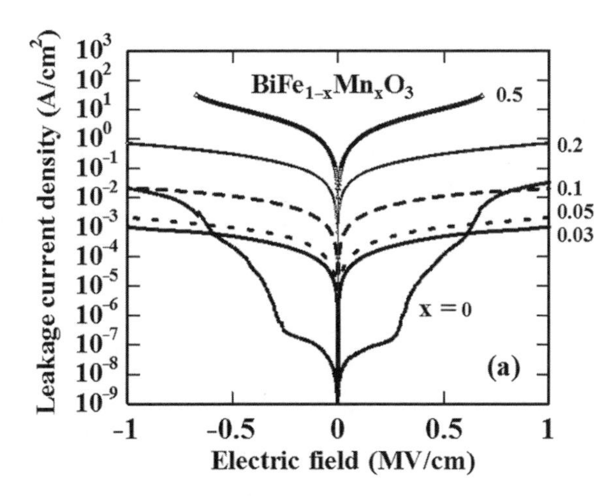

Fig. 12.7 *J-E* characteristics of $BiFe_{1-x}Mn_xO_3$ (x = 0 – 0.5) films on a Pt/Ti/SiO_2/Si(100) structure measured at RT.

the current conduction mechanism at a low applied voltage is ohmic when ohmic contact is formed as an electron injection source. Under this condition, traps are partially filled and the Fermi level is virtually fixed. With an increase in the applied voltage, all traps are filled at a certain voltage. This voltage is known as the trap-filled-limit voltage and is proportional to trap density. When the applied voltage is further increased, beyond the trap-filled-limit voltage, the Fermi level shifts upwards rapidly and the current increases sharply. Finally, the current reaches the space-charge-limited current, which increases in proportion to the square of the applied voltage in a solid.

Therefore, if it is assumed that some Mn atoms in a BFMO film substitute for Fe atoms to form a $BiMnO_3$ phase, and some remain in interstitial sites or grain boundaries, the following conduction model holds in the BFMO film. Since $BiMnO_3$ has a perovskite structure and its energy band gap is narrower than that of BFO, it is speculated that the band gap of the BFO–$BiMnO_3$ mixed crystal is narrower than that of BFO and the intrinsic carrier concentration is higher. This speculation well explains the experimental result that the current density of the BFMO film is higher in the ohmic region. On the other hand, the Mn atoms in the interstitial site or in grain boundaries form deep trap levels, and thus the trap-filled-limit voltage increases with an increase of the Mn concentration. As the result, the limit voltage below which ohmic current flows shifts upwards in the BFMO film [15].

Figure 12.8 shows comparison of *P-E* hysteresis loops between pure and 5 at% Mn-substituted BFO films. In this measurement, the applied voltage was swept linearly with a frequency of 1 kHz. As can be seen from Figure 12.8 (a), the loops in the pure BFO film are

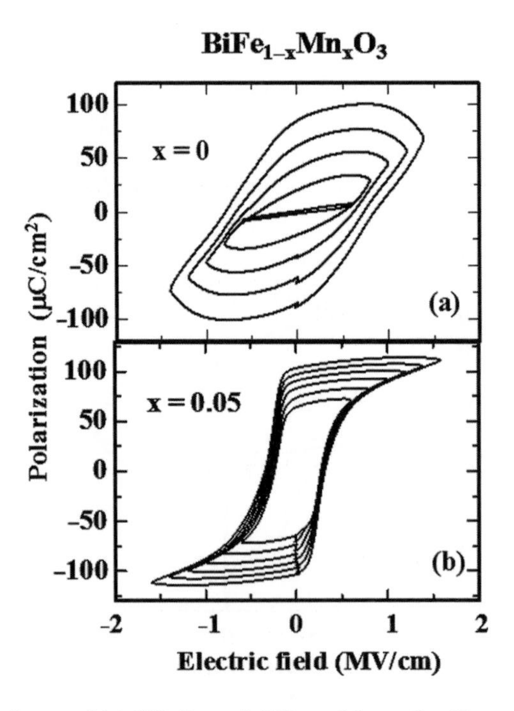

Fig. 12.8 *P-E* hysteresis loops of (a) $BiFeO_3$ and $BiFe_{0.95}Mn_{0.05}O_3$ films on a $Pt/Ti/SiO_2/Si(100)$ structure.

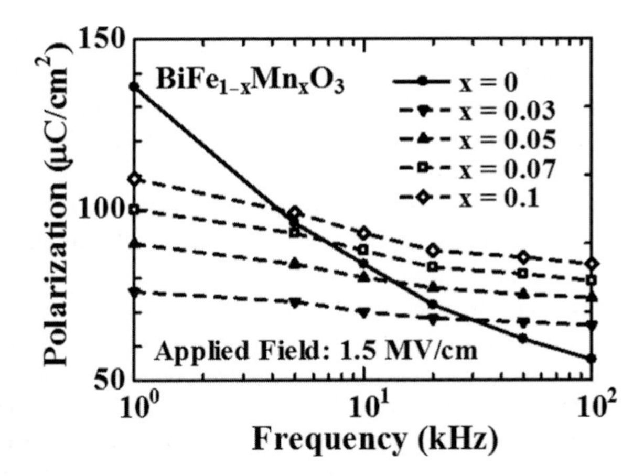

Fig. 12.9 *P-E* hysteresis loops of 5 at% Mn-substituted $BiFeO_3$ films measured at RT using triangular voltage pulses with 1–100 kHz frequency.

rounded, because of the high leakage current density at a high electric field. On the other hand, well-saturated hysteresis loops were obtained in the 5 at% Mn-substituted BFO film, as shown in Figure 12.8 (b). In this film, the remanent polarization and coercive field at 1.6 MV/cm were 100 μC/cm^2 and 0.33 MV/cm, respectively. In the 10 at% Mn-substituted film, the leakage current density became high again and rounded hysteresis loops were obtained. These results clearly show that a decrease in the leakage current in the high electric field region is essential in obtaining saturated hysteresis loops in the *P-E* measurement.

Figure 12.9 shows the frequency-dependent remanent polarizations in a 600 nm-thick BFO film and 450 nm-thick BFMO films [16]. As shown in the figure, the frequency dependence of P_r is very large in the pure BFO film, which is probably due to the large leakage current. In BFMO films, on the other hand, the frequency dependences are weaker, particularly at frequencies higher than 10 kHz. This figure clearly shows that the large P_r values measured in the BFMO films at RT are not due to the leakage current, but are due to the ferroelectricity of the films. The enhanced P_r in BFMO films might be due to an increase of the tetragonality in the crystal structure. Similar Mn substitution effects have also been observed in RF-sputtered BFO films [15].

12.3 Cell structure and operation principle of capacitor-type FeRAMs

12.3.1 Cell structure of 1T1C(2T2C)-type FeRAMs

A 1T1C-type FeRAM cell is composed of a ferroelectric capacitor and an MOSFET used for cell selection, as shown schematically in Figure 12.2. In the case of a 2T2C-type FeRAM, two pairs of the capacitor and FET are used in each cell, by which the voltage difference between data "0" and "1" can be kept largely in the "read" operation. The 2T2C-type cell was successfully used in the early stages of commercialized FeRAMs, because the characteristics of ferroelectric capacitors were less reliable for repeated polarization reversal at that stage. However, since each cell area is twice the size of that of a 1T1C cell, and since the ferroelectric capacitors and the current detection system have become more reliable by improvement in materials, fabrication processes, and circuit technology, the 1T1C-type cell is usually used in present-day high-density FeRAMs.

In a planar capacitor cell, a ferroelectric capacitor is formed on a field oxide film and it is connected to the drain of the FET using the upper electrode, as shown in Figure 12.10 (a). In the fabrication process, after fabricating the FET, the chip surface is covered with the interlayer oxide and planarized by chemical mechanical polishing. Then, the Pt bottom electrode with a Ti or TiO_2 sticking layer to SiO_2, the ferroelectric film, and the Pt top electrode are successively blanket-deposited and the capacitor structure is formed by etching the films using two or three different masks. In FeRAMs, since a plate line is connected to the individual capacitors, it is necessary to separate ferroelectric capacitors cell by cell, which is different from DRAM cells. In the case of PZT capacitors, stacked layers such as $Pr/IrO_2/TiO_2$, $Pr/IrO_2/Ir/TiO_2$, and $SrRuO_3/Pt/TiO_2$ are used as the bottom electrode, instead of a Pt/TiO_2 layer, for improving fatigue and imprint characteristics. For the same purpose, Ir/IrO_2 and $Pt/SrRuO_3$ are often

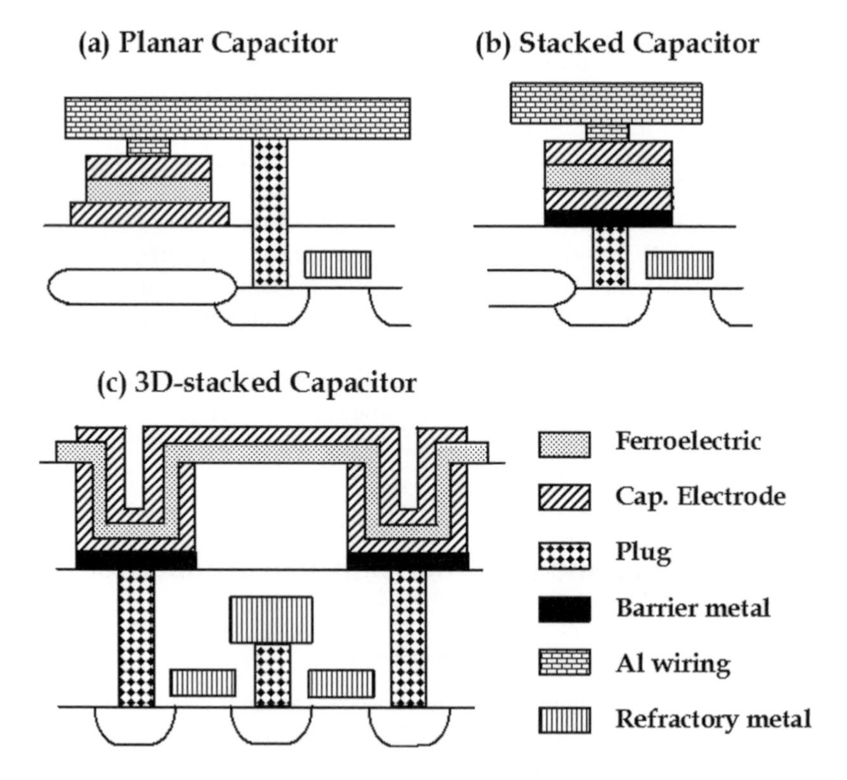

Fig. 12.10 Classification of cell structures. (a) Planar capacitor cell, (b) stacked capacitor cell, and (c) 3D-stacked capacitor cell.

used as the top electrode. Figure 12.11 shows a schematic drawing of a cross-section of actual FeRAM cells with planar ferroelectric capacitors.

In a stacked capacitor cell, shown in Figure 12.10 (b), the ferroelectric capacitor is formed on the FET and the bottom electrode of the capacitor is connected to the drain of the FET using a plug. A key technology in fabricating this structure is the electrical connection between the plug and the bottom electrode, because plug materials such as poly-Si and W are easily oxidized and disconnected electrically through the crystallization process of the ferroelectric film. To

Fig. 12.11 A schematic drawing of a cross-section of actual FeRAM cells with planar ferroelectric capacitors. Courtesy of Fujitsu Semiconductor Ltd. This figure is reproduced in color in the color plate section.

solve this problem, a barrier metal layer such as Ir/IrO_2 or $Ir/IrO_2/TiAlN$ is inserted between the bottom electrode and the plug. In this cell structure, it is possible to etch the stacked films continuously using a single mask. This method has the advantage that the capacitor area can be reduced, particularly when the etching angle is close to 90°. Under such an etching condition, however, a short-circuit problem in the capacitor often occurs, because of deformation of the edge of the Pt electrode. Thus, typical etching angles are 70° to 80°.

In future high-density memories, it is important to further shrink the cell size without reducing the stored charge. One approach towards this aim is to develop a novel ferroelectric material with a large remanent polarization, as discussed in 12.2.3. Another approach is to fabricate ferroelectric capacitors in three-dimensions, as shown in Figure 12.10 (c). In fabrication of this structure, the MOCVD technique is needed to deposit a ferroelectric film uniformly on the side wall of the holes as well as the bottom face. After fabrication of the capacitors, the wafer surface is again planarized by depositing a SiO_2 film. During this process, since SiH_4 gas is decomposed, hydrogen gas is inevitably generated. Furthermore, it has been found that H_2 gas is decomposed to hydrogen atoms by the catalytic action of Pt, and the ferroelectric properties of the film are severely degraded by penetration of hydrogen atoms. Thus, to minimize degradation of the ferroelectric properties of the capacitor, formation of a hydrogen barrier layer such as an Al_2O_3 layer is needed prior to deposition of the SiO_2 film.

12.3.2 Operation principle of 1T1C(2T2C)-type FeRAMs

Figure 12.12 shows the simplified circuit configurations of (a) a 2T2C cell and (b) a 1T1C cell, in which the plate lines (PL) are run parallel to the word line (WL). In a 2T2C cell, the opposite data are stored in the two ferroelectric capacitors. To write a "1" datum in the cell, the bit line (BL) and the PL are raised to V_{DD} (power supply voltage), while the $\overline{BL}$ is kept grounded. Then, the WL is raised to V_{PP} ($V_{DD} + V_T$ or the higher voltage) so that the voltage drop across the FET is negligible, where V_T is the threshold voltage of the FET. At this time, the ferroelectric film in the right capacitor is polarized upwards, because of the voltage difference between the PL and the $\overline{BL}$, while the polarization direction of the film in the left capacitor is unchanged, because the voltages of the PL and the BL are equal. Next, the voltage of the PL is driven back to zero, keeping the voltage of the BL at V_{DD}. At this time, the ferroelectric film in the left capacitor is polarized downwards, while the polarization direction in the right capacitor is unchanged. Finally, the BL and the WL are driven back to zero. To write a "0" datum, the roles of the BL and the $\overline{BL}$ are exchanged. The time sequence diagram is shown schematically in Figure 12.12 (c).

To read the stored data, the PL is raised to V_{DD}, a sense amplifier connected between the BL and the $\overline{BL}$ is turned on so that the difference in currents flowing out to the BL and the $\overline{BL}$ is detected. If the stored datum is "1", polarization of the ferroelectric film is reversed only in the left capacitor and thus a current flowing out to the BL is larger than that to the $\overline{BL}$. This small imbalance in the currents is amplified by the positive feedback function of the sense amplifier and the voltages of the BL and the $\overline{BL}$ reach V_{DD} and zero, respectively in a short time. The voltage difference is transferred to the periphery circuit as the datum "1" signal. After the voltages of the sense amplifier reach constant values, the voltage of the PL is driven back to

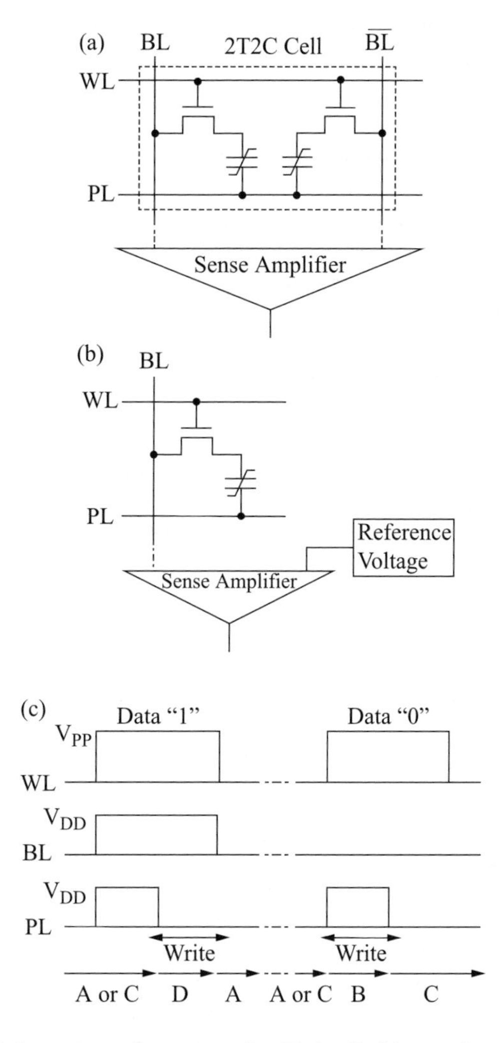

Fig. 12.12 (a) A simplified circuit configuration of a 2T2C cell, (b) a configuration of a 1T1C cell, and (c) a schematic time sequence diagram for applying various voltage pulses in the "write" operation. A, B, C, and D correspond to those in Fig.12.3.

zero, by which the polarization of the ferroelectric film in the left capacitor returns fully to the downward direction (rewrite operation).

In a 1T1C cell, one terminal of the sense amplifier is connected to a reference voltage, instead of the $\overline{BL}$, as shown in Figure 12.12 (b). To generate the reference voltage, which is requested to be kept in the middle of the cell voltages corresponding to "1" and "0" data, a ferroelectric capacitor with a larger area is used, and its polarization is reversed whenever a "read" or "write" operation is conducted. In this case, the reference voltage gradually changes by fatigue of the ferroelectric capacitor, and the change in the reference voltage roughly coincides with that of the cell voltages. For this reason, correct operation can be expected in a 1T1C-type cell, even if the ferroelectric film is somewhat fatigued.

In WL//PL architecture, shown in Figure 12.12, an entire row that shares the same WL and PL pair is accessed at once and all stored data are destroyed by raising the voltage of the PL during a "write/read" operation. Thus, a "rewrite" operation using the sense amplifier is needed for all non-selected cells along the row. Since no "refresh" operation is needed in a FeRAM, the "rewrite" operation for the non-selected cells is disadvantageous from the viewpoint of fatigue of the ferroelectric film. One solution to this problem is to use BL//PL architecture, in which the PL is run parallel to the BL. However, since the BL direction is usually longer than the WL direction in a chip, the operation speed of FeRAM is decreased by the *RC* delay originating from the PL resistance. Various architectures, including staircase PL architecture [17] have been proposed and those that are appropriately matched with the technology development have been adopted.

12.3.3 Other capacitor-type FeRAMs

To decrease the cell area of a 1T1C-type cell and to increase stability in the "write/read" operation, a chain FeRAM has been proposed and its operation has successfully been demonstrated [18]. Figure 12.13 shows the circuit diagram (a) and the cross-sectional structure (b) of a chain cell block. As shown in the figure, a ferroelectric capacitor and a MOSFET are connected in

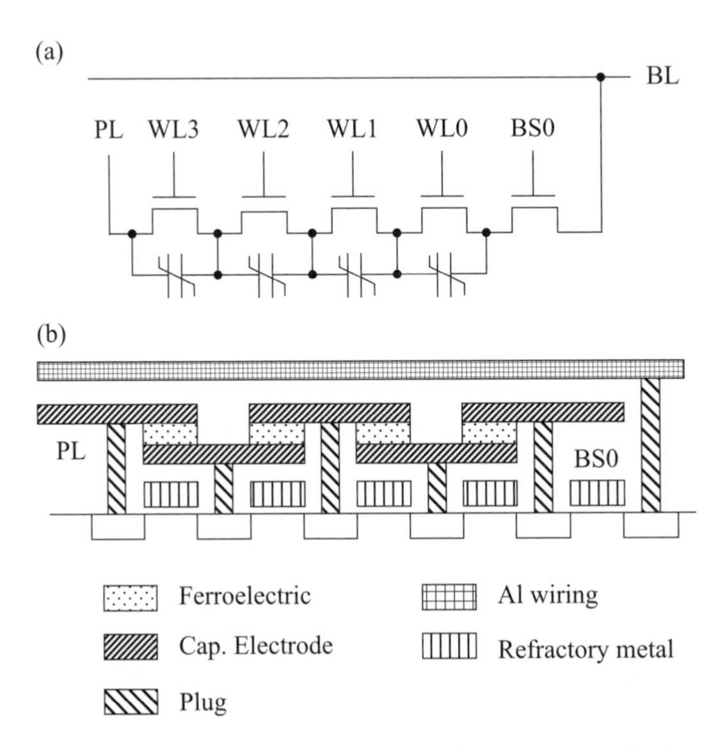

Fig. 12.13 (a) A circuit configuration and (b) a cross-sectional structure of a cell block in a chain FeRAM.

parallel in each cell and the cells are connected in series, forming a chain cell block. During the stand-by period, the gate voltage of the FET (BS0) for selecting the cell block is grounded, while all word lines are boosted to V_{PP}, so that all ferroelectric capacitors are short-circuited by the FETs, by which means, the possibility that the polarization of the ferroelectric capacitors is reversed by noise signals becomes very low.

In "write/read" operation, the gate voltage of the selected BS0 is raised to V_{DD} and the FET in the selected cell is turned off by pulling down the voltage of the selected WL. Under this condition, since the BL voltage is applied only to the ferroelectric capacitor in the selected cell, the "write/read" operation can be conducted in a similar manner to that for a 1T1C-type cell. Additionally, high-speed operation can be expected, because the voltage is not applied to the unnecessary ferroelectric capacitors. Figure 12.13 (b) shows a cross-sectional structure of a chain cell block with the minimum cell area. As can be seen from the figure, integration with the $4F^2$ cell size ($2F$: line pitch) can be expected by stacking the ferroelectric capacitor on the gate electrode of the FET.

Another group within capacitor-type FeRAMs is NVSRAMs (non-volatile static RAMs) using ferroelectric capacitors, which can be used as embedded non-volatile memories in high-speed system LSI (large-scale integration) circuits. The prototype was proposed in 1988 as a shadow SRAM [4]. In this circuit, ferroelectric capacitors are connected to the storage nodes of a SRAM cell through pass transistors, as shown in Figure 12.14(a). The circuit usually operates as a SRAM and when electricity is turned off, the voltages at the storage nodes are transformed to the polarization direction of the two ferroelectric capacitors by making the pass transistors active and by driving the PL ("store" operation). When electricity is turned on, the data stored in the ferroelectric capacitors are returned to the SRAM ("recall" operation) by driving the PL. In this circuit, since the ferroelectric capacitors are used only when electricity is turned off or turned on, degradation of the ferroelectric film can be neglected in the circuit operation. The disadvantage of the shadow SRAM is its larger cell area than the usual SRAM.

To decrease the cell area of the shadow SRAM, a 6T4C-type NVSRAM, shown in Figure 12.14(b) has been proposed and is already produced commercially [19]. In this circuit, since the four ferroelectric capacitors are stacked on the SRAM circuit, the cell area is almost the same as that of a usual volatile SRAM. The "store" operation is conducted by driving both PL1 and PL2 from high to low, by which the polarization of the two ferroelectric capacitors connecting to the high voltage node is directed from the storage node to the PL, and that of the two capacitors connecting to the low voltage node is directed from the PL to the storage node. In the "recall" operation, the PL1 is raised to a high voltage, keeping the PL2 grounded, by which the polarization in the two ferroelectric capacitors (C_1/C_4 pair or C_2/C_3 pair) is reversed and the voltages of the storage nodes are determined by the difference of the polarization reversal current.

During "write/read" operation, both PL1 and PL2 are kept grounded. Under this condition, the polarization of the ferroelectric capacitors is directed from the storage node to the PL, when the voltage of the node is high, and this polarization direction never changes, even if the node voltage becomes low or it again becomes high. That is, the maximum number of polarization reversals is two in the time interval from turn-on of electricity to its turn-off. Since the polarization direction hardly changes, the operation speed of this circuit is as fast as that of a normal SRAM, and there is no limitation to "write/read" cycles.

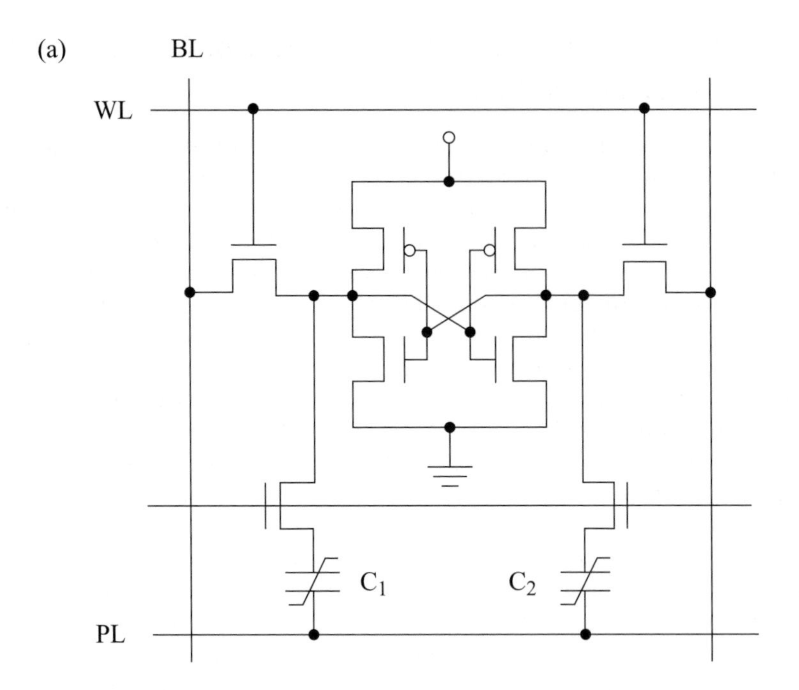

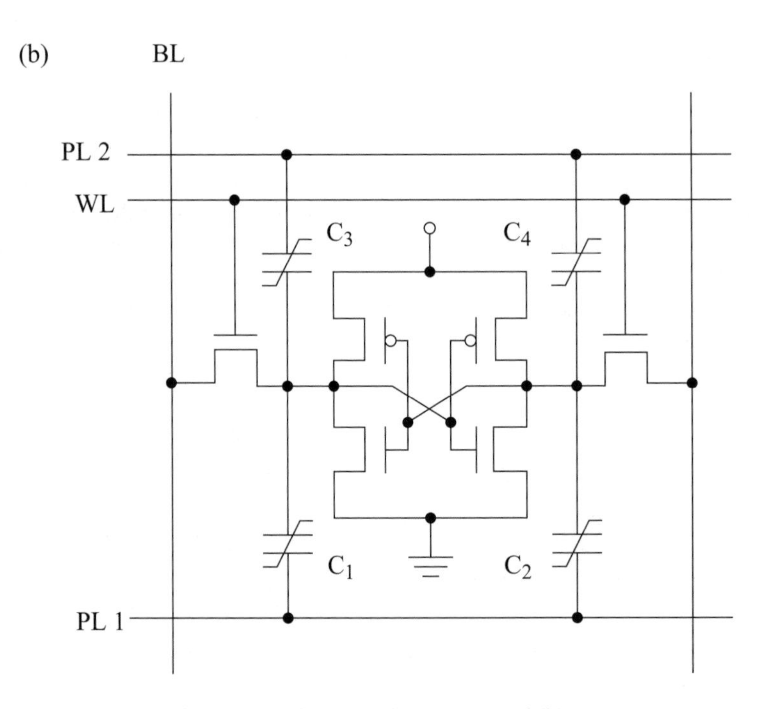

Fig. 12.14 Circuit configurations of (a) a shadow SRAM and (b) a 6T4C-type NVSRAM.

12.4 Cell structure and operation principle of FET-type FeRAMs

12.4.1 Optimization of FeFET structure

One-transistor-type (1T-type) FeRAM, shown in Figure 12.2 (b), has the potential to be integrated in high-density, because each memory cell is composed of a single ferroelectric-gate FET (FeFET) and because the FET can be scaled down using the proportionality rule. In a FeFET, electrons or holes are accumulated at the surface of the semiconductor according to the polarization direction of the gate ferroelectric film, and drain current flows between the source and drain regions, only when one type of carrier is accumulated at the interface. Thus, 1T-type FeRAM has another advantage, that stored data can be non-destructively read out using the drain current of the FET. Concerning the remanent polarization of the gate ferroelectric film, a large value is unnecessary, because the surface carrier density necessary for operation of MOSFETs is on the order of 10^{12} electrons(holes)/cm^2 (0.16 μC/cm^2).

However, it is very difficult to fabricate FeFETs with excellent electrical properties, because of inter-diffusion of the constituent elements in the film and the substrate. That is, when a ferroelectric film such as PZT or SBT is deposited directly on a Si substrate, constituent elements diffuse amongst each other during crystallization annealing. To avoid degradation due to inter-diffusion, an insulating buffer layer is often inserted between the ferroelectric film and the Si substrate. Even in the structure with the buffer layer, carriers are induced on the semiconductor surface by polarization of the ferroelectric film, as long as the charge neutrality condition is satisfied at the interface between the ferroelectric film and the insulating buffer layer. It is also possible to insert a floating gate electrode between the ferroelectric film and the buffer layer. The resultant gate structure is either an MFIS (M; metal, F; ferroelectric, S; semiconductor) structure (Figure 12.15 (a)) or an MFMIS (I; insulator) structure (Figure 12.15 (b)).

In these structures, however, new problems arise such that the data retention time is short and the operation voltage is high. The reason why the data retention time is short is as follows. In a FeFET, when the power supply is off and the gate terminal of the FET is grounded, the top and bottom electrodes of the two capacitors are short-circuited. At the same time, electric charges $\pm Q$ remain on the electrodes of both capacitors due to the remanent polarization of the ferroelectric film and due to the charge neutrality condition at a node between the two capacitors. The $Q - V$ (charge versus voltage) relation in the dielectric capacitor is $Q = CV$ (C:

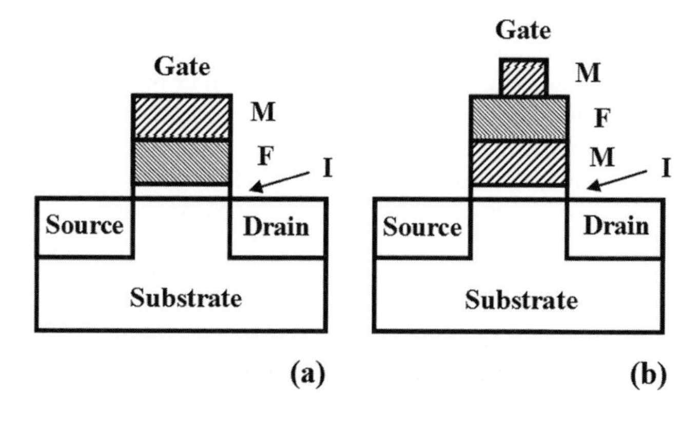

Fig. 12.15 Schematic drawings of MFIS (a) and MFMIS (b) gate structures.

capacitance of the dielectric layer), and thus the relation in the ferroelectric capacitor becomes $Q = -CV$ under the short-circuited condition. This relation means that the direction of the electric field in the ferroelectric film is opposite to that of the polarization. This field is known as the depolarization field and it reduces the data retention time significantly.

In order to make the depolarization field low, C must be as large as possible. That is, a thin buffer layer with a high dielectric constant is desirable. It is also desirable in the MFMIS structure that the area of the MIS part is made larger than that of the MFM part, as shown in Figure 12.15 (b). Another important point is to reduce the leakage current in both the ferroelectric film and the buffer layer. If the charge neutrality at a node between the two capacitors is destroyed by the leakage current, electric charges on the electrodes of the buffer layer capacitor disappear, which means that carriers on the semiconductor surface disappear and the stored data can not be read out by the drain current of the FET, even if the polarization of the ferroelectric film is retained. Thus, it is very important to reduce the leakage current across both the ferroelectric film and the buffer layer [20].

Based on these considerations, various buffer layer materials have been investigated experimentally. Several MFIS devices with buffer layers such as Si_3N_4 [21], $LaAlO_3$ [22], HfAlO [23], and HfO_2 [24] have exhibited good diode and FET characteristics. Particularly, FETs with HfAlO and HfO_2 layers have shown excellent data retention characteristics, as discussed in the next section. As well as studies on buffer layer materials, studies on ferroelectric materials with low dielectric constants have also been conducted. When the dielectric constant of a ferroelectric film is low, the external voltage is more effectively applied to the ferroelectric film and thus a wider memory window in drain current versus gate voltage (I_D-V_{GS}) characteristics is expected. Typical ferroelectric materials used for this purpose are $Sr_2(Ta,Nb)_2O_7$ [25] and $Pb_5Ge_3O_{11}$ [26].

12.4.2 Data retention characteristics of FeFETs

MFIS diodes and FETs have been fabricated on a Si substrate using HfO_2 as a buffer layer and using SBT or $(Bi,La)_4Ti_3O_{12}$ (BLT) as a ferroelectric film [24]. The buffer layer was deposited by vacuum evaporation of sintered HfO_2 targets at room temperature and subsequently annealed in an O_2 atmosphere at 800 °C to 1 min for obtaining an excellent insulator–semiconductor interface. Then, an SBT or BLT film was deposited by spin-coating, dried and calcined in air, and annealed in an O_2 atmosphere at 750 °C for 30 min for crystallization. Finally, Pt top electrodes were deposited either by vacuum evaporation or by RF-sputtering. It has been observed from cross-section TEM images that the buffer layer formed under typical fabrication conditions has a stacked structure composed of an 8 nm-thick HfO_2 layer and a 5 nm-thick SiO_2 layer after the crystallization annealing of the ferroelectric film.

Figure 12.16 shows I_D-V_{GS} characteristics of FeFETs with SBT(400 nm)/HfO_2(8 nm) and BLT(400 nm)/HfO_2(8 nm) gate structures [27]. As can be seen from the figure, I_D-V_{GS} characteristics show clockwise hysteresis and the drain current on/off ratio at a gate voltage of 0.8 V is as large as 10^5 in the SBT/HfO_2 sample. The memory window width in the hysteresis loop is about 1.0 V in the SBT/HfO_2 sample and about 0.5 V in the BLT/HfO_2 sample. It has also been found in a BLT/HfO_2 sample that the drain current on/off ratio is about 200 for a "write" pulse width as narrow as 20 ns. These results clearly show that the hysteresis loops in the I_D-V_{GS} characteristics are due to ferroelectricity of the gate films.

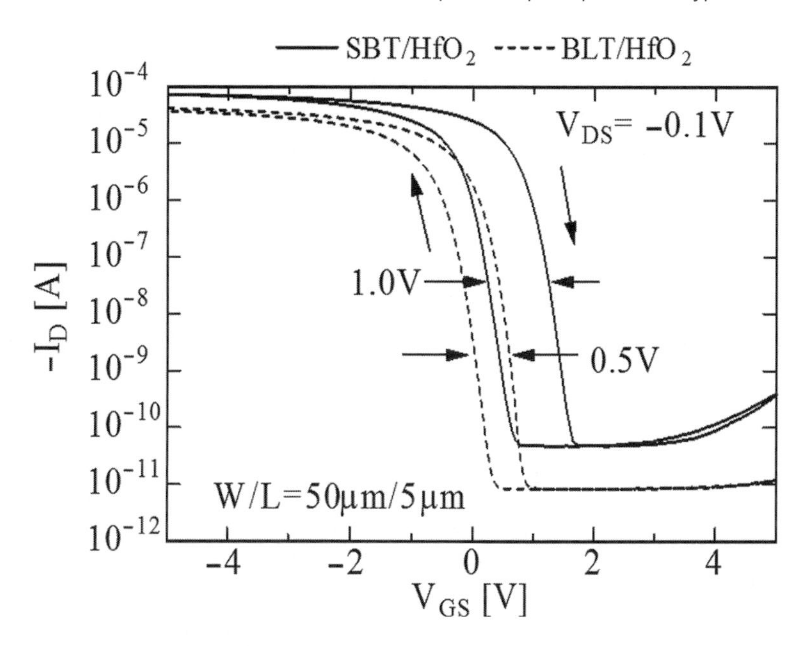

Fig. 12.16 $I_D - V_{GS}$ characteristics of FeFETs with SBT/HfO_2 and BLT/HfO_2 gate structures.

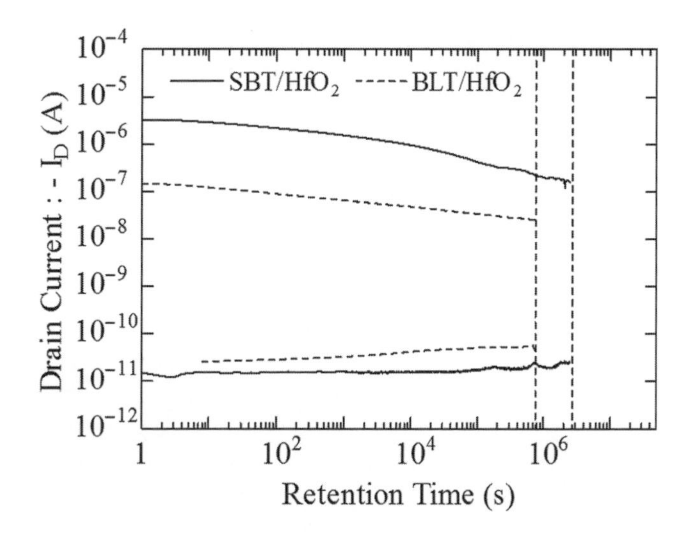

Fig. 12.17 Data retention characteristics of FeFETs with SBT/HfO_2 and BLT/HfO_2 gate structures. Two vertical lines in the figure correspond to 10 days and 30 days.

Figure 12.17 shows data retention characteristics of FeFETs with the $Pt/SBT/HfO_2/Si$ and $Pt/BLT/HfO_2/Si$ gate structures. In these measurements, "write" pulses of ± 10 V in amplitude and 1 μs in width were initially applied to the gate, and variation of the drain currents with time was measured, keeping the gate voltages at 0.6 V and 0.3 V in the SBT/HfO_2 and BLT/HfO_2 samples, respectively. In the SBT/HfO_2 sample, the drain current on/off ratio was larger than

10^3, even after 30 days had elapsed. Furthermore, if the experimental data are simply extrapolated towards a longer time scale, the current on/off ratio at 10 years (3×10^8 sec) is expected to be much larger than 100. Concerning the fatigue endurance of the gate ferroelectric film, the memory window width in $I_D - V_{GS}$ characteristics hardly changed, even after application of 2.2×10^{11} bipolar pulses.

These results show that HfO_2 is one of the best buffer layer materials to be inserted between the ferroelectric film and Si substrate, and to prevent interdiffusion of constituent elements in MFIS FETs. However, the "write" voltage in these devices is still high and thus a further decrease of the operation voltage is needed. Recently, it has also been shown in a FeFET with a HfAlO buffer layer that the data retention time is not seriously degraded, even if the operation temperature is increased to 85 °C [28].

12.4.3 Cell array structures

To increase the packing density of FET-type FeRAMs, it is desirable that each memory cell be composed of a single FeFET. In this 1T-type cell array, however, the stored data in non-

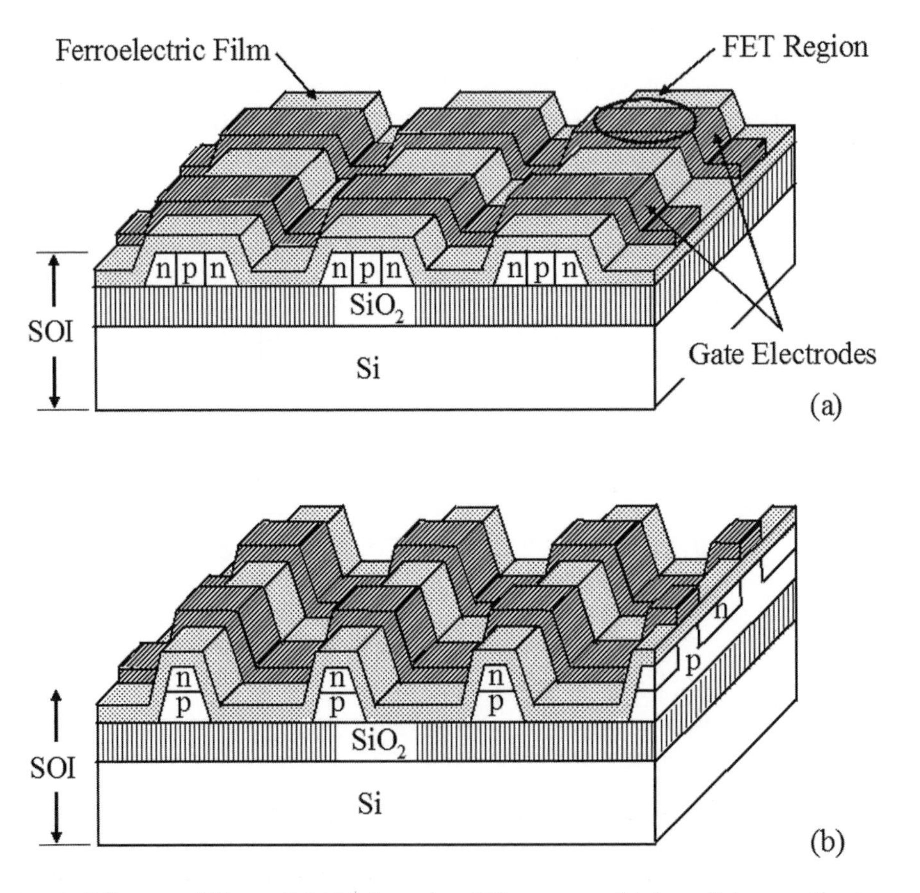

Fig. 12.18 Cell arrays of 1T-type FeRAMs formed on SOI structures. (a) A parallel connection type and (b) a series connection type.

selected cells are often reversed through repetition of "write/read" operation. Thus, the array structure and the "write/read" method for minimizing the data disturbance phenomenon are important. Typical 1T-type cell arrays are shown in Figure 12.18 (a) and (b), in which memory cells are fabricated on an SOI (silicon-on-insulator) structure. In Figure 12.18 (a), Si stripes with a lateral npn structure are placed on an insulating substrate; they are covered with a uniform ferroelectric film, and then metal stripes are placed on the film perpendicular to the Si stripes. Thus, each Si stripe represents a parallel connection of FeFETs and no via hole through the ferroelectric film exists in the array area [29]. In Figure 12.18 (b), on the other hand, the surface n layer is disconnected under the individual gate electrodes, as shown in the cross-section along the Si stripe. Thus, each Si stripe represents a series connection of FeFETs and the packing density of $4F^2$ ($2F$: line pitch) can be expected in this cell array.

To write a datum "1" in a selected cell in the array, the so-called $V/3$ rule is used, in which V and $V/3$ are applied to the selected and non-selected metal electrodes, respectively, while 0 and $2V/3$ are applied to the selected and non-selected Si stripes, as shown in Figure 12.19 (a). In the both cell arrays shown in Figure 12.18, "write" voltage pulses can be applied to the ferroelectric film of any cell by using the body p-layer in the stripe as a lower electrode. In this method, the disturbance voltage in most non-selected cells is $-V/3$, and it is $V/3$ in the non-selected cells placed along the selected column and row. Although the disturbance voltage is 1/3 of the "write" voltage, it has been found from an experimental result using ferroelectric capacitors, that depolarization due to the disturbance voltage cannot be neglected in the worst case [29]. This problem has almost been solved by applying compensation voltage pulses in the next timing, as shown in Figure 12.19 (b).

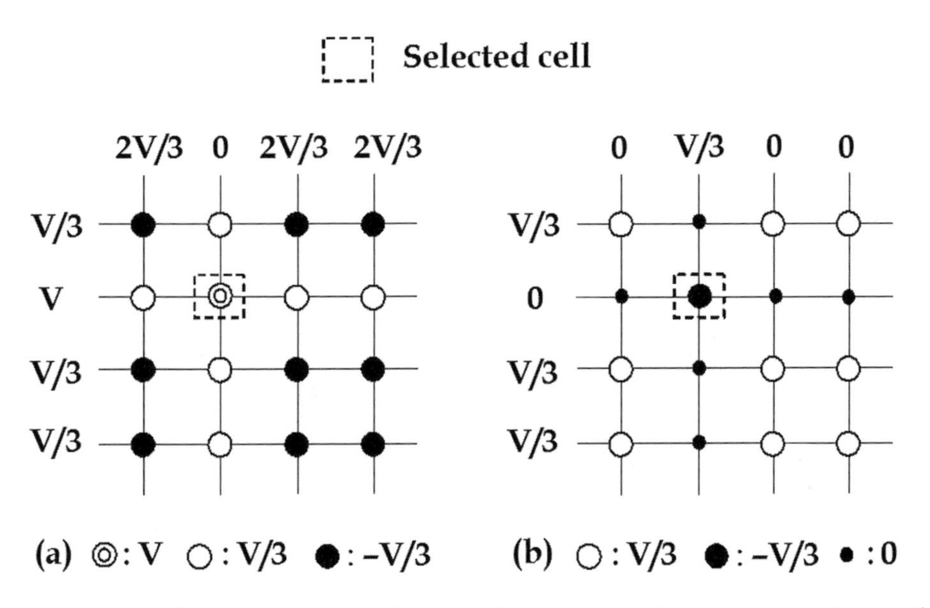

Fig. 12.19 (a) "write" operation to a selected cell and (b) the compensation operation to minimize the data disturbance phenomenon.

To read out the stored data in the 1T-type cell arrays, it is necessary to measure the drain current of the selected FeFET. First, from a data retention viewpoint, the center of the hysteresis loops in $I_D - V_{GS}$ characteristics should be kept at 0 V. In this case, it is unnecessary to apply any voltage to the gate of the selected cell for distinguishing data "0" and "1". On the contrary, it is necessary to apply a voltage to the gate of the non-selected cells. In the case of the parallel connection shown in Figure 12.18 (a), the resistance of all non-selected FETs is requested to be sufficiently high for both cases of data "0" and "1". Thus, it is necessary to apply a negative voltage to the gate of the non-selected FETs during the "read" operation, if FETs are n-channel. Similarly, when they are connected in series, it is necessary to apply a positive voltage to the gate of the non-selected FETs to sufficiently decrease their resistance. In both cases, since there is a possibility that data disturbance is generated during the "read" operation, it is important to determine the optimum voltages to stably conduct "write/read" operation.

References

[1] Ross, I.M. (1957). Semiconductor translating device. US Patent No.2791760.

[2] Wu, S.Y. (1974). A new ferroelectric memory device, metal-ferroelectric-semiconductor transistor. *IEEE Trans. Electron Devices*, **ED-21**, 499–504.

[3] Sugibuchi, K., Kurogi, Y., and Endo, N. (1975). Ferroelectric field-effect memory device using $Bi_4Ti_3O_{12}$ film. *J. Appl. Phys.*, **46**, 2877–2881.

[4] Eaton, S.S., Butler, D.B., Parris, M., Wilson, D., and McNeille, H. (1988). A ferroelectric nonvolatile memory. *IEEE Intern. Solid State Circuits Conf. Digest of Technical Papers*, pp. 130–131.

[5] Evans, J. and Womack, K. (1988). An experimental 512-bit nonvolatile memory with ferroelectric storage cell. *IEEE Solid State Circuits*, **23**, 1171–1175.

[6] Fujisaki, S., Ishiwara, H., and Fujisaki, Y. (2007). Low-voltage operation of ferroelectric poly(vinylidene fluoride-trifluoroethylene) copolymer capacitors and metal-ferroelectric- insulator-semiconductor diodes *Appl. Phys. Lett.* **90**, 162902 (3 pages).

[7] Arimoto, Y. and Ishiwara, H. (2004). Current status of ferroelectric random-access memory, *MRS Bulletin*, **29**, 823–828.

[8] Kijima, T., Aoyama, T., Miyazawa, H. *et al.* (2005). Novel Si codoped $Pb(Zr, Ti, Nb)O_3$ thin film for high-density ferroelectric random access memory. *Jpn. J. Appl. Phys.* **44**, 267–274.

[9] Matsuzaki, T. and Funakubo, H. (1999). Preparation and characterization of $Pb(Nb,Ti)O_3$ thin films by metalorganic chemical vapor deposition. *J. Appl. Phys.* **86**, 4559–4564.

[10] Koo, C-Y., Cheon, J-H., Yeom, J-H., *et al.* (2006). Electrical properties of $BiFeO_3$ doped PZT thin films for embedded FeRAM devices. *J. Korean Phys. Soc.* **49**, S514–S517.

[11] Tang, M-H., Dong, G-J., Sugiyama, Y., and Ishiwara, H. (2010). Frequency-dependent electrical properties in $Bi(Zn_{0.5}Ti_{0.5})O_3$ doped $Pb(Zr_{0.4}Ti_{0.6})O_3$ thin film for ferroelectric memory application. *Semiconductor Science & Technology* **25**, 035006 (4 pages).

[12] Noguchi, T., Hase, T., and Miyasaka, Y. (1996). Analysis of the dependence of ferroelectric properties of strontium bismuth tantalate (SBT) thin films on the composition and process temperature. *Jpn. J. Appl. Phys.* **35**, 4900–4905.

[13] Wang, J., Neaton, J.B., Zheng, H. *et al.* (2003). Epitaxial $BiFeO_3$ multiferroic thin film heterostructures. *Science*, **299**, 1719–1722.

[14] Singh, S.K., Ishiwara, H., and Maruyama, K. (2006). Room temperature ferroelectric properties of Mn-substituted $BiFeO_3$ thin films deposited on Pt electrodes using chemical solution deposition *Appl. Phys. Lett.* **88**, 262908 (3 pages).

[15] Kim, J-H., Funakubo, H., Sugiyama, Y., and Ishiwara, H. (2009). Characteristics of undoped and Mn-doped $BiFeO_3$ films formed on Pt and $SrRuO_3$/Pt electrodes by radio-frequency sputtering. *Jpn. J. Appl. Phys.* **48**, 09KB02 (4 pages).

[16] Singh, S.K., Ishiwara, H., Sato, K., and Maruyama, K. (2007). Microstructure and frequency dependent electrical properties of Mn-substituted $BiFeO_3$ thin films *J. Appl. Phys.* **102**, 094109 (5 pages).

[17] Kawashima, S., Endo, T., Yamamoto, A., *et al.* (2002). Bitline GND sensing technique for low-voltage operation FeRAM. *IEEE J. Solid-State Circuits* **37**, 592–598.

[18] Takashima, D. and Kunishima, I. (1998). High-density chain ferroelectric random access memory (chain FRAM). *IEEE J. Solid-State Circuits*, **33**, 787–792.

[19] Masui, S., Ninomiya, T., Ohkawa, T., *et al.* (2004). Design and application of ferroelectric memory based nonvolatile SRAM. *Inst. Electron. Inform. Commun. Eng. Trans. Electron.* **E87-C**, 1769–1776.

[20] Ishiwara, H. (2003). Recent progress in ferroelectric-gate FETs. *Mater. Res. Soc. Sympo. Proc.* **748**, 297–304.

[21] Kim, K-H., Han, J-P., Jung, S-W., and Ma, T-P. (2002). Ferroelectric DRAM (FEDRAM) FET with metal/$SrBi_2Ta_2O_9$/SiN/Si gate structure. *IEEE Electron Device Lett.* **23**, 82–84.

[22] Park, B-E. and Ishiwara, H. (2004). Fabrication and characterization of $(Bi,La)_4Ti_3O_{12}$ films using $LaAlO_3$ buffer layers for MFIS structures. *Integrated Ferroelectrics* **62**, 141–147.

[23] Sakai, S. and Ilangovan, R. (2004). Metal-ferroelectric-insulator-semiconductor memory FET with long retention and high endurance. *IEEE Electron Device Lett.* **25**, 369–371.

[24] Aizawa, K., Park, B-E., Kawashima, Y., Takahashi, K., and Ishiwara, H. (2004). Impact of HfO_2 buffer layers on data retention characteristics of ferroelectric-gate field effect transistors. *Appl. Phys. Lett.* **85**, 3199–3201.

[25] Fujimori, Y., Nakamura, T., and Kamisawa, A. (1999). Properties of ferroelectric memory FET using $Sr_2(Ta, Nb)_2O_7$ thin film. *Jpn. J. Appl. Phys.* **38**, 2285–2288.

[26] Li, T., Hsu, S-T., Ulrich, B.D., *et al.* (2002). One transistor ferroelectric memory with Pt/$Pb_5Ge_3O_{11}$/Ir/poly-Si/SiO_2/Si gate-stack. *IEEE Electron Device Lett.* **23**, 339–341.

[27] Takahashi, K., Aizawa, K., Park, B-E., and Ishiwara, H. (2005). Thirty-day-long data retention in ferroelectric-gate field-effect transistors with HfO_2 buffer layers. *Jpn. J. Appl. Phys.* **44** 6218–6220.

[28] Li, Q-H and Sakai, S. (2006). Characterization of Pt/$SrBi_2Ta_2O_9$/Hf-Al-O/Si field-effect transistors at elevated temperatures. *Appl. Phys. Lett.* **89**, 222910 (3 pages).

[29] Ishiwara, H., Simamura, T., and Tokumitsu, E. (1997). Proposal of a single-transistor-type ferroelectric memory using an SOI structure and experimental study on interference problem in write operation. *Jpn. J. Appl. Phys.*, **36**, 1655–1658.

13
$LaAlO_3/SrTiO_3$-based device concepts

Daniela F. Bogorin, Patrick Irvin, Cheng Cen, and Jeremy Levy

13.1 Introduction

The two-dimensional electron gas (2DEG) that forms at the interface between two heterogeneous semiconductors or between a semiconductor and an oxide is the basis for some of the most useful and prevalent electronic devices. Silicon-based metal-oxide-semiconductor field-effect transistors, GaAs-based high-mobility transistors, solid-state lasers, and photodetectors are but a few examples of technologies that have emerged from semiconductor interfaces. Herb Kroemer, when receiving the Nobel Prize in Physics for the invention of the semiconductor heterostructure, stated that "the interface is the device" [1].

Complex oxides differ in many ways from traditional semiconductors that are used to form 2DEG devices [2]. Both $LaAlO_3$ and $SrTiO_3$ are compatible with a large class of materials that, collectively, exhibits a remarkable range of behaviors that include ferroelectricity, magnetism, multiferroic behavior, and superconductivity. There are many other complex oxide systems with exceptional properties, specifically ZnO/MgZnO heterostructures grown with high mobility [3].

The development of complex oxides over the past fifteen years has raised the prospect for new classes of electronic devices [2, 4]. In 2004, Ohtomo and Hwang published their seminal discovery that a high-mobility 2DEG can form at the interface between $LaAlO_3$ and $SrTiO_3$ [5], which forms the backdrop for this chapter. Since that time a number of striking properties of this interface have been discovered and explored. Here we will highlight the most important features that may lead to novel devices that can challenge or disrupt current technologies and perhaps create new ones. One feature in particular that we will focus on is the ability to reversibly create conducting nanostructures at the $LaAlO_3/SrTiO_3$ interface using a conductive atomic-force microscope (c-AFM). Again, it seems appropriate to quote from Kroemer's Nobel Lecture, in which he introduced his "*Lemma of New Technology*" [1]:

The principal applications of any sufficiently new and innovative technology always have been – and will continue to be – applications created by that technology.

In this chapter we will describe features and functionality that *resemble* existing technologies, but we hope that the reader will be inspired to think about what kind of *fundamentally new* technologies may emerge.

13.1.1 Semiconductor 2DEGs

A 2DEG is a structure in which the electrons are restricted to move only in a two-dimensional plane. Quantum confinement along the growth direction leads to the formation of "sub-bands," in which the lowest one is typically occupied for a 2DEG. Semiconductor 2DEGs are traditionally formed in Si metal-oxide-semiconductor field-effect transistors (MOSFETs) or modulation-doped GaAs/AlGaAs heterostructures. In modulation-doped structures the carriers are physically separated from the doping centers, enabling very high mobilities up to 31,000,000 cm^2/Vs [6].

Graphene is another 2D system of contemporary interest [7]. A fascinating property of the Dirac fermions in graphene is their ability to "Klein tunnel" through large barriers without scattering [8]. This property leads to anomalously large mobilities at room temperature.

13.1.2 2DEG at $LaAlO_3/SrTiO_3$ interface

$LaAlO_3$ and $SrTiO_3$ both have the perovskite crystal structure. $SrTiO_3$, a non-polar oxide, is composed of alternating, stacked layers of $(SrO)^0$ and $(TiO_2)^0$(Fig. 13.1). $SrTiO_3$ is pseudo-cubic at room temperature with a lattice constant 3.905 Å. It is a band-insulator with a bandgap of 3.25 eV, is chemically inert, and has been a substrate of choice for the growth of many other oxides and high-T_C superconductors [4, 9, 10]. $LaAlO_3$, a polar oxide, is composed of alternating stacked layers of $(LaO)^+$ and $(AlO_2)^-$. $LaAlO_3$ has a pseudo-cubic structure with a lattice constant of 3.791 Å at room temperature. It is a Mott insulator [11] with a wide bandgap of 5.6 eV. Films of $LaAlO_3$ thinner than ~15 monolayers (ML) [12] can be grown coherently strained to $SrTiO_3$. The manageably small lattice mismatch between $LaAlO_3$ and $SrTiO_3$ (about 3%) enables high-quality epitaxial growth of $LaAlO_3$ on $SrTiO_3$ via pulsed laser deposition (PLD) [13] or molecular-beam epitaxy (MBE) [14].

13.1.3 Polar catastrophe model

Epitaxial growth of $LaAlO_3$ on $SrTiO_3$ [5] can lead to an unusual and energetically unstable charge distribution. Two situations are shown in Fig. 13.2, the difference being the final termination of the $SrTiO_3$ layer. We first consider TiO_2-terminated $SrTiO_3$ (Fig. 13.2 (a)). The first-grown LaO layer will have a positive charge (+1/unit cell), after which the charge of each layer alternates between −1 and +1. This alternating charge density can be integrated to reveal both the internal electric field E and voltage V across the $LaAlO_3$ layer. The magnitude of the voltage increases linearly with the thickness of the $LaAlO_3$ layer. The divergence of the electrostatic energy can lead to a "polar catastrophe" and an associated electronic reconstruction in which the polarization is screened (in part) by the formation of a 2DEG at the $LaAlO_3/SrTiO_3$ interface (Figure 13.2 (c)). The other possible growth condition (Fig. 13.2 (b)) occurs when the

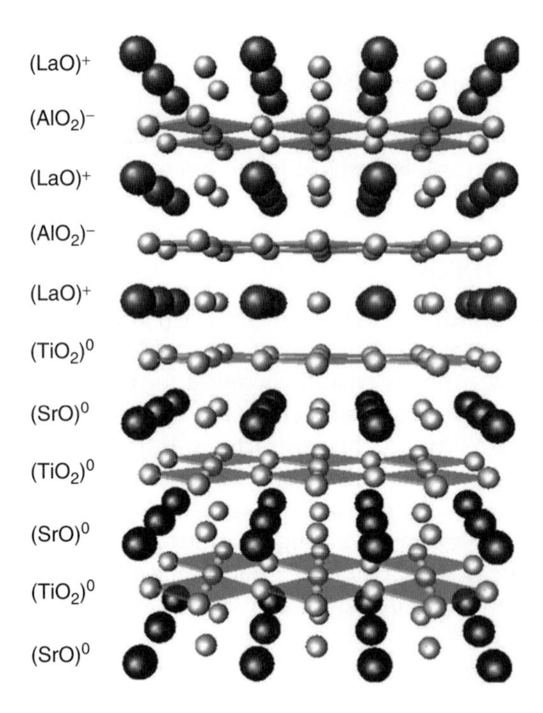

Fig. 13.1 $LaAlO_3/(TiO_2^-)SrTiO_3$ structure. Courtesy J. Mannhart.

$SrTiO_3$ substrate is SrO-terminated. This situation also leads to a "polar catastrophe," although it is predicted to result in a positively charged interface.

One missing component of this model is a discussion of what happens at the top $LaAlO_3$ surface. The absence of charge neutrality (indicated in Fig. 13.2 (c,d)) leads to a much more severely catastrophic situation than the existence of a finite, but diverging, polarization. To achieve charge neutrality, the top surface must also be compensated with charge. One such scenario that was invoked to explain metastable writing at the $LaAlO_3/SrTiO_3$ interface (described in more detail below) is the formation of oxygen vacancies at the top $LaAlO_3$ surface. Another scenario, described by Son *et al.* [16], involves the adsorption of hydrogen at the top AlO_2 surface. The hydrogen could be formed by spontaneous dissociation of H_2O molecules or through direct adsorption of H_2 in gas form. The charged top surface is expected to retain its insulating properties, but donation of electrons to the $LaAlO_3/SrTiO_3$ interface can be achieved for TiO_2-terminated $SrTiO_3$.

13.1.4 Metal–insulator transition in $LaAlO_3/SrTiO_3$

One predicted consequence of the polar discontinuity between $LaAlO_3$ and $SrTiO_3$ is an interfacial insulator-to-metal transition that is dependent on the $LaAlO_3$ thickness. This effect, first reported in 2006 by Thiel *et al.* [13] and confirmed by many groups, occurs in structures of $LaAlO_3$ grown on TiO_2-terminated $SrTiO_3$. When $LaAlO_3$ is grown with greater than a

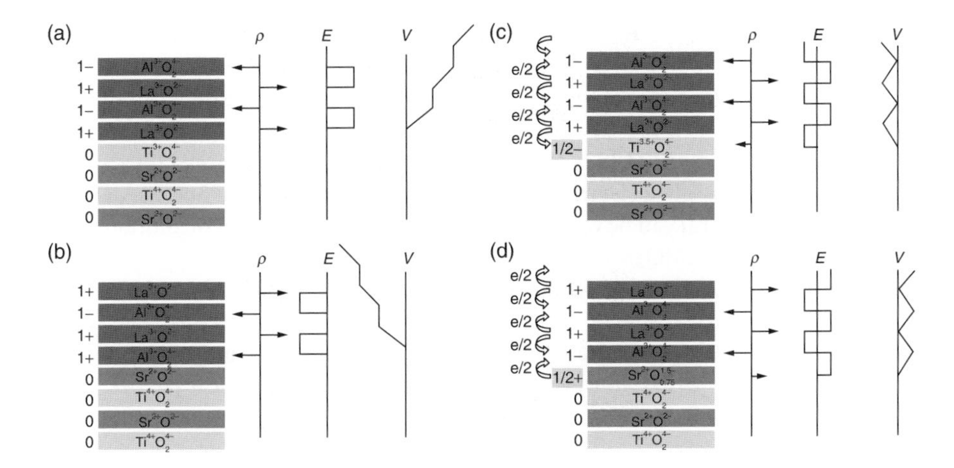

Fig. 13.2 Illustration of charge redistribution due to polar catastrophe at interfaces of $LaAlO_3/SrTiO_3$ grown with different termination conditions. (a) Non-reconstructed LaO/TiO_2 interface. A positive potential exists in the $LaAlO_3$ layer and diverges with film thickness. (b) A similar but negative potential occurs in the case of an AlO_2/SrO interface. (c) At the LaO/TiO_2 interface, adding 1/2 electron per unit cell to the top TiO_2 layer can minimize the potential in $LaAlO_3$. (d) At the AlO_2/SrO interface, adding 1/2 hole per unit-cell to the top SrO layer can minimize the potential in $LaAlO_3$. Reprinted by permission from Macmillan Publishers Ltd: Nature Materials [15], copyright 2006.

critical thickness $d_c = 3$ unit-cell (uc), the interface between $LaAlO_3$ and $SrTiO_3$ is found to be conducting. When the thickness of $LaAlO_3$ is smaller than d_c the interface remains insulating. In samples grown with approximately 3 uc of $LaAlO_3$ (normally insulating), the interface can be switched between the insulating and conducting states by applying a voltage to the back of the $SrTiO_3$ substrate [13]. This transition is a hysteretic function of the applied electric field.

13.1.5 Inconsistencies with the polar catastrophe model

While the polar catastrophe model helps to explain a number of empirical observations in the $LaAlO_3/SrTiO_3$ system, several experiments have identified inconsistencies with this model. Typical mobilities of a 2DEG formed at the $LaAlO_3/SrTiO_3$ interface is on the order of 100–1000 cm^2/Vs. However, in the original paper by Ohtomo and Hwang [5], the carrier density of the highest mobility $LaAlO_3/SrTiO_3$ structure ($\mu = 10^4$ cm^2/Vs at 2 K) was found to be two orders of magnitude larger than that required to screen the polarization in the $LaAlO_3$. The most likely explanation is that the majority of carriers in this system were donated by other sources (e.g. oxygen vacancies). Subsequent investigations with samples grown under different conditions have yielded carrier densities that are lower but consistent with the polar catastrophe model.

The effect of oxygen vacancies has been investigated explicitly by several groups. During growth of $LaAlO_3$ by PLD, the growth temperature, oxygen pressure, and post-deposition annealing conditions can greatly influence the distribution of oxygen vacancies [17, 18]. In addition, substrate preparation methods [19] used to produce a TiO_2-terminated surface can lead to the formation of oxygen vacancies. Under certain conditions, the density of oxygen vacancies can be sufficient to lead to three-dimensional electronic properties (i.e. no confinement) near the $LaAlO_3/SrTiO_3$ interface [20–24]. Post-deposition annealing can significantly reduce the density of oxygen vacancies near the interface, as has been measured directly using cross-sectional conducting AFM [22]. Because oxygen vacancies are charged they are capable of redistribution under applied electric fields. The hysteretic switching of 3uc-$LaAlO_3/SrTiO_3$ observed by Thiel *et al.* [13] can be explained by the field-induced hysteretic motion of oxygen vacancies at or near the $LaAlO_3/SrTiO_3$ interface.

Another experimental observation that is difficult to reconcile with the polar catastrophe model is the absence of clear signature in X-ray photoemission spectroscopy (XPS) measurements [25–28]. A number of alternate theoretical explanations have been given to describe the interface confinement [29], however at this point a complete physical picture has yet to emerge.

13.2 Field-effect devices

The variety of electronic phenomena exhibited at the $LaAlO_3/SrTiO_3$ interface depends on the ability to modulate the carrier density at the interface through a combination of electronic reconstruction, doping, and field effects. Electronic reconstruction was discussed above and can drive the $LaAlO_3/SrTiO_3$ interface close to or through the metal–insulator transition. Direct chemical doping of $SrTiO_3$ with Nb can produce high-mobility 2DEGs that become superconducting at low temperatures [30, 31]. Oxygen vacancies can donate electrons to the interface if they are located at or near the interface itself. The top $LaAlO_3$ surface, or even defects within the $LaAlO_3$, can also provide a kind of modulation doping of the $LaAlO_3/SrTiO_3$ interface.

13.2.1 $SrTiO_3$-based channels

A variety of field-effect devices that use $SrTiO_3$ as the channel layer have been reported [32–37]. The $SrTiO_3$ channel can be either intrinsic or heavily doped. One device utilized the ferroelectric field effect from a $Pb(Zr_{0.2}Ti_{0.8})O_3$ (PZT) to modulate the superconducting transition temperature of a 400 Å, Nb-doped $SrTiO_3$ using the metallic tip of an AFM as a mobile gate electrode [35]. Similar approaches were reported in [32, 34]. FET devices were created by combining two p-type copper oxide superconductors gated by the $SrTiO_3$ dielectric layer. A metal-insulator-semiconductor field effect transistor (MISFET) was demonstrated by Ueno [33] using an undoped $SrTiO_3$ substrate as a conduction channel and Al_2O_3 as a gate insulator. Another FET device using $SrTiO_3$ as a channel and $CaHfO_3$ as an insulator was reported by Shibuya with amorphous [38] and epitaxial [37] interfaces.

13.2.2 Electrical gating of $LaAlO_3/SrTiO_3$ structures

There are two common methods for electrical gating of $LaAlO_3/SrTiO_3$ structures. The first uses a top gate to modulate the 2DEG carrier density (Figure 13.3 (a)). This method of gating was used by Jany *et al.* to create Schottky-like devices [39]. As shown in Figure 13.4, the interface of a 3–4 uc-$LaAlO_3/SrTiO_3$ heterostructure is directly contacted with a buried contact; an adjacent area is top-gated through the $LaAlO_3$. When forward biased

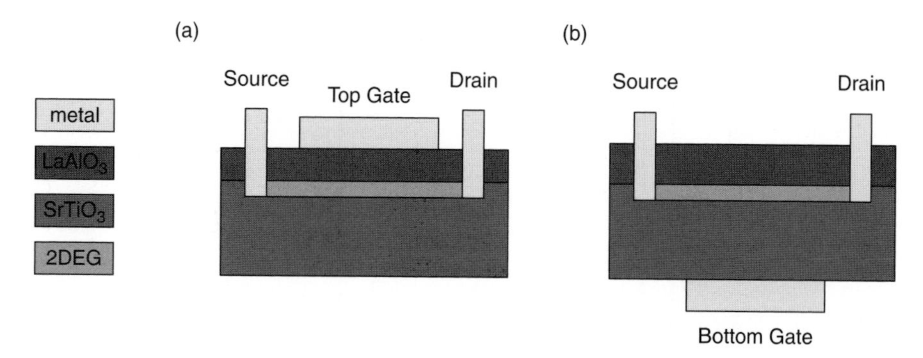

Fig. 13.3 Two methods for electrical gating of $LaAlO_3/SrTiO_3$ heterostructures. (a) Top gating, where the electric field is applied across the thin ($\sim$ 1 nm) $LaAlO_3$ layer. (b) Bottom gating, where the electric field is applied across the thick ($\sim$ 1 mm) $SrTiO_3$ layer.

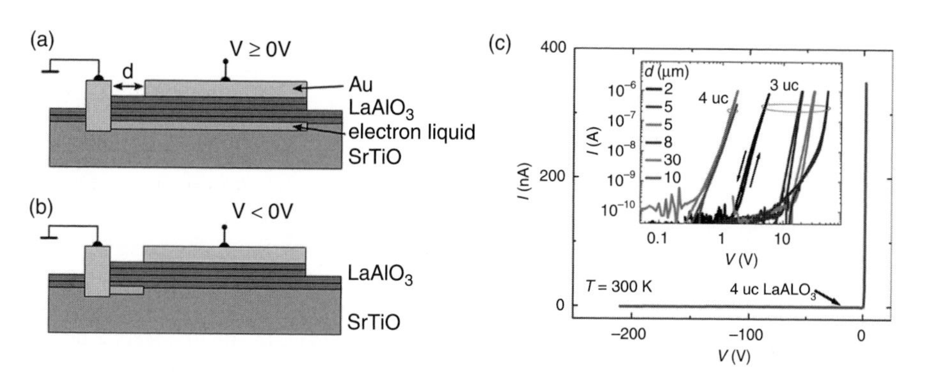

Fig. 13.4 Illustration of a $LaAlO_3/SrTiO_3$ field-effect device exhibiting Schottky-diode behavior. (a) Enhancement mode ($V \geq 0$) in which there is large conductance between the interface contact and top gate. (b) Depletion mode ($V < 0$) in which the interface conduction has been extinguished. (c) *I-V* curve. Reprinted with permission from [39]. Copyright 2010, American Institute of Physics.

(Figure 13.4 (a)), the interface becomes conducting and a large current flows. Under negative gate bias (Figure 13.4 (b)), the 2DEG is depleted and the conduction between the gate and interface contact becomes very low. The top electric field effect switches the interface between insulating and conducting states, leading to highly rectifying behavior (Figure 13.4 (c)).

A second approach to electrical gating is to apply a voltage to the bottom of the relatively thick ($\sim$1 mm) $SrTiO_3$ substrate (Figure 13.3 (b)). Although the effect of a back gate is diminished by the spatial separation from the 2DEG layer, it is simultaneously enhanced by the larger dielectric constant of $SrTiO_3$ ($\sim$300) compared to $LaAlO_3$ ($\sim$25). The transport properties of the $LaAlO_3/SrTiO_3$ 2DEG can be modified by applying large voltages ($\pm$100 V) to the bottom of the $SrTiO_3$ substrate. At room temperature, this type of gating can lead to hysteretic switching of 3uc-$LaAlO_3/SrTiO_3$ structures (Figure 13.5) [13]. The discovery of interfacial

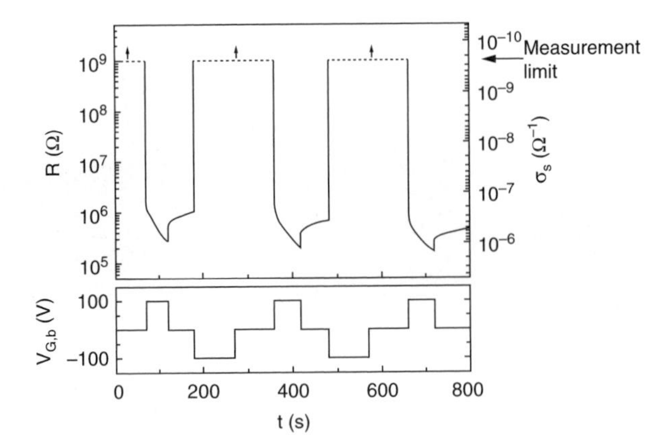

Fig. 13.5 Hysteretic switching of the resistance of 3 uc-$LaAlO_3/SrTiO_3$ using a back gate. Positive gate biases produce a conducting interface, even after the bias is restored to zero. Similarly, negative gate biases produce an insulating interface. From [13]. Reprinted with permission from AAAS.

superconductivity at the $LaAlO_3/SrTiO_3$ interface was subsequently found to be electric-field tunable using the same type of back gating [40].

13.2.3 $LaAlO_3/SrTiO_3$-based field-effect devices

One of the earliest attempts to laterally define conducting microstructures at the $LaAlO_3/SrTiO_3$ interface focused on lithographically modifying the growth process in order to spatially modulate the number of *crystalline* $LaAlO_3$ layers [41]. In the first step, 2 uc $LaAlO_3$ are grown epitaxially on $SrTiO_3$ (Figure 13.6 (a)). A lithographic mask is then used to define areas on which to grow *amorphous* $LaAlO_3$ (Figure 13.6 (b)). Following liftoff, an additional 3 uc of *crystalline* $LaAlO_3$ are grown (Figure 13.6 (c)). The result is conducting interfaces beneath the regions with 5 uc $LaAlO_3/SrTiO_3$ surrounded by insulating regions with 2 uc $LaAlO_3/SrTiO_3$, as shown in Figure 13.6 (d). Using a combination of optical and electron-beam lithography, conducting features as small as 200 nm were demonstrated.

13.3 Reconfigurable nanoscale devices

A powerful method for creating nanoscale devices at the $LaAlO_3/SrTiO_3$ interface involves metastable charging of the top $LaAlO_3$ surface with a conducting AFM probe. By locally and reversibly controlling a metal–insulator transition, the creation of both isolated and continuous conducting features has been demonstrated with length scales smaller than 2 nm. These structures can be erased and rewritten numerous times. As a result of the enormous flexibility in controlling electronic properties at near-atomic dimensions, a variety of nanoscale devices can be realized. Here we describe the nanoscale writing technique and illustrate some of these prototype devices and their properties.

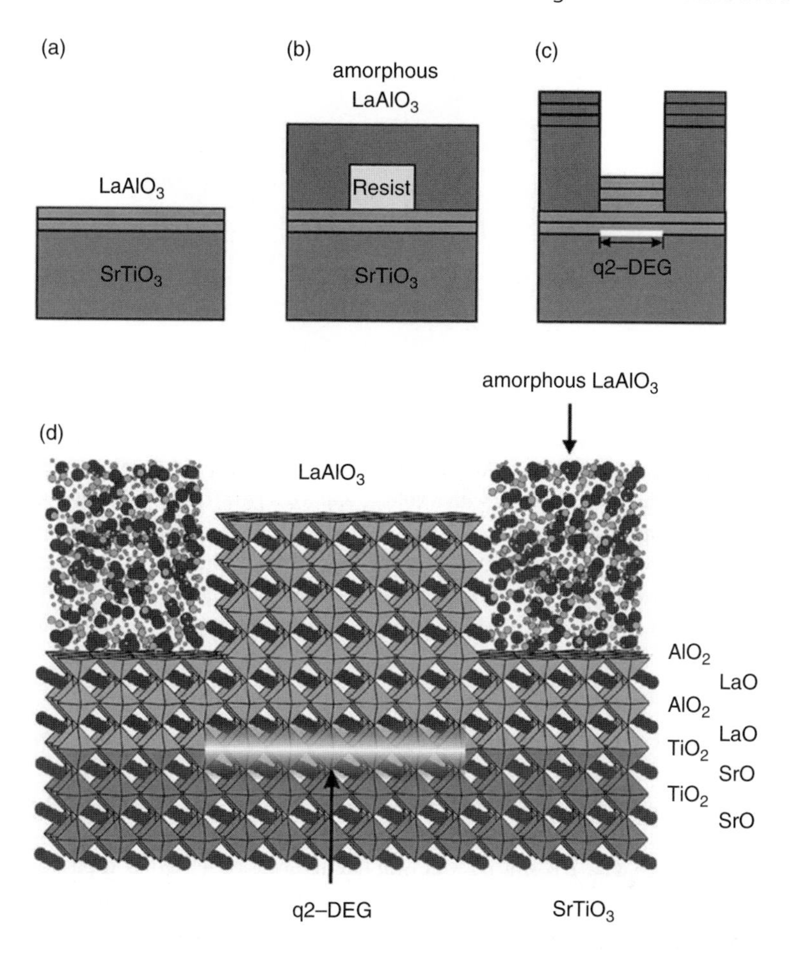

Fig. 13.6 Lithographic method for producing conducting microstructures at the $LaAlO_3/SrTiO_3$ interface. (a) Two unit-cells of $LaAlO_3$ are grown on TiO_2-terminated $SrTiO_3$. (b) Selected areas are masked off by photoresist, and amorphous $LaAlO_3$ is grown. (c) After liftoff, $LaAlO_3$ grows epitaxially on areas that were masked off but not on the amorphous regions. (d) Final structure, illustrating patterning of the 2DEG. Reprinted with permission from [41]. Copyright 2006, American Institute of Physics.

13.3.1 Nanoscale writing and erasing

To create conducting nanostructures, a conducting AFM tip is placed in contact with the top $LaAlO_3$ surface and biased at V_{tip} with respect to the interface, which is held at electrical ground (Figure 13.7). Positive tip voltages locally produce a metallic interface, while negative tip voltages locally restore the insulating state. During the writing and erasing process, the conductance is monitored between buried Au electrodes that directly contact the interface. Figure 13.8 (a) illustrates the result of writing with V_{tip} = +3 V. A sudden increase in

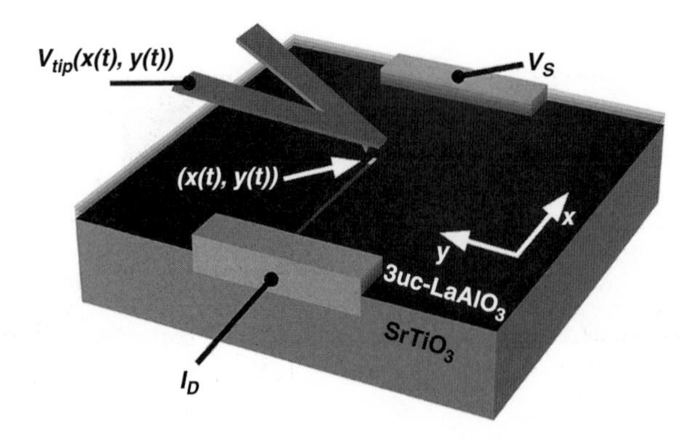

Fig. 13.7 Schematic illustrating of the nanowriting process at the $LaAlO_3/SrTiO_3$ interface. Au electrodes are electrically contacted to the $LaAlO_3/SrTiO_3$ interface. The AFM tip with an applied voltage is scanned once between the two electrodes with a voltage applied $V_{tip}(x(t), y(t))$. Positive voltages locally switch the interface to a conducting state, while negative voltages locally restore the insulating state. Here, a conducting nanowire is being written from the left electrode. The conductance between the two electrodes is monitored by applying a small voltage bias on one of the two gold electrodes (V_S) and reading the current at the second electrode (I_D). Reprinted with permission from [42]. Copyright 2010, American Institute of Physics.

conductance is observed when a conducting path is obtained between the two monitored electrodes ($y = 0$ nm).

To provide a measure of the transverse dimension of the conducting wire, and to demonstrate that the writing process is reversible, the wire can be "cut" with a reverse voltage $V_{tip} = -3$ V (Figure 13.8 (b)). A sharp reduction in current is observed, comparable in abruptness to the one found for the writing process. Assuming that the erasure process has a resolution comparable to the writing process, one can infer the nanowire width from the deconvolved differential profile full width at half maximum (FWHM).

The writing and erasing process can be repeated hundreds of times without noticeable degradation of the nanowire. Figure 13.8 (c) illustrates the repeated erasing and writing of a nanowire using 100 ms voltage pulses. The inset shows that a conducting and insulating state is consistently achieved.

13.3.2 "Water cycle"

One possible explanation for the switching of the 2DEG interface between the insulating and conducting states is the selective removal of OH^- or H^+ in the water layer naturally adsorbed at the $LaAlO_3$ top surface. We can assume that the top surface has water naturally adsorbed on the top AlO_2 layer in the atmosphere. The water is dissociated into OH^- and H^+, as illustrated in Figure 13.9. The large binding energy makes the first layer stable even in ultra-high vacuum. When the positively charged AFM tip scans the top surface, the OH^- will be removed and leave behind a positive H^+. Scanning with a negative voltage will remove the remaining H^+ and thus restore the neutral state of the surface. The neutral surface will easily adsorb new water molecules.

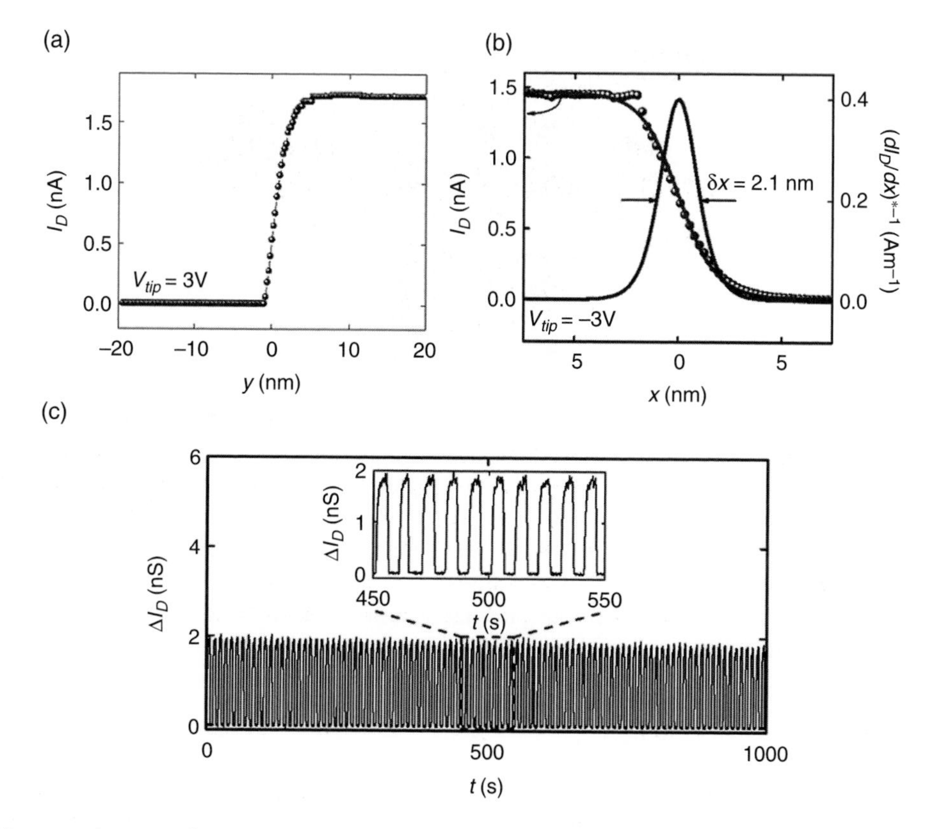

Fig. 13.8 Writing and erasing nanostructures at the $LaAlO_3/SrTiO_3$ interface (a) Conductance between the two electrodes measured with a lock-in amplifier as a function of the tip position while writing a conducting wire with $V_{tip} = +3$ V. A steep increase in conductance shows when the tip reaches the second electrode $\delta x = 3.3$ nm. (b) Conductance drop when the wire is cut with $V_{tip} = -3$ V. (c) Repeated cutting and restoring the conductance of a 12 nm-wide nanowire using $V_{tip} = \pm 10$ V. From [43]. Reprinted with permission from AAAS.

13.3.3 $LaAlO_3/SrTiO_3$ as a floating-gate transistor network

The properties of $LaAlO_3/SrTiO_3$ heterostructures are similar to enhancement-mode MOSFET devices used to form logic gates in microprocessors and other digital electronics. Figure 13.10 (a, b) illustrates the "off" and "on" states of an n-type, enhancement-mode MOSFET. In the off state, electrons cannot form a conducting path from source to drain because the region below the gate is p-type. However, application of a positive bias to the gate can create a channel at the oxide–silicon interface (Figure 13.10 (b)), thus closing the switch. The switch is volatile, in that when the externally applied voltage is turned off, the switch returns to its open state. A 3 uc-$LaAlO_3/SrTiO_3$ heterostructure can behave in a similar fashion. Figure 13.10 (c, d) illustrates the "off" and "on" states of a $LaAlO_3/SrTiO_3$ heterostructure. The main difference is that the top gate is "floating" and can store charge metastably. This charge can be

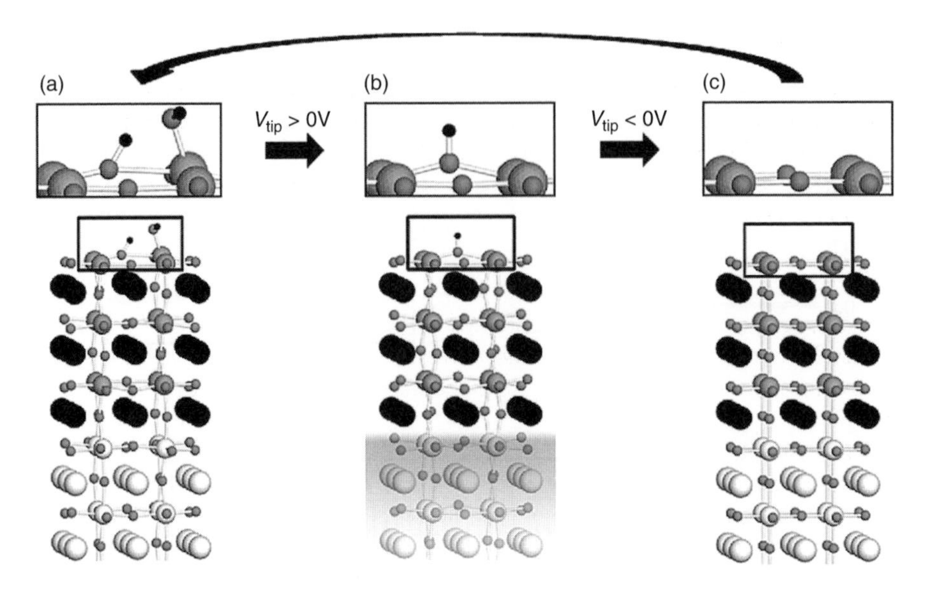

Fig. 13.9 "Water cycle" mechanism for writing and erasing structures at the $LaAlO_3/SrTiO_3$ interface. (a) H_2O adsorbs, dissociates on $LaAlO_3$ surface. (b) Positive tip removes OH^-, leaving H^+ on surface and producing a conducting interface. (c) Negative tip removes H^+, restoring the insulating state. (Courtesy, C. S. Hellberg.)

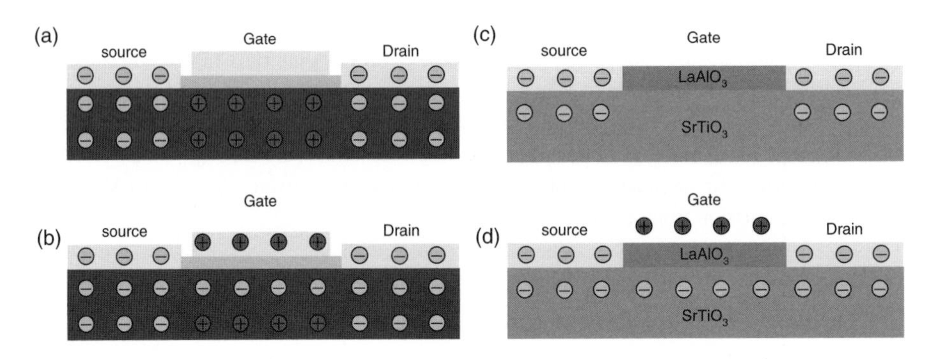

Fig. 13.10 Enhancement-mode MOSFET in (a) "off" and (b) "on" states and $LaAlO_3/SrTiO_3$ device in (c) "off" and (d) "on" states.

added and removed with extremely high spatial resolution using a scanning probe microscope. The 3 uc-$LaAlO_3/SrTiO_3$ system can thus be regarded as a two-dimensional network of non-volatile field-effect transistors. As will be seen below, the patterns of charge applied to the top $LaAlO_3$ surface can be used to create conducting lines, isolated islands, transistors, and other devices.

13.3.4 Quasi-0D structures

It is also possible to write isolated conducting islands or "dots" by applying voltage pulses with fixed amplitude but varying duration at a fixed tip position. A monotonically increasing island size with pulse duration is observed in writing experiments that create a chain of dots spaced a fixed distance apart. The critical spacing for transitioning between isolated dots and a continuous line can vary from ~1 nm to >35 nm depending on the pulse duration [43].

13.3.5 Designer potential barriers

The high degree of control over the energy landscape within the $LaAlO_3/SrTiO_3$ 2DEG allows for the development of a variety of nonlinear devices such as nanoscale junctions [43]. The shape of the barrier can determine whether the transport is reciprocal ($I(V)$ = -I(-V)) or rectifying.

The controlled creation of rectifying structures is further described below. Non-reciprocal nanostructures can be created using a slightly different c-AFM manipulation. In this approach, spatial variations in the conduction-band profile are created by a precise sequence of erasure steps. In a first experiment, a conducting nanowire is created using V_{tip} = +10 V. The initial *I-V* curve (Figure 13.11 (a), closed symbols) is highly linear and reciprocal. This nanowire is then cut by scanning the AFM tip across the nanowire at a speed v_y = 100 nm/s using V_{tip} = −2 mV at a fixed location (x = 20 nm) along the length of the nanowire. This erasure process increases the conduction-band minimum $E_c(x)$ locally by an amount that scales monotonically with the number of passes N_{cut} (Figure 13.11 (a), inset); the resulting nanostructure exhibits a crossover from conducting to activated to tunneling behavior [42]. Here we focus on the symmetry of

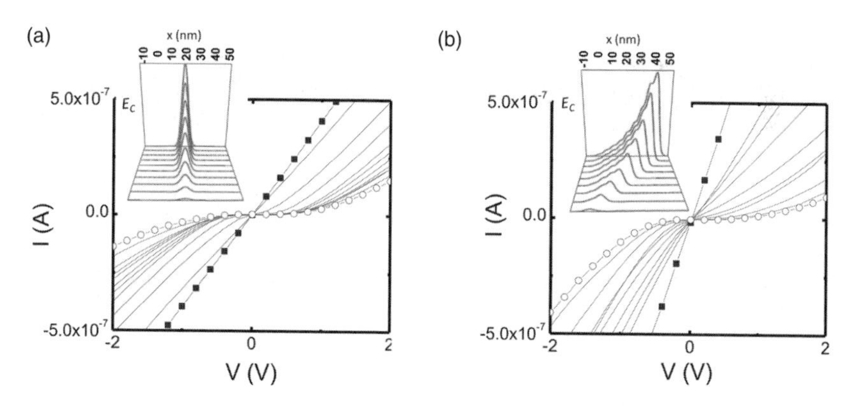

Fig. 13.11 (a) $I-V$ plots for a nanowire cut at the same location multiple times with an AFM tip bias V_{tip} = −2 mV. Intermediate *I-V* curves are shown after every alternate cut. As the wire is cut, the potential barrier increases (inset) and the zero-bias conductance decreases; however, the overall *I-V* curve remains highly reciprocal. (b) *I-V* plots for a nanowire subject to a sequence of cuts $N_{cut}(x)$ at nine locations spaced 5 nm apart along the nanowire (inset). The asymmetry in $N_{cut}(x)$ results in a non-reciprocal *I-V* curve. Curves with closed (open) symbols indicate $I-V$ curve before first (last) cut. Reprinted with permission from [42]. Copyright 2010, American Institute of Physics.

the full *I-V* curve. As N_{cut} increases, the transport becomes increasingly nonlinear; however, the *I-V* curve remains highly reciprocal. The canvas is subsequently erased and a uniform conducting nanowire is written in a similar fashion as before ($V_{tip} = +10$ V, $v_x = 400$ nm/s). A similar erasure sequence is performed; however, instead of cutting the nanowire at a single x coordinate, a sequence of cuts is performed at nine adjacent x coordinates along the nanowire (separated by $D_x = 5$ nm). The number of cuts at each location along the nanowire $N_{cut}(x)$ increases monotonically with x, resulting in a conduction band profile $E_c(x)$ that is asymmetric by design (Figure 13.11 (b), inset). The resulting *I-V* curve for the nanostructure evolves from being highly linear and reciprocal before writing (Figure 13.11 (b), closed symbols) to highly nonlinear and non-reciprocal (Figure 13.11 (b), open symbols).

Nanoscale control over asymmetric potential profiles at the interface between $LaAlO_3$ and $SrTiO_3$ can have many potential applications in nanoelectronics and spintronics. Working as straightforward diodes, these junctions can be used to create half-wave and full-wave rectifiers for AC–DC conversion or for RF detection and conversion to DC. Cascading two or more such junctions, with a third gate for tuning the density in the intermediate regime could form the basis for low-leakage transistor devices. The ability to control the potential along a nanowire could be used to create wires with built-in polarizations similar to those created in heterostructures that lack inversion symmetry [44].

13.3.6 SketchFET

Here we describe the creation of a sketch-defined transistor or "SketchFET". A transistor is a three-terminal device and one begins with a uniformly non-conductive region that spans three electrodes (Figure 13.12 (a)). A "T-junction" is written which will form the source (S), gate (G), and drain (D) leads of the SketchFET (Figure 13.12 (b)). The central junction is then erased over a $\sim$1 μm area (Figure 13.12 (c)), and a smaller junction is created using a narrower line width (Figure 13.12 (d)). A close-up of this region is shown in Figure 13.12 (e). Finally, the horizontal nanowire is cut and the gate electrode is moved further away ($\sim$25 nm).

Transport measurements of a SketchFET can be performed by monitoring the drain current I_D as a function of the source and gate voltages (V_{SD} and V_{GD}, respectively). Both V_{SD} and V_{GD} are referenced to the drain, which is held at virtual ground. At zero gate bias, the *I-V* characteristic between source and drain is highly nonlinear and non-conducting at small $|V_{SD}|$ (Figure 13.13 (b)). A positive gate bias $V_{GD} > 0$ lowers the potential barrier for electrons in the source and gate leads. With V_{GD} large enough ($\geq$ 5V in this specific device), the barrier eventually disappears. In this regime, ohmic behavior between source and drain is observed. The field effect in this case is non-hysteretic, in contrast to field effects induced by the AFM writing procedure. When a sufficiently large gate bias is applied, a small gate leakage current I_{GD} can contribute to the total drain current I_D. The amount of leakage current can be adjusted by placing the gate electrode at different distances away from the source-drain junction. Conductance changes of up to four orders of magnitude have been observed in SketchFET devices, which make them interesting components for logic devices. We note that complementary logic families directly analogous to CMOS are not possible since only electron gases have been demonstrated thus far. A significantly different logic device must be constructed that can simultaneously switch its logic state and consume minimal power while holding either of two states.

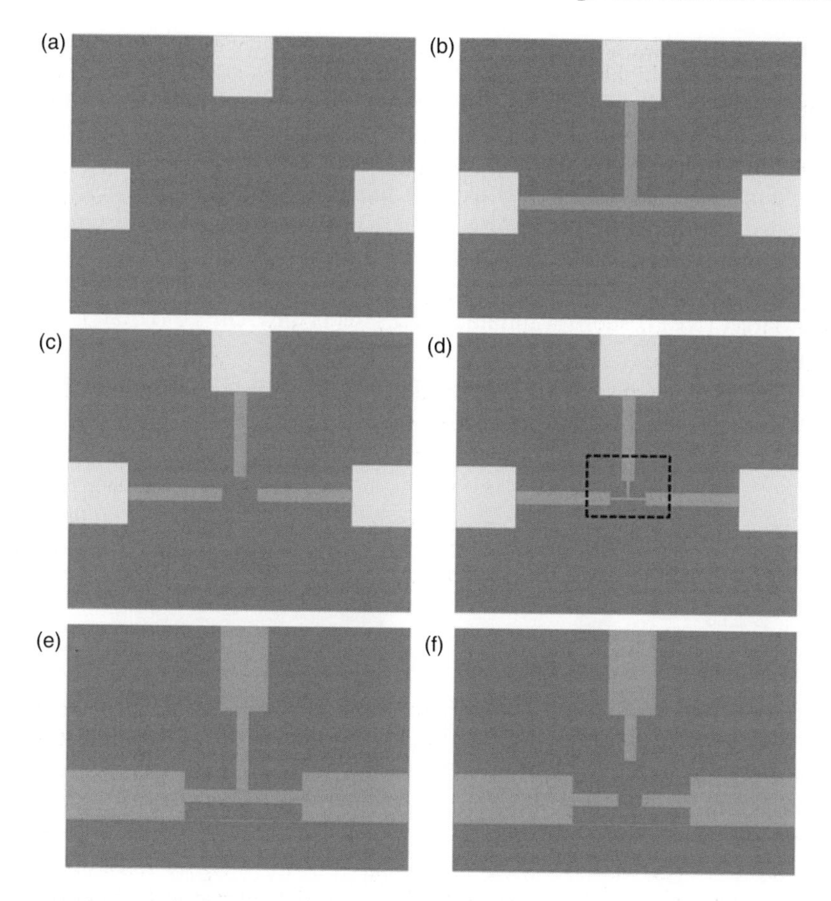

Fig. 13.12 Schematic illustration of step-by-step creation of a SketchFET.

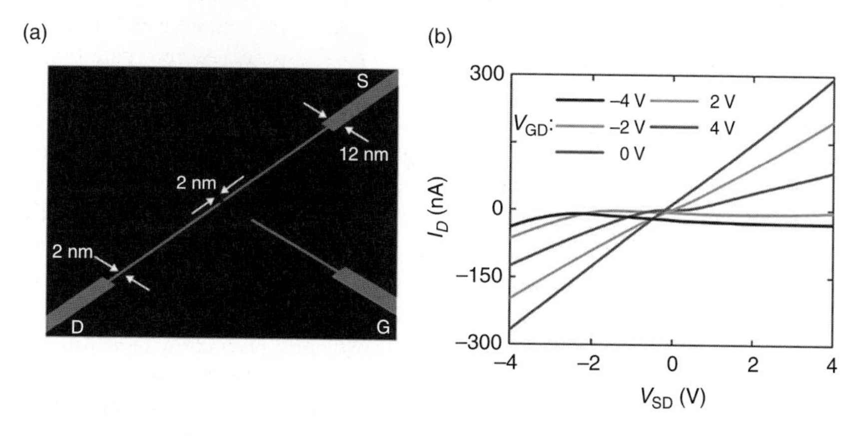

Fig. 13.13 SketchFET device characteristics. (a) Schematic diagram of SketchFET structure. (b) *I-V* characteristic between source and drain for different gate biases V_{GD}. From [43]. Reprinted with permission from AAAS.

13.3.7 Nanoscale photodetectors

In addition to the variety of electronic devices described above, one can also use AFM patterning of the $LaAlO_3/SrTiO_3$ interface to create rewritable, nanoscale photodetectors [45]. Nanophotonic devices seek to generate, guide, and/or detect light using structures whose nanoscale dimensions are closely tied to their functionality [46, 47]. Nanoscale photodetectors created at the interface of $LaAlO_3/SrTiO_3$ possess an electric-field tunable spectral response spanning the visible-to-near-infrared regime. Following illumination with up to kW/cm^2 of optical intensity they are still able to be erased and rewritten. An analysis of noise equivalent target (NET)) shows a minimum NET of 11 $mW/cm^2/\sqrt{Hz}$ (at T = 80 K and $\lambda = 735$ nm).

Optical properties of nanostructures are spatially mapped using scanning photocurrent microscopy (SPCM) [48, 49] (Figure 13.14 (a)). The intensity of a laser source is modulated by

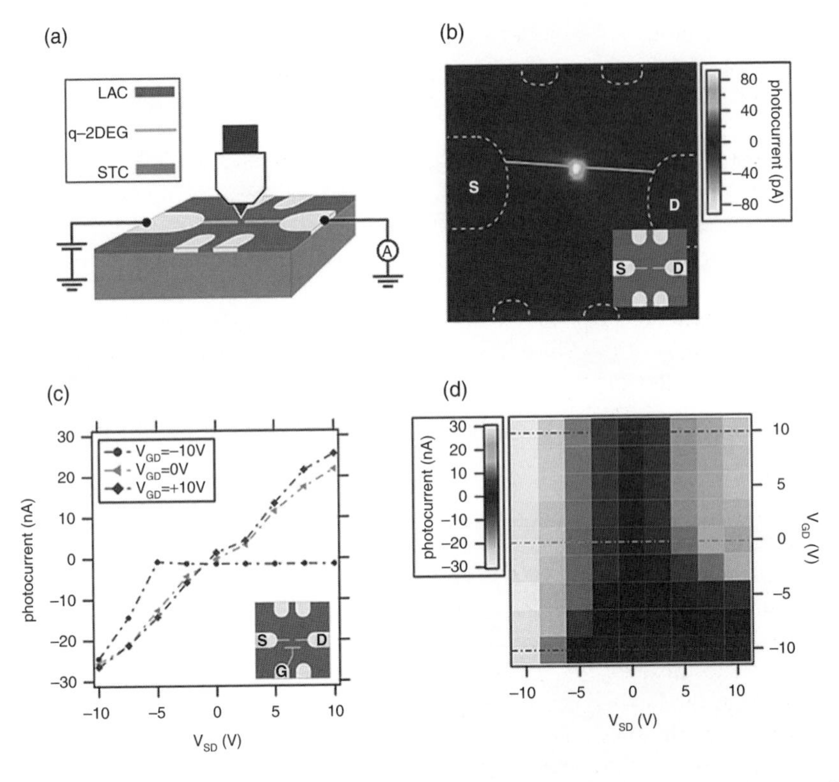

Fig. 13.14 (a) Diagram of photocurrent measurement. (b) Scanning photocurrent microscopy (SPCM) image of two-terminal device shown in inset. $I \sim 30$ kW/cm^2 ($NA = 0.73$), $V_{SD} = 0.1$ V, $T = 300$ K. (c) Photocurrent versus V_{SD} of the three-terminal device shown in the inset. $I \sim 20$ W/cm^2 ($NA = 0.13$), $T = 80$ K. (d) Intensity map of photocurrent of three-terminal device as a function of V_{SD} and V_{GD}. Adapted from [45].

an optical chopper at frequency f_R and the resulting photocurrent is measured with a lock-in amplifier at f_R. When the light overlaps with the device a sharp increase in the photocurrent is observed. An SPCM image shows spatially localized photocurrent detected only in the region of the junction (Figure 13.14 (b)). In a three-terminal geometry (née SketchFET) the gate-drain bias (V_{GD}) can be used to modify the source-drain conductance, enabling conduction between source and drain for positive V_{GD} and inhibiting it for negative V_{GD} (Figure 13.14 (c)).

The spectral response is sensitive to both V_{SD} and V_{GD}. At positive V_{SD} the photodetector response red-shifts as the gate bias is increased. A similar Stark shift is observed when sweeping the source bias. This evidence of a Stark effect, along with finite element analysis showing that the electric field is predominantly confined to the gap region, indicates that the photo-induced absorption is highly localized.

13.3.8 Integration of $LaAlO_3/SrTiO_3$ with silicon

Because of its ubiquitous use in modern microelectronics, silicon is the most desirable platform on which to base multifunctional oxide-based nanoelectronic devices. As a result, one of the major challenges of oxide-based electronic device research is integrating with Si-based electronic circuits and subsequently scaling to a commercially relevant available large wafer process compatible with present Si-process. Using a combination of oxide growth techniques, the integration of oxide heterostructures with silicon can be achieved. Oxide molecular beam epitaxy (MBE) growth of virtual $SrTiO_3$ substrate layers can be achieved on Si substrates and can be followed by pulse laser deposition (PLD) or laser-MBE for growth of oxide heterostructures [50].

Figure 13.15 shows cross-sectional TEM images of oxide heterostructures consisting of 5 uc of $LaAlO_3$ grown on a 100 nm-thick $SrTiO_3$ template grown on Si substrate. The 100 nm-thick epitaxial (001) $SrTiO_3$ layers were deposited on (001) Si wafers in an 8 inch-diameter molecular beam epitaxy (MBE) machine [50]. The termination of the epitaxial $SrTiO_3$ thin films was controlled by halting the film growth at SrO−, TiO_2−, or undefined-termination. During the STO deposition, if surface termination is not controlled, it may have both SrO and TiO_2 terminations. To improve the crystalline quality and surface morphology, all $SrTiO_3$-on-Si templates were annealed at 900°C for 2 hrs in an O_2 atmosphere [51].

The temperature dependence of the sheet resistance, carrier concentration, and mobility of an unpatterned heterointerface between 10 uc $LaAlO_3$ and annealed TiO_2-$SrTiO_3$ on Si are measured by the van der Pauw method (Figure 13.16). The room-temperature properties are comparable to those for heterointerfaces between $LaAlO_3$ and TiO_2-terminated $SrTiO_3$ [5, 53]. Although the low-temperature mobility is lower than $LaAlO_3/SrTiO_3$ single crystal, presumably due to the sensitivity to the defect structure of the heterointerface, we believe that the 2DEG at the $LaAlO_3/SrTiO_3$ heterointerface on Si is useful for room-temperature nanoelectronic devices. A comparison between the room-temperature thickness-dependent electrical properties of unpatterned $LaAlO_3/SrTiO_3$ on Si and $LaAlO_3$ on $SrTiO_3$ bulk single crystal grown under identical conditions shows the same 4 uc critical thickness. The electrical properties of the $SrTiO_3$ layer on Si shows insulating behavior relative to the interfacial 2DEG for $LaAlO_3$ thicknesses above the critical thickness. The bare Si substrate was determined

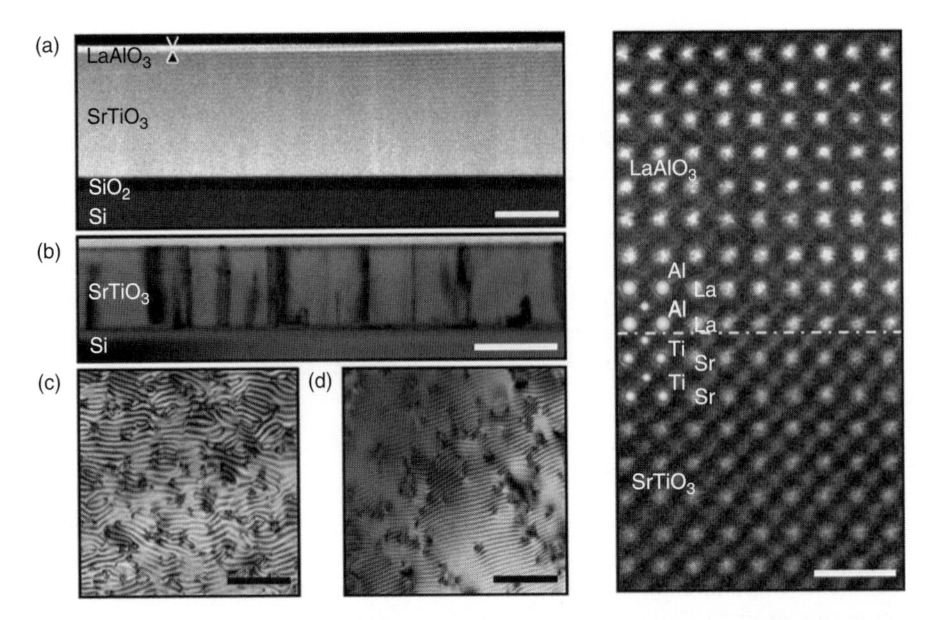

Fig. 13.15 TEM investigation of $LaAlO_3/SrTiO_3$ heterointerface on Si. (s) Cross-sectional high-angle dark field (HAADF) image of 5 nm-thick $LaAlO_3$/annealed TiO_2-$SrTiO_3$ heterostructure grown on Si. Arrows indicate the thickness of the $LaAlO_3$ layer. The scale bar corresponds to 50 nm. (b) Cross-sectional bright field TEM image of the same sample showing the existence of threading dislocations in the $SrTiO_3$ layer. Planar view TEM images of (c) as-grown and (d) annealed $SrTiO_3$ films showing moiré patterns. The scale bars in (b), (c), and (d) correspond to 100 nm. (e) High-resolution HAADF image showing an atomically sharp interface between $LaAlO_3$ film and annealed TiO_2- $SrTiO_3$ on Si. The dimension of the scale bar is 1 nm. (Adapted from [52].)

to be *p*-type, with a room-temperature mobility $\mu \sim 350\ cm^2/Vs$ and carrier concentration $n \sim 0.8 \times 10^{15}\ cm^{-3}$, significantly different from the electrical properties of the interfacial 2DEG for $LaAlO_3/SrTiO_3$ on Si.

Nanostructure writing experiments show that reversible patterning of the $LaAlO_3/SrTiO_3$ interface can be achieved in essentially the same way as for $LaAlO_3/SrTiO_3$ interfaces formed on bulk $SrTiO_3$ substrates. The ability to sustain 2DEG behavior depends on the interface preparation. 2DEG behavior has not been found for the heterointerfaces between 3 uc $LaAlO_3$ and SrO- or undefined-terminated $SrTiO_3$ templates on Si. These results are consistent with previous reports that a well-defined TiO_2-terminated $SrTiO_3$ surface is critical for the formation of the 2DEG. Even the heterointerface between $LaAlO_3$ on as-grown TiO_2-$SrTiO_3$/Si did not exhibit 2DEG behavior [52] and it is expected that his result is due to the defective surface of the as-grown $SrTiO_3$ template on Si substrate. In contrast, the annealed $SrTiO_3$ template has relatively lower dislocation density through the annihilation of dislocations with contrary Burgers vectors and the dissociation of two whole dislocations into partial dislocations during high-temperature post-annealing [54]. Further optimization of the $SrTiO_3$/Si substrate quality is needed for useful device applications.

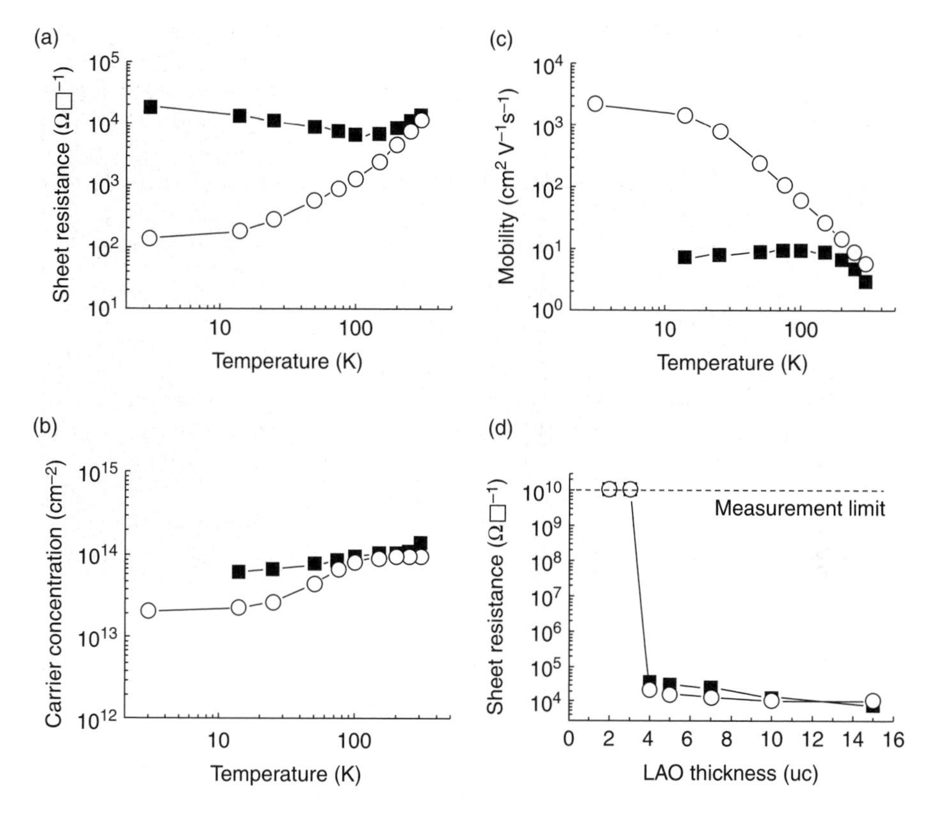

Fig. 13.16 Electrical transport properties of $LaAlO_3/SrTiO_3$ heterointerface on Si. Temperature dependence of (a) sheet resistance, (b) carrier concentration, and (c) mobility of the heterointerface between unpatterned 10 uc $LaAlO_3$ and TiO_2-$SrTiO_3$ on Si (closed squares). The $SrTiO_3$ layer on the Si substrate without the $LaAlO_3$ layer showed highly insulating behavior, indicating that the measured transport properties of $LaAlO_3/SrTiO_3$ on Si originated only from the 2DEG. For comparison, the electrical transport properties of unpatterned 10 uc $LaAlO_3$ on $SrTiO_3$ single crystal were also measured (open circles). (d) $LaAlO_3$ thickness dependence of sheet resistance of unpatterned $LaAlO_3/SrTiO_3$ heterointerface on Si. (Adapted from [52].)

13.4 Future prospects

It is still far too premature to make definitive proclamations about the applicability of this novel interface for technology, but there are many interesting features that are certainly worth exploring. There are well-defined possibilities regarding ultra-high-density transistor and memory elements, though stability and scaling issues are just as clearly defined. In keeping with Kroemer's lemma, it is important to keep an open mind to entirely new classes of devices that are enabled by the unique combination of properties present in the $LaAlO_3/SrTiO_3$ system. Several possible directions are discussed below.

13.4.1 Room-temperature devices

Semiconductors are the workhorse of modern electronics industry. To obtain ever higher device integration densities, the size of metal-oxide-semiconductor field-effect transistors (MOSFET) has successfully followed Moore's Law for over four decades. Now, due to the intrinsic limitations, the scaling of MOSFET devices is truly reaching fundamental limits [55–57]. In addition to being a possible avenue for scaling beyond Moore's law, oxide-based 2DEG devices may also provide opportunities in terms of adding new functionality.

13.4.2 Information processing

SketchFET structures are interesting candidates for post-CMOS transistors. The quantum tunneling-dominated field emission operates in a distinct manner compared with traditional MOSFET devices. Owing to the fact that the gate, source, drain, and junction areas are essentially made of the same material, SketchFETs are less affected by the "short channel" effect which is a major limit of the MOSFET scaling [58].

AFM lithography eliminates defects that can be introduced during conventional chemical lithography steps and also obtains the separation of carriers from dopants through modulation doping processes. These achievements, coupled with future improvements of material quality and device layout, have the potential to produce very high carrier mobility and lead to SketchFETs with fast switching speeds. Further enhancements in the room-temperature mobility of devices may be possible through advances in materials growth techniques.

Scaling to large numbers of SketchFET transistors will require either a scalable array of AFM probes (e.g. IBM's Millipede project [59]) or other techniques for creating the desired $LaAlO_3$ surface state. Nanoimprint lithography [60] may prove useful for the latter approach. Still, it will be important to integrate oxide materials with conventional silicon-based electronics.

Using an AFM to lithographically pattern oxide interfaces allows the fabrication and integration of devices with different functionalities on a single chip without the need to incorporate many different materials. Junctions between nanowires can be formed with various potential profiles and can be cascaded to produce, for example, nanodiodes with different polarities and turn-on voltages. In the presence of mid-bandgap states such as oxygen vacancies, junctions can be placed and electrically gated to form wide-bandwidth nanoscale photodetectors, allowing on-chip optical signal coupling. The nanoscale footprint of such photodetectors could be especially useful in near-field optics. In addition, nanoscale control over inversion symmetry breaking in junction profiles could in principle be used to produce nonlinear optical frequency conversion (i.e. second-harmonic generation or difference frequency mixing), thus providing a means for the generation of nanoscale sources of light or THz radiation.

13.4.3 Spintronics

Taking advantage of the large Rashba coupling in these 2DEG structures may enable new classes of spintronic devices. Intrinsic magnetic effects have already been reported at temperatures as high as 35 K [53, 61, 62], and large spin-orbit splittings have been measured through magnetotransport [44, 63, 64]. The resulting effective magnetic fields could allow control over spin precession along two orthogonal axes [65] and thus exert full three-dimensional control over electron spin [66] in a nanowire. Another approach to spintronics could be in the development of a Datta–Das spin transistor [67], which would require spin injection and filtering components. The development of spin-based oxide nanostructures could lead to hybrid magnetoelectric components with low power requirements.

13.4.4 Quantum Hall regime

For almost three decades, the integer and fractional quantum Hall effects has been observed almost exclusively in high-mobility silicon or III-V heterostructures. Recently, Tsukazaki *et al.* have observed both the integer [3] and fractional quantum Hall effect [68] in high mobility $ZnO/Mg_xZn_{1-x}O$ heterostructures. In the case of a planar, well-oxidized $LaAlO_3/SrTiO_3$ 2DEG, the observation of quantum Hall phenomena seems *a priori* precluded by its low carrier mobility and high carrier density. However, the carrier density can be much lower in AFM-written nanowires than in planar conductive $LaAlO_3/SrTiO_3$ interfaces. Secondly, quasi-one-dimensional confinement from the nanowire is expected to suppress elastic backscattering and increase carrier mobility [69]. A third effect arises from the peculiar dielectric properties of $SrTiO_3$ at low temperature: the large local electric field inside the nanowire formed by the writing process can lead to a dielectric constant in the nanowire that is significantly less than the surrounding region. For this geometry, Jena *et al.* predict an order-of-magnitude mobility enhancement in two dimensions and additional enhancements for one-dimensional geometries [70]. Nanowires created via conducting-AFM with controllable carrier concentration and mobilities may be suitable for quasiparticle interference experiments in experimentally realizable topological quantum computing geometries [71, 72].

13.4.5 Superconducting devices

$SrTiO_3$ was one of the first semiconductors discovered to exhibit superconducting behavior at low temperature [73, 74]. Recently, superconductivity in $SrTiO_3$ has experienced a renaissance due to the discovery of 2D superconductivity at the $LaAlO_3/SrTiO_3$ interface [75]. The measured transition temperature T_C is approximately 200–400 mK (which is comparable to bulk superconducting $SrTiO_3$) and is $LaAlO_3$-thickness dependent [75]. Subsequent investigations have shown that the application of an electric field can tune the carrier density and induce a superconductor–metal quantum phase transition [40].

In addition to constructing superconducting wires and Josephson couplings, new types of interactions may be possible that take advantage of both the dimensionality of those structures and other intrinsic properties like the strong spin-orbit coupling in these systems [44].

13.4.6 Solid-state Hubbard simulators

One long-range goal might be the creation of a platform for quantum simulation—a *solid-state* Hubbard toolbox—using the $LaAlO_3/SrTiO_3$ interface whose metal–insulator transition can be controlled on scales approaching the underlying unit cell (Figure 13.17). The ground states of strongly correlated systems have generally eluded exact analytical solution. For example, the phase diagram of the 2D Hubbard model away from half filling, which has been proposed to explain the mechanism for high-temperature superconductivity, is still poorly understood. Numerical methods remain limited in their ability to explore the full relevant phase space. Some novel approaches to this problem involve the use of physical simulations using atomic gases trapped in optical lattices [76–80]. Using a so-called "cold atom Hubbard toolbox"

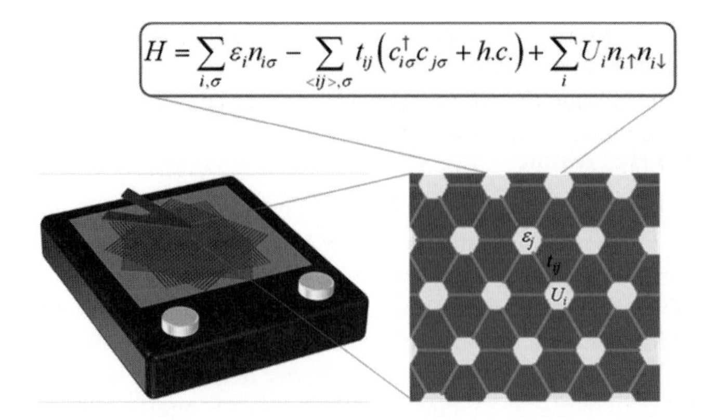

Fig. 13.17 Illustration of $LaAlO_3/SrTiO_3$ as a rewritable nanoelectronic interface. Here, a conducting AFM creates a hexagonal array of conducting islands separated by ultrathin tunnel barriers. Using a demonstrated nanoscale electronic confinement with 2 nm resolution, artificial condensed matter systems such as this may be constructed and probed. For these systems, electronic correlations play a critical role in determining their properties (metallic, insulating, and superconducting).

[81, 82], it may be possible to explore parameter regimes or measurement classes not accessible in conventional solid-state systems. Pioneering experiments such as the observation of a reversible quantum phase transition between a Bose–Einstein condensate to a Mott insulator [83] lend credence to the idea that quantum systems can simulate one another. In the language of quantum information, optical lattices can be regarded as approaching the universal quantum simulator first conceived by Feynman [84, 85].

References

[1] Kroemer H. (2001). Nobel Lecture: Quasielectric fields and band offsets: teaching electrons new tricks. Rev Mod Phys **73**, 783–93.

[2] Mannhart J. and Schlom D. G. (2010). Oxide Interfaces—An Opportunity for Electronics. Science **327**, 1607–11.

[3] Tsukazaki A., Ohtomo A., Kita T., *et al.* (2007). Quantum Hall Effect in Polar Oxide Heterostructures. Science **315**, 1388–91.

[4] Mannhart J., Blank D. H. A., Hwang H. Y., Millis A. J., and Triscone J. M. (2008). Two-Dimensional Electron Gases at Oxide Interfaces. MRS Bull **33**, 1027–34.

[5] Ohtomo A. and Hwang H. Y. (2004). A high-mobility electron gas at the $LaAlO_3/SrTiO_3$ heterointerface. Nature **427**, 423–6.

[6] Eisenstein J. P., Cooper K. B., Pfeiffer L. N., and West K. W. (2002). Insulating and Fractional Quantum Hall States in the First Excited Landau Level. Phys Rev Lett **88**, 076801.

[7] Geim A. K. and Novoselov K. S. (2007). The rise of graphene. Nat Mater **6**, 183–91.

[8] Young A. F. and Kim P. (2009). Quantum interference and Klein tunnelling in graphene heterojunctions. Nat Phys **5**, 222–6.

[9] Christen H. M., Kim D. H., and Rouleau C. M. (2008). Interfaces in perovskite heterostructures. Appl Phys A: Mat Sci **93**, 807–11.

[10] Pauli S. A. and Willmott P. R. (2008). Conducting interfaces between polar and non-polar insulating perovskites. J Phys: Cond Mat **20**, 264012.

[11] Okamoto S., Millis A. J., and Spaldin N. A. (2006). Lattice relaxation in oxide heterostructures: $LaTiO_3/SrTiO_3$ superlattices. Phys Rev Lett **97**, 056802.

[12] Ohtomo A., Muller D. A., Grazul J. L., and Hwang H. Y. (2002). Artificial charge-modulation in atomic-scale perovskite titanate superlattices. Nature **419**, 378–80.

[13] Thiel S., Hammerl G., Schmehl A., Schneider C. W., and Mannhart J. (2006). Tunable quasi-two-dimensional electron gases in oxide heterostructures. Science **313**, 1942–5.

[14] Jalan B., Moetakef P., and Stemmer S. (2009). Molecular beam epitaxy of $SrTiO_3$ with a growth window. Appl Phys Lett **95**, 032906.

[15] Nakagawa N., Hwang H. Y., and Muller D. A. (2006). Why some interfaces cannot be sharp. Nat Mater **5**, 204–9.

[16] Son W.-j., Cho E., Lee J., and Han S. (2010). Hydrogen adsorption and carrier generation in $LaAlO_3$–$SrTiO_3$ heterointerfaces: a first-principles study. J Phys: Cond Mat **22**, 315501.

[17] Huijben M., Brinkman A., Koster G., *et al.* (2009). Structure-Property Relation of $SrTiO_3/LaAlO_3$ Interfaces. Adv Mater **21**, 1665–77.

[18] Copie O., Garcia V., Bödefeld C., *et al.* (2009). Towards Two-Dimensional Metallic Behavior at $LaAlO_3/SrTiO_3$ Interfaces. Phys Rev Lett **102**, 216804.

[19] Kawasaki M., Takahashi K., Maeda T., *et al.* (1994). Atomic Control of the $SrTiO_3$ Crystal-Surface. Science **266**, 1540–2.

[20] Herranz G., Basletic M., Bibes M., *et al.* (2006). High Mobility in $LaAlO_3/SrTiO_3$ Heterostructures: Origin, Dimensionality, and Perspectives. Phys Rev Lett **98**, 216803.

[21] Herranz G., Basleticacute M., Bibes M., *et al.* (2007). High Mobility in $LaAlO_3/SrTiO_3$ Heterostructures: Origin, Dimensionality, and Perspectives. Phys Rev Lett **98**, 216803.

[22] Basletic M., Maurice J. L., Carretero C., *et al.* (2008). Mapping the spatial distribution of charge carriers in $LaAlO_3/SrTiO_3$ heterostructures. Nat Mater **7**, 621–5.

[23] Kalabukhov A., Gunnarsson R., Borjesson J., *et al.* (2007). Effect of oxygen vacancies in the $SrTiO_3$ substrate on the electrical properties of the $LaAlO_3/SrTiO_3$ interface. Phys Rev B **75**, 121404.

[24] Herranz G., Basletić M., Copie O., *et al.* (2009). Controlling high-mobility conduction in $SrTiO_3$ by oxide thin film deposition. Appl Phys Lett **94**, 012113.

[25] Segal Y., Ngai J. H., Reiner J. W., Walker F. J., and Ahn C. H. (2009). X-ray photoemission studies of the metal-insulator transition in $LaAlO_3/SrTiO_3$ structures grown by molecular beam epitaxy. Phys Rev B **80**, 241107.

[26] Chambers S. A., Engelhard M. H., Shutthanandan V., *et al.* (2010). Instability, Intermixing and Electronic Structure at the Epitaxial $LaAlO_3/SrTiO_3(001)$ Heterojunction. Surf Sci Rep **65**, 317–52.

[27] Yoshimatsu K., Yasuhara R., Kumigashira H., and Oshima M. (2008). Origin of metallic states at the heterointerface between the band insulators $LaAlO_3$ and $SrTiO_3$. Phys Rev Lett **101**, 026802.

[28] Janicka K., Velev J. P., and Tsymbal E. Y. (2009). Quantum Nature of Two-Dimensional Electron Gas Confinement at $LaAlO_3/SrTiO_3$ Interfaces. Phys Rev Lett **102**, 106803.

[29] Pentcheva R. and Pickett W. E. (2010). Electronic phenomena at complex oxide interfaces: insights from first principles. J Phys: Cond Mat **22**, 043001.
[30] Bell C., Harashima S., Kozuka Y., *et al.* (2009). Dominant Mobility Modulation by the Electric Field Effect at the $LaAlO_3/SrTiO_3$ Interface. Phys Rev Lett **103**, 226802.
[31] Kozuka Y., Hikita Y., Bell C., and Hwang H. Y. (2010). Dramatic mobility enhancements in doped $SrTiO_3$ thin films by defect management. Appl Phys Lett **97**, 012107.
[32] Mannhart J. (1996). High-T_C transistors. Supercond Sci Technol **9**, 49–67.
[33] Ueno K., Inoue I. H., Akoh H., *et al.* (2003). Field-effect transistor on $SrTiO_3$ with sputtered Al_2O_3 gate insulator. Appl Phys Lett **83**, 1755–7.
[34] Matthey D., Gariglio S., and Triscone J. M. (2003). Field-effect experiments in $NdBa_2Cu_3O_{7-\delta}$ ultrathin films using a $SrTiO_3$ single-crystal gate insulator. Appl Phys Lett **83**, 3758.
[35] Takahashi K. S., Matthey D., Jaccard D., *et al.* (2004). Electrostatic modulation of the electronic properties of Nb-doped $SrTiO_3$ superconducting films. Appl Phys Lett **84**, 1722–4.
[36] Shibuya K., Ohnishi T., Lippmaa M., Kawasaki M., and Koinuma H. (2004). Single crystal $SrTiO_3$ field-effect transistor with an atomically flat amorphous $CaHfO_3$ gate insulator. Appl Phys Lett **85**, 425–7.
[37] Shibuya K., Ohnishi T., Uozumi T., *et al.* (2006). Field-effect modulation of the transport properties of nondoped $SrTiO_3$. Appl Phys Lett **88**, 212116.
[38] Shibuya K., Ohnishi T., Lippmaa M., Kawasaki M., and Koinuma H. (2004). Domain structure of epitaxial $CaHfO_3$ gate insulator films on $SrTiO_3$. Appl Phys Lett **84**, 2142–4.
[39] Jany R., Breitschaft M., Hammerl G., *et al.* (2010). Diodes with breakdown voltages enhanced by the metal-insulator transition of $LaAlO_3$–$SrTiO_3$ interfaces. Appl Phys Lett **96**, 183504.
[40] Caviglia A. D., Gariglio S., Reyren N., *et al.* (2008). Electric field control of the $LaAlO_3/SrTiO_3$ interface ground state. Nature **456**, 624–7.
[41] Schneider C. W., Thiel S., Hammerl G., Richter C., and Mannhart J. (2006). Microlithography of electron gases formed at interfaces in oxide heterostructures. Appl Phys Lett **89**, 122101.
[42] Bogorin D. F., Bark C. W., Jang H. W., *et al.* (2010). Nanoscale rectification at the $LaAlO_3/SrTiO_3$ interface. Appl Phys Lett **97**, 013102.
[43] Cen C., Thiel S., Mannhart J., and Levy J. (2009). Oxide Nanoelectronics on Demand. Science **323**, 1026–30.
[44] Caviglia A. D., Gabay M., Gariglio S., *et al.* (2010). Tunable Rashba Spin-Orbit Interaction at Oxide Interfaces. Phys Rev Lett **104**, 126803.
[45] Irvin P., Ma Y., Bogorin D. F., *et al.* (2010). Rewritable nanoscale oxide photodetector. Nature Photonics **4**, 849–52.
[46] Sirbuly D. J., Law M., Yan H., and Yang P. (2005). Semiconductor Nanowires for Subwavelength Photonics Integration. J Phys Chem B **109**, 15190–213.
[47] Agarwal R. and Lieber C. M. (2006). Semiconductor nanowires: optics and optoelectronics. Appl Phys A: Mat Sci **85**, 209–15.
[48] Balasubramanian K., Burghard M., Kern K., Scolari M., and Mews A. (2005). Photocurrent Imaging of Charge Transport Barriers in Carbon Nanotube Devices. Nano Lett **5**, 507–10.

[49] Xia F., Mueller T., Golizadeh-Mojarad R., *et al.* (2009). Photocurrent Imaging and Efficient Photon Detection in a Graphene Transistor. Nano Lett **9**, 1039–44.

[50] Goncharova L. V., Starodub D. G., Garfunkel E., *et al.* (2006). Interface structure and thermal stability of epitaxial $SrTiO_3$ thin films on Si (001). J Appl Phys **100**, 014912.

[51] Park J. W., Baek S. H., Bark C. W., Biegalski M. D., and Eom C. B. (2009). Quasi-single-crystal (001) $SrTiO_3$ templates on Si. Appl Phys Lett **95**, 061902.

[52] Park J. W., Bogorin D. F., Cen C., *et al.* (2010). Creation of a two-dimensional electron gas at an oxide interface on silicon. Nat Commun **1**, 94.

[53] Brinkman A., Huijben M., Van Zalk M., *et al.* (2007). Magnetic effects at the interface between non-magnetic oxides. Nat Mater **6**, 493–6.

[54] Su D., Yamada T., Sherman V. O., *et al.* (2007). Annealing effect on dislocations in $SrTiO_3/LaAlO_3$ heterostructures. J Appl Phys **101**, 064102.

[55] Zhirnov V. V., Ralph K., Cavin I., Hutchby J. A., and Bourianoff G. I. (2003). Limits to Binary Logic Switch Scaling—A Gedanken Model. Proc IEEE **91**, 1934–9.

[56] Thompson S. E., Chau R. S., Ghani T., *et al.* (2005). In Search of "Forever," Continued Transistor Scaling One New Material at a Time. IEEE Trans Semicond Manuf **18**, 26–36.

[57] Haensch W., Nowak E. J., Dennard R. H., *et al.* (2006). Silicon CMOS devices beyond scaling. IBM J Res Dev **50**, 339–61.

[58] Björkqvist K. and Arnborg T. (1981). Short Channel Effects in MOS-Transistors. Physica Scripta **24**, 418–21.

[59] Vettiger P., Cross G., Despont M., *et al.* (2002). The "millipede" - Nanotechnology entering data storage. IEEE Trans Nanotechnol **1**, 39–55.

[60] Chou S. Y., Krauss P. R., and Renstrom P. J. (1996). Nanoimprint lithography. J Vac Sci Technol B **14**, 4129–33.

[61] Ben Shalom M., Tai C. W., Lereah Y., *et al.* (2009). Anisotropic magnetotransport at the $SrTiO_3$ / $LaAlO_3$ interface. Phys Rev B **80**, 140403.

[62] Seri S. and Klein L. (2009). Antisymmetric magnetoresistance of the $SrTiO_3$ / $LaAlO_3$ interface. Phys Rev B **80**, 180410.

[63] Caviglia A. D., Gariglio S., Cancellieri C., *et al.* (2010). Two-dimensional quantum oscillations of the conductance at $LaAlO_3/SrTiO_3$ interfaces. Phys Rev Lett **105**, 236802.

[64] Ben Shalom M., Sachs M., Rakhmilevitch D., Palevski A., and Dagan Y. (2010). Tuning Spin-Orbit Coupling and Superconductivity at the $SrTiO_3/LaAlO_3$ Interface: A Magnetotransport Study. Phys Rev Lett **104**, 126802.

[65] Bychkov Y. A. and I. R. E. (1984). Oscillatory effects and the magnetic susceptibility of carriers in inversion layers. J Phys C: Sol Stat **17**, 6039–45.

[66] Kato Y., Myers R. C., Driscoll D. C., *et al.* (2003). Gigahertz Electron Spin Manipulation Using Voltage-Controlled g-Tensor Modulation. Science **299**, 1201–4.

[67] Datta S. and Das B. (1990). Electronic analog of the electro-optic modulator. Appl Phys Lett **56**, 665–7.

[68] Tsukazaki A., Akasaka S., Nakahara K., *et al.* (2010). Observation of fractional quantum Hall effect in an oxide. Nat Mater **9**, 889–93.

[69] Sakaki H. (1980). Scattering suppression and high-mobility effect of size-quantized electrons in ultrafine semiconductor wire structures. Jap J Appl Phys **19**, L735–8.

[70] Jena D. and Konar A. (2007). Enhancement of carrier mobility in semiconductor nanostructures by dielectric engineering. Phys Rev Lett **98**, 136805.

[71] Das Sarma S., Freedman M., and Nayak C. (2005). Topologically Protected Qubits from a Possible Non-Abelian Fractional Quantum Hall State. Phys Rev Lett **94**, 166802.
[72] Sau J. D., Lutchyn R. M., Tewari S., and Das Sarma S. (2010). Generic New Platform for Topological Quantum Computation Using Semiconductor Heterostructures. Phys Rev Lett **104**, 040502.
[73] Schooley J. F., Hosler W. R., and Cohen M. L. (1964). Superconductivity in Semiconducting $SrTiO_3$. Phys Rev Lett **12**, 474–5.
[74] Schooley J. F., Hosler W. R., Ambler E., *et al.* (1965). Dependence of the Superconducting Transition Temperature on Carrier Concentration in Semiconducting $SrTiO_3$. Phys Rev Lett **14**, 305–7.
[75] Reyren N., Thiel S., Caviglia A. D., *et al.* (2007). Superconducting interfaces between insulating oxides. Science **317**, 1196–9.
[76] Guidoni L., Triche C., Verkerk P., and Grynberg G. (1997). Quasiperiodic optical lattices. Phys Rev Lett **79**, 3363–6.
[77] Garcia-Ripoll J. J., Martin-Delgado M. A., and Cirac J. I. (2004). Implementation of spin Hamiltonians in optical lattices. Phys Rev Lett **93**, 250405.
[78] Fertig C. D., O'Hara K. M., Huckans J. H., *et al.* (2005). Strongly inhibited transport of a degenerate 1D Bose gas in a lattice. Phys Rev Lett **94**, 120403.
[79] Lewenstein M., Sanpera A., Ahufinger V., *et al.* (2007). Ultracold atomic gases in optical lattices: mimicking condensed matter physics and beyond. Adv Phys **56**, 243–379.
[80] Spielman I. B., Phillips W. D., and Porto J. V. (2007). Mott-insulator transition in a two-dimensional atomic Bose gas. Phys Rev Lett **98**, 080404.
[81] Jaksch D. and Zoller P. (2005). The cold atom Hubbard toolbox. Ann Phys **315**, 52–79.
[82] Bloch I. (2008). Quantum Gases. Science **319**, 1202–3.
[83] Greiner M., Mandel O., Esslinger T., Hansch T. W., and Bloch I. (2002). Quantum phase transition from a superfluid to a Mott insulator in a gas of ultracold atoms. Nature **415**, 39–44.
[84] Feynman R. (1982). Simulating physics with computers. Int J Theor Phys **21**, 467–88.
[85] Lloyd S. (1996). Universal quantum simulators. Science **273**, 1073–8.

Index

Page numbers in *italic* refer to Figures and Tables.